Synthetic Antibodies

Methods and Protocols

Edited by

Thomas Tiller

MorphoSys AG, Discovery Alliances & Technologies, Planegg, Germany

Humana Press

Editor
Thomas Tiller
MorphoSys AG, Discovery Alliances & Technologies
Planegg, Germany

ISSN 1064-3745 ISSN 1940-6029 (electronic)
Methods in Molecular Biology
ISBN 978-1-4939-6855-8 ISBN 978-1-4939-6857-2 (eBook)
DOI 10.1007/978-1-4939-6857-2

Library of Congress Control Number: 2017933077

Printed on acid-free paper

This Humana Press imprint is published by Springer Nature
The registered company is Springer Science+Business Media LLC
The registered company address is: 233 Spring Street, New York, NY 10013, U.S.A.

Dedication

Dedicated to Isi, Josh & Family.
...and Rea.
Thanks for inspirational, wise, and humorous support.

Preface

Antibodies are important tools that are used extensively in basic biomedical research, in diagnostics, and in the treatment of diseases.

Traditionally, the production of antibodies relies on the immunization of an animal. For example, for the generation of monoclonal antibodies by the hybridoma technology, usually mice and rats are preferred. For polyclonal antibody production, larger mammals (e.g., rabbits, sheep, and goats) are used as the relatively huge amount of serum that can be collected from these animals serves as a rich source for antibody purification. These antibodies are all based on an immunoglobulin scaffold and are derived from a genuine in vivo immune response. Despite their widespread applications as detection, diagnostic, and therapeutic agents, in vivo-generated polyclonal and monoclonal antibodies bear some limitations. For example, polyclonal antibodies as detection reagents are not only prone to batch-to-batch variability but also contain significant amounts of nonspecific antibodies. Furthermore, due to their inadequate characterization, it is not surprising that many experimental results that are obtained with polyclonal antibodies are often not reproducible. In contrast, hybridoma-derived monoclonal antibodies are considered to be perfectly defined reagents with unique specificities. Very often, however, they secrete additional light and/or heavy chains, which makes it cumbersome to evaluate if the binding behavior of the hybridoma-derived mAb is intrinsic to the mAb from the target B cell or due to artificial chain combinations caused by the presence of the additional chains derived from the fusion cell line. Furthermore, hybridoma cells can lose expression, are prone to mutations, and thus require frequent retesting.

The restrictions of these traditional in vivo-generated antibodies have been overcome by modern synthetic recombinant in vitro antibody technologies.

One of the most significant difference between naturally occurring and *synthetic immunoglobulins* per se is the way these two groups are generated. Naturally occurring immunoglobulins are generated in vivo by processes of V(D)J recombination and somatic hypermutation of the B cell antigen receptor during B cell development and differentiation and its secretion as soluble immunoglobulin by plasma cells. *Synthetic antibodies* on the other hand can be defined in general as affinity reagents engineered entirely in vitro, thus completely eliminating animals from the production process. (Although this definition might get blurred, e.g., by processes such as antibody humanization, which basically is the replacement of frameworks of a murine antibody generated in vivo with their human counterparts by recombinant genetic engineering in vitro. Therefore, a humanized antibody could be considered as "semisynthetic").

Synthetic affinity reagents include recombinantly produced *immunoglobulin antibodies derived from combinatorial antibody libraries* (i.e., antibody libraries built on in silico-designed and chemically defined diversity on the basis of synthetic oligonucleotides) and so-called *antibody mimetics* that are based on alternative protein/polypeptide scaffolds.

In addition, the term "*synthetic antibody*" is also often used to describe affinity reagents that are different from protein/polypeptides but share typical antibody characteristics such as diversity and specific binding affinities. For example, *aptamers* as a class of small nucleic acid ligands are composed of RNA or single-stranded DNA oligonucleotides. Like antibodies, *aptamers* interact with their corresponding targets with high specificity and affinity.

An example of synthetic "plastic antibodies" are *molecularly imprinted polymers* (MIPs), which are polymeric matrices obtained by a technique called *molecular imprinting technology* to design artificial receptors with a predetermined selectivity and specificity for a given analyte. MIPs are able to mimic natural recognition entities, such as antibodies and biological receptors.

This volume on Synthetic Antibodies aims to present a set of protocols useful for research in the field of recombinant immunoglobulin and alternative scaffold engineering, aptamer development, and generation of MIPs. Part I includes methods that deal with amino acid-based synthetic antibodies. Brief protocols about the generation of antibody libraries are detailed, as well as techniques for antibody selection, characterization, and validation. This section is completed by a brief description of a bioinformatics platform that supports antibody engineering during Research and Development. Part II contains basic procedures about the selection and characterization of aptamer molecules, and Part III describes fundamental processes of MIP generation and application.

I would like to express my sincere thanks to all contributing authors for sharing their research expertise. Without their support, this volume would not have been possible. Many thanks to John M. Walker for the invitation to edit this volume on "Synthetic Antibodies" and to Monica Suchy and Patrick Marton from Springer for helpful advice and for publishing this book.

Planegg, Germany ***Thomas Tiller***

Contents

Contributors

ROCIO ARANDA-RODRIGUEZ • *Environmental Health Science and Research Bureau, Health Canada, Ottawa, ON, Canada*
XUELIAN BAI • *Department of Life Science, Ewha Womans University, Seoul, Korea*
ELYSE D. BERNARD • *Environmental Health Science and Research Bureau, Health Canada, Ottawa, ON, Canada*
BERNADETTE TSE SUM BUI • *CNRS Enzyme and Cell Engineering Laboratory, Sorbonne Universités, Université de Technologie de Compiègne, Compiègne Cedex, France*
FRANCESCO CANFAROTTA • *MIP Diagnostics Ltd., University of Leicester, Leicester, UK*
GUIDO CAPPUCCILLI • *Genedata AG, Basel, Switzerland*
LINDA CHIO • *Department of Chemical and Biomolecular Engineering, University of California, Berkeley, CA, USA*
AGNES CIBIEL • *Commissariat à l'Energie Atomique et aux Energies Alternatives (CEA), Département de la Recherche Fondamentale (DRF), Institut d'Imagerie Biomédicale (I2BM), Molecular Imaging Center (MIRCen), CNRS UMR 9199, Neurodegenerative Diseases Laboratory (LMN), Université Paris-Sud, Université Paris-Saclay, Fontenay-aux-Roses, France*
STEPHEN J. DEMAREST • *Eli Lilly Biotechnology Center, San Diego, CA, USA*
MARIA C. DEROSA • *Chemistry Department, Carleton University, Ottawa, ON, Canada*
THIRUMALAI DIRAVIYAM • *College of Veterinary Medicine, Northwest Agriculture and Forestry University, Yangling, Shaanxi, China; Department of Microbiology, Karpagam University, Coimbatore, Tamil Nadu, India*
FRÉDÉRIC DUCONGÉ • *Commissariat à l'Energie Atomique et aux Energies Alternatives (CEA), Département de la Recherche Fondamentale (DRF), Institut d'Imagerie Biomédicale (I2BM), Molecular Imaging Center (MIRCen), CNRS UMR 9199, Neurodegenerative Diseases Laboratory (LMN), Université Paris-Sud, Université Paris-Saclay, Fontenay-aux-Roses, France*
BLANCA I. ESCUDERO-ABARCA • *Department of Food, Bioprocessing, and Nutrition Sciences, North Carolina State University, Raleigh, NC, USA*
GRIGORY S. FILONOV • *Essen Bioscience, Ann Arbor, MI, USA (current); Department of Pharmacology, Weill Medical College, Cornell University, New York, NY, USA*
JONATHAN R. FITCHETT • *Eli Lilly Biotechnology Center, San Diego, CA, USA*
MELISSA L. GEDDIE • *Merrimack Pharmaceuticals, Inc., Cambridge, MA, USA*
SUSANNE GRÄSLUND • *Structural Genomics Consortium, Department of Biochemistry and Biophysics, Karolinska Institutet, Solna, Sweden*
LAURENCE GUGLIELMI • *IRCM, Institut de Recherche en Cancérologie de Montpellier, Montpellier, France; INSERM, U1194, Montpellier, France; Université de Montpellier, Montpellier, France; Institut régional du Cancer de Montpellier, Montpellier, France*
YOSHIRO HANYU • *Structure Physiology Research Group, Biomedical Research Institute, National Institute of Advanced Industrial Science and Technology (AIST), Tsukuba, Japan*

Karsten Haupt • *CNRS Enzyme and Cell Engineering Laboratory, Sorbonne Universités, Université de Technologie de Compiègne, Compiègne Cedex, France*
Viola Horvath • *Department of Inorganic and Analytical Chemistry, Budapest University of Technology and Economics, Budapest, Hungary*
Chuixiu Huang • *School of Pharmacy, University of Oslo, Blindern, Oslo, Norway*
Lee-Ann Jaykus • *Department of Food, Bioprocessing, and Nutrition Sciences, North Carolina State University, Raleigh, NC, USA*
Mieko Kato • *Bio-Peak Co., Ltd., Takasaki, Japan*
Neeraj Kohli • *Merrimack Pharmaceuticals, Inc., Cambridge, MA, USA*
Anthony A. Kossiakoff • *Department of Biochemistry and Molecular Biology, The University of Chicago, Chicago, IL, USA; Institute for Biophysical Dynamics, The University of Chicago, Chicago, IL, USA*
Stephanie Kunath • *CNRS Enzyme and Cell Engineering Laboratory, Sorbonne Universités, Université de Technologie de Compiègne, Compiègne Cedex, France*
Markita Landry • *Department of Chemical and Biomolecular Engineering, University of California, Berkeley, CA, USA; Landry Lab, California Institute for Quantitative Biosciences, QB3, University of California, Berkeley, CA, USA*
Johan Lengqvist • *Centre for Molecular Medicine, Rheumatology Unit, Department of Medicine, Karolinska Institutet, Karolinska University Hospital, Solna, Sweden*
Richard A. Lerner • *Department of Cell and Molecular Biology, The Scripps Research Institute, La Jolla, CA, USA*
Xiuling Li • *Department of Cancer Biology, The Scripps Research Institute, Jupiter, FL, USA*
Peter A. Lieberzeit • *Department of Analytical Chemistry, University of Vienna, Vienna, Austria*
Edyta Marcon • *Terrence Donnelly Center for Cellular & Biomolecular Research, University of Toronto, Toronto, ON, Canada*
Pierre Martineau • *IRCM, Institut de Recherche en Cancérologie de Montpellier, Montpellier, France; INSERM, U1194, Montpellier, France; Université de Montpellier, Montpellier, France; Institut régional du Cancer de Montpellier, Montpellier, France*
Marigona Mazlami • *Department of Biochemistry, University of Zurich, Zurich, Switzerland*
Anna Miodek • *Commissariat à l'Energie Atomique et aux Energies Alternatives (CEA), Département de la Recherche Fondamentale (DRF), Institut d'Imagerie Biomédicale (I2BM), Molecular Imaging Center (MIRCen), CNRS UMR 9199, Neurodegenerative Diseases Laboratory (LMN), Université Paris-Sud, Université Paris-Saclay, Fontenay-aux-Roses, France*
Matthew D. Moore • *Department of Food, Bioprocessing, and Nutrition Sciences, North Carolina State University, Raleigh, NC, USA*
Kathy C. Nguyen • *Environmental Health Science and Research Bureau, Health Canada, Ottawa, ON, Canada*
Maria Panagiotopoulou • *CNRS Enzyme and Cell Engineering Laboratory, Sorbonne Universités, Université de Technologie de Compiègne, Compiègne Cedex, France*
Marcin Paduch • *Department of Biochemistry and Molecular Biology, The University of Chicago, Chicago, IL, USA*

Helena Persson • *Science for Life Laboratory, Drug Discovery and Development Platform & School of Biotechnology, KTH-Royal Institute of Technology, Solna, Sweden*
Sergey Piletsky • *Department of Chemistry, University of Leicester, Leicester, UK*
Andreas Plückthun • *Department of Biochemistry, University of Zurich, Zurich, Switzerland*
Charlotta Preger • *Structural Genomics Consortium, Department of Biochemistry and Biophysics, Karolinska Institutet, Solna, Sweden*
Nam Nguyen Quang • *Commissariat à l'Energie Atomique et aux Energies Alternatives (CEA), Département de la Recherche Fondamentale (DRF), Institut d'Imagerie Biomédicale (I2BM), Molecular Imaging Center (MIRCen), CNRS UMR 9199, Neurodegenerative Diseases Laboratory (LMN), Université Paris- Sud, Université Paris-Saclay, Fontenay-aux-Roses, France*
Christoph Rader • *Department of Cancer Biology, The Scripps Research Institute, Jupiter, FL, USA; Department of Molecular Therapeutics, The Scripps Research Institute, Jupiter, FL, USA*
Emilie Renaud • *IRCM, Institut de Recherche en Cancérologie de Montpellier, Montpellier, France; INSERM, U1194, Montpellier, France; Université de Montpellier, Montpellier, France; Institut régional du Cancer de Montpellier, Montpellier, France*
Tibor Renkecz • *Department of Inorganic and Analytical Chemistry, Budapest University of Technology and Economics, Budapest, Hungary*
Chak Sangma • *Faculty of Science, Department of Chemistry, Center for Advanced Studies in Nanotechnology and Its Applications in Chemical, Food and Agricultural Industries, Kasetsart University, Chatuchak, Bangkok, Thailand*
Romana Schirhagl • *University Medical Center Groningen, Groningen University, Groningen, The Netherlands*
Annika Schmid • *MorphoSys AG, Planegg, Germany*
Hyunbo Shim • *Department of Life Science, Ewha Womans University, Seoul, Korea; Department of Bioinspired Science, Ewha Womans University, Seoul, Korea*
Xiantao Shen • *Key Laboratory of Environment and Health, Ministry of Education & Ministry of Environmental Protection, School of Public Health, Tongji Medical College, Huazhong University of Science and Technology, Wuhan, Hubei, China*
Eric V. Shusta • *Department of Chemical and Biological Engineering, University of Wisconsin-Madison, Madison, WI, USA*
Katarzyna Smolinska-Kempisty • *Department of Chemistry, University of Leicester, Leicester, UK*
Yvonne Stark • *MorphoSys AG, Planegg, Germany*
Andrea Sterner • *MorphoSys AG, Planegg, Germany*
Wannisa Sukjee • *Department of Chemistry, Kasetsart University, Chatuchak, Bangkok, Thailand*
Azam F. Tayabali • *Environmental Health Science and Research Bureau, Health Canada, Ottawa, ON, Canada*
Michael W. Traxlmayr • *Department of Chemistry, BOKU-University of Natural Resources and Life Sciences, Vienna, Austria*
Sophie Venet • *MorphoSys AG, Planegg, Germany*
Wouter P.R. Verdurmen • *Department of Biochemistry, University of Zurich, Zurich, Switzerland*
Maria Wendt • *Genedata AG, Basel, Switzerland*

XIUFENG WU • *Eli Lilly Biotechnology Center, San Diego, CA, USA*
DARWIN YANG • *Department of Chemical and Biomolecular Engineering, University of California, Berkeley, CA, USA*
LEI YE • *Division of Pure and Applied Biochemistry, Lund University, Lund, Sweden*
CAROLIN ZEHETMEIER • *MorphoSys AG, Planegg, Germany*
HONGKAI ZHANG • *Department of Cell and Molecular Biology, The Scripps Research Institute, La Jolla, CA, USA*
KAI ZHANG • *Eli Lilly Biotechnology Center, San Diego, CA, USA*
XIAOYING ZHANG • *College of Veterinary Medicine, Northwest Agriculture and Forestry University, Yangling, Shaanxi, China*

The original version of this book was revised. An erratum to this book can be found at DOI 10.1007/978-1-4939-6857-2_27

Part I

Amino Acid-Based Synthetic Antibodies

Part I

Amino Acid-Based Synthetic Scaffolds

Chapter 1

Antibody Mimetics, Peptides, and Peptidomimetics

Xiaoying Zhang and Thirumalai Diraviyam

Abstract

In spite of their widespread applications as therapeutic, diagnostic, and detection agents, the limitations of polyclonal and monoclonal antibodies have enthused scientists to plan for next-generation biomedical agents, the so-called antibody mimetics, which offer many advantages compared to traditional antibodies. Antibody mimetics could be designed through protein-directed evolution or fusion of complementarity-determining regions with intervening framework regions. In the recent decade, extensive progress has been made in exploiting human, butterfly (*Pieris brassicae*), and bacterial systems to design and select mimetics using display technologies. Notably, some of the mimetics have made their way to market. Numerous limitations lie ahead in developing mimetics for different biomedical usage, particularly for which conventional antibodies are ineffective. This chapter presents a brief overview of the current characteristics, construction, and applications of antibody mimetics.

Key words Antibody mimetics, Protein engineering, Monoclonal antibodies (mAbs), Therapeutics, Diagnostics

1 Introduction

A revolution has been made in the biological science through the development of the hybridoma technique to generate monoclonal antibodies (mAbs) [1]. In the meantime, advancements in genetic engineering revolutionized the methods to select, humanize and produce recombinant antibodies. The accomplishment of fabricating antibody fragments in different host systems (e.g., bacteria and yeast) and selection technologies, such as phage and ribosome display, permitted the production of antibody-based reagents for varied applications. On the other hand, animal-sourced antibodies faced some challenges such as ethical concerns to use animals for experiments, the penetration difficulty for large sized antibodies in solid tumors, immunogenicity [2], presence of six hypervariable loops that are difficult to manipulate at once, if generation of a large synthetic library is required [3], complex multi-chain architecture and glycosylation of the heavy chains [4]. Besides, some studies reported that, some antibodies have lost their activity when

Thomas Tiller (ed.), *Synthetic Antibodies: Methods and Protocols*, Methods in Molecular Biology, vol. 1575, DOI 10.1007/978-1-4939-6857-2_1, © Springer Science+Business Media LLC 2017

used in microarrays [5], are required in high doses to achieve clinical efficacy [6], exhibit poor pharmacokinetic behavior and costly manufacturing processes [7].

The tremendous advancements of biotechnology and cutting-edge protein engineering have made it possible to synthesize antibody-like molecules, the so-called antibody mimetics. The process of producing antibody mimetics upholds the precepts of 3Rs (replacement, reduction, and refinement) for using laboratory animals [8]. They mimic natural antibodies and functionally exhibit many advantages than conventional antibodies. To date, several antibody mimetics such as, affibodies, anticalins, avimers, bicycles, DARPins, fynomers, iBodies, and nanofitins, have been developed and many more are under development. These novel approaches are gaining acceptance by offering versatile advantages to combat with clinically important diseases such as cancer, autoimmune diseases, and acquired immunodeficiency syndrome.

2 Steps Involved in Constructing Antibody Mimetics

Antibody mimetics are mainly constructed by two methods, protein-directed evolution and fusion of complementary determining regions (CDRs) through cognate framework regions (FRs) in different sequences.

Presently, the protein-directed evolution is employed to harness the power of natural selection to evolve proteins with preferred properties. In principle, it involves four key steps as illustrated in Fig. 1a: (1) Identification: the sequence of interest is chosen on the basis of its perceived proximity to the desired function and its evolvability [9]; (2) Diversification: the parent sequence is

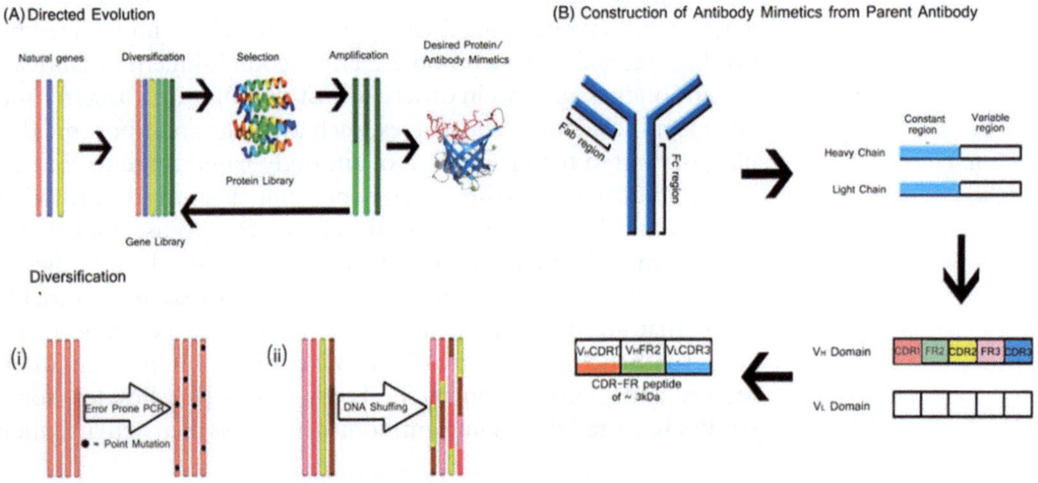

Fig. 1 Construction strategies of antibody mimetics

subjected to diversification by error-prone PCR and DNA shuffling; (3) Selection: the screening or selection is used to test the presence of mutants in the generated library; and (4) Amplification: the variants are screened, selected and replicated many-fold to harvest a variant with the desired properties. Massive combinatorial libraries have been constructed by randomizing amino acid positions in structurally variable loops of proteins [10] or by exon shuffling and phage display [11]. The mutant libraries specific against desired antigens are screened by phage display or ribosome display selection (Fig. 1).

By fusing two CDRs through a cognate framework region (FR) the CDR-FR peptides are constructed. Protein antigens are generally recognized by all six CDRs from both the VL and VH domains of the intact antibody combining site. The CDRH3 loop is considered the most indispensable part of the mimetic, as it is often the most accessible of the CDR loops, and is almost always involved in antigen binding to the greatest extent due to its greater sequence diversity. The C-terminus of the selected CDR1 or CDR2 loop and the N-terminus of the selected CDRH3 loop are joined with a FR chosen from VH or VL [12] (Fig. 1b). On the basis of these principles numerous antibody mimetics have been developed.

3 Antibody Mimetics as Therapeutic Agents

Human epidermal growth factor receptors (HER1, HER2, HER3, and HER4) dysregulation and overexpression may cause different types of cancers, and therefore the HER proteins are considered as reliable biomarkers for cancer progression and treatment [13]. The FDA approved anti-HER2 mAb Trastuzumab (Herceptin) is successfully used for the treatment of breast cancer; however, Trastuzumab application may also result in side effects such as cardiac dysfunction in some patients [14]. Small, non-immunogenic, stable, and specific affibody molecules named ZHER2:342 with good tissue penetration are successfully used as imaging and treatment agents as alternatives to mAbs [15]. ZHER2:342 has been fused with a truncated form of Pseudomonas exotoxin A, and the fusion protein was found to bind successfully to HER2-expressing cells [16]. It was also evaluated whether albumin binding domain (ABD) conjugation with the anti-HER2 affibody could improve its pharmacokinetics and enable radionuclide therapy for small tumors expressing HER2. This conjugation strategy [177Lu-CHX-A00-DTPA-ABD-(ZHER2:342)2] exhibited significantly enhanced half-life and reduced the kidney uptake [17]. The affibody molecule ZEGFR was encapsulated in liposomes to prevent degradation from metabolizing enzymes and was successfully delivered to EGFR-expressing cells [18]. The effects of two other affibodies (Z05416 and Z05417) were investigated against HER3 on different cell lines and these molecules completely inhibited heregulin (HRG)-induced cancer cell growth in an in vitro assay. The antiproliferative effect of these affibodies on cells was caused by blocking the physiological interaction between HER3 and HRG [19].

It is well known that, targeting cytotoxic T lymphocyte associated antigen-4 (CTLA-4) has opened new avenues in immunotherapy of cancer, HIV as well as other infectious diseases. A novel engineered antibody mimetic anticalin (lipocalin), derived from neutrophil gelatinase- associated lipocalin (NGAL), is a potential candidate for immunotherapy of cancer and infectious diseases by blocking the activity of CTLA-4. A combinatorial library of ~2×10^{10} variants was constructed by randomizing the positions of 20-aa in a structurally variable loop of NGAL. The mutant library was then subjected to phage display selection. Lipocalin (Lcn) selected by phage display competitively inhibited physiological interaction between CTLA-4 and B7.1/B7.2, and interestingly, selected lipocalins showed no cross-reactivity with CD28, a structurally related T-cell coreceptor [10]. The anticalin complex with its target CTLA-4 is shown in Fig. 2.

PRS-190, a bi-specific anticalin (Duocalin), was developed with the dual specificity to target IL-17 and IL-23 (members of Th17 cytokine family involved in autoimmunity and inflammation). DigA16 (H86N) anticalin functions as a digoxin antidote when administrated intravenously in rats, dramatically decreasing the free digoxin concentration in plasma and rapidly reducing its toxic effects [20]. Anticalins are also demonstrated to be suitable candidates for treatment of digitalis intoxications [21]. The other anticalin programs from Pieris Pharmaceuticals such as PRS-050, PRS-110, PRS-080 are developed to target VEGF-A, c-Met oncogene, and chemotherapy-induced anemia (CIA) and chronic kidney disease (CKD), respectively. The PRS-060 is an advanced anticalin program developed to target IL-4 for treating asthma (http://www.pieris.com/).

The E7 protein is well known for inactivation of pRb (tumor suppressor) and is a strong element involved in rampant growth of cervical cancer [22]. It is identified that inhibition and functional knockout of E7 protein leads to arrest of cell proliferation and/or cell growth and apoptosis. The intracellular protein E7 was the target for inhibition by anti-E7 affilin molecules, which were able to arrest cellular growth and were confirmed to be highly specific for E7+ mammalian cells [23].

It was demonstrated that different cytokines including IL-5, IL-6, IL-13, and TNFa, produced by cultured human mast cells, were cleaved by chymase [24]. Fynomers bind chymase with a KD of 0.9 nM and k_{off} of $1.1 \times 10^{-3}s^{-1}$ selectively inhibiting chymase activity with an IC_{50} value of 2 nM [25]. The D3 fynomer was discovered from a Fyn Src Homology3 (SH3) phage library that

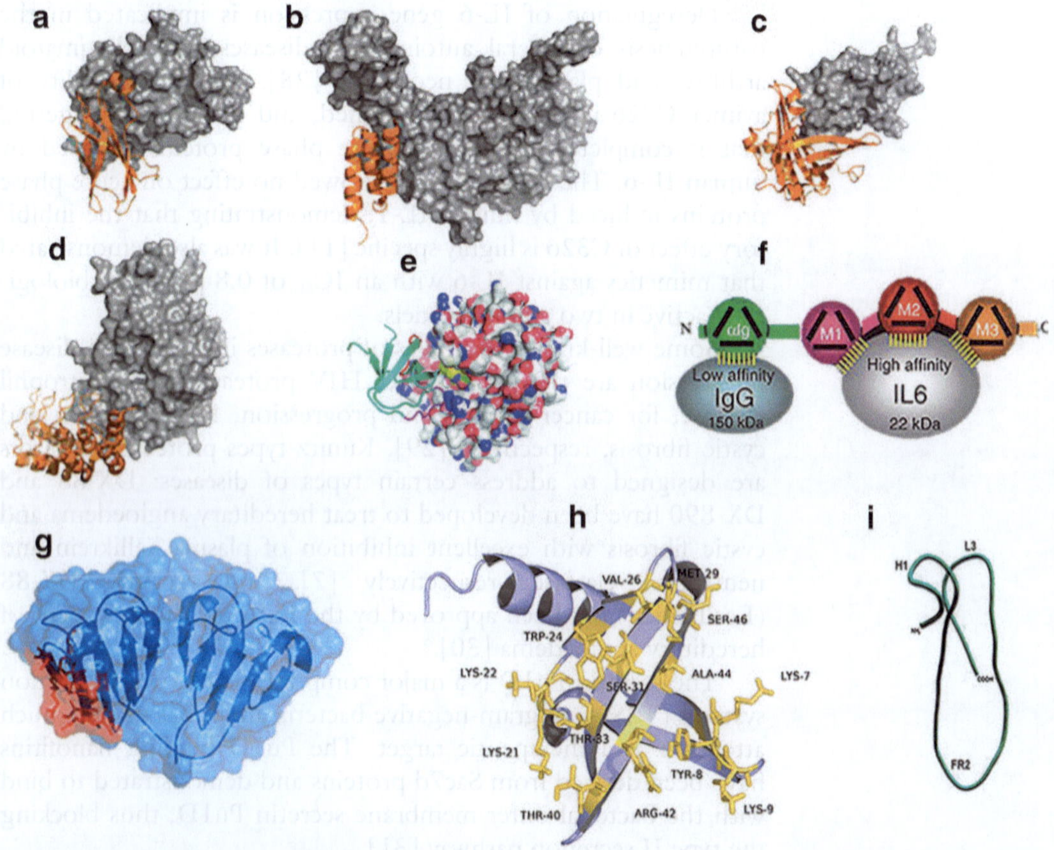

Fig. 2 Structures of antibody mimetics and their parent proteins. (**a–f**) Antibody mimetics in complex with their targets. (**g** and **h**) Parent proteins of the antibody mimetics. (**a–d**) the *orange color* represents the antibody mimetics while the targets are shown as *gray surfaces*. (**a**) Engineered Adnectin/Monobody (10Fn3) in complex with human estrogen receptor alpha binding domain. PDB: 2OCF. (**b**) Affibody molecule in complex with HER2 extra domain cellular region. PDB: 3MZW. (**c**) Anticalin in complex with the extracellular domain of Human CTLA-4. PDB: 3BX7. (**d**) DARPin in complex with aminoglycoside phosphotransferase. PDB: 2BKK. (**e**) Fynomer 4C-A4 (ribbon diagram) in complex with human chymase (space filling model). For Fynomer: The *magenta* represents RT-loop and *red* represents n-src-loop. The accession numbers for six Fynomer-chymase complexes in PDB are: 4afq, 4afs, 4afu, 4afz, 4ag1, and 4ag2. (**f**) Schematic representation of anti-IL6 Avimer (C326), in a tetramer construct. The first domain binds monovalently with human IgG in serum to prolong half-life while the remaining three domains bind to various epitopes on the surface of IL-6. (**g**) Representation of bovine g-B-crystallin, which is used to model the human g-B-crystallin scaffold (Affilins). The *red color* shows the eight selected amino acid residues (Positions 2, 4, 6, 15, 17, 19, 38, and 38) used to construct the library (PDB: 1AMM). The bovine molecule consists of 174 amino acid residues with a molecular weight of 20 kDa. (**h**) Schematic representation of wild-type Sac7d, the parent protein of nanofitins (PDB: 1AZP). (**i**) Ribbon diagram (model) of CDR-FR mimetics. The VHFR2 that links VHCDR1 and VHCDR2 in native Fab here plays the role of connecting VHCDR1 and VLCDR3, keeping them in a "quasi-physiological" binding site orientation (refer: 11, 12, 26, 27, 33, 46, and 47)

binds extra-domain B (EDB) but no other structurally related proteins [26]. The COVA301, a dual TNF/IL-17A inhibitor has also been developed, in which the fynomer against IL-17A was fused with an approved anti-TNF antibody [27].

Deregulation of IL-6 gene expression is implicated in the pathogenesis of several autoimmune diseases, e.g., rheumatoid arthritis and plasma cell neoplasias [28]. The functionality of avimer C326 in vivo was determined, and the results suggested that it completely abrogated acute phase proteins induced by human IL-6. The same mimetic showed no effect on acute phase proteins induced by human IL-1, demonstrating that the inhibitory effect of C326 is highly specific [11]. It was also demonstrated that mimetics against IL-6 with an IC_{50} of 0.8 pM were biologically active in two animal models.

Some well-known examples of proteases implicated in disease progression are the proteasome, HIV proteases and neutrophil elastase, for cancer growth and progression, HIV infection and cystic fibrosis, respectively [29]. Kunitz-types protease inhibitors are designed to address certain types of diseases: DX-88 and DX-890 have been developed to treat hereditary angioedema and cystic fibrosis with excellent inhibition of plasma kallikrein and neutrophil elastase, respectively [7], and recently DX-88 (Ecallantide) has been approved by the FDA for the treatment of hereditary angioedema [30].

The secretin PulD is a major component of Type II secretion systems (T2SSs) of gram-negative bacteria and it has gained much attention as a therapeutic target. The PulD binding nanofitins have been derived from Sac7d proteins and demonstrated to bind with the bacterial outer membrane secretin PulD, thus blocking the type II secretion pathway [31].

HUMIRA (Adalimumab), a human monoclonal antibody directed against TNFa was approved by the FDA to treat rheumatoid arthritis in 2002 and later for some other diseases. However, as HUMIRA suppresses the immune response; consequently, patients receiving HUMIRA treatment are also more prone to diseases like hepatitis B infections, allergic reactions, nervous system problems, heart failure and psoriasis [32]. Adnectins have been developed against the same pharmacological target but without aforementioned side-effects. Adnectins are mainly selected by phage, mRNA and yeast display technologies, and yeast two-hybrid techniques [33]. As another example, Adnectin Ct-322 binding to vascular endothelial growth factor receptor 2 (VEGFR-2) displays antitumor activities and results also suggest that adnectins can be developed for the treatment of various others diseases [34]. Adnectins are generated and selected to target Src SH3, Abelson (Abl) kinase SH2 domain lysozyme, TNF-α, and estrogen receptor a ligand [33]. Polyethylene glycol (PEG) has been used as a flexible scaffold molecule to link two Fabs together to generate Fab-PEG-Fab (FpF) molecule that is capable to act as IgG mimetic. Anti-VEGF and anti-Her2 FpFs molecules have successfully been prepared and evaluated. The prepared FpFs displayed similar affinities to their parent IgG molecule. In vitro antiangiogenic properties of anti-VEGF FpFs were evaluated and it was found that these properties were comparable to or even better than bevacizumab (monoclonal antibody used to treat various cancers) [35].

The CDR-FR peptides retain the antigen recognition function of their intact parent molecule IgG but have superior capacity to penetrate solid tumors. The mimetics that are fused with the C-terminus of bacterial toxin Colicin Ia are called pheromonicins. Therapeutic efficacy of such fusion proteins was tested for their killing effects against Epstein-Barr virus (EBV)-induced BL, AIDS-related body-cavity lymphoma and nasopharyngeal cancer cells, and results showed the killing effects of PMC-EBV within solid tumors bearing specific surface antigens. The bacterial toxin used as a payload has many significant advantages such as solubility, heat stability and absence of cystine residues; through indirect ELISA and assessment in normal mice, it was also shown that the cancer killing toxin was non-immunogenic [12]. The mimetic-Fc small antibodies were generated by using CDR and FR sequences from trastuzumab, a humanized anti-HER2 monoclonal antibody, fused with the Fc domain of IgG. The designed fully functional mimetic-Fc small antibody called HMTI-Fc successfully inhibited the binding of trastuzumab with HER2-overexpressing SK-BR3 cells, thus showing its potential to treat cancer [36]. The Fc part of the antibody participates in recruiting the immune cells in antibody-dependent cell-mediated cytotoxicity (ADCC) [37], and intriguingly the HMTI-Fc effectively mediated ADCC against HER2-positive breast-cancer cells [36]. Other anti-HER2 [38] and anti-CD4 [39] antibody mimetics known as DARPins have also been developed, and anti-CD4 DARPins with pM affinity blocked the entry of HIV into cells by competing with binding of gp120 to CD4. The CD4+ cells are a type of white blood cell (lymphocyte) and are critical to the immune system. The MP0112 DARPin is perhaps the most advanced program, and has been developed as a VEGF-A inhibitor (IC_{50} less than 10 pM) to treat ocular neovascularization. MP0112 has been demonstrated to be safe and well tolerated in wet age-related macular degeneration (wet AMD) and diabetic macular edema (DME). The therapeutic effect of MP0112 lasted for 16 weeks and several studies have revealed that MP0112 is long-acting and highly efficacious [40].

4 Applications of Antibody Mimetics Diagnosis and Imaging

Antibody mimetics could be labeled and used to image metabolite pathways, intracellular targets such as kinases and polymerases, and other proteins associated with cancers. Studies have revealed that affibodies are promising among the tracers for HER2-specific molecular imaging [41]. ZHER2:342 affibody molecules with the chelator sequence of maEE were synthesized and labeled with technetium-99 m. The synthesized molecule 99mTc-(mercaptoacetyl-Glu-Glu-Glu) maEEE-ZHER2:342 appeared to be a better tracer for clinical imaging of HER2 overexpression in tumors and metastases [42]. Affibodies have been used in protein capture microarrays and due to their high specificity they can be used for affinity capture in analyses of complex samples, e.g., human serum or plasma [5]. Anticalins, due to their small size when conjugated with radioactive isotopes, can be used for in vivo diagnostics [3] and images of high contrast have been obtained soon after administration [43]. Anticalin C26 was developed with high binding affinity for rare-earth metal–chelate complexes, and further improvement in this anticalin by in vitro selection yielded CL31 with fourfold slower dissociation (more than 2 h). Oncofetal isoform of extracellular matrix protein fibronectin carries the EDB and is exclusively expressed in neovasculature, and has gained significant interest for tumor diagnosis. The human Lcn2 has been employed as a small non-immunoglobulin scaffold to selecting EDB-specific anticalins, and anticalins showing low nanomolar affinities for EDB were isolated and biochemically characterized. When these isolated anticalins were used in immunofluorescence microscopy, they showed specific staining of EDB positive tumor cells, and the analysis of BIAcore affinity data showed that they recognized distinct epitopes of EDB, suggesting that these EDB specific anticalins could provide potential biomolecules both in research and biomedical drug development [44].

The CDR-FR peptides have been used for in vivo fluorescence imaging and these antibody mimetics also conferred enhanced intracellular delivery, thus rendering the mimetics potent candidates for cancer diagnostic applications [12]. The nanofitins have been designed to selectively bind to a wide range of targets and have been reliable tools for targeting (immunolocalization, in vivo neutralization), capture (affinity chromatography, protein removal) and detection (immunoassays, western blot). The DARPins H6-2-B3 and H6-2-A7 have been used for in vivo tumor imaging and tumor targeting and were shown to localize at the tumor [38].

5 Future Prospects for Antibody Mimetics

Due to their high target retention, rapid tissue penetration and blood clearance, antibody mimetics are gaining importance both in therapeutics and diagnostics, especially in tumor targeting and treatment. Antibody mimetics can be generated against a range of biomarkers associated with specific diseases for the development of electronic and other formats of multiplex biosensors, reagents for detection in routine immunological analysis such as ELISA and Western blot. Other small molecules called aptamers (about 10 kDa) [45] have attracted the attention of scientific community due to their merits such as thermal stability, cost-effectiveness, and unlimited applications. Therapeutic efficacy and continuing advances in the production of human-derived molecules suggest a promising future for antibody mimetics; however, some questions remain relating to both therapeutic and diagnostic uses, principally their short half-life. The mimetics exhibit shorter half-lives because they lack the Fc region and have much lower molecular weights. However, when antibody mimetics are engineered with some functional antibody part such as Fc [36], they better mimic the real antibody and combine the advantages of both natural antibodies and antibody mimetics. With the recent advancements of bioengineering, the biological activity of mimetics can be increased by many-fold. Despite their reduced size and increased affinity, the effects of mimetics in treating diseases other than solid tumors and autoimmune diseases still need to be further assessed.

Acknowledgements

This work has been supported by China Nature Science Foundation (grant no. 31572556), Ph.D. Programs Foundation of Ministry of Education (grant no. 20130204110023), and The Key Construction Program (grant no. 2015SD0018) of International Cooperation Base in S&T, Shaanxi Province, China.

References

1. Köhler G, Milstein C (1975) Continuous cultures of fused cells secreting antibody of predefined specificity. Nature 256:495–497
2. Murali R, Greene MI (2012) Structure based antibody-like peptidomimetics. Pharmaceuticals 5:209–235
3. Skerra A (2001) 'Anticalins': a new class of engineered ligand-binding proteins with antibody-like properties. Rev Mol Biotechnol 74:257–275
4. Jefferis R (2009) Glycosylation as a strategy to improve antibody-based therapeutics. Nat Rev Drug Discov 8:226–234
5. Renberg B, Nordin J, Merca A, Uhlén M, Feldwisch J, Nygren PÅ, Eriksson Karlström A (2007) Affibody molecules in protein capture microarrays: evaluation of multidomain ligands and different detection formats. J Proteome Res 6:171–179
6. Chames P, Van Regenmortel M, Weiss E, Baty D (2009) Therapeutic antibodies: successes, limitations and hopes for the future. Br J Pharmacol 157:220–233
7. Hey T, Fiedler E, Rudolph R, Fiedler M (2005) Artificial, non-antibody binding proteins for pharmaceutical and industrial applications. Trends Biotechnol 23:514–522
8. Russell WMS, Burch RL, Hume CW (1959) The principles of humane experimental technique. Methuen & Co., London
9. Romero PA, Arnold FH (2009) Exploring protein fitness landscapes by directed evolution. Nat Rev Mol Cell Biol 10:866–876
10. Schönfeld D, Matschiner G, Chatwell L, Trentmann S, Gille H, Hülsmeyer M, Brown N, Kaye PM, Schlehuber S, Hohlbaum AM, Skerra A (2009) An engineered lipocalin specific for CTLA-4 reveals a combining site with structural and conformational features similar to antibodies. Proc Natl Acad Sci 106:8198–8203
11. Silverman J, Lu Q, Bakker A, To, W, Duguay A, Alba BM, Smith R, Rivas A, Li P, Le H, Whitehorn E (2005) Multivalent avimer proteins evolved by exon shuffling of a family of human receptor domains. Nat Biotechnol 23:1556–1561
12. Qiu XQ, Wang H, Cai B, Wang LL, Yue ST (2007) Small antibody mimetics comprising two complementarity-determining regions and

a framework region for tumor targeting. Nat Biotechnol 25:921–929
13. Ahlgren S, Wållberg H, Tran TA, Widström C, Hjertman M, Abrahmsén L, Berndorff D, Dinkelborg LM, Cyr JE, Feldwisch J, Orlova A (2009) Targeting of HER2-expressing tumors with a site-specifically 99mTc-labeled recombinant affibody molecule, ZHER2: 2395, with C-terminally engineered cysteine. J Nucl Med 50:781–789
14. Seidman A, Hudis C, Pierri MK, Shak S, Paton V, Ashby M, Murphy M, Stewart SJ, Keefe D (2002) Cardiac dysfunction in the trastuzumab clinical trials experience. J Clin Oncol 20:1215–1221
15. Wållberg H, Löfdahl PÅ, Tschapalda K, Uhlén M, Tolmachev V, Nygren PÅ, Ståhl S (2011) Affinity recovery of eight HER2-binding affibody variants using an anti-idiotypic affibody molecule as capture ligand. Protein Expr Purif 76:127–135
16. Zielinski R, Lyakhov I, Jacobs A, Chertov O, Kramer-Marek G, Francella N, Stephen A, Fisher R, Blumenthal R, Capala J (2009) Affitoxin–a novel recombinant, HER2-specific, anti-cancer agent for targeted therapy of HER2-positive tumors. J Immunother 32:817–825 (Hagerstown, Md: 1997)
17. Tolmachev V, Orlova A, Pehrson R, Galli J, Baastrup B, Andersson K, Sandström M, Rosik D, Carlsson J, Lundqvist H, Wennborg A (2007) Radionuclide therapy of HER2-positive microxenografts using a 177Lu-labeled HER2-specific Affibody molecule. Cancer Res 67:2773–2782
18. Beuttler J, Rothdiener M, Müller D, Frejd FY, Kontermann RE (2009) Targeting of epidermal growth factor receptor (EGFR)-expressing tumor cells with sterically stabilized affibody liposomes (SAL). Bioconjug Chem 20:1201–1208
19. Göstring L, Malm M, Höidén-Guthenberg I, Frejd FY, Ståhl S, Löfblom J, Gedda L (2012) Cellular effects of HER3-specific affibody molecules. PLoS One 7:e40023
20. Eyer F, Steimer W, Nitzsche T, Jung N, Neuberger H, Müller C, Schlapschy M, Zilker T, Skerra A (2012) Intravenous application of an anticalin dramatically lowers plasma digoxin levels and reduces its toxic effects in rats. Toxicol Appl Pharmacol 263:352–359
21. Rusconi CP, Roberts JD, Pitoc GA, Nimjee SM, White RR, Quick G, Scardino E, Fay WP, Sullenger BA (2004) Antidote-mediated control of an anticoagulant aptamer in vivo. Nat Biotechnol 22:1423–1428
22. Balsitis S, Dick F, Dyson N, Lambert PF (2006) Critical roles for nonpRb targets of human papillomavirus type 16 E7 in cervical carcinogenesis. Cancer Res 66:9393–9400
23. Mirecka EA, Hey T, Fiedler U, Rudolph R, Hatzfeld M (2009) Affilin molecules selected against the human papillomavirus E7 protein inhibit the proliferation of target cells. J Mol Biol 390:710–721
24. Loefblom J, Frejd FY (2011) Alternative scaffolds as bispecific antibody mimetics. In: Kontermann RE (ed) Bispecific antibodies. Springer-Verlag, Berlin Heidelberg, pp 115–133
25. Schlatter D, Brack S, Banner DW, Batey S, Benz J, Bertschinger J, Huber W, Joseph C, Rufer AC, van der Klooster A, Weber M (2012) Generation, characterization and structural data of chymase binding proteins based on the human Fyn kinase SH3 domain. Landes Biosci 4:497–508
26. Grabulovski D, Kaspar M, Neri D (2007) A novel, non-immunogenic Fyn SH3-derived binding protein with tumor vascular targeting properties. J Biol Chem 282:3196–3204
27. Covagen, Advanced Biopharmaceuticals. Covagen utilizes the unique versatility of Fynomers to create next generation biologic. http://www.covagen.com/index.php?id¼118. Accessed 21 Jun 2013
28. Hirano T (1992) Interleukin-6 and its relation to inflammation and disease. Clin Immunol Immunopathol 62:S60–S65
29. Krehenbrink M, Chami M, Guilvout I, Alzari PM, Pécorari F, Pugsley AP (2008) Artificial binding proteins (Affitins) as probes for conformational changes in secretin *PulD*. J Mol Biol 383:1058–1068
30. Scott CJ, Taggart CC (2010) Biologic protease inhibitors as novel therapeutic agents. Biochimie 92(11):1681–1688
31. Mouratou B, Schaeffer F, Guilvout I, Tello-Manigne D, Pugsley AP, Alzari PM, Pecorari F (2007) Remodeling a DNA binding protein as a specific in vivo inhibitor of bacterial secretin PulD. Proc Natl Acad Sci 104:17983–17988
32. Lee CM, Shrieve DC, Zempolich KA, Lee RJ, Hammond E, Handrahan DL, Gaffney DK (2005) Correlation between human epidermal growth factor receptor family (EGFR, HER2, HER3, HER4), phosphorylated Akt (P-Akt), and clinical outcomes after radiation therapy in carcinoma of the cervix. Gynecol Oncol 99:415–421
33. Lipovšek D (2011) Adnectins: engineered target-binding protein therapeutics. Protein Eng Des Sel 24:3–9
34. Mamluk R, Carvajal IM, Morse BA, Wong HK, Abramowitz J, Aslanian S, Lim AC, Gokemeijer J, Storek MJ, Lee J, Gosselin M

(2010) Anti-tumor effect of CT-322 as an adnectin inhibitor of vascular endothelial growth factor receptor-2. Landes Biosci 2:199–208

35. Khalili H, Godwin A, Choi JW, Lever R, Khaw PT, Brocchini S (2013) Fab-PEG-Fab as a potential antibody mimetic. Bioconjug Chem 24:1870–1882
36. Gao J, Li B, Li H, Zhang X, Zhang D, Zhao L, Wang C, Fang C, Qian W, Hou S, Kou G (2009) Development and characterization of a fully functional small anti-HER2 antibody. BMB Rep 42:636–641
37. Ladner RC (2007) Antibodies cut down to size. Nat Biotechnol 25:875–877
38. Stumpp MT, Amstutz P (2007) DARPins: a true alternative to antibodies. Curr Opin Drug Discov Dev 10:153–159
39. Schweizer A, Rusert P, Berlinger L, Ruprecht CR, Mann A, Corthésy S, Turville SG, Aravantinou M, Fischer M, Robbiani M, Amstutz P (2008) CD4-specific designed ankyrin repeat proteins are novel potent HIV entry inhibitors with unique characteristics. PLoS Pathog 4:e1000109
40. Wurch T, Pierré A, Depil S (2012) Novel protein scaffolds as emerging therapeutic proteins: from discovery to clinical proof-of-concept. Trends Biotechnol 30:575–582
41. Löfblom J, Feldwisch J, Tolmachev V, Carlsson J, Ståhl S, Frejd FY (2010) Affibody molecules: engineered proteins for therapeutic, diagnostic and biotechnological applications. FEBS Lett 584:2670–2680
42. Tran T, Engfeldt T, Orlova A, Sandström M, Feldwisch J, Abrahmsén L, Wennborg A, Tolmachev V, Karlström AE (2007) 99mTc-maEEE-ZHER2: 342, an Affibody molecule-based tracer for the detection of HER2 expression in malignant tumors. Bioconjug Chem 18:1956–1964
43. Skerra A (2008) Alternative binding proteins: anticalins–harnessing the structural plasticity of the lipocalin ligand pocket to engineer novel binding activities. FEBS J 275:2677–2683
44. Gebauer M, Schiefner A, Matschiner G, Skerra A (2013) Combinatorial design of an Anticalin directed against the extra-domain B for the specific targeting of oncofetal fibronectin. J Mol Biol 425:780–802
45. Cai S, Singh BR (2007) Strategies to design inhibitors of clostridium botulinum neurotoxins. Infect Disord Drug Targets 7:47–57
46. Arnoux B, Ducruix A, Prange T (2002) Anisotropic behaviour of the Cterminal Kunitz-type domain of the 3 chain of human type VI collagen at atomic resolution (0.9 A). Acta Crystallographica D Biol Crystallograph 58:1252–1254
47. Ebersbach H, Fiedler E, Scheuermann T, Fiedler M, Stubbs MT, Reimann C, Proetzel G, Rudolph R, Fiedler U (2007) Affilin–novel binding molecules based on human g-b-crystallin, an all b-sheet protein. J Mol Biol 372:172–185

Chapter 2

Construction of a scFv Library with Synthetic, Non-combinatorial CDR Diversity

Xuelian Bai and Hyunbo Shim

Abstract

Many large synthetic antibody libraries have been designed, constructed, and successfully generated high-quality antibodies suitable for various demanding applications. While synthetic antibody libraries have many advantages such as optimized framework sequences and a broader sequence landscape than natural antibodies, their sequence diversities typically are generated by random combinatorial synthetic processes which cause the incorporation of many undesired CDR sequences. Here, we describe the construction of a synthetic scFv library using oligonucleotide mixtures that contain predefined, non-combinatorially synthesized CDR sequences. Each CDR is first inserted to a master scFv framework sequence and the resulting single-CDR libraries are subjected to a round of proofread panning. The proofread CDR sequences are assembled to produce the final scFv library with six diversified CDRs.

Key words Antibody library, scFv, Phage display, Non-combinatorial CDR diversity, Synthetic CDR diversity, Synthetic antibody library

1 Introduction

Synthetic antibody libraries are a powerful and versatile tool for the generation of target-specific human monoclonal antibodies suitable for therapeutic and other demanding applications. Since the construction of the first synthetic antibody library using degenerate random oligonucleotides [1], significant technological advancements in library designs and synthetic methodologies have been made, and many highly sophisticated antibody libraries with synthetic diversity have been successfully constructed and utilized to generate therapeutic antibodies in various stages of preclinical and clinical development.

A current focus of the synthetic library design is the optimization of the antibody sequences for such aspects as lower immunogenicity, higher levels of expression and stability, lower aggregation propensity, and the minimization of undesirable posttranslational modifications (PTM) [2–4]. For example, the heavy and light

Thomas Tiller (ed.), *Synthetic Antibodies: Methods and Protocols*, Methods in Molecular Biology, vol. 1575, DOI 10.1007/978-1-4939-6857-2_2, © Springer Science+Business Media LLC 2017

chain framework variable gene segments for the library construction can be selected and/or designed based on their favorable biophysical properties, either individually [2] or as pairs [3]. The complementarity determining regions (CDRs), upon which most of the library's sequence diversity is concentrated, are typically designed to mimic the sequences and amino acid usage of natural human antibodies, and synthesized by the random concatenation of mononucleotide or trinucleotide units. The codon-based random combinatorial syntheses of CDR sequences [3, 5] enable the precise implementation of intricate CDR sequence designs, resulting in the antibody libraries with a high degree of humanness. However, the intrinsically random nature of the synthetic CDR diversity generation inevitably produces a significant number of sequences that are problematic or unnatural, i.e., the emulation of the natural amino acid frequency at each position does not always produce a natural amino acid *sequence*, and sequences that contain undesirable PTM motifs can also be produced.

A novel approach to eliminate or reduce these limitations of the random synthetic sequence diversity has recently been reported by our laboratory [6]. The *SCIEN* (*S*imulation of *C*DRs *I*nspired by and *E*mulating *N*ature) principle is based on the simulation of natural rearranged and hypermutated CDRs and the parallel synthesis of thousands of oligonucleotides encoding the predefined CDR sequences. Because the CDR sequence diversity is predefined and synthesized without relying on random combinatorial events, the incorporation of undesired sequences to the library can be prevented in the design stage. Although the diversity of each of the six CDRs is low because of the non-combinatorial synthetic approach (500 ~ 8000 unique sequences per CDR in the current example; *see* ref. 6), the combination of six such regions provides a total diversity that is large enough to construct a highly functional antibody library.

In this article, we will provide a detailed method for the generation of the antibody library based on the *SCIEN* principle, starting with the oligonucleotide mixtures that contain the simulated CDR sequences prepared by array synthesis. While this chapter describes the construction of an antibody library with non-combinatorial synthetic CDR diversity, the protocol may also be applied to the preparation of synthetic antibody libraries with other methods of CDR diversification such as random degenerate oligonucleotides [1] or trinucleotide-directed mutagenesis (TRIM) [2, 5, 7].

2 Materials

Reagents and equipment suggested here can be substituted with equivalent products from different vendors. If not listed below, standard molecular biology laboratory equipment and molecular biology grade chemicals/reagents can be used.

1. Nuclease-free water.
2. Oligonucleotide mixtures: Thousands of predefined oligonucleotide sequences can be synthesized in parallel on a microarray chip and chemically cleaved into solution. OligoMix® (LC Sciences, Houston, TX, USA), for example, provides up to 3918 individual oligonucleotide sequences of up to 100-mer length in pools.
3. scFv framework gene: The human germline variable gene segments DP47, DPK22, and DPL3 were used as the frameworks for library construction. The template scFv genes (DP47-linker-DPK22 and DP47-linker-DPL3) were codon-optimized and synthesized by GenScript Inc, (Pascataway, NJ, USA) (*see* **Note 1**), with the $(GGGGS)_3$ linker sequence and two asymmetric *Sfi*I restriction sites for cloning into pComb3X phagemid vector [8]. For the fourth framework regions (FR4) of V_H, V_λ, and V_κ, JH1, JL2, and JK1 were used, respectively, and a short dummy CDR-H3 sequence (CARNKLWFDY) was used for V_H. The scFv constructs were cloned and supplied in pUC57 vector. The master framework scFv sequences are shown in Fig. 1.
4. Oligonucleotide primers for polymerase chain reaction: *see* Table 1.
5. DNA polymerases: Taq polymerase (New England Biolabs, Ipswich, MA, USA) and Pfu polymerase (Promega, Madison, WI, USA) with vendor-provided reaction buffers.
6. dNTP mixture (New England Biolabs).
7. T4 DNA ligase (Invitrogen, Carlsbad, CA, USA).
8. *Sfi*I (New England Biolabs).
9. pComb3XTT phagemid vector (Addgene, Cambridge, MA, USA. Plasmid #63891).
10. DH5α *E. coli* chemically competent cells: Prepared according to Inoue method [9].
11. LB medium: Dissolve 10 g tryptone, 5 g yeast extract, and 10 g NaCl in 1 L water. Autoclave and store at room temperature.
12. SB medium: Dissolve 20 g yeast extract, 30 g tryptone, and 10 g 3-(N-morpholino) propanesulfonic acid (MOPS) in 1 L water. Adjust pH to 7.0 and autoclave. Store at room temperature.
13. QIAprep spin miniprep kit (QIAGEN, Hilden, Germany).
14. LBAG (LB-ampicillin-glucose) agar plates: Dissolve 10 g tryptone, 5 g yeast extract, and 10 g NaCl in 1 L water. Add 18 g bacteriological agar and autoclave. When the autoclaved solution cools down below 50 °C, add 1 mL of filter-sterilized

(a)

```
GAAGTGCAGCTGCTG GAAAGTGGAGGTGGA CTGGTGCAGCCTGGC GGCAGCCTGCGCCTG AGCTGTGCCGCCAGC
 E  V  Q  L  L   E  S  G  G  G   L  V  Q  P  G   G  S  L  R  L   S  C  A  A  S
GGATTCACCTTCAGC AGCTATGCCATGAGC TGGGTTCGCCAAGCA CCTGGCAAAGGCCTG GAATGGGTGAGCGCC
 G  F  T  F  S   S  Y  A  M  S   W  V  R  Q  A   P  G  K  G  L   E  W  V  S  A
ATCAGCGGCAGCGGC GGCAGCACCTATTAT GCCGATAGCGTGAAA GGCCGCTTTACCATC AGCCGCGATAACAGC
 I  S  G  S  G   G  S  T  Y  Y   A  D  S  V  K   G  R  F  T  I   S  R  D  N  S
AAAAACACCCTGTAT CTGCAGATGAACAGC CTGCGCGCCGAGGAC ACCGCAGTCTACTAC TGTGCCAAAAACAAA
 K  N  T  L  Y   L  Q  M  N  S   L  R  A  E  D   T  A  V  Y  Y   C  A  K  N  K
CTGTGGTTTGATTAT TGGGGACAAGGTACT CTGGTGACCGTGAGC AGCGGTGGAGGAGGT AGCGGAGGTGGTGGA
 L  W  F  D  Y   W  G  Q  G  T   L  V  T  V  S   S  G  G  G  G   S  G  G  G  G
TCTGGAGGTGGAGGT AGTGAAATCGTGCTG ACCCAGAGCCCTGGC ACCCTGAGCCTGAGC CCTGGCGAACGCGCA
 S  G  G  G  G   S  E  I  V  L   T  Q  S  P  G   T  L  S  L  S   P  G  E  R  A
ACACTGTCATGCCGC GCCAGCCAGAGCGTG AGCAGCAGCTATCTG GCCTGGTATCAGCAG AAACCAGGTCAGGCT
 T  L  S  C  R   A  S  Q  S  V   S  S  S  Y  L   A  W  Y  Q  Q   K  P  G  Q  A
CCACGTCTGCTGATC TATGGCGCCAGCAGC CGCGCCACCGGCATC CCTGATCGCTTCTCA GGATCTGGAAGCGGT
 P  R  L  L  I   Y  G  A  S  S   R  A  T  G  I   P  D  R  F  S   G  S  G  S  G
ACCGATTTTACCCTG ACCATCAGCCGCCTG GAACCTGAGGACTTT GCCGTGTATTATTGT CAGCAGTATGGCAGC
 T  D  F  T  L   T  I  S  R  L   E  P  E  D  F   A  V  Y  Y  C   Q  Q  Y  G  S
AGCCCTTGGACCTTC GGTCAGGGCACTAAA GTGGAAATCAAA
 S  P  W  T  F   G  Q  G  T  K   V  E  I  K
```

(b)

```
GAAGTGCAGCTGCTG GAAAGTGGAGGTGGA CTGGTGCAGCCTGGC GGCAGCCTGCGCCTG AGCTGTGCCGCCAGC
 E  V  Q  L  L   E  S  G  G  G   L  V  Q  P  G   G  S  L  R  L   S  C  A  A  S
GGATTCACCTTCAGC AGCTATGCCATGAGC TGGGTTCGCCAAGCA CCTGGCAAAGGCCTG GAATGGGTGAGCGCC
 G  F  T  F  S   S  Y  A  M  S   W  V  R  Q  A   P  G  K  G  L   E  W  V  S  A
ATCAGCGGCAGCGGC GGCAGCACCTATTAT GCCGATAGCGTGAAA GGCCGCTTTACCATC AGCCGCGATAACAGC
 I  S  G  S  G   G  S  T  Y  Y   A  D  S  V  K   G  R  F  T  I   S  R  D  N  S
AAAAACACCCTGTAT CTGCAGATGAACAGC CTGCGCGCCGAGGAC ACCGCAGTCTACTAC TGTGCCAAAAACAAA
 K  N  T  L  Y   L  Q  M  N  S   L  R  A  E  D   T  A  V  Y  Y   C  A  K  N  K
CTGTGGTTTGATTAT TGGGGACAAGGTACT CTGGTGACCGTGAGC AGCGGTGGAGGAGGT AGCGGAGGTGGTGGA
 L  W  F  D  Y   W  G  Q  G  T   L  V  T  V  S   S  G  G  G  G   S  G  G  G  G
TCTGGAGGTGGAGGT AGTCAGAGCGTGCTG ACCCAGCCTCCTAGC GCCTCCGGTACACCA GGACAGCGCGTGACT
 S  G  G  G  G   S  Q  S  V  L   T  Q  P  P  S   A  S  G  T  P   G  Q  R  V  T
ATTAGCTGTAGCGGC AGCAGCAGCAACATC GGCAGCAACTATGTG TATTGGTACCAGCAA CTGCCTGGAACTGCA
 I  S  C  S  G   S  S  S  N  I   G  S  N  Y  V   Y  W  Y  Q  Q   L  P  G  T  A
CCTAAGCTGCTGATC TATCGCAACAACCAG CGCCCTAGCGGCGTG CCTGATCGCTTTAGC GGTAGCAAATCAGGC
 P  K  L  L  I   Y  R  N  N  Q   R  P  S  G  V   P  D  R  F  S   G  S  K  S  G
ACCAGCGCCAGCCTG GCCATCAGCGGCCTT CGCTCCGAAGATGAA GCCGATTATTATTGT GCCGCCTGGGATGAT
 T  S  A  S  L   A  I  S  G  L   R  S  E  D  E   A  D  Y  Y  C   A  A  W  D  D
AGCCTGAGCGGCTAT GTGTTTGGTGGCGGT ACCAAGCTGACCGTG CTG
 S  L  S  G  Y   V  F  G  G  G   T  K  L  T  V   L
```

Fig. 1 The master scFv framework sequences for (**a**) VH-linker-VK (DP47 and DPK22 for VH and VK, respectively), and (**b**) VH-linker-VL (DP47 and DPL3 for VH and VL, respectively)

ampicillin (100 mg/mL) and 50 mL of 40% (w/v) filter-sterilized glucose. Mix evenly with gently stirring and pour on 100 mm diameter polystyrene petri dishes (20 mL per dish). Cool down at room temperature until agar solidifies, and keep the plates at 4 °C.

15. Agarose electrophoresis gel: For 1% agarose gel, use 1 g of agarose and 100 mL TAE buffer (40 mM Tris, 20 mM acetic acid, and 1 mM EDTA, pH 8.0). Change the amount of agarose as needed to make 1.5 or 2% gels.
16. QIAquick Gel Extraction Kit (QIAGEN).
17. Minimal media agar plate: Add 5.6 g 5 × M9 salt and 15 g agar to 500 mL of deionized water and autoclave. When cooled to

Table 1
List of primers used in this protocol

Name*	Sequence
H1-f (H1-rc-b)	CAGCGGATTCACCTTCAGC
H1-b (H1-rc-f)	AGGTGCTTGGCGAACCCA
H2-f (H2-rc-b)	GGCCTGGAATGGGTGAGC
H2-b (H2-rc-f)	GCGGCTGATGGTAAAGCG
H3-f (H3-rc-b)	GGACACCGCAGTCTACTACT
H3-b (H3-rc-f)	CACCAGAGTACCTTGTCCC
K1-f (K1-rc-b)	CGCGCAACACTGTCATGC
K1-b (K1-rc-f)	TGGAGCCTGACCTGGTTTC
K2-f (K2-rc-b)	CCAGGTCAGGCTCCACGT
K2-b (K2-rc-f)	ACCGCTTCCAGATCCTGAG
K3-f (K3-rc-b)	CTGGAACCTGAGGACTTTG
K3-b (K3-rc-f)	ACTTTAGTGCCCTGACCG
L1-f (L1-rc-b)	GCGCGTGACTATTAGCTGT
L1-b (L1-rc-f)	AGGTGCAGTTCCAGGCAGT
L2-f (L2-rc-b)	GCCTGGAACTGCACCTAAG
L2-b (L2-rc-f)	GCCTGATTTGCTACCGCTA
L3-f (L3-rc-b)	CTTCGCTCCGAAGATGAAG
L3-b (L3-rc-f)	GTCAGCTTGGTACCGCCA
FR4 (FR4-b)	CTGGTGACCGTGAGCAGC
kFR1-f (kFR1-b)	GAAATCGTGCTGACCCAG
lFR3-f (lFR3-b)	CTGGCCATCAGCGGCCTTC
pC3X-f	GCACGACAGGTTTCCCGAC
pC3X-b	AACCATCGATAGCAGCACCG
pUC57-b	TTC GCC ATT CAG GCT GCG
pC3-seq	GTGAGCGGATAACAATTGA
dp-seq	AGAAGCGTAGTCCGGAACG

*Reverse-complement (rc) sequences of these primers were also prepared and used. Names of the reverse-complement primers are given in parentheses

~45 °C, add 1 mL of 1 M $MgSO_4$ (autoclaved), 0.1 mL of 1 M $CaSO_4$ (autoclaved), 5 mL of 40% glucose (filter-sterilized), and 0.25 mL of 1% thiamine HCl (filter-sterilized). Mix evenly with gently stirring and pour on 100 mm diameter polystyrene petri dishes (20 mL per dish). Cool down at room temperature until agar solidifies, and keep the plates at 4 °C.

18. Electrocompetent TG1 *E. coli* cells (Lucigen, Middleton, WI, USA). Recovery medium is supplied with the competent cells. Also streak 1 μL of the electrocompetent TG1 *E. coli* cells on a minimal media plate for panning experiments.
19. Phosphate Buffered Saline (PBS): Dissolve 80 g NaCl, 2.0 g KCl, 17 g Na_2HPO_4, and 1.63 g KH_2PO_4 in 0.95 L deionized water. Set pH to 7.4 with HCl, and add water to 1 L.
20. PBS-Tween-20 (PBST): Add 0.05% (v/v) Tween 20 to PBS. Mix well.
21. 3% milk-PBST (mPBST): Dissolve 3% (w/v) nonfat dried milk in PBST.
22. Electroporation cuvette: 1 mm gap (Bio-Rad, Hercules, CA, USA).
23. Electroporator (Micropulser™, Bio-Rad).
24. Immunotube (Thermo Scientific, Waltham, MA, USA).
25. 5× polyethyleneglycol (PEG) precipitation buffer: 20% PEG-8000 (w/v) and 15% NaCl (w/v) in deionized water.
26. 100 mM triethylamine (TEA) solution: Mix 140 μL of TEA in 10 mL of deionized water.
27. VCSM13 helper phage (Agilent Technologies, Santa Clara, CA, USA).

3 Methods

3.1 Amplification of Oligonucleotides by Polymerase Chain Reaction

1. Dissolve the CDR oligonucleotide mixtures in 25 μL of nuclease-free water. The amount of DNA from the array synthesis is typically small, and the solution will contain a few ng/μL of DNA.
2. Add the followings to nuclease-free water to make a final volume of 100 μL in a PCR tube on ice: 2 ng of template DNA (CDR oligonucleotide mixture), 0.6 μM final concentration each of forward and backward primers, dNTP mixture (0.2 mM final concentration of each dNTP), 10 μL of Pfu polymerase buffer, 1 μL of Taq polymerase (5 units), and 0.2 μL of Pfu polymerase (0.6 units). Primer sequences are shown in Table 1, and primer pairs for the amplification of CDRs are shown in Table 2.
3. Perform PCR with following thermal cycle: initial melting at 94 °C for 2 min; 25 cycles of 94 °C for 30 s, 56 °C for 30 s, and 72 °C for 30 s; final extension at 72 °C, 7 min.

Table 2

Primer pairs for the amplification of CDRs from the oligonucleotide mixture

CDR	Forward primer	Backward primer
H1	H1-f	H1-b
H2	H2-f	H2-b
H3	H3-f	H3-b
K1 (kappa L1)	K1-f	K1-b
K2 (kappa L2)	K2-f	K2-b
K3 (kappa L3)	K3-f	K3-b
L1 (lambda L1)	L1-f	L1-b
L2 (lambda L2)	L2-f	L2-b
L3 (lambda L3)	L3-f	L3-b

4. PCR products are loaded onto 2% agarose gel and electrophoresed, and the gel is inspected under UV light. Gel bands near ~100 bp length are excised, and the amplified DNA fragments are extracted from the agarose gel using the DNA gel extraction kit, according to the manufacturer's protocol.

3.2 Construction of Single-CDR scFv Libraries

1. Amplify parts of the master framework sequences (DP47-linker-DPK22 [κscFv] and DP47-linker-DPL3 [λscFv]) in pUC57 by PCR according to Table 3, using primers shown in Table 1. Use the following PCR mixture: 200 ng of template DNA, 0.6 μM of the forward and backward primers, 0.2 mM each of dNTP, 10 μL of Pfu polymerase buffer, 0.5 μL of Taq polymerase (2.5 units), 0.2 μL of Pfu polymerase (0.6 units), and nuclease-free water added to 100 μL. The amplification thermal cycle is: initial melting at 94 °C for 2 min; 25 cycles of 94 °C for 30 s, 56 °C for 30 s, and 72 °C for 1.5 min; final extension at 72 °C, 7 min. Purify the amplified PCR products by 1.5% agarose gel electrophoresis as described above.
2. Assemble single-CDR scFv libraries by overlap extension PCR, according to the scheme shown in Table 4. Use the same reaction mixture and thermal cycle as in Subheading 3.2, **step 1**, and purify the PCR products by 1% agarose gel electrophoresis as described above.
3. Digest the PCR products and pComb3X vector with *Sfi*I (*see* **Note 2**), and purify the digested DNA by 1% agarose gel electrophoresis as described above.
4. Ligate the *Sfi*I-digested single-CDR scFv libraries and pComb3X vector at room temperature for ~16 h by mixing 1 μg each of the scFv and vector DNA, 5 μL of T4 DNA ligase

Table 3
PCR scheme for the amplification of framework regions for single-CDR scFv library construction

Template	Forward primer	Reverse primer	Product name
λscFv-pUC57	pUC57-b	H1-rc-b	H1-N
		H2-rc-b	H2-N
		H3-rc-b	H3-N
		L1-rc-b	L1-N
		L2-rc-b	L2-N
		L3-rc-b	L3-N
	H1-rc-f	pC3X-f	H1-C
	H2-rc-f		H2-C
	H3-rc-f		H3-C
	L1-rc-f		L1-C
	L2-rc-f		L2-C
	L3-rc-f		L3-C
κscFv-pUC57	pUC57-b	K1-rc-b	K1-N
		K2-rc-b	K2-N
		K3-rc-b	K3-N
	K1-rc-f	pC3X-f	K1-C
	K2-rc-f		K2-C
	K3-rc-f		K3-C

buffer (10×), and 2 μL of T4 DNA ligase (800 units) in 50 μL final volume adjusted with nuclease-free water.

5. After ligation, precipitate DNA by adding 5 μL of 3 M sodium acetate (pH 5.2) and 128 μL of absolute ethanol to the ligation mixture and incubating it at −20 °C for >2 h (70% final ethanol concentration by volume). Subsequently, the precipitated DNA was spun down (14,000 × *g*, 15 min at 4 °C) and washed twice with cold 70% ethanol. The DNA pellets were briefly dried and dissolved in 10 μL of 10% glycerol solution.
6. Mix the ligated DNA (10 μL) with 50 μL of electrocompetent TG1 *E. coli* cells, and add the mixture to an electroporation cuvette (1 mm gap). Incubate the cuvette on ice for 1 min, and transform the bacteria by electroporation (a single 2.50 kV pulse).
7. After electroporation, immediately add 1 mL of warm (37 °C) recovery medium to the cuvette and pipet up and down several times to resuspend the cells. Repeat the procedure once again,

Table 4
Assembly of single-CDR scFv libraries by overlap extension PCR

Library	Template 1[a]	Template 2[b]	Template 3[a]	Fwd primer	Rev primer
H1-scFv	H1-N	H1	H1-C	pUC57-b	pC3X-f
H2-scFv	H2-N	H2	H2-C		
H3-scFv	H3-N	H3	H3-C		
L1-scFv	L1-N	L1	L1-C		
L2-scFv	L2-N	L2	L2-C		
L3-scFv	L3-N	L3	L3-C		
K1-scFv	K1-N	K1	K1-C		
K2-scFv	K2-N	K2	K2-C		
K3-scFv	K3-N	K3	K3-C		

[a]Templates 1 and 3 are PCR products shown in Table 3
[b]Templates 2 are amplified synthetic CDR oligonucleotide mixtures shown in Table 2

and combine the bacterial suspensions (2 mL). Incubate the transformed cells at 37 °C for 1 h with shaking at 250 rpm.

8. To estimate the transformation titer, plate 100 μL of 10^{-3} and 10^{-4} dilutions of the cells on LBAG agar plates (*see* **Note 3**). Centrifuge the remaining cells (2000 × *g*, 15 min), and resuspend the pellet in 200 μL of LB medium. Plate the resuspended bacteria on a 150 mm diameter LBAG agar plate and incubate overnight at 37 °C.
9. Next morning, add 5 mL of SB medium to the 150 mm diameter agar plates, and scrape the bacterial growth using flame-sterilized glass spreader. Add 0.5 volume of sterile 50% glycerol (16.7% final glycerol concentration), mix well, freeze 1 mL aliquots with liquid nitrogen or dry ice–acetone bath, and store at −80 °C.

3.3 Proofreading of Synthetic CDRs

The synthetic CDRs were proofread by one round of panning of the single-CDR scFv libraries against surfaced-immobilized anti-HA monoclonal antibody. Because the scFv constructs have an HA-tag at C-terminus, CDR sequences with stop codons or frameshifts can be selectively depleted. Follow the protocol below for each single-CDR scFv library.

1. Add 50 μL of the frozen single-CDR library *E. coli* stock from Subheading 3.2, **step 9** to 20 mL of SB medium with 100 μg/mL ampicillin, and grow at 37 °C for 2 h with shaking at 200 rpm.

2. When the culture become slightly turbid (OD600 = ~0.5–1), add VCSM13 helper phage (10^{11} pfu) and incubate for 1 h at 37 °C with gentle shaking (120 rpm).
3. Add kanamycin to 70 μg/mL and shake overnight at 30 °C, 200 rpm.
4. On a minimal media plate, streak TG1 *E. coli* cells (*see* **Note 4**). Incubate the plate overnight at 37 °C.
5. Next morning, centrifuge the overnight culture at 14,000 × *g* for 15 min. Take the phage-containing supernatant and add 5 mL of 5× polyethyleneglycol (PEG) precipitation solution. Incubate the mixture on ice for 30 min.
6. While precipitating phage, coat an immunotube with anti-HA monoclonal antibody. Dilute the antibody in 1 mL PBS at 1 μg/mL concentration and add the solution to an immunotube. Incubate the tube at 37 °C for 1 h. Also inoculate 10 mL of SB medium with a single colony of TG1 *E. coli*.
7. Centrifuge the precipitated phage at 14,000 × *g* at 4 °C for 20 min. Discard the supernatant and dissolve the phage pellet in 300 μL PBS. Centrifuge the phage suspension again at 14,000 × *g* for 15 min to clear cell debris.
8. Take the phage-containing supernatant and add 700 μL of mPBST to block the phage for 1 h at room temperature.
9. Remove the coating solution from the immunotube after 1 h, and rinse the tube with deionized water three times. Block the tube by adding mPBST to the brim and incubating at room temperature for 1 h.
10. Remove the blocking solution (mPBST), and add the blocked phage to the tube. Incubate the tube at 37 °C for 1.5 h with shaking (200 rpm). Save a few microliters of the blocked phage for input titration (*see* below).
11. Discard the phage solution and wash unbound phages with PBST. Add 1 mL PBST to the tube, vortex briefly, further fill the tube with PBST to the brim, discard the wash solution, and rinse with deionized water. Repeat the wash cycle three times.
12. Elute captured phages by adding 1 mL of 100 mM triethylamine solution and incubating at room temperature for 10 min (*see* **Note 5**). Vortex briefly, and transfer the eluted phage to a conical tube. Add 0.5 mL of 1 M Tris (pH 7.0) to neutralize the solution.
13. Add 8.5 mL of mid-log phase (OD600 = ~0.5–1.0) TG1 *E. coli* cells. Gently shake the culture (120 rpm) at 37 °C for 1 h to allow phage infection. Also prepare a 10^{-7} dilution (in SB medium) of the input phage from Subheading 3.3, **step 10**.

Mix 1 μL of the diluted phage with 50 μL of the mid-log phase TG1 and incubate at room temperature for 1 h for input titration.

14. After 1 h, take 1 μL of the infected culture and dilute it in 1 mL SB medium. Plate 10 and 100 μL of the diluted bacteria on LB-ampicillin agar plates for output titration. Also plate the input titration mixture from Subheading 3.3, **step 13** on an LB-ampicillin agar plate. Incubate the plates overnight at 37 °C (*see* **Note 6**).
15. Centrifuge the remaining infected bacteria at 3000 × *g* for 15 min. Discard the supernatant and resuspend the pellet in 500 μL of SB medium. Plate the resuspended bacteria on a 150 mm diameter LBAG plate. Incubate the plate overnight at 37 °C.
16. Next morning, collect bacteria from the 150 mm plate and stock 500 μL frozen aliquots as described in Subheading 3.2, **step 9**.

3.4 Construction of the Final scFv Library

Perform the following protocol for each of the sub-libraries (in this example, scFv libraries with a kappa or a lambda light chain repertoire are constructed, amplified, and rescued separately).

1. Take out one 500 μL frozen stock of each proofread single-CDR scFv library from Subheading 3.3, **step 16**. Centrifuge the bacteria at 14,000 × *g* for 15 min.

Table 5

Amplification of the proofread CDR and adjoining framework regions by PCR

Template	Forward primer	Reverse primer	Product Name
H1-scFv (proofread)	pC3X-f	H2-rc-b	H1-PR
H2-scFv (proofread)	H1-rc-f	H3-rc-b	H2-PR
H3-scFv (proofread)	H2-rc-f	FR4-b	H3-PR
L1-scFv (proofread)	L1-f	L2-rc-b	L1-PR
L2-scFv (proofread)	L1-rc-f	lFR3-b	L2-PR
L3-scFv (proofread)	lFR3-f	pC3X-b	L3-PR
K1-scFv (proofread)	kFR1-f	K2-rc-b	K1-PR
K2-scFv (proofread)	K1-rc-f	K3-rc-b	K2-PR
K3-scFv (proofread)	K2-rc-f	pC3X-b	K3-PR
λscFv-pUC57	FR4-f	L1-rc-b	Lambda linker
κscFv-pUC57	FR4-f	kFR1-b	Kappa linker

2. Discard supernatant, and extract plasmid DNA from the pellets by Miniprep following the manufacturer's protocol.
3. Using primers shown in Table 1 and PCR scheme shown in Table 5, amplify the proofread CDR and adjoining framework regions (initial melting at 94 °C for 2 min; 25 cycles of 94 °C for 30 s, 56 °C for 30 s, and 72 °C for 30 s; final extension at 72 °C, 7 min). Use the PCR mixture shown in Subheading 3.2, **step 1**.
4. Separate and extract DNA bands from 1% agarose gel using the DNA gel extraction kit, following the manufacturer's protocol.
5. Assemble V_H and linker-V_L fragments, each with three diversified and proofread CDRs, by overlap extension PCR following the scheme shown in Table 6 (initial melting at 94 °C for 2 min; 25 cycles of 94 °C for 30 s, 56 °C for 30 s, and 72 °C for 1 min; final extension at 72 °C, 7 min). Use the PCR mixture shown in Subheading 3.2, **step 1** except the amounts of the template DNA. Separate and extract DNA bands from 1% agarose gel using the DNA gel extraction kit, following the manufacturer's protocol.
6. Assemble the final scFv repertoires with six diversified and proofread CDRs by overlap-extension PCR of V_H and linker-V_L fragments using pC3X-f and pC3X-b primers. Use the same thermal cycle and reaction mixture as in Subheading 3.2, **step 1**, except the amounts of the template DNA (50 ng each of V_H and linker-V_L fragments). Perform eight 100-μL reactions in parallel.
7. Pool the PCR products (800 μL total) and add 80 μL of 3 M sodium acetate (pH 5.2) and 2.0 mL of absolute ethanol. Mix well and incubate at −20 °C for >2 h. Centrifuge the mixture at 14,000 × *g* for 15 min, and dissolve the precipitate DNA

Table 6
Assembly of VH, linker-VL, and scFv by overlap extension PCR

Template 1	Template 2	Template 3 (if any)	Fwd primer	Rev primer	Product name
H1-PR	H2-PR	H3-PR	pC3X-f	FR4-b	VH
L1-PR	H2-PR	L3-PR	L1-f	pC3X-b	VL
K1-PR	K2-PR	K3-PR	kFR1-f	pC3X-b	VK
VL	Lambda linker		FR4-f	pC3X-b	Linker-VL
VK	Kappa linker		FR4-f	pC3X-b	Linker-VK
VH	Linker-VL		pC3X-f	pC3-seq	λ library
VH	Linker-VK		pC3X-f	dp-seq	κ library

pellet in 50 μL of nuclease-free water. Separate and extract DNA bands from 1% agarose gel using the DNA gel extraction kit, following the manufacturer's protocol.

8. Digest the purified PCR product and pComb3X vector with *Sfi*I restriction endonuclease. Incubate 10 μg of DNA with supplied buffer and 40 units of *Sfi*I in 50 μL reaction volume at 50 °C for 16 h. Separate and extract DNA bands from 1% agarose gel using the DNA gel extraction kit, following the manufacturer's protocol.
9. Set up ligation reaction. Mix 2 μg each of the *Sfi*I-digested PCR product and vector, 5 μL of 10× T4 ligase buffer, 2 μL of T4 DNA ligase (800 units), and nuclease-free water to 50 μL final volume. Incubate the reaction mixture overnight at room temperature.
10. Next morning, inactivate the ligase by incubating the reaction mixture for 5 min at 70 °C, then precipitate the ligated DNA and dissolve the pellet in 10 μL of 10% glycerol, as described in Subheading 3.2, **step 5**.
11. Transform electrocompetent TG1 *E. coli* cells with the ligated DNA as described in Subheading 3.2 **steps 6** and 7.
12. For transformation titration, plate 100 μL of 10^{-3} and 10^{-4} dilutions of the cells on LBAG agar plates (*see* **Note 3**). Add the remaining culture to 400 mL of SB medium supplemented with 100 μg/mL ampicillin and 2% glucose. Incubate the culture overnight at 37 °C with shaking (200 rpm).
13. Next morning, harvest the cells by centrifugation (3000 × *g*, 15 min), resuspend the pellet in 10 mL SB medium, and add 0.5 volume of 50% glycerol (16.7% final glycerol concentration). Mix well, freeze 1 mL aliquots with liquid nitrogen or dry ice–acetone bath, and store at −80 °C.
14. Phage antibody library can be rescued from the frozen *E. coli* stocks and used for the generation of antigen-binding clones, following the protocol described in [6].

4 Notes

1. For specific amplification of CDRs, some nucleotides in the codon-optimized master framework sequences may need to be changed to minimize cross-priming while maintaining the same amino acid sequences. Antibody variable gene segments have relatively high sequence similarities with one another, and careful inspection of the primer sequences for possible nonspecific priming or primer dimer formation is necessary.

2. Incubation of the reaction mixture at 50 °C in a water bath or a dry block heater may cause evaporation and condensation of water droplets on the cap of the microcentrifuge tube, which changes the concentrations of reaction components and affects the digestion efficiency and/or accuracy. To minimize this, a heater with heated lid can be used. Alternatively, the reaction tubes can be put in a 50 mL conical tube, which is then submerged using a flask weight in a water bath set at 50 °C.
3. The transformation titer is calculated as:

   [No. of colonies × 2 (mL recovery medium culture) × 1000 (μL/mL)]/[100 (μL plated) × dilution fold (10^{-3} or 10^{-4})].
4. TG1 *E. coli* has a genotype of:

   K-12 *glnV44 thi-1 Δ(lac-proAB) Δ(mcrB-hsdSM)5, (rK– mK–)* F′ [*traD36 proAB+ lacIq lacZΔM15*].

   Growth of TG1 cells on a minimal media agar plate (with thiamine and without proline) selects for bacteria with F' factor which harbors *proA* and *proB* genes required for proline biosynthesis. Also the amber stop codon (UAG) is suppressed by the *glnV44* mutation, which is required for the display of a foreign protein on the surface of M13 phage when using pComb3X or some other phagemid vectors that carry an amber codon at the 5' end of *gIII*.
5. The pH of 100 mM triethylamine is about 11.5. At pH 11, M13 bacteriophage retains the infectivity for at least 20 min [10], suggesting that the elution time of 10 min would probably not adversely affect the outcome of panning.
6. The input titer is calculated as:

[No. of colonies × 1000 (μL blocked phage)]/[dilution fold (10^{-7}) × 1 (μL of diluted phage added to TG1)].

The output titer is calculated as:

[No. of colonies × 10 (mL infected culture) × 1000 (μL/mL)]/[dilution fold (10^{-3}) × plated volume in μL (10 or 100 μL)].

Acknowledgment

This work was supported by the National Research Foundation of Korea (NRF) grant for Medical Bioconvergence Research Center (NRF-2013M3A6A4044991) and the Bio & Medical Technology Development Program of the NRF funded by the Korean government, MSIP (NRF-2015M3A9B6029138) to H.S.

References

1. Barbas CF 3rd, Bain JD, Hoekstra DM, Lerner RA (1992) Semisynthetic combinatorial antibody libraries: a chemical solution to the diversity problem. Proc Natl Acad Sci U S A 89:4457–4461
2. Prassler J, Thiel S, Pracht C, Polzer A, Peters S, Bauer M, Norenberg S, Stark Y, Kolln J, Popp A, Urlinger S, Enzelberger M (2011) HuCAL PLATINUM, a synthetic Fab library optimized for sequence diversity and superior performance in mammalian expression systems. J Mol Biol 413:261–278
3. Tiller T, Schuster I, Deppe D, Siegers K, Strohner R, Herrmann T, Berenguer M, Poujol D, Stehle J, Stark Y, Hessling M, Daubert D, Felderer K, Kaden S, Kolln J et al (2013) A fully synthetic human Fab antibody library based on fixed VH/VL framework pairings with favorable biophysical properties. MAbs 5:445–470
4. Zhai W, Glanville J, Fuhrmann M, Mei L, Ni I, Sundar PD, Van Blarcom T, Abdiche Y, Lindquist K, Strohner R, Telman D, Cappuccilli G, Finlay WJ, Van den Brulle J, Cox DR et al (2011) Synthetic antibodies designed on natural sequence landscapes. J Mol Biol 412:55–71
5. Knappik A, Ge L, Honegger A, Pack P, Fischer M, Wellnhofer G, Hoess A, Wolle J, Pluckthun A, Virnekas B (2000) Fully synthetic human combinatorial antibody libraries (HuCAL) based on modular consensus frameworks and CDRs randomized with trinucleotides. J Mol Biol 296:57–86
6. Bai X, Kim J, Kang S, Kim W, Shim H (2015) A novel human scFv library with non-combinatorial synthetic CDR diversity. PLoS One 10:e0141045
7. Rothe C, Urlinger S, Lohning C, Prassler J, Stark Y, Jager U, Hubner B, Bardroff M, Pradel I, Boss M, Bittlingmaier R, Bataa T, Frisch C, Brocks B, Honegger A et al (2008) The human combinatorial antibody library HuCAL GOLD combines diversification of all six CDRs according to the natural immune system with a novel display method for efficient selection of high-affinity antibodies. J Mol Biol 376:1182–1200
8. Andris-Widhopf J, Rader C, Steinberger P, Fuller R, Barbas CF 3rd (2000) Methods for the generation of chicken monoclonal antibody fragments by phage display. J Immunol Methods 242:159–181
9. Sambrook J, Russell DW (2001) In molecular cloning: a laboratory manual. Cold Spring Harbor Laboratory Press, Cold Spring Harbor, NY, p. 1.112
10. Branston SD, Stanley EC, Ward JM, Keshavarz-Moore E (2013) Determination of the survival of bacteriophage M13 from chemical and physical challenges to assist in its sustainable bioprocessing. Biotechnol Bioprocess Eng 18:560–566

Chapter 3

Enzymatic Assembly for scFv Library Construction

Mieko Kato and Yoshiro Hanyu

Abstract

Recombinant monoclonal antibodies can be established by displaying single-chain variable fragment (scFv) antibody libraries on phages and then biopanning against the target. For constructing superior scFv libraries, antibody light-chain variable region (V_L) and heavy-chain variable region (V_H) fragments must be assembled into scFvs without loss of diversity. A high-quality scFv library is a prerequisite for obtaining strong binders from the scFv library. However, the technical challenges associated with the construction of a diverse library have been the bottleneck in the establishment of recombinant antibodies through biopanning. Here, we describe a simple and efficient method for assembling V_L and V_H fragments through the concerted action of λ-exonuclease and *Bst* DNA polymerase. We successfully used this method to construct a diverse chicken scFv library.

Key words scFv, Phage display library, λ-exonuclease, Assembly, Antibody

Abbreviations

scFv	Single-chain variable fragment
V_L	Light-chain variable region
VH	Heavy-chain variable region
CDR	Complementarity-determining region

1 Introduction

Recombinant monoclonal antibodies can be established by screening antibody libraries [1]. Because these antibodies are screened in vitro, the assays can be specifically tailored to the properties of the antigen and the desired antibody [2]. Efficient establishment of high-affinity antibodies requires monoclonal antibody selection by using genetic engineering techniques such as phage display [3]. Antibody heavy-chain and light-chain variable region (V_H and V_L) genes are isolated from the lymph tissue of immunized animals and linked to create scFv libraries, which are incorporated into a phagemid and expressed in phages. Subsequently, scFvs exhibiting

Thomas Tiller (ed.), *Synthetic Antibodies: Methods and Protocols*, Methods in Molecular Biology, vol. 1575,
DOI 10.1007/978-1-4939-6857-2_3, © Springer Science+Business Media LLC 2017

antigen affinity are isolated through panning to establish monoclonal antibodies. The selected phages contain the DNA encoding specific scFvs, and the complementarity-determining regions (CDRs) from the scFvs can be recloned into a vector to express full-size recombinant monoclonal antibodies. Currently, many researchers use synthetic [4] and naïve antibody libraries [5], which can be derived e.g., from B cells isolated from lymph nodes of non-immunized animals. However, an immune library, which on the other hand is derived from an immunized animal, is the most reliable source for establishing monoclonal antibodies exhibiting high affinity and specificity.

For selecting clones displaying scFvs with high affinity for target antigens, one must construct scFv libraries that reflect the diversity of the parental antibody library [6]. To construct such scFv libraries, antibody V_H and V_L genes are amplified, joined by using an appropriate linker sequence, and incorporated into expression vectors. When creating scFv libraries, V_L and V_H genes must be incorporated into expression vectors in a specific order to prevent mutations, deletions, and frameshifts. Moreover, because these genes contain a region that exhibits considerable diversity (CDR; the antigen-recognition region), scFv libraries must be constructed in a manner that maintains this diversity.

Two approaches can be used to construct scFvs: assembly of V_L and V_H domains through either successive cloning [7] or overlap-extension PCR [8]. The first method constructs scFvs through successive cloning of V_L and V_H domains into a phagemid; however, this method, although reliable, requires several steps and is very time-consuming. In the second approach, V_L and V_H domains can be assembled using overlap-extension PCR. Primers for scFv-fragment assembly and amplification are designed to include a 10–15-residue-long overlapping linker sequence, and the primers are used for separately amplifying V_L and V_H domains. A second overlap-extension PCR is then performed using a mixture of V_L and V_H fragments to connect these fragments, followed by cloning of the resulting PCR products into a phagemid vector [9]. V_H and V_L libraries comprise genes containing extremely diverse antibody recognition sites. Therefore, if the diversity region includes a sequence resembling the original primer-binding site, then uniform PCR-amplification will be inhibited. Consequently, clones featuring certain sequences will be selectively amplified, which will substantially reduce library diversity.

We developed an efficient method for constructing scFv libraries through enzymatic assembly of V_L and V_H domains by using the concerted action of λ-exonuclease and *Bst* DNA polymerase [10]. Our strategy for assembling V_H and V_L gene libraries outlined herein (Fig. 1) enables simple construction of scFv libraries: The V_L reverse primer and V_H forward primer contain the overlapping linker sequence GGSSRSS [11, 12]. V_L and V_H are PCR-amplified

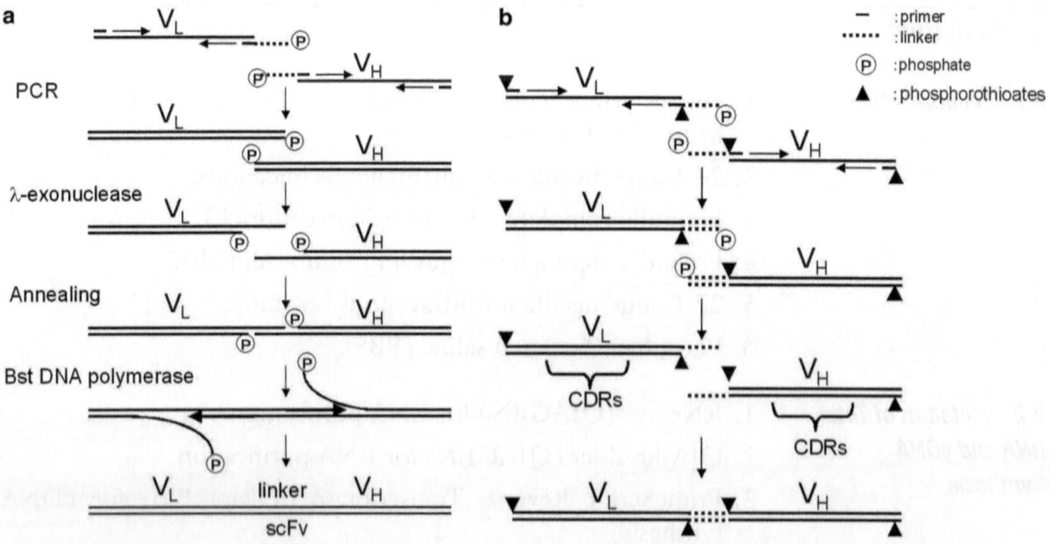

Fig. 1 Schematic diagram of enzymatic assembly of V_L and V_H domains. (**a**) V_L and V_H fragments were PCR-amplified. The V_L reverse primer and V_H forward primer contain the overlapping linker sequence and their 5′-ends were phosphorylated. The PCR products were mixed and treated with λ-exonuclease. Following digestion with the enzyme, the overlapping region formed the single-stranded DNA overhangs, which annealed to generate the scFv. *Bst* polymerase was used to polymerize the new complementary single strand while simultaneously displacing the original single strand. (**b**) Two consecutive phosphorothioates were introduced at the 3′-end of the V_L reverse primer and V_H forward primer to prevent further digestion. Two consecutive phosphorothioates were also introduced at the 5′-end of the forward primer of V_L and reverse primer of V_H to block nonspecific digestion of the non-phosphorylated 5′-ends by λ-exonuclease (reproduced from [10] with permission from Elsevier)

using these primers, whose 5′-ends are phosphorylated (Fig. 1a), and the PCR products are mixed and treated with λ-exonuclease to remove nucleotides from the phosphorylated 5′-ends of the double-stranded DNA molecules [13]. After λ-exonuclease digestion, the overlapping region forms single-stranded DNA overhangs that anneal to generate the scFv. Lastly, *Bst* polymerase is used to polymerize the new complementary single strand while concurrently displacing the original single strand [14]. Phosphorothioates are used in the primers to limit the digestion by λ-exonuclease (Fig. 1b); 3′-end phosphorothioation of the phosphorylated primer can regulate the length of λ-exonuclease-mediated DNA digestion, which in turn will prevent the conversion of the highly diverse CDR into a single strand and thus substantially reduce background reactions.

We constructed an scFv library from a chicken IgG immune library and analyzed the library's diversity through DNA fingerprinting. Our method can also be employed for highly specific assembly of two DNA molecules and could serve as a useful molecular engineering tool [15].

2 Materials

2.1 Immunization

1. Female white leghorn chickens (15–16 weeks old), obtained from a local supplier.
2. 24-Gauge needle for subcutaneous injections.
3. Freund's complete adjuvant (Sigma-Aldrich).
4. Freund's incomplete adjuvant (Sigma-Aldrich).
5. 27-Gauge needle for intravenous boosting.
6. Phosphate-buffered saline (PBS).

2.2 Isolation of Total RNA and cDNA Synthesis

1. RNeasy® (QIAGEN) for RNA purification.
2. QIAshredder (QIAGEN) for RNA purification.
3. PrimeScript Reverse Transcriptase (Takara Bio) for cDNA synthesis.
4. RNase AWAY™ (Thermo Scientific).
5. Experion™ automated electrophoresis system (Bio-Rad).

2.3 PCR-Amplification of V_L and V_H Fragments

1. KOD FX polymerase (Toyobo) (*see* **Note 1**).
2. Primers (Table 1) (*see* **Note 2**).
3. SeaKem GTG Agarose (Lonza).
4. NucleoSpin Gel and PCR Clean-up kit (Macherey-Nagel).

Table 1
Oligonucleotide primers used for PCR of antibody genes

Primers	5′ to 3′ sequence
V_L forward	GTGGCCCAGCCGGCCCTGACTCAGCCGTCCTCGGTGTC
V_L forward with phosphorothioates	gaTCGTGGCCCAGCCGGCCCTGACTCAGCCGTCCTCGGTGTC
V_L reverse	GGAAGATCTAGAGGAACCACCTAGGACGGTCAGG
V_L reverse with phosphorothioates	GGAAGATCTAGAGGAACCACCTAGGACGGTCagG
V_H forward	GGTGGTTCCTCTAGATCTTCCGCCGTGACGTTGGACGAG
V_H forward with phosphorothioates	GGTGGTTCCTCTAGATCTTCCGCCGTGACGTtgGACGAG
V_H reverse	CCTTTTGCGGCCGCACTAGTGGAGGAGACGATGACTTCGGTCC
V_H reverse with phosphorothioates	ccTTTTGCGGCCGCACTAGTGGAGGAGACGATGACTTCGGTCC

Nucleotide with phosphorothioates is indicated as small character (reproduced from [10] with permission from Elsevier)

2.4 Construction of scFv Library Through Enzymatic Assembly

1. λ-exonuclease (New England Biolabs).
2. *Bst* polymerase (New England Biolabs).

2.5 Transformation

1. *Sfi*I and *Not*I (New England Biolabs).
2. pCAMTAB5E phagemid (GE Healthcare).
3. T4 DNA ligase (New England Biolabs).
4. Electroporation cuvette, 0.1-cm gap (Bio-Rad).
5. Electrocompetent *Escherichia coli* TG1 (Stratagene).
6. Micro Pulser (Bio-Rad).
7. SOC medium: 20 g Bacto tryptone, 5 g Bacto yeast extract, 0.5 g NaCl, 20 mM glucose, 2.5 mM KCl, 10 mM $MgCl_2$, 10 mM $MgSO_4$, make up to 1 L.
8. SOBAG plates: 20 g Bacto tryptone, 5 g Bacto yeast extract, 0.5 g NaCl, 10 mL of 1 M $MgCl_2$, 1.5% agar, 2% glucose, 1 mL of 50 mg/mL carbenicillin, make up to 1 L.

2.6 Fingerprinting of Enzymatically Assembled scFvs

1. Ex Taq DNA polymerase (Takara Bio).
2. *Bst*OI enzyme (Promega).

2.7 Phage Display of scFv Libraries

1. SOB medium: 20 g Bacto tryptone, 5 g Bacto yeast extract, 0.5 g NaCl, 10 mL of 1 M $MgCl_2$, make up to 1 L.
2. M13-KO7 helper phage (New England Biolabs).
3. Carbenicillin, 100 mg/mL stock.
4. Kanamycin, 100 mg/mL stock.
5. Polyethylene glycol (PEG) 6000 (Fluka) (*see* **Note 3**).
6. Bovine serum albumin (BSA).

3 Methods

3.1 Immunization

1. Prepare an emulsion of antigen with Freund's adjuvant (*see* **Note 4**).
2. Inject the antigen emulsion (50–200 μg) subcutaneously twice into 15–16-week-old female white leghorn chickens; the first injection is in Freund's complete adjuvant, the second in Freund's incomplete adjuvant in a volume of 0.2 mL (*see* **Note 5**).
3. Perform test bleeds after immunization procedure.
4. Boost intravenously with 200 μg of antigen in normal PBS at least 4 weeks after the second subcutaneous injection.
5. Remove the spleens aseptically from the chickens after sacrificing them at 3 days after the intravenous boost.

3.2 Isolation of Total RNA and cDNA Synthesis

1. Cut spleen tissue into pieces, suspend in RLT buffer (RNeasy), and homogenize with QIAshredder (*see* **Note 6**).
2. Purify total RNA from chicken spleens by using RNeasy according to manufacturer instructions (*see* **Note 7**).
3. Elute the purified total RNA by adding 50 μL of RNase-free water. The RNA is ready for cDNA synthesis (*see* **Note 8**).
4. Add 4 μg of total RNA, 2 μL of 50 μM oligo(dT) primer, 2 μL of 100 μM random hexamer primer, 16 μL of 5× PrimeScript Buffer, and 2 μL of PrimeScript Enzyme Mix to a sterile 0.2-mL tube and adjust the volume to 80 μL with RNase-free water.
5. Incubate at 37 °C for 15 min.
6. Terminate the reaction by incubating the sample at 85 °C for 5 s (*see* **Note 9**).

3.3 PCR-Amplification of V_L and V_H Fragments from Antibody Library

1. Use the primers listed in Table 1; primers are dissolved at a concentration of 50 μM (*see* **Note 10**).
2. Amplify V_L and V_H by performing PCR with the synthesized cDNA as the template. Set up the following PCR reactions:

PCR mix for VL	
cDNA	0.5 μL
V_L Forward primer (50 μM)	1 μL
V_L Reverse primer (50 μM)	1 μL
2× PCR buffer for KOD FX polymerase	25 μL
dNTPs (2 mM each)	10 μL
KOD FX polymerase	1 μL
H_2O	11.5 μL

PCR mix for VH	
cDNA	0.5 μL
V_H Forward primer (50 μM)	1 μL
V_H Reverse primer (50 μM)	1 μL
2× PCR for KOD FX buffer	25 μL
dNTPs (2 mM each)	10 μL
KOD FX polymerase	1 μL
H_2O	11.5 μL

3. Run PCR program (*see* **Note 11**):

1 cycle:	98 °C for 2 min
25 cycles:	98 °C for 10 s, 58 °C for 30 s, 68 °C for 1 min
1 cycle:	68 °C for 3 min

4. Purify the PCR-amplified V_L and V_H fragments on 2% agarose gels by using a NucleoSpin Gel and PCR Clean-up kit. Elute in 50 μL of DNase-free water (*see* **Note 12**).

3.4 Construction of scFv Library Through Enzymatic Assembly

1. Treat V_L and V_H fragments with λ-exonuclease (*see* **Note 13**). Set up the following λ-exonuclease reactions:

V_H or V_L PCR product	1 μg
10× reaction buffer	3 μL
λ-exonuclease	2 μL (10 U)
Adjust the volume to 30 μL with DNase-free water	

2. Treat the mixture for 1 min at 37 °C (*see* **Note 14**).
3. Heat-denature at 75 °C for 10 min (*see* **Note 15**).
4. Mix V_H and V_L at a 1:1 molar ratio. Add 4 μL of 10× reaction buffer and 2 μL of *Bst* polymerase to the mixture of V_H and V_L, and adjust the volume to 40 μL with DNase-free water.
5. Treat the annealed fragment with *Bst* DNA polymerase for 30 min at 65 °C to form a complete double-stranded scFv (*see* **Note 16**).
6. Analyze 5 μL of reaction products by performing 1.5% agarose-gel electrophoresis to check product size and purity. If the assembly reaction is successful, a single strong band should be visible; the expected size for a single-chain clone is about 750 base pairs (bp). (*see* **Note 17**).
7. Purify the assembled scFv genes on 1.5% agarose gels by using a NucleoSpin Gel and PCR Clean-up kit. Elute in 40 μL of Elution buffer.

3.5 Transformation

1. Set up restriction-digestion reactions for the assembled scFv genes. Mix 1 μL of *Sfi*I, 5 μL of 10× enzyme buffer, 3 μL of DNase-free water, and 40 μL of assembled scFv genes for 3–4 h at 50 °C, and then add 1 μL of *Not*I and incubate the mixture for 2 h at 37 °C. Purify the digested scFv genes with a NucleoSpin Gel and PCR Clean-up kit.
2. Digest the pCAMTAB5E phagemid vector with *Sfi*I and *Not*I in the same manner. Purify the digested vector with a NucleoSpin Gel and PCR Clean-up kit (*see* **Note 18**).
3. Ligate 100 ng of scFv fragments with the pCAMTAB5E phagemid vector (vector-to-insert molar ratio = 1:5) by using 2 μL of T4 DNA ligase in the presence of 1× T4 DNA ligase buffer in a total volume of 50 μL. Incubate for 1 h at 16 °C (*see* **Note 19**).
4. Purify the ligated pCAMTAB5E/scFv expression-vector library by using a NucleoSpin Gel and PCR Clean-up kit, and

then elute the purified library from the column by adding 30 μL of DNase-free water (*see* **Note 20**).

5. Place electrocompetent *E. coli* TG1 cells and a 0.1 cm electroporation cuvette on ice.
6. Mix 40 μL of the *E. coli* TG1 cell suspension with 1 μL of the pCAMTAB5E/scFv expression-vector library. Mix well and incubate on ice for 1 min. Transfer the mixture of cells and DNA to the cold electroporation cuvette.
7. Place the cuvette in the chamber in a Micro Pulser and apply the electroporation pulse by using the program Ec1 (1.8 kV and one pulse).
8. Remove the cuvette from the chamber and add 1 mL of SOC medium to the cuvette immediately. Gently resuspend the cells by using a Pasteur pipette (*see* **Note 21**).
9. Incubate the cultures for 1 h at 37 °C with shaking at 200 rpm.
10. Dilute the transformed cultures 100–100,000-fold with prewarmed SOC medium and plate 100 μL of each dilution on SOBAG plates and culture overnight at 30 °C. Calculate the library size from the total number of transformants (*see* **Note 22**).

Figure 2 shows the results of enzymatic gene assembly for the construction of the scFv library. The cDNA was synthesized using RNA from the spleen of a chicken immunized with rabbit IgG. After λ-exonuclease and *Bst* polymerase treatment, V_L and V_H fragments (350 and 400 bp, respectively) were assembled into the larger scFv fragment (750 bp). Assembly efficiency was higher with phosphorothioate-containing primers than with primers lacking phosphorothioates.

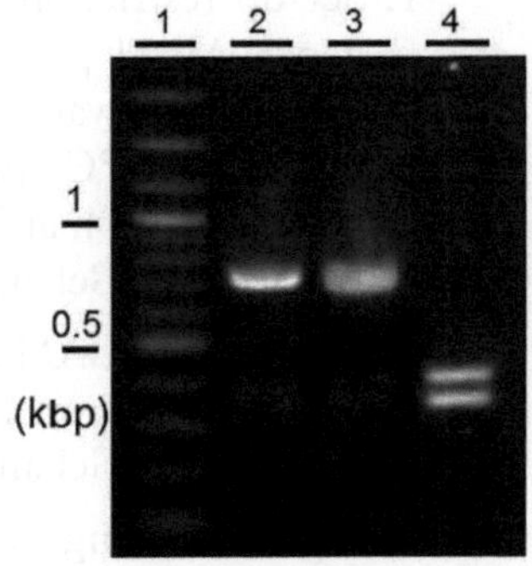

Fig. 2 Agarose-gel electrophoresis analysis of the assembly of the V_L and V_H library. Lane 1, 100-bp DNA ladder marker; lane 2, reaction products from the PCR performed with phosphorylated primers; lane 3, reaction products from the PCR performed with the phosphorylated primers that included phosphorothioates; lane 4; mixture of the PCR products of the V_L and V_H library (reproduced from [10] with permission from Elsevier)

3.6 Fingerprinting of scFvs Prepared Through Enzymatic Assembly

1. For each clone to be analyzed, amplify the scFvs through colony-PCR by using Ex Taq DNA polymerase with the V_L forward and V_H reverse primers (Table 1) at 50 μM each.

   Colony-PCR master mix (for n clones):

n × 0.5 μL	V_L forward primer (GTGGCCCAGCCGGCCCTG ACTCAGCCGTCCTCGGTGTC)
n × 0.5 μL	V_H reverse primer (CCTTTTGCGGCCGCACTA GTGGAGGAGACGATGACTTCGGTCC)
n × 3 μL	10× Ex Taq buffer
n × 2 μL	2.5 mM dNTPs
n × 0.25 μL	Ex Taq DNA polymerase
n × 23.75 μL	DNase-free water

2. Add 30 μL of the PCR master mix to each PCR tube.
3. Use sterile tips to pick clones from SOBAG plates and introduce *E. coli* cells into the PCR tubes.
4. Run PCR program:

1 cycle:	98 °C for 2 min
25 cycles:	98 °C for 10 s, 55 °C for 30 s, 72 °C for 1 min

5. Analyze 5 μL of each PCR product in 1.5% agarose gels to check their size and purity (*see* **Note 23**).
6. Prepare a *Bst*OI enzyme master mix containing 2 μL of 10× reaction buffer, 0.2 μL of 100× BSA solution, 3 U (0.3 μL) of *Bst*OI, and DNase-free water to a final volume of 10 μL per reaction.
7. Add 10 μL of the master mix and 10 μL of each colony-PCR reaction to a PCR tube. Incubate for 2 h for at 60 °C.
8. Analyze fragment pattern on 2–3% agarose gels.

To validate the library prepared from the chicken immunized with the rabbit IgG, 48 clones from the library were analyzed using colony-PCR. All the clones carried full-length scFv inserts. Fingerprinting analysis through *Bst*OI digestion was used to demonstrate clonal diversity (Fig. 3). All of the tested clones were different from each other, which indicated that the library contained a broad representation of the V-domain repertoire.

3.7 Phage Display of scFv Libraries

1. Collect cells from SOBAG plates and suspend in 100 mL of SOB medium (adjust OD_{600} to <0.1), supplement with 2% glucose and 0.1 mg/mL carbenicillin.
2. Allow cells to grow at 37 °C with shaking at 200 rpm until OD_{600} = 0.5.

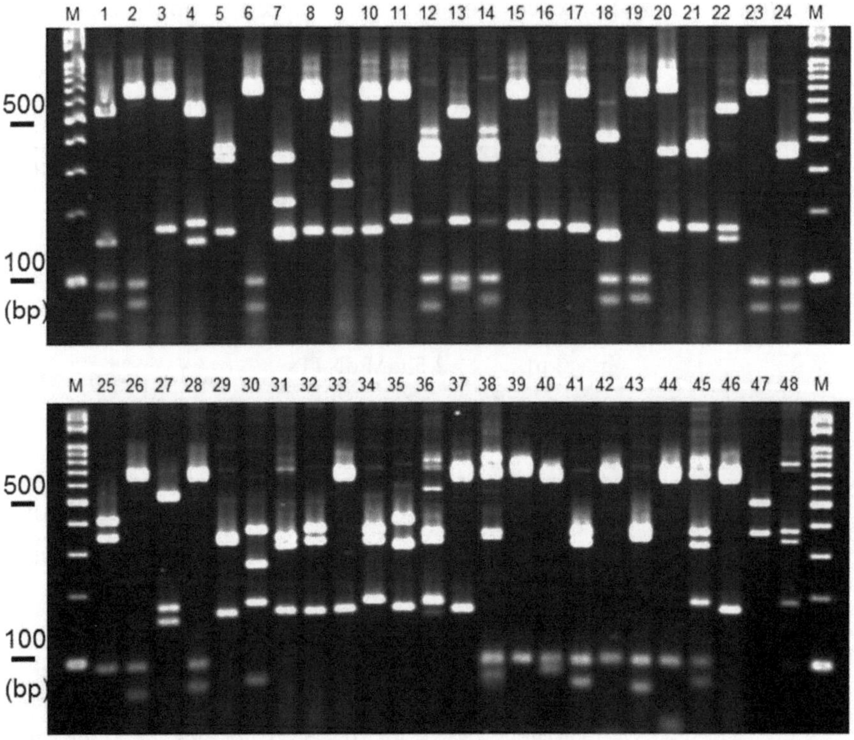

Fig. 3 Validation of scFv libraries prepared through enzymatic assembly. Fingerprinting analysis of randomly selected single clones: The 750-bp V-region-coding DNA from random clones was PCR-amplified, digested with *Bst*OI, and analyzed using 2% agarose-gel electrophoresis (reproduced from [10] with permission from Elsevier)

3. Add 1 mL of M13-KO7 helper phage (10^{11} pfu/mL). Incubate at 37 °C for 1 h with slow agitation (100 rpm).
4. Centrifuge the cultures at 2000 × *g* for 10 min. Resuspend the cell pellet in 100 mL of SOB medium, supplement with 0.1 mg/mL carbenicillin and kanamycin. Grow cultures overnight at 26 °C with shaking at 200 rpm.
5. Pellet cells from overnight cultures through centrifugation at 2000 × *g*. Transfer the supernatant to an appropriate centrifuge tube and add 1/4 volume of 20% PEG 6000/2.5 M NaCl. Mix gently and incubate on ice for 30 min.
6. Centrifuge at 8000 × *g* for 20 min at 4 °C.
7. Resuspend the phage pellet in PBS containing 0.5% BSA (*see* **Note 24**).

4 Notes

1. The scFv library must be constructed to maintain high antibody diversity; thus, a DNA polymerase appropriate for high-fidelity PCR is required.

2. For PCR-amplification of V_H and V_L genes, a linker is added to the 3′-end of one primer and the 5′-end of the other and phosphate is added to both 5′-ends (Fig. 1a). We designed primers to construct scFvs in the $V_L - V_H$ format [11, 12], but the $V_H - V_L$ format is also possible. Primer must be HPLC-purified for precise PCR-amplification. ScFv libraries from mouse [9] or rat [16] can also be constructed, and in these cases, degenerate primers must be used. Because λ-exonuclease cannot digest phosphorothioated DNA, the 5′-end of the non-phosphorylated primer is phosphorothioated to prevent nonspecific digestion by λ-exonuclease (Fig. 1b). Moreover, phosphorothioation at the 3′-end of the phosphorylated primer can regulate the length of the DNA digested by λ-exonuclease, which will prevent the conversion of the highly diverse CDR into a single strand and markedly reduce background reactions.
3. PEG solution must be prepared fresh.
4. Emulsification is critical for generating a high-quality antibody library. Check the quality of the emulsion by dropping it into water before immunization. If the emulsion disperses in water, the formation of the emulsion is inadequate.
5. Injecting antigens multiple times occasionally increases the titer. However, this is not essential for obtaining a high-quality antibody library.
6. Keep all reagents RNase free. Avoid RNase contamination from the bench, centrifuge, pipette, and hands. Dust is also a major source of contamination. Treat all materials with RNase AWAY.
7. Highly pure total RNA must be prepared for constructing the scFv library precisely. Spleen contains larger amounts of DNA than other organs. To obtain highly pure total RNA, the amount of spleen tissue used per column must not exceed 30 mg. RNA quality can be checked with a Experion. RNA Quality Indicator (RQI) must be higher than 9.
8. To obtain highest-quality cDNA, the cDNA should be prepared immediately after purification of total RNA. Storage of total RNA is not recommended, even at −80 °C.
9. The cDNA is ready and can be used for PCR or stored at −80 °C for up to 2 months. If the cDNA is stored at −80 °C, repeated freeze-thaw should be avoided.
10. In this reaction, a linker featuring the sequence GGSSRSS is used [11, 12]. A long linker such as GGSSRSSSSGGGGSGGGG or $(G_4S)_3$ might be more favorable because shorter linkers (<10-mer) occasionally cause scFv multimerization [17]. If a $(G_4S)_3$ linker is used, the middle $(G_4S)_3$ could be used for the overlapping region.

11. To avoid contamination, use pipette tips with filters and predispensed DNase-free water. Contamination makes the genes susceptible to deletion, insertion, replacement, or frameshifts during PCR-amplification, which in turn makes it difficult to obtain functional scFvs. Addressing this problem is critical for generating high-quality libraries.
12. Use highest-quality agarose to prevent contamination in the following enzymatic reactions. Because the scFv library must be constructed to maintain extremely high antibody diversity, special care must be devoted to conducting the experiment precisely.
13. The phosphorylated 5′-ends of the 2 DNA fragments are digested with λ-exonuclease. The overlapping parts of the V_L and V_H fragments are recessed by λ-exonuclease, which yields single-stranded overhangs that specifically anneal between the V_L and V_H fragments (Fig. 1a).
14. Even though λ-exonuclease cannot digest phosphorothioated DNA, the reaction time used must not exceed 1 min; this time restriction is crucial for ensuring that unwanted digestion is avoided.
15. To precisely control λ-exonuclease digestion, the reaction is performed in a programmed thermal cycler.
16. The remaining complementary strand is unzipped in the 3′direction by *Bst* DNA polymerase and a new complementary strand is synthesized (strand-displacement synthesis); this generates a complete double strand in which V_H and V_L are connected by a linker.
17. V_H (400 bp) and V_L (350 bp) assembly generates a band corresponding to scFv (750 bp). Product purity is critical for obtaining a high-quality library. A high yield of the strand-displacement synthesis is crucial for constructing a good library. If the 750-bp band is not clear, repeat the reaction, because even if the obtained scFv band is purified from the gel, the assembly reaction will not be suitable for library construction. The reaction period for the strand-displacement synthesis must be optimized.
18. The restriction enzyme's digestion efficiency critically affects the library quality: Excess reaction will lead to nonspecific digestion, but insufficient reaction will result in the retention of uncut fragments. Thus, restriction-enzyme reactions must be optimized. We recommend checking the number of colonies formed following deficient cutting and self-ligation. Moreover, vector dephosphorylation should be avoided because it lowers the efficiency of ligation between scFvs and the vector.
19. The vector-to-insert ratio should be optimized by conducting a small-scale test before performing large-scale ligation.

20. *E. coli* transformation efficiency is a major factor that determines library size, and thus high-efficiency transformations must be performed using techniques such as electroporation. For effective electroporation, salt must be excluded completely.
21. The time taken to transfer the cells to the growth medium critically affects recovery of transformant clones: A delay of 1 min before adding the SOC medium can cause a threefold reduction in transformation efficiency.
22. To maintain a large antibody library, each enzymatic reaction and *E. coli* transformation must be efficient.
23. A single strong band should be visible; the expected size for a single-chain clone is about 750 bp. For effective analysis of the smaller DNA fragments in the next step (*Bst*OI digest), additional bands should not be amplified in this reaction.
24. The library can now be panned for screening and selecting scFvs that bind antigens.

References

1. Hoogenboom HR (2005) Selecting and screening recombinant antibody libraries. Nat Biotechnol 23:1105–1116
2. Kato M, Hanyu Y (2015) Screening technologies for recombinant antibody libraries. Med Res Arch 2:12–18
3. McCafferty J, Griffiths AD, Winter G et al (1990) Phage antibodies: filamentous phage displaying antibody variable domains. Nature 348:552–554
4. Chen G, Sidhu SS (2014) Design and generation of synthetic antibody libraries for phage display. Methods Mol Biol (Clifton, N.J.) 1131:113–131
5. Chan CEZ, Chan AHY, Lim APC et al (2011) Comparison of the efficiency of antibody selection from semi-synthetic scFv and non-immune Fab phage display libraries against protein targets for rapid development of diagnostic immunoassays. J Immunol Methods 373:79–88
6. Griffiths AD, Williams SC, Hartley O et al (1994) Isolation of high affinity human antibodies directly from large synthetic repertoires. EMBO J 13:3245–3260
7. Müller D (2010) scFv by two-step cloning. In: Kontermann R, Dübel S (eds) Antibody engineering. Springer, Berlin Heidelberg, pp 55–59
8. Krebber A, Bornhauser S, Burmester J et al (1997) Reliable cloning of functional antibody variable domains from hybridomas and spleen cell repertoires employing a reengineered phage display system. J Immunol Methods 201:35–55
9. Schaefer JV, Honegger A, Plückthun A (2010) Construction of scFv fragments from hybridoma or spleen cells by PCR assembly. In: Kontermann R, Dübel S (eds) Antibody engineering. Springer, Berlin Heidelberg, pp 21–44
10. Kato M, Hanyu Y (2013) Construction of an scFv library by enzymatic assembly of VL and VH genes. J Immunol Methods 396:15–22
11. Andris-Widhopf J, Rader C, Steinberger P et al (2000) Methods for the generation of chicken monoclonal antibody fragments by phage display. J Immunol Methods 242:159–181
12. Andris-Widhopf J, Steinberger P, Fuller R et al (2001) Generation of antibody libraries: PCR amplification and assembly of light- and heavy-chain coding sequences. In: Barbas Iii CF, Burton DR, Scott JK et al (eds) Phage display: a laboratory manual. Cold Spring Harbor Laboratory Press, Cold Spring Harbor, New York, pp 102–109
13. Subramanian K, Rutvisuttinunt W, Scott W et al (2003) The enzymatic basis of processivity in Lambda exonuclease. Nucleic Acids Res 31:1585–1596
14. Notomi T, Okayama H, Masubuchi H et al (2000) Loop-mediated isothermal amplification of DNA. Nucleic Acids Res 28:E63
15. Gibson DG (2011) Enzymatic assembly of overlapping DNA fragments. Methods Enzymol 498:349–361
16. Sepulveda J, Shoemaker CB (2008) Design and testing of PCR primers for the construction of scFv libraries representing the immuno-

globulin repertoire of rats. J Immunol Methods 332:92–102

17. Kortt AA, Lah M, Oddie GW et al (1997) Single-chain Fv fragments of anti-neuraminidase antibody NC10 containing five- and ten-residue linkers form dimers and with zero-residue linker a trimer. Protein Eng 10:423–433

Chapter 4

Directed Evolution of Protein Thermal Stability Using Yeast Surface Display

Michael W. Traxlmayr and Eric V. Shusta

Abstract

Yeast surface display is a powerful protein engineering technology that has been used for many applications including engineering protein stability. Direct screening for improved thermal stability can be accomplished by heat shock of yeast displayed protein libraries. Thermally stable protein variants retain binding to conformationally specific ligands, and this binding event can be detected by flow cytometry, facilitating recovery of yeast clones displaying stabilized protein variants. In early efforts, the major limitation of this approach was the viability threshold of the yeast cells, precluding the application of significantly elevated heat shock temperatures (>50 °C) and therefore limited to the engineering of intrinsically unstable proteins. More recently, however, techniques for stability mutant gene recovery between sorting rounds have obviated the need for yeast growth amplification of improved mutant pools. The resultant methods allow significantly higher denaturation temperatures (up to 85 °C), thereby enabling the engineering of a broader range of protein substrates. In this chapter, a detailed protocol for this stability engineering approach is presented.

Key words Thermal stability, Conformationally specific ligand, Thermal denaturation, Yeast surface display, Protein engineering, Directed evolution

1 Introduction

Yeast surface display has been widely used in directed evolution for binder identification and affinity maturation [1–3], epitope mapping [4, 5], enzyme evolution [6–8], and stability engineering of proteins [9, 10]. The yeast display platform employed for the aforementioned studies is based on the a-agglutinin mating proteins [1]. In this system, the protein of interest is expressed as a fusion protein with the a-agglutinin subunit Aga2p. Aga2p, in turn, is linked via two disulfide bonds to the glycosylphosphatidylinositol (GPI) anchored subunit Aga1p [11], ultimately resulting in yeast surface anchorage of the target protein (Fig. 1).

Key to directed evolution experiments is the genotype–phenotype linkage. The challenge for stability engineering screens

Thomas Tiller (ed.), *Synthetic Antibodies: Methods and Protocols*, Methods in Molecular Biology, vol. 1575,
DOI 10.1007/978-1-4939-6857-2_4,

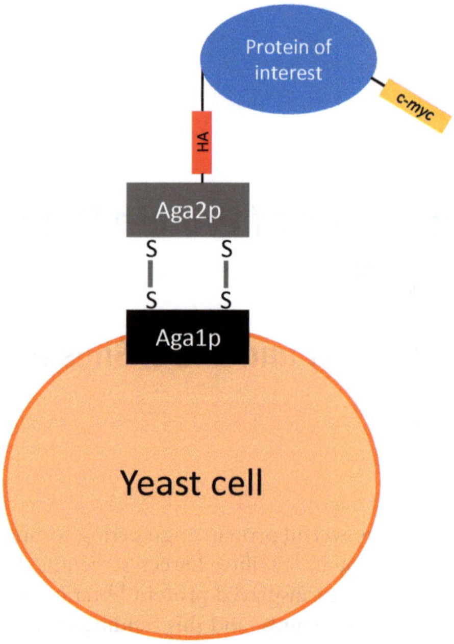

Fig. 1 Molecular architecture of the Aga2p-based yeast display system. The following fusion protein is expressed: Aga2p–HA-tag–protein of interest–c-*myc*-tag. Aga2p is linked via two disulfide bonds to the yeast surface protein Aga1p, resulting in surface anchorage of the entire protein complex. HA- and c-*myc*-tags can be detected with appropriate antibodies, allowing expression level normalization

therefore is to link the thermal stability of the displayed protein to a binding event that can be combined with cell sorting technologies. In general, two such yeast surface display-based screening strategies have been developed. The first approach is based on the correlation between the thermal stability and protein expression levels [9, 12]. Since more stable proteins are often expressed at higher levels in yeast, screening yeast-displayed protein libraries for higher expression levels can often yield thermally stabilized mutants. This approach has been successfully applied for stabilizing a variety of proteins, such as T cell receptors (TCRs) [9, 12–15], class I [16] and class II [17, 18] major histocompatibility complexes (MHC), the extracellular domain of epidermal growth factor receptor (EGFR-ECD) [19], the extracellular domain of the p55 tumor necrosis factor receptor (TNFr-ECD) [20], and binding loop-inserted GFP scaffolds [21]. However, while a straightforward strategy, there are also limitations. For example, some unstable proteins are displayed at high levels on the surface of yeast, but in a misfolded form [17, 19, 22]. Thus, screening for improved stability should be performed by using a conformationally specific ligand that can read-out a properly folded protein variant in addition to assaying for full length protein expression using an antibody directed against a terminally expressed epitope tag (Fig. 1). Moreover, the correlation between thermal stability and expression is not always observed for proteins with a higher basal stability. For example, the thermally stabilized IgG1-Fc (crystallizable fragment of human immunoglobulin G subclass 1) variant K370L showed similar expression levels compared to its wild type counterpart which already had a high basal stability [23]. Similar results were obtained for the de novo engineered three-helix bundle protein α3D [22]. This suggests that there may be a stability threshold above which all proteins are efficiently displayed,

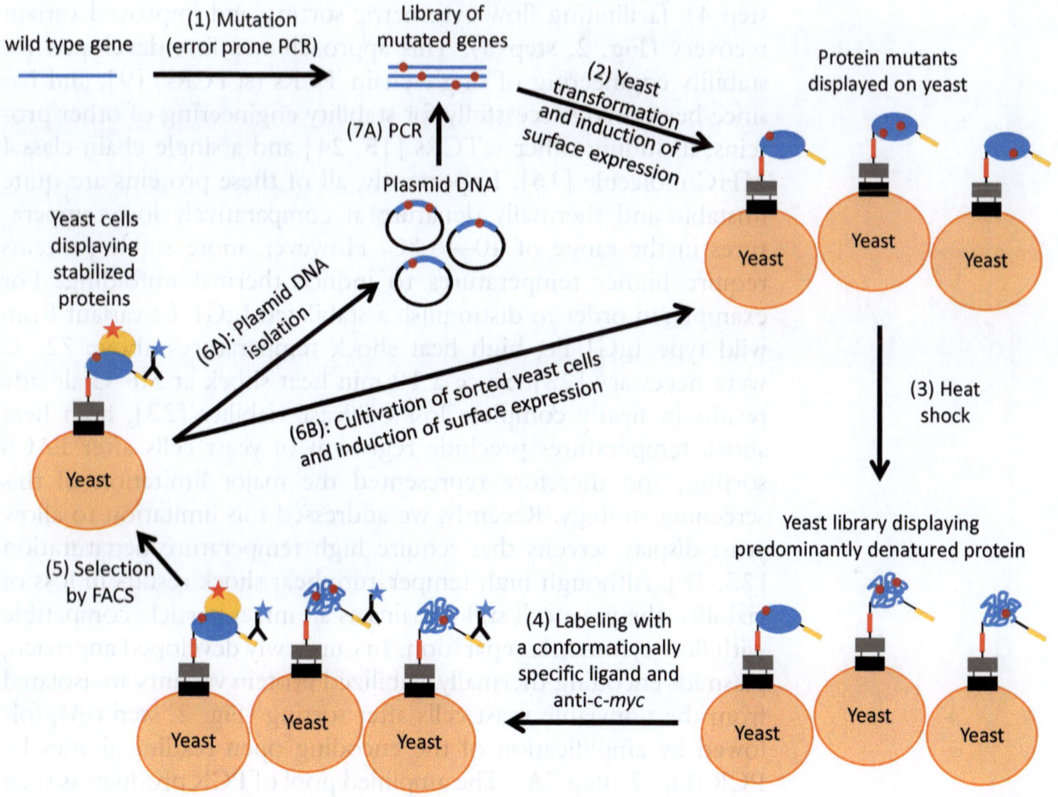

Fig. 2 Experimental procedure for stability engineering by yeast surface display. The wild-type gene is mutated as indicated by *red dots* (*step 1*). For many applications, error prone PCR will be the facile method of choice for random mutagenesis. Yeast libraries are created by homologous recombination of co-transformed linearized vector backbone and mutated genes (*step 2*). After induction of surface display, yeast-displayed protein libraries are subjected to heat shock (*step 3*), resulting in thermal denaturation (indicated by *blue random coils*). Stabilized variants resist thermal denaturation and remain in a native fold. As a consequence they still bind to fluorescently labeled conformationally specific ligands (*yellow spheres* with *red asterisk*, *step 4*), enabling flow cytometric sorting of these cells (*step 5*). The next step depends on the viability of the yeast cells. If the yeast cells did not survive the heat shock ($T > 46$ °C), plasmid DNA is isolated directly after flow cytometric sorting (*step 6A*), followed by insert amplification by PCR (*step 7A*) and transformation of yeast with this newly generated pool of genes. Alternatively, if the heat shock temperature was low enough to yield viable yeast cells ($T < 46$ °C), there is no need for plasmid isolation and therefore the sorted cells can be regrown in yeast media (*step 6B*). The process is repeated until an enriched pool of stability mutants is achieved

rendering expression level screening ineffective for identifying protein variants having improved stability.

The second screening strategy for yeast surface display-based stability engineering directly screens for thermal stability and therefore avoids the potential confounding factors described above. Briefly, after induction of yeast display, the library is subjected to heat shock appropriately chosen in temperature and duration to result in the irreversible thermal denaturation of the wild-type protein (Fig. 2, step 3). Protein variants that are more resistant to irreversible thermal denaturation remain capable of binding to their conformationally specific ligands after the heat shock (Fig. 2, step 4), facilitating flow cytometric sorting and improved variant recovery (Fig. 2, step 5). This approach was first developed for stability engineering of single-chain TCRs (scTCRs) [9], and has since been used successfully for stability engineering of other proteins, including other scTCRs [15, 24] and a single chain-class I MHC molecule [16]. Importantly, all of these proteins are quite unstable and thermally denature at comparatively low temperatures in the range of 40–46 °C. However, more stable proteins require higher temperatures to induce thermal unfolding. For example, in order to distinguish a stabilized IgG1-Fc variant from wild type IgG1-Fc, high heat shock temperatures above 72 °C were necessary [23]. Since a 10 min heat shock at 50 °C already results in nearly complete loss of yeast viability [23], high heat shock temperatures preclude regrowth of yeast cells after FACS sorting, and therefore represented the major limitation of this screening strategy. Recently, we addressed this limitation to allow yeast display screens that require high temperature denaturation [23, 25]. Although high temperature heat shock results in loss of viability, the yeast cell still remains as an intact particle compatible with flow cytometric separation. In our newly developed approach, plasmids encoding thermally stabilized protein variants are isolated from the nonviable yeast cells after sorting (Fig. 2, step 6A), followed by amplification of the encoding open reading frames by PCR (Fig. 2, step 7A). The amplified pool of PCR products is then used for constructing an enriched yeast library for the next screening round. Generally speaking, there are three prerequisites for this approach: (1) a conformationally specific ligand capable of discriminating correctly folded variants that resist thermal denaturation; (2) an irreversible denaturation process in order to avoid refolding after the heat shock and artifactual escape from the screening pressure; (3) a maximum heat shock temperature of 76 and 85 °C for 30 and 10 min incubation times, respectively. Above these temperatures the yeast cells lyse, preventing sorting and recovery of improved variants [23]. Importantly, for most proteins these temperatures are sufficiently high to induce at least partial thermal denaturation.

This enhanced stability engineering strategy has been successfully applied for stabilization of loop-inserted GFP scaffolds [21, 25], wild type IgG1-Fc [23] and of a Her2-binding IgG1-Fc variant [26]. Moreover, this screening strategy has also been combined with deep sequencing, resulting in a stability landscape of the entire CH3 domain of IgG1-Fc showing mutational tolerances at single residue resolution [27]. Thus, this approach can be easily adapted to any given protein, as long as the prerequisites mentioned above are fulfilled. The detailed experimental procedure for the enhanced stability engineering strategy is described in the protocol below.

2 Materials

2.1 Plasmids and Yeast and Bacterial Strains

1. pCTCON2: Cloning of the gene of interest into this plasmid results in yeast surface expression at the C-terminus of Aga2p after induction with galactose (Fig. 1) [1, 28, 29]. The plasmid is available at Addgene (plasmid #41843; insert: CD20).
2. EBY100: This yeast strain carries a genomic insertion of Aga1 under a GAL promoter. EBY100 must be cultivated in complex media (YPD). Cells that have been transformed with the yeast display plasmid pCTCON2 grow in minimal media (SD-CAA or SG-CAA, containing all essential amino acids except for tryptophan), because pCTCON2 contains *TRP1* [28, 29]. EBY100 can be purchased through ATCC (American Type Culture Collection, catalogue no. MYA-4941).
3. *E. coli* top ten chemically competent cells (Thermo Fisher Scientific); these chemically competent cells are delivered together with the *lacZ+* pUC19 plasmid that can be used as an internal standard plasmid.

2.2 Yeast Media and Buffers

1. SD-CAA medium (pH 4.5, *see* **Note 1**): 20 g/L dextrose (D-glucose), 6.7 g/L yeast nitrogen base, 5 g/L casamino acids (−ade, −ura, −trp), 11.85 g/L sodium citrate dihydrate, and 7.4 g/L citric acid monohydrate are filled up with ddH_2O to the final volume and filter-sterilized.
2. SD-CAA agar plates (pH 6.0): 20 g/L dextrose (D-glucose), 6.7 g/L yeast nitrogen base, and 5 g/L casamino acids (−ade, −ura, −trp) are dissolved in 0.1× final volume ddH_2O and sterile-filtered. 5.40 g/L Na_2HPO_4 anhydrous, 8.56 g/L NaH_2PO_4 monohydrate, 15 g/L agar, and 182 g/L sorbitol are filled up with ddH_2O to 0.9× final volume and autoclaved. After cooling the autoclaved mixture to approximately 50–60 °C, the filtered mixture is added.
3. SG-CAA medium: 20 g/L D-galactose, 2 g/L dextrose (D-glucose), 6.7 g/L yeast nitrogen base, 5 g/L casamino acids

(−ade, −ura, −trp), 5.40 g/L Na_2HPO_4 anhydrous and 8.56 g/L NaH_2PO_4 monohydrate are filled up with ddH_2O to the final volume and filter sterilized.

4. YPD medium: 20 g/L dextrose (D-glucose), 10 g/L yeast extract, 20 g/L peptone are filter-sterilized.
5. YPD agar plates: 10 g/L yeast extract, 20 g/L peptone and 15 g/L agar are autoclaved, followed by addition of 20 g/L filter-sterilized dextrose (D-glucose).
6. Phosphate buffered saline (PBS), pH 7.4: 10 mM Na_2HPO_4, 2 mM KH_2PO_4, 137 mM NaCl and 2.7 mM KCl.
7. PBSA: 1 g/L bovine serum albumin (BSA) is added to PBS.

2.3 Reagents and Materials for Cloning and Transformation

1. Deoxynucleotide (dNTP) mix (New England Biolabs).
2. 8-oxo-2′-deoxyguanosine-5′-triphosphate (8-oxo-dGTP) (TriLink BioTechnologies).
3. 2′-deoxy-*p*-nucleoside-5′-triphosphate (dPTP) (TriLink BioTechnologies).
4. Taq DNA Polymerase with *Thermopol* Buffer (New England Biolabs).
5. Primers:

   pCT_forward: 5′-GGCTCTGGTGGAGGCGGTAGCGGAGGCGGAGGGTCGGCTAGC-3′

   pCT_reverse: 5′- CTATTACAAGTCCTCTTCAGAAATAAGCTTTTGTTCGGATCC-3′ (the underlined regions in these primers facilitate homologous recombination with *Nhe*I/*Bam*HI-digested pCTCON2 after yeast transformation).
6. T4 DNA Ligase (New England Biolabs).
7. *Nhe*I-HF (20,000 U/mL, New England Biolabs).
8. *Bam*HI (20,000 U/mL, New England Biolabs).
9. *Sal*I (20,000 U/mL, New England Biolabs).
10. 3 M sodium acetate, pH 5.2.
11. 100 mM lithium acetate.
12. Plasmid Maxi Kit (Qiagen).
13. QIAquick PCR Purification Kit, QIAEX II Gel Extraction Kit or QIAquick Gel Extraction Kit (all from Qiagen).
14. Zymoprep Yeast Plasmid Miniprep II kit (Zymo Research).
15. Electroporation cuvettes, 2 mm gap, 400 μL (VWR).
16. Frozen-EZ Yeast Transformation II Kit (Zymo Research).

2.4 FACS Reagents

1. Mouse anti-c-*myc*, clone 9E10 (Thermo Fisher Scientific).
2. Alexa Fluor 488-conjugated goat anti-mouse IgG (Thermo Fisher Scientific).

3. Conformationally specific ligand, for example an antibody or natural ligand recognizing the native conformation of the protein of interest.
4. An appropriate fluorophore-conjugated secondary reagent is needed for detection of the conformationally specific ligand. If the secondary reagent for ligand detection is a mouse IgG antibody, the c-*myc* detection has to be adapted accordingly (*see* **Note 2**).

2.5 Instruments

1. Bio-Rad Gene Pulser Xcell (Bio-Rad).
2. FACS sorting machine; since only two fluorescence channels are needed for this screening procedure, most FACS sorters will suffice.

3 Methods

3.1 Library Construction

3.1.1 Mutagenesis by Error Prone PCR

A number of mutagenesis strategies have been described in the literature, including error prone PCR using nucleotide analogues [30–32], parsimonious mutagenesis [26, 33], as well as saturation mutagenesis by using oligonucleotides containing fully randomized codons (e.g., NNN, NNK or NNS), among others. While error prone PCR distributes point mutations randomly over the entire gene fragment that is located between the two primers, usage of degenerated oligonucleotides allows high mutation rates within short regions of the gene. Thus, the choice of the most appropriate mutagenesis strategy depends on the aim of the project. In this protocol only error prone PCR using nucleotide analogues will be described, since this randomization strategy is often the method of choice for screening stabilizing point mutations in a given protein (*see* **Note 3**).

1. Ligate the gene of interest into the vector pCTCON2 using the restriction sites *Nhe*I and *Bam*HI using standard molecular biology techniques, resulting in the following fusion construct: Aga2p—HA-tag—(Gly$_4$Ser)$_3$linker—protein of interest—c-*myc*-tag (Fig. 1) (*see* **Notes 4** and **5**). Use the resulting vector as the template for error prone PCR.
2. Amplify the gene of interest by error prone PCR using nucleotide analogues. The primers used in this protocol flank the gene of interest, resulting in the incorporation of random point mutations distributed over the entire gene (Table 1).

Cycling parameters can be seen in Table 2.

3. Purify the PCR-product by excision of the correctly sized band after agarose gel electrophoresis. This step ensures that the majority of the library members will contain the correct insert (*see* **Note 6**).

Table 1
Master mix composition for error prone PCR

	Stock concentration	Final concentration	Volume (μL)
ddH_2O			35.5
Thermopol Reaction Buffer	10×	1×	5
pCT_forward	10 μM	0.5 μM	2.5
pCT_reverse	10 μM	0.5 μM	2.5
dNTP mix (dATP, dCTP, dGTP, and dTTP)	10 mM each	0.2 mM each	1
8-oxo-dGTP	100 μM	2 μM	1
dPTP	100 μM	2 μM	1
Template DNA	10 ng/μL	0.2 ng/μL	1
Taq DNA Polymerase	5 U/μL	0.05 U/μL	0.5

Table 2
PCR cycling parameters

Cycle	Step	Temperature, °C	Time
1 (×1)	Denaturation	94	3 min
2 (×15)	Denaturation	94	45 s
	Annealing	60	30 s
	Extension	72	90 s (for 500 bp insert)
3 (x1)	Final extension	72	10 min

3.1.2 Amplification of Insert by PCR

Further amplify the resulting gel-purified insert by PCR in order to generate sufficient amounts for yeast transformation. The following PCR reaction (200 μL) is sufficient for one electroporation reaction.

1. Adjust the template DNA concentration (the gel-purified error prone PCR product) to 10 ng/μL.
2. Prepare the PCR reactions according to Table 3. The cycling parameters for this PCR are identical to the error prone PCR above, except for an increased number of cycles (30 instead of 15).

3.1.3 Ethanol Precipitation

The PCR-product is purified and concentrated by ethanol precipitation.

1. Transfer the DNA-sample into a 1.5 mL microcentrifuge tube (*see* **Note 7**).

Table 3

Master mix composition for insert amplification PCR

	Stock concentration	Final concentration	Volume (μL)
ddH_2O			150
Thermopol Reaction Buffer	10×	1×	20
pCT_forward	10 μM	0.5 μM	10
pCT_reverse	10 μM	0.5 μM	10
dNTP mix (dATP, dCTP, dGTP, and dTTP)	10 mM each	0.2 mM each	4
Template DNA	10 ng/μL	0.2 ng/μL	4
Taq DNA Polymerase	5 U/μL	0.05 U/μL	2

2. Adjust the pH of the DNA-sample by adding 10% volume of 3 M sodium acetate, pH 5.2.
3. Add at least 2× volume of 100% ethanol and incubate at room temperature for 2 min.
4. Centrifuge at high speed (12,000 × *g*, 5 min).
5. Remove the supernatant and add 500 μL of 70% ethanol.
6. Mix/vortex briefly.
7. Centrifuge at high speed (12,000 × *g*, 5 min).
8. Remove the supernatant and add 500 μL of 100% ethanol.
9. Mix/vortex briefly.
10. Centrifuge at high speed (12,000 × *g*, 5 min).
11. Remove the supernatant and air dry until the pellet is dry (this usually takes a couple of hours; alternatively, the pellet can also be dried overnight).
12. Dissolve dried pellet in 10 μL H_2O.

3.1.4 Vector Preparation

Digest the pCTCON2-CD20 vector by using the restriction enzymes *Nhe*I, *Bam*HI, and *Sal*I according to the manufacturer's protocols (*see* **Note 8**). A detailed protocol for the digestion has been described previously [29]. Finally, purify and concentrate the digested vector to a final concentration of approximately 1 μg/μL by ethanol precipitation as described in Subheading 3.1.3.

3.1.5 Electroporation of Yeast

Yeast can be efficiently transformed by electroporation. The electroporation protocol described here has been optimized for the Bio-Rad Gene Pulser Xcell (Bio-Rad) electroporation system. Yeast cells are electroporated with linearized vector and insert. The insert is flanked by regions that are homologous to the linearized vector (about 30–50 bp on each side), facilitating homologous recombination in the yeast cell after electroporation. The following transformation protocol typically yields 2×10^7 to 1×10^8 transformants per electroporation.

1. Prepare an EBY100 starter culture in YPD and grow overnight at 30 °C with shaking at 250 rpm.
2. Inoculate YPD at OD_{600} = 0.2 and grow (30 °C, shaking at 250 rpm) until an OD_{600} = 1.3–1.5 is reached. This typically takes about 4–6 h (*see* **Note 9**).
3. Centrifuge in 50 mL conical tubes (2000 × g, 3 min). A fifty-milliliter culture volume is enough for two electroporations. Thus, the total volume depends on the number of transformations to be performed. Allow for additional cells for the two recommended control electroporations (*see* **Note 11**).
4. Remove supernatant and resuspend cell pellets with 100 mM lithium acetate to achieve half of the original YPD volume (25 mL per 50 mL tube). Add sterile 1 M dithiothreitol to the cells to a final concentration of 10 mM. Incubate the 50 mL tubes with the cell suspensions at 30 °C with shaking for 10 min.
5. Centrifuge (2000 × *g*, 3 min) and remove supernatant.
6. Place cells on ice; all of the remaining steps before electroporation will be performed on ice and chilled reagents will be used.
7. Resuspend cells in half the original volume (25 mL per 50 mL tube) chilled sterile ddH_2O.
8. Centrifuge (2000 × *g*, 3 min, 4 °C) and remove supernatant.
9. Resuspend the cell pellet in 250 μL of chilled sterile H_2O per 50 mL tube, resulting in a total volume of approximately 500 μL per tube.
10. Chill 2 mm electroporation cuvettes on ice.
11. Add 4 μg of linearized vector to the insert in a microcentrifuge tube. Both, linearized vector and insert were purified and concentrated by ethanol precipitation (*see* Subheading 3.1.3) (*see* **Note 10**).
12. Add 250 μL of cells to the premixed vector and insert DNA and transfer the mixture to a prechilled electroporation cuvette (*see* **Note 11**).
13. Use the Bio-Rad Gene Pulser Xcell for electroporation. The square wave protocol is performed with a single pulse at 500 V and a pulse duration of 15 ms. "Droop" readings that are reported by the machine are typically in the range of 5–6%.
14. Add 1 mL of YPD medium to electroporated cells and transfer the cell suspension to a 15 mL conical tube (Falcon). Rinse

the cuvette with an additional 1 mL of YPD medium and add these cells to the same 15 mL tube. Incubate at 30 °C for 1 h (without shaking).

15. Centrifuge (2000 × *g*, 3 min) and resuspend in at least 100 mL SD-CAA medium.
16. Prepare serial dilutions and plate cell suspensions on SD-CAA agar plates in order to determine the library diversity. The protocol presented here typically yields between 2×10^7 and 1×10^8 transformants.
17. Incubate the cell suspensions at 30 °C with shaking at 250 rpm overnight.
18. On the next day the library can be either induced in SG-CAA medium (see Subheading 3.2), stored at 4 °C or frozen for long term storage (*see* **Notes 12–14**).

3.2 Induction of Surface Expression

1. Incubate the library at 30 °C in SD-CAA medium while shaking overnight.
2. Dilute the culture to an OD_{600} of 1 in SD-CAA medium (usually on the next morning).
3. Pellet at least 10× (100× if possible, *see* **Note 15**) the library diversity (2000 × *g*, 3 min) when the culture reaches an OD_{600} of 3–5 (usually after about 4–6 h).
4. Resuspend cells to an OD_{600} of 1 in SG-CAA medium for induction of surface expression. Usually, yeast surface display libraries are induced at 20 °C while shaking overnight (for 12–24 h). Alternatively, libraries can also be induced at 30 °C, which, however, results in stronger proliferation compared to 20 °C. Moreover, induction at 37 °C has been shown to result in increased screening pressure towards stability for some proteins, such as single chain T cell receptors (scTCRs) and loop-inserted GFP scaffolds [9, 10, 21].

3.3 FACS Screening of Yeast Cells Displaying Thermally Stable Protein Variants

1. Measure OD_{600} of the culture.
2. Pellet the appropriate amount of culture (2000 × *g*, 3 min, 4 °C) to cover at least 10× diversity (100× if possible; *see* **Note 15**).
3. Resuspend cell pellet in PBSA to achieve an OD_{600} of 3.
4. Split the samples into 700 μL aliquots in 1.5 mL microcentrifuge tubes (*see* **Note 16**). Important: Pipet cell suspensions into the bottom of the tubes and do NOT invert tubes before heat shock. Depending on the thermal cycler or thermomixer that is used, the temperature of the lid may differ from that on the bottom of the tube. Thus, the cells that are located underneath the lid after inverting the tube would be subjected to differential thermal stress.
5. Incubate aliquots on ice for at least 10 min (*see* **Note 17**).

6. Incubate aliquots in a thermomixer while shaking at 300 rpm for 10 min at an appropriate heat shock temperature. Note that the choice of the duration and temperature of the heat shock is of critical importance for the success of the screening experiment (*see* **Note 18**).
7. Put samples on ice immediately after the heat shock for at least 5 min.
8. If more than one tube was used for one library, pool all aliquots of the same library; in all following steps in this protocol it is assumed that the sample consists of only one 700 μL aliquot at an OD_{600} of 3 (corresponding to about 2.1×10^7 cells); if the library consists of several aliquots that are pooled in this step, the following volumes need to be adjusted accordingly.
9. Centrifuge the cells (2000 × *g*, 3 min, 4 °C) and discard the supernatant.
10. Resuspend cells (still being in 1.5 mL microcentrifuge tubes) in PBSA containing a conformationally specific ligand (*see* **Notes 19** and **20**) and mouse anti-c-*myc* (clone 9E10, 1:100) (*see* **Note 2**).
11. Incubate at 4 °C, shaking for 30 min.
12. Centrifuge (2000 × *g*, 3 min, 4 °C), discard supernatant and wash with 1 mL PBSA.
13. Centrifuge (2000 × *g*, 3 min, 4 °C) and discard supernatant.
14. Resuspend the cell pellet in secondary reagents labeled with compatible fluorophores. For example, mouse anti-c-*myc* can be detected by using Alexa Fluor 488-conjugated goat anti-mouse IgG (1:200, *see* **Note 21**).
15. Incubate at 4 °C, shaking for 30 min.
16. Centrifuge (2000 × *g*, 3 min, 4 °C), discard supernatant and wash with 1 mL PBSA.
17. Centrifuge (2000 × g, 3 min, 4 °C) and discard supernatant.
18. Store cell pellet on ice and resuspend immediately before the FACS sort in order to minimize dissociation of the ligand.
19. Sort the yeast clones that are most positive for binding to the conformationally specific ligand by FACS. By normalizing for expression density (c-*myc* signal), stabilized variants can be separated more efficiently from their unstable counterparts by using a diagonal sorting gate that is placed just above the main population (Fig. 3).

3.4 Plasmid Isolation from Sorted Yeast Cells Directly After FACS

If the heat shock temperature was low enough to keep the yeast cells alive (Fig. 2, step 6B; *see* **Note 22**), yeast cells can be sorted directly into SD-CAA medium and cultivated at 30 °C while shaking. However, if the yeast cells do not survive the heat shock,

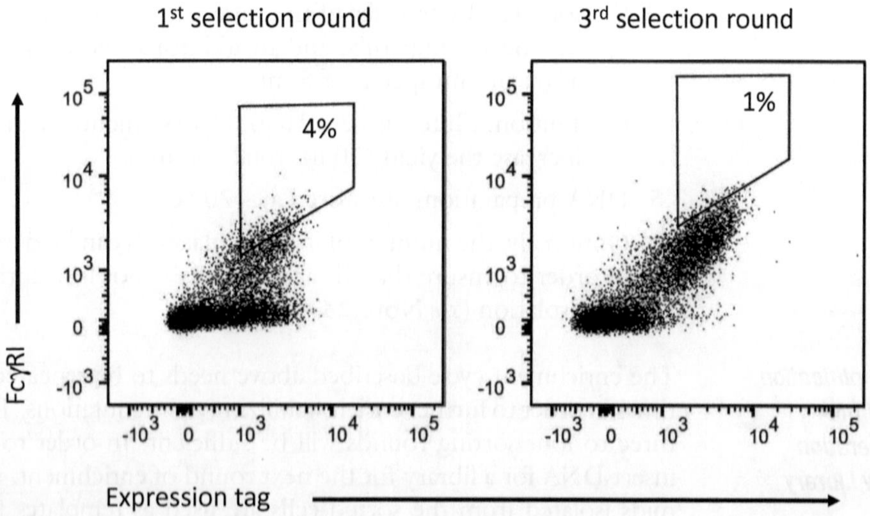

Fig. 3 Typical sort gates for yeast display screening. These FACS *dot plots* represent yeast displayed IgG1-Fc libraries that were screened for FcγRI-binding after a heat shock at 79 °C for 10 min [23]. Simultaneous staining with a conformationally specific ligand (in this case FcγRI) and an antibody recognizing an expression tag (in this case the Xpress-tag, located between Aga2p and IgG1-Fc, similarly to the HA-tag in the pCTCON2 vector described in the present protocol) allows for expression normalization by using a diagonal sort gate. That is, cells that display more protein also need to bind more ligand. This expression normalization enables more precise separation of clones with only slightly different thermal stabilities. Usually, in early sorting rounds (*left*) a higher percentage of cells (typically 2–5%) is selected in order to avoid loss of rare clones. In later sorting rounds (*right*), when clones are already enriched, the gate can be set more stringently (typically 0.1–1%) in order to improve the enrichment rate and screen for the very best clones. A more detailed discussion on the stringency of sort gates is published elsewhere [37]

plasmid DNA has to be isolated directly after the FACS screen (Fig. 2, step 6A) as described in the following steps.

1. Sort cells directly into 1.5 mL microcentrifuge tubes filled with 500 μL PBSA.
2. Centrifuge the cells (2000 × *g*, 15 min, 4 °C). If the number of sorted cells is less than 5×10^5, carrier yeast (non-transformed EBY100, 5×10^6 cells) is added in order to facilitate the formation of a cell pellet (*see* **Note 23**).
3. Pipet off the supernatant. Be careful to not pipet off the cells. Dependent on the number of cells, the pellet might be hardly visible.
4. Plasmids are isolated from yeast cells by using the Zymoprep Yeast Plasmid Miniprep II kit according to the manufacturer's manual with the following modifications:
   - 6 μL instead of 3 μL Zymolyase are added.
   - After addition of Solution 3 the suspension is centrifuged at maximum speed for 10 min instead of 3 min,

followed by transfer of the supernatant into a new 1.5 mL microcentrifuge tube and an additional centrifugation step at maximum speed for 5 min.

- Elution: Elute with 2 × 10 μL H_2O sequentially in order to increase the yield (20 μL total volume).

5. DNA-preparations are stored at −20 °C.
6. Optionally, the number of isolated plasmids can be determined in order to ensure that library diversity is not lost during plasmid isolation (*see* **Note 24**).

3.5 Amplification of Insert DNA and Generation of a New Library

The enrichment cycle described above needs to be repeated several times in order to further enrich stabilizing point mutations. Typically, three to four sorting rounds will be sufficient. In order to prepare insert DNA for a library for the next round of enrichment, the plasmids isolated from the sorted cells are used as templates for PCR according to the protocol presented in Subheading 3.1.2. In order to avoid loss of library diversity, the entire purified plasmid sample needs to be amplified. Two 200 μL PCR reactions are performed with 10 μL isolated plasmid DNA each. Subsequently, PCR products from the two reactions are pooled, purified and concentrated by ethanol precipitation (see Subheading 3.1.3), and used for electroporation of yeast together with linearized vector DNA (see Subheading 3.1.5) to yield a library for the next sorting round.

After the last sorting round insert DNA is amplified again and a new yeast library is generated by electroporation. Subsequently, plasmids are isolated from those yeast cells, followed by transformation into *E. coli*, plasmid DNA preparation of single clones and subsequent sequencing.

3.6 Clonal Evaluation of Stability Properties

The thermal stability of enriched protein variants can be evaluated directly on the surface of yeast on a single clone basis. The basic experimental procedure of such experiments consists of irreversible heat denaturation of surface displayed protein variants, followed by flow cytometric analysis to determine the fraction of displayed protein that can still bind to a conformationally specific ligand. Depending on the design of the experiment, two different parameters can be obtained.

On the one hand, subjecting the yeast displayed protein variant to a range of temperatures yields a denaturation curve as a function of temperature and therefore allows calculation of the temperature of half-maximum irreversible denaturation ($T_{1/2}$). These $T_{1/2}$ values have been shown to correlate very well with $T_{1/2}$ values obtained from experiments with soluble protein [34]. Moreover, although $T_{1/2}$ values obtained from yeast display experiments did not match exactly the T_m values obtained from differential scanning calorimetry (DSC) experiments (which is not unexpected given the different experimental procedures of those two methods, i.e., constant heating rate in DSC experiments vs. prolonged incubation time at a fixed temperature during the yeast display heat shock), the ranking among different mutants was highly consistent between the two methods [35]. This yeast display heat denaturation method has even been used for determining the average thermal stability of entire library pools [35].

On the other hand, subjecting the yeast suspension to a fixed temperature for a range of incubation times allows detailed analysis of denaturation kinetics. Fitting the data to a first order exponential function yields the denaturation rate constant, as well as the half-life at the given denaturation temperature. Again, kinetic denaturation data obtained from such yeast display experiments have been shown to correlate very well with values measured with the soluble protein counterparts [21].

1. Transform EBY100 with the pCTCON2 plasmids of enriched single clones by using the Frozen-EZ Yeast Transformation II Kit according to the manufacturer's protocol. Plate transformed yeast on SD-CAA agar plates and incubate at 30 °C for 2–4 days.
2. Induce yeast surface expression as described in Subheading 3.2.
3. Harvest induced yeast cells and subject them to a heat shock as described in Subheading 3.3. As mentioned above in this section, depending on the type of stability parameter that is to be measured, each yeast displayed protein variant has to be exposed either to a range of temperatures for a fixed time period (yielding $T_{1/2}$ values) or to a fixed temperature for a range of incubation times (yielding kinetic denaturation parameters).
4. Label yeast cells with a conformationally specific ligand and with anti-c-*myc* as described in Subheading 3.3 and analyze on a flow cytometer.
5. When analyzing the data, gate on the displaying yeast population (c-*myc*-positive) and calculate the total median fluorescence intensity for binding to the conformationally specific ligand (MFI_{tot}).
6. Plot MFI_{tot} as a function of temperature (T) to fit the temperature of half-maximum irreversible denaturation ($T_{1/2}$). The data should fit the following equation:

$$MFItot = \frac{MFI0 - MFIbg}{1 + \exp(\frac{T - T1/2}{b})} + MFIbg$$

where MFI_0 is the fluorescence intensity without heat shock, MFI_{bg} is the background fluorescence intensity, and b is a constant determining the slope of the transition.

7. For determination of the first order exponential denaturation rate constant and half-life, MFI_{tot} is plotted as a function of heat shock time (t). The data should fit the following equation:

$$MFItot = (MFI0 - MFIbg) \times \exp(-k \times t) + MFIbg$$

where k is the denaturation rate constant.

4 Notes

1. Low pH (4.5) discourages bacterial growth. Optionally, bacterial growth can be further prevented by addition of penicillin and streptomycin.
2. If mouse IgG is used as a secondary reagent for detection of the conformationally specific ligand, chicken anti-c-*myc* (Gallus Immunotech, 1:200) has to be used instead of mouse anti-c-*myc* in order to avoid cross-reactivity of the secondary antibody. In this case the chicken antibody is detected with goat anti-chicken IgY Alexa Fluor 488 (Thermo Fisher Scientific, 1:200).
3. The mutation rate can be adjusted by changing the nucleotide analogue concentration or the number of cycles. The error prone PCR protocol described in this manuscript is expected to yield, on average, 2–5 amino acid mutations per 1000 bp.
4. The C-terminal tag is not absolutely necessary. In cases where a C-terminal tag might interfere with protein function or folding, it may be omitted. However, if possible, a C-terminal tag is recommended, because it allows for recovery of full-length proteins.
5. In most yeast display studies the proteins of interest were fused to the C-terminus of Aga2p and therefore this orientation is described in this protocol. However, N-terminal fusions are also possible. pCHA (Addgene, plasmid #44874) is an appropriate vector for fusing the protein of interest to the N-terminus of Aga2p [36].
6. After 15 PCR-cycles the band on the agarose gel might be hardly visible. Usage of a ladder or a control insert with higher concentration as a standard in a separate lane facilitates excision of the correctly sized PCR-product.
7. Optional: Add 2 μL of Pellet Paint® Co-Precipitant (Novagen, EMD Millipore) to visualize the DNA pellet.
8. The usage of *Sal*I is not absolutely necessary, as it cuts in the insert. However, it further reduces the amount of uncut vector and therefore reduces the contamination of the library with yeast cells carrying the parental plasmid.

9. The doubling time of EBY100 in YPD medium is about 1.5–2 h.
10. It is recommended that the vector is added to the tube that was used for ethanol precipitation of the insert. Subsequently, the cells are added to the same tube, mixed with vector and insert DNA and finally transferred to the prechilled electroporation cuvette. This procedure minimizes the loss of insert DNA.
11. Two control electroporations are recommended: (a) Cells without any DNA in order to verify that the untransformed yeast strain does not grow on the selective media and (b) cells with linearized vector but without insert DNA in order to estimate the background in the library due to non-linearized vector (typically <1% of the library diversity).
12. Yeast cultures contain approximately 10^7 cells/mL at an OD_{600} of 1. Thus, for example, 5 mL of yeast culture at an OD_{600} of 3 contain about 1.5×10^8 cells. For all subsequent steps (freezing, passaging, induction of surface expression) it has to be ensured that the library diversity is covered at least ten times (e.g., 10^9 cells should be passaged for a 10^8 library).
13. Optional: The library can be passaged on the day after electroporation in order to reduce the number of untransformed cells.
14. For long term storage yeast cultures are frozen in SD-CAA medium + 15% glycerol. In brief, cells are centrifuged and the cell pellets are resuspended at high density in SD-CAA medium + 30% glycerol to achieve a final concentration of 15%. Subsequently, the tubes are stored at −80 °C. In order to minimize the loss in library diversity, ten times the library diversity should be frozen. Glycerol can also be added directly to the yeast culture to yield a final concentration of 15% (without centrifugation). This procedure yields less cells per tube, but might be sufficient if the library diversity is low.
15. Note that during plasmid isolation and amplification of insert DNA there will always be some loss of diversity. Thus, instead of covering the library diversity 10× as is usually done, we recommend to cover it 100 times during FACS-screening, if possible.
16. Other aliquot volumes are possible and the heat shock can also be performed in PCR tubes in thermal cyclers. However, if one library is split into aliquots and pooled again after the heat shock, the volumes in all tubes must be the same, because different volumes will heat up at different heating rates, resulting in differential thermal stress. The same must be taken into consideration if different samples are compared by FACS after

heat shock. In general, it is advisable to always use the same volume throughout all experiments and samples.

17. This incubation step ensures that all samples and aliquots start the heat shock at exactly the same temperature.
18. The heat shock temperature has to be high enough to result in partial thermal denaturation of the displayed protein. Otherwise, it is impossible to discriminate stabilized protein variants from their wild type counterparts [23]. On the other hand, if the heat shock temperature is too high, even the stabilized protein variants will completely denature, not allowing enrichment of stable mutants. Thus, the optimal heat shock duration and temperature need to be determined in initial experiments, where various yeast samples displaying the wild type protein are subjected to a range of temperatures and subsequently tested for binding to a conformationally specific ligand. In general, a temperature where the yeast-displayed wild type protein shows considerable denaturation, but still residual binding to the conformationally specific ligand, will be appropriate. The heat shock temperature is also limited by the stability of the yeast cell. Above 85 °C the cells lyse after a 10-min incubation, resulting in gelatinization of the solution making it unsuitable for FACS analysis. A detailed discussion on the choice of the optimal heat shock temperature is available elsewhere [10]. Moreover, if the denaturation rate constant of the wild-type protein at a given temperature is known, the appropriate denaturation screening time using flow cytometric methods can be directly calculated using a straightforward approach as described by Boder and Wittrup [37].
19. It is critical for the success of the screening procedure that the ligand that is used is indeed conformationally specific, which means that it only binds to correctly folded protein and therefore is able to specifically recognize stabilized protein variants that did not denature during the heat shock. The best and easiest way to test whether a ligand is conformationally specific is to express the protein of interest on yeast and compare ligand binding between yeast populations that have been subjected to an appropriate heat shock or not. Differential binding between heat-treated and untreated cells confirms (a) that this ligand is indeed conformationally specific and (b) that the protein denatured irreversibly, which is another prerequisite for the success of the screening experiment. It should be noted that the yeast surface display approach tends to result in irreversible (not equilibrium) thermal denaturation and hence is an imperfect albeit effective facsimile of true thermal stability properties.

20. In contrast to affinity maturation, where the screening pressure is directed towards increased affinity, the ligand concentration should not be limiting in the stability screen described here. Instead, the ligand should be present at a saturating concentration so that the screening pressure is directed towards thermal stability instead of affinity to the ligand. One exemption is the usage of the antigen as the conformationally specific ligand for stabilizing the respective binder. In this case the antigen should not be at a saturation concentration in order to avoid loss of affinity during the stability engineering process. An example of such a strategy was the generation of stabilized Her2-binding Fc-mutants by using Her2 for screening [26].
21. It is recommended to perform all steps after addition of the fluorophores in the dark in order to minimize photobleaching.
22. For the conditions described in this protocol (700 μL aliquots in 1.5 mL microcentrifuge tubes on a Thermomixer) the viability threshold for the yeast EBY100 is approximately 48 and 46 °C for a 10 and 30 min incubation, respectively. Detailed temperature-dependent survival curves of EBY100 can be found in Traxlmayr et al., Supplemental Fig. 1 [23].
23. In our experience it is key that at least 5×10^5 yeast cells are sorted in order to obtain a cell pellet. If the number of sorted cells is lower, we add 5×10^6 non-transformed yeast and therefore plasmid-negative EBY100 cells, which results in efficient formation of a cell pellet and plasmid DNA extraction.
24. Since the majority of isolated DNA will be genomic yeast DNA, an estimation of the number of plasmids from the total DNA concentration measured by absorption at 260 nm is not possible. An appropriate method for estimating the number of isolated plasmids has been described elsewhere [23]. Briefly, *E. coli* cells are transformed with a mix containing a certain fraction of the isolated yeast surface display plasmids (pCTCON2) and a known concentration of *lacZ*+ pUC19 plasmids. When plated on agar plates containing ampicillin and X-gal, pCTCON2 and pUC19 plasmids will yield white and blue colonies, respectively. Since the concentration and therefore the number of pUC19 plasmids in the transformation reaction is known, the number of pCTCON2 plasmids can be calculated from the ratio of white and blue colonies.

Acknowledgments

This work was supported by the Austrian Science Fund (FWF; Erwin-Schrödinger-Fellowship J3496-N28) and the National Science Foundation USA (CBET-1403350).

References

1. Boder ET, Wittrup KD (1997) Yeast surface display for screening combinatorial polypeptide libraries. Nat Biotechnol 15:553–557
2. Hackel BJ, Kapila A, Dane Wittrup K (2008) Picomolar affinity fibronectin domains engineered utilizing loop length diversity, recursive mutagenesis, and loop shuffling. J Mol Biol 381:1238–1252
3. Feldhaus MJ, Siegel RW, Opresko LK et al (2003) Flow-cytometric isolation of human antibodies from a nonimmune *Saccharomyces cerevisiae* surface display library. Nat Biotechnol 21:163–170
4. Cochran JR, Kim YS, Olsen MJ et al (2004) Domain-level antibody epitope mapping through yeast surface display of epidermal growth factor receptor fragments. J Immunol Methods 287:147–158
5. Chao G, Cochran JR, Dane Wittrup K (2004) Fine epitope mapping of anti-epidermal growth factor receptor antibodies through random mutagenesis and yeast surface display. J Mol Biol 342:539–550
6. Lam SS, Martell JD, Kamer KJ et al (2015) Directed evolution of APEX2 for electron microscopy and proximity labeling. Nat Methods 12:51–54
7. Lipovšek D, Antipov E, Armstrong KA et al (2007) Selection of horseradish peroxidase variants with enhanced enantioselectivity by yeast surface display. Chem Biol 14:1176–1185
8. Antipov E, Cho AE, Wittrup KD et al (2008) Highly L and D enantioselective variants of horseradish peroxidase discovered by an ultrahigh-throughput selection method. Proc Natl Acad Sci U S A 105:17694–17699
9. Shusta EV, Holler PD, Kieke MC et al (2000) Directed evolution of a stable scaffold for T-cell receptor engineering. Nat Biotechnol 18:754–759
10. Traxlmayr MW, Obinger C (2012) Directed evolution of proteins for increased stability and expression using yeast display. Arch Biochem Biophys 526:174–180
11. Dranginis AM, Rauceo JM, Coronado JE et al (2007) A biochemical guide to yeast adhesins: glycoproteins for social and antisocial occasions. Microbiol Mol Biol Rev 71:282–294
12. Shusta EV, Kieke MC, Parke E et al (1999) Yeast polypeptide fusion surface display levels predict thermal stability and soluble secretion efficiency. J Mol Biol 292:949–956
13. Kieke M, Shusta E, Boder E et al (1999) Selection of functional T cell receptor mutants from a yeast surface-display library. Proc Natl Acad Sci U S A 96:5651–5656
14. Buonpane RA, Moza B, Sundberg EJ et al (2005) Characterization of T cell receptors engineered for high affinity against toxic shock syndrome toxin-1. J Mol Biol 353:308–321
15. Weber KS, Donermeyer DL, Allen PM et al (2005) Class II-restricted T cell receptor engineered in vitro for higher affinity retains peptide specificity and function. Proc Natl Acad Sci U S A 102:19033–19038
16. Jones LL, Brophy SE, Bankovich AJ et al (2006) Engineering and characterization of a stabilized alpha1/alpha2 module of the class I major histocompatibility complex product Ld. J Biol Chem 281:25734–25744
17. Esteban O, Zhao H (2004) Directed evolution of soluble single-chain human class II MHC molecules. J Mol Biol 340:81–95
18. Starwalt SE, Masteller EL, Bluestone JA et al (2003) Directed evolution of a single-chain class II MHC product by yeast display. Protein Eng 16:147–156
19. Kim Y, Bhandari R, Cochran JR et al (2006) Directed evolution of the epidermal growth factor receptor extracellular domain for expression in yeast. Proteins 62:1026–1035
20. Schweickhardt RL, Jiang X, Garone LM et al (2003) Structure-expression relationship of tumor necrosis factor receptor mutants that increase expression. J Biol Chem 278:28961–28967
21. Pavoor TV, Cho YK, Shusta EV (2009) Development of GFP-based biosensors possessing the binding properties of antibodies. Proc Natl Acad Sci U S A 106:11895–11900
22. Park S, Xu Y, Stowell XF et al (2006) Limitations of yeast surface display in engineering proteins of high thermostability. Proteins 19:211–217

23. Traxlmayr MW, Faissner M, Stadlmayr G et al (2012) Directed evolution of stabilized IgG1-Fc scaffolds by application of strong heat shock to libraries displayed on yeast. Biochim Biophys Acta Proteins Proteomics 1824:542–549
24. Aggen DH, Chervin AS, Insaidoo FK et al (2011) Identification and engineering of human variable regions that allow expression of stable single-chain T cell receptors. Protein Eng Des Sel 24:361–372
25. Pavoor TV, Wheasler JA, Kamat V et al (2012) An enhanced approach for engineering thermally stable proteins using yeast display. Protein Eng Des Sel 25:625–630
26. Traxlmayr MW, Lobner E, Antes B et al (2013) Directed evolution of Her2/neu-binding IgG1-Fc for improved stability and resistance to aggregation by using yeast surface display. Protein Eng Des Sel 26:255–265
27. Traxlmayr MW, Hasenhindl C, Hackl M et al (2012) Construction of a stability landscape of the CH3 domain of human IgG1 by combining directed evolution with high throughput sequencing. J Mol Biol 423:397–412
28. Chao G, Lau WL, Hackel BJ et al (2006) Isolating and engineering human antibodies using yeast surface display. Nat Protoc 1:755–768
29. Angelini A, Chen TF, De Picciotto S et al (2015) Protein engineering and selection using yeast surface display. Methods Mol Biol 1319:3–36
30. Colby DW, Kellogg BA, Graff CP et al (2004) Engineering antibody affinity by yeast surface display. Methods Enzymol 388:348–358
31. Zaccolo M, Gherardi E (1999) The effect of high-frequency random mutagenesis on in vitro protein evolution: a study on TEM-1 beta-lactamase. J Mol Biol 285:775–783
32. Zaccolo M, Williams DM, Brown DM et al (1996) An approach to random mutagenesis of DNA using mixtures of triphosphate derivatives of nucleoside analogues. J Mol Biol 255:589–603
33. Balint RF, Larrick JW (1993) Antibody engineering by parsimonious mutagenesis. Gene 137:109–118
34. Orr BA, Carr LM, Wittrup KD et al (2003) Rapid method for measuring ScFv thermal stability by yeast surface display. Biotechnol Prog 19:631–638
35. Hasenhindl C, Traxlmayr MW, Wozniak-Knopp G et al (2013) Stability assessment on a library scale: a rapid method for the evaluation of the commutability and insertion of residues in C-terminal loops of the CH3 domains of IgG1-Fc. Protein Eng Des Sel 26:675–682
36. Mata-Fink J, Kriegsman B, Yu HX et al (2013) Rapid conformational epitope mapping of anti-gp120 antibodies with a designed mutant panel displayed on yeast. J Mol Biol 425:444–456
37. Boder ET, Wittrup KD (1998) Optimal screening of surface-displayed polypeptide libraries. Biotechnol Prog 14:55–62

Chapter 5

Whole Cell Panning with Phage Display

Yvonne Stark, Sophie Venet, and Annika Schmid

Abstract

Phage display has emerged as one of the leading technologies for the selection of highly specific monoclonal antibodies, offering a number of advantages over traditional methods of antibody generation. While there are various possibilities to conduct phage display (e.g., solution panning, solid-phase panning), whole cell panning is an elegant way to present membrane embedded target antigens in their natural environment and conformation to antibody-bearing phages. Here, a whole cell panning procedure using a Fab-based antibody library including primary cell based screening for selectivity is described.

Key words Antibody fragment, Phage display, Antibody library, Whole cell panning (WCP)

1 Introduction

The outstanding success story of antibodies as therapeutics started as early as in the 1890ies when a group of researchers demonstrated that the injection of a treated animal's immune serum led to the elimination of the effects of diphtheria toxin [1]. Despite the exact source of cure was unknown, this is the first description of the healing potential of antibodies. Since then, several alternative techniques have been developed for the discovery of target specific antibodies, substituting animal hosts. Amongst the new techniques, one proved to be in particular very efficient: the isolation of antibodies from large antibody combinatorial libraries after in vitro selection. Such libraries rely on a support, such as phage [2, 3], yeast [4], mammalian cells [5], bacteria [6], eukaryotic virus particles [7], or ribosomes [8], which allows maintaining a linkage between the antibody fragments phenotype and their corresponding encoding DNA sequence (genotype). These techniques have been described in a number of excellent reviews in recent years, e.g., [9–11]. In the context of phage display, which has become one of the leading technologies, phage particles presenting antibody fragments are mixed with an antigen of interest for specific interaction to occur. After washing, only phage displaying antibody

Thomas Tiller (ed.), *Synthetic Antibodies: Methods and Protocols*, Methods in Molecular Biology, vol. 1575, DOI 10.1007/978-1-4939-6857-2_5, © Springer Science+Business Media LLC 2017

fragments specific to the antigen are recovered and the antibody candidate may be identified via the corresponding DNA trapped in the phage capsid. Various sources of diversity have been used for the generation of fully human combinatorial libraries bearing either natural [12] or synthetic [13–15] sequences. Natural sequences have been proofread in vivo for proper folding but contain in part rare frameworks and are pre-cleared for self-antigens. Synthetic diversity, in turn, allows focusing on sequences of highest interest, e.g., the most common human frameworks to reduce immunogenicity or the most stable framework combinations for developability. To obtain a high quality synthetic library, it is essential to use appropriate techniques such as TRIM (trinucleotide-directed mutagenesis) or a ligation-based strategy for chemical gene synthesis to introduce CDR (complementarity determining regions) diversity [15, 16].

The effectiveness of in vitro selection relies heavily on large and high quality combinatorial libraries, but also on the quality of the antigen used for panning. If possible, such antigen should reflect the three dimensional conformation of the target protein. Such condition eventually increases the probability to isolate antibodies which can recognize the target molecule in its natural form. Recombinant proteins have largely been used in pannings and most often led to the identification of specific antibodies. However, certain classes of proteins, which contain transmembrane domains, are extremely challenging to produce as soluble recombinant molecules, e.g., G protein-coupled receptors (GPCRs) and ion channels. Despite major progress has been made in stabilizing such proteins [17, 18], their most common source remains their expression at the surface of cells (naturally or genetically modified). This presentation mode better reflects their native conformation and natural context, which might include, for instance, interaction partners. Whole cell pannings (WCP) are however challenging and often associated with the amplification of many unspecific antibodies. These hurdles are often due to a low level of expression of the target protein and the availability of negative cells for an effective deselection of antibodies specific to other cell surface molecules. In this chapter, phage display using a fully human Fab based antibody library will be presented applying an optimized whole cell panning approach. For additional information on the Ylanthia library and corresponding vectors used in this chapter, please refer to Tiller et al. [13].

2 Materials

For this protocol about the isolation of antibody fragments from a phage display library by WCP, we used a phagemid vector based library harboring a chloramphenicol-resistance gene. Fab heavy and light chain gene as well as gene pIII expression is inducible and regulated by an IPTG-inducible lac promoter region. All solutions prepared should be performed in double-distilled water. Solutions and media should be autoclaved before use if not otherwise indicated. Waste disposal regulations should be considered.

2.1 Blocking of Phage Prior to Selection

1. FBS (fetal bovine serum) (e.g., Sigma-Aldrich, mycoplasma negative, heat-inactivated).
2. PBS (Dulbecco's PBS, 1× with Ca^{2+} Mg^{2+}, e.g., Gibco® Life Technologies).
3. 10% FBS/PBS (v/v) (with Ca^{2+} Mg^{2+}).
4. 2 mL low binding tubes (e.g., Biozym, siliconized).
5. Library phage preparation (e.g., Ylanthia® Library Phages, MorphoSys).
6. 2-YT medium: 31 g yeast extract/tryptone medium (e.g., Difco) in 1 L ddH_2O autoclaved.
7. *E. coli* competent cells (e.g., TG1F+, Stratagene/Agilent Technologies, supE thi-1 Δ(lac-proAB). Δ(mcrB-hsdSM)5 (rK– mK–) [F′ traD36 proAB lacIqZΔM15] for propagation of phage particles).

2.2 Preparation of Target and Adsorption Cells

1. T175 cell culture flasks (e.g., Thermo Scientific).
2. Target cells (transfected or antigen positive).
3. Adsorption cells (mock-transfected or antigen negative).
4. PBS (Dulbecco's PBS, 1× without Ca^{2+} Mg^{2+}, e.g., Gibco® Life Technologies™).
5. Versene (e.g., Gibco® Life Technologies™).
6. 5% FBS/PBS (v/v) (without Ca^{2+} Mg^{2+}).
7. 50 mL Falcon tube (e.g., Corning b.V. Life Sciences).
8. Trypan Blue 0.4% for cell culture (e.g., Sigma).

2.3 Selection Process: Binding of Phage to Specific Antigen Expressing Target Cells Followed by Acidic Elution of Selected Phage

1. 100 mM Glycine-HCl, pH 2.2: 100 mM glycine, 0.5 M NaCl, in ddH_2O. After dissolving, adjust pH with 37% HCl from initial pH (~6.2) to 2.2; sterilized by filtration (0.22 μm).
1. 2 M Tris base, unbuffered: 24.22 g Tris powder in 100 mL ddH_2O; sterilized by filtration (0.22 μm).
2. 7.5% FBS/PBS (v/v) (without Ca^{2+} Mg^{2+}).
3. Sterile glass beads (e.g., Roth, Karlsruhe, Germany).
4. 34 mg/mL chloramphenicol stock solution (e.g., Sigma-Aldrich); sterilized by filtration (0.22 μm); aliquots stored at −20 °C.
5. 50 mg/mL kanamycin stock solution (e.g., Invitrogen); sterilized by filtration (0.22 μm); aliquots stored at −20 °C.

6. LB/Kan agar plate small: Ø 9.4 cm (prepare 1 L containing 50 μg/mL kanamycin and 1% glucose).
7. LB/Cam agar plates small: Ø 9.4 cm; large: Ø 14.5 cm (prepare 1 L containing 34 μg/mL chloramphenicol and 1% glucose).
8. Ethanol 80% and RBS 3%, for workbench cleaning.

2.4 Recovery of Panning Output

1. Pasteur pipette (e.g., VWR). Drygalski spatula.
2. 100% glycerol.
3. 40% glucose: 44 g glucose monohydrate in 100 mL ddH_2O; sterilized by filtration (0.22 μm).
4. Freezing medium #1: 2-YT liquid medium containing final conc. 34 μg/mL chloramphenicol, 1% glucose and 15% glycerol for freezing of stock cultures.
5. 13 mL disposable snap cap tube (e.g., Sarstedt).
6. Cryovials, 2 mL screw cap micro tubes (e.g., Sarstedt).
7. Growth medium: 2-YT liquid medium containing final conc. 34 μg/mL chloramphenicol and 1% glucose.

2.5 Phage Production and Precipitation

1. Helper phage (Hyperphage-Progen # PRHYPE 1*E*+12 tu/mL; resuspended in 2 mL ddH_2O).
2. 1 M IPTG stock solution (e.g., Carl Roth); sterilized by filtration (0.22 μm); aliquots stored at −20 °C.
3. Induction medium #1: 2-YT liquid medium supplemented with 34 μg/mL chloramphenicol, 50 μg/mL kanamycin, and 0.25 mM IPTG final concentration.
4. PEG/NaCl: 20% PEG 6000/Lipoxol (e.g., Staub), 2.5 M NaCl in ddH_2O; precooled to −20 °C (ice-cold).
5. 10× PBS according to Gibco: 1.36 M NaCl, 26.8 mM KCl, 81 mM Na_2HPO_4, 14.6 mM KH_2PO_4.
6. 1× PBS (pH 7.2) (diluted with ddH_2O from 10× stock).

2.6 Subcloning of Selected Phage Pools into Expression Vector

1. DNA preparation kit (e.g., Qiagen).
2. Restriction enzymes: *Xba*I, *Eco*RI-HF, 10× CutSmart TM buffer (e.g., New England Biolabs).
3. 10× gel loading buffer (e.g., Invitrogen).
4. DNA ladder (e.g., Thermo Scientific).
5. Agarose (e.g., Invitrogen).
6. DNA gel extraction kit/ DNA purification kit (e.g., Wizard® SV Gel and PCR Clean-Up System; Promega).
7. T4 DNA ligase and ligase buffer (e.g., Invitrogen).

8. 1.5 mL low binding tubes (e.g., Biozym, siliconized)
9. 2-butanol (e.g., Sigma-Aldrich).
10. 70% ethanol (precooled to −20 °C).
11. Glycogen (e.g., Roche).
12. *E. coli* competent cells (e.g., DH5α, Invitrogen, or TG1F-, for expression of soluble Fabs, supE thi-1Δ(lac-proAB) Δ(mcrB-hsdSM)5 (rK− mK−)).
13. Electroporation system (e.g., Gene Pulser II, Bio-Rad).
14. Electroporation cuvette (e.g., VWR).
15. SOB medium: 2% (w/v) tryptone, 0.5% yeast extract, 10 mM NaCl, 2.5 mM KCl, 10 mM $MgCl_2$, 10 mM $MgSO_4$, in H_2O.
16. Expression vector DNA with purification tags (e.g., pYBex10_h_Fab_FH).
17. Sterile ddH_2O.

2.7 Preparation of Selection Plates for Subsequent Primary Cell Screening

1. Toothpicks (autoclaved).
2. 384-well microtiter plates flat bottom (e.g., VWR).
3. Growth medium: 2-YT medium supplemented with 34 μg/mL chloramphenicol and 1% glucose (for master plate).
4. Aluminum foil seal/gas-permeable foil seal (e.g., Thermo Scientific).
5. Freezing medium #2: 2-YT/60% glycerol (v/v) /34 μg/mL chloramphenicol/1% glucose (v/v).
6. Induction medium #2: 2-YT liquid medium supplemented with 34 μg/mL chloramphenicol, 0.1% glucose and 0.5 mM IPTG final concentration (for expression plate).
7. 0.5 M EDTA solution (pH 8.0): 186.12 g EDTA $C_{10}H_{14}N_2Na_2O_8{\bullet}2H_2O$ and 22 g NaOH pellets in 1 L ddH_2O; adjusted to pH 8.0 using 10 M NaOH (if required); sterilize by filtration (0.22 μm).
8. 2× BBS: 0.4 M H_3BO_3, 0.32 M NaCl, 66 mM NaOH, pH 8.0; 24.7 g H_3BO_3 and 18.7 g NaCl in 1 L ddH_2O; adjusted to pH 8.0 using 10 M NaOH; autoclaved.
9. 2× lysis buffer: For 100 mL lysis buffer dissolve 250 mg lysozyme in 2× BBS; add 0.8 mL from 0.5 M EDTA stock solution and 5 μL Benzonase® (e.g., Merck; >250 U/μL); adjust to 100 mL using 2× BBS (final concentrations: EDTA = 4 mM, lysozyme = 2.5 mg/mL, Benzonase® = >12.5 U/mL). Lysozyme and Benzonase® should be added only shortly prior to use.
10. Blocking buffer: 12.5% (v/v) FBS in 1× PBS (with Ca^{2+} Mg^{2+}).

2.8 Primary Multiplex Cell Screening

1. Calcein-AM (e.g., Invitrogen).
2. Flow cytometer (e.g., HTFC®, IntelliCyt Corporation).
3. Sodium Azide (e.g., Sigma-Aldrich, dissolved in 25 mL H_2O, 20% Sodium Azide stock solution).
4. FACS buffer: 1× PBS (without Ca^{2+} Mg^{2+}), 3% FBS, 0.02% Sodium Azide.
5. 384 well V-bottom PP-plate (e.g., Greiner).
6. Alexa Fluor 647 labeled anti-human Fab detection antibody (e.g., Jackson Immuno Research).
7. Alexa Fluor 647 labeled anti-reference antibody detection antibody. Specificity has to be selected depending on the species origin of the reference antibody.

3 Methods

3.1 Isolation of Antibody Fragments from a Phage Display Library by Whole Cell Panning

3.1.1 Blocking of Phage Prior to Selection

1. Mix the required amount of phages of the respective library (recommended: covering 1000-fold the diversity) 1:1 with 10% FBS/PBS for a final volume of 1 mL in 2 mL low binding tubes. Incubate the phages for at least 1 h at 4 °C on a rotator (*see* **Notes 1** and **2**).
2. For each panning inoculate ~30 mL 2-YT medium with *E. coli* (e.g., TG1) in a phage-free working space (*see* **Note 3**). Use a phage-free or disposable flask for *E. coli* incubation to avoid any contamination with library phage or helper phage. This *E. coli* culture will be used for infection with the selected phage. Incubate the culture at 37 °C shaking at 250 rpm, until an OD_{600nm} of 0.6–0.8 is reached. Keep the *E. coli* on ice until required for infection with the eluted phage. Avoid prolonged incubation (>60–90 min) of cells on ice to prevent reduced efficiency of *E. coli*.

3.1.2 Preparation of Target and Adsorption Cells

For whole cell pannings, antigen-expressing cells serve as panning matrix. A prerequisite for whole cell pannings is the availability of cells expressing the antigen in good quality and sufficient copy number. Transfected cells or antigen-overexpressing (tumor) cell lines may be used. Optimal antigen densities are >1*E*+06 molecules per cell (*see* **Note 4**). However, a successful selection also strongly depends on the nature and the accessibility of the antigen, including e.g., antigen glycosylation, steric hindrance by neighboring proteins, size and folding of the extracellular domain. Hence, long protruding structures, which provide optimal accessibility for antibody displaying phage, are most advantageous.

In order to select clones binding to the target antigen but not to any of the other cell surface proteins, it is indispensable to have adsorption cells at hand which display the same overall expression pattern as the target cells but do not express the target antigen (*see* **Note 5**). These cells are used for either post- or pre-adsorption steps which deplete Fab phage binding to cell surface molecules other than the target antigen (*see* **Note 6**). If transfected cells are used for panning, mock-transfected cells (identical growth conditions preferred) would be most suitable for this purpose.

All volumes indicated in this section refer to T175 cell culture flasks.

1. Grow a sufficient amount of antigen-expressing cells and adsorption cells. If necessary, add agents which stimulate antigen expression. Discard cultivation medium and wash cells once with 10 mL PBS. If required, detach cells (e.g., with 3 mL versene) for ~5–8 min at 37 °C. Monitor detachment under microscope. Add 7 mL of ice cold 5% FBS/PBS and collect detached cells in 50 mL falcon tube (*see* **Notes 7** and **8**).
2. Some antibodies may stick to dead cells and intracellular structures. To reduce enrichment of such not antigen-specific antibodies, the cells utilized for panning should have a high percentage of viable cells. This should be checked by vital staining prior to each selection round: e.g., use trypan blue for staining of dead cells by standard cell culture techniques.
3. Wash the target cells (transfected or antigen positive) and adsorption cells (mock-transfected or antigen negative) 3× with 5% FBS/PBS.
4. Count the target cells and prepare aliquots of 0.5–1*E*+07 cells in 1 mL 5% FBS/PBS in 2 mL low binding tubes for each selection.
5. Count the adsorption cells and adjust to 0.5–1*E*+07 cells in 1 mL 5% FBS/PBS for each post-adsorption step. Prepare several aliquots in 2 mL low binding tubes according to the number of post-absorption steps.
6. Keep the cells on ice until use for phage selection process (*see* **Notes 9** and **10**).

3.1.3 Selection Process: Binding of Phage to Specific Antigen Expressing Target Cells Followed by Acidic Elution of Selected Phage

1. Centrifuge the target cells for 2 min at 100–200 × *g* at 4 °C and resuspend the cell pellet in 1 mL blocked phages. Incubate for 2 h at 4 °C on a rotator (*see* **Note 11**).
2. Spin cells for 2 min at 100–200 × *g* at 4 °C. Remove the supernatant carefully and discard it. Carefully resuspend the cell pellet in 1 mL 5% FBS/PBS and incubate the cell suspension for 5 min on a rotator.
3. Repeat the washing step twice (in total three washing steps). For the last washing step, transfer the cells into a new tube

after resuspending in 5% FBS/PBS to get rid of phage which bound unspecifically to the selection tube.

4. Elute selected phage by adding 1 mL of 100 mM glycine-HCl, pH 2.2 and incubate for 10 min at room temperature.
5. Spin cells for 2 min at 100–200 × *g* at 4 °C and transfer supernatant into a new low binding tube. Neutralize with ~25 μL of 2 M Tris-base (*see* **Notes 12** and **13**). Use the eluted phage for post-adsorption as described below.

3.1.4 Post-adsorption of Selected Phage

1. Add 0.5 mL 7.5% FBS/PBS to the eluted, neutralized phage.
2. Centrifuge three aliquots of adsorption cells for 2 min at 100–200 × *g* and remove the supernatant. Keep pellets at 4 °C or on ice.
3. Carefully resuspend a pellet of adsorption cells in the eluted phage. Incubate for 30 min at 4 °C on a rotator.
4. Spin cells for 2 min at 100–200 × *g* at 4 °C and use the supernatant for the next post-adsorption step.
5. Repeat the post-adsorption twice (in total three post-adsorption steps).

3.1.5 Bacterial Infection

1. After three post-adsorption steps, transfer the supernatant containing the selected phage into one 50 mL Falcon tube with 30 mL *E. coli* culture at OD_{600nm} of 0.6–0.8 for infection. If the *E. coli* culture was stored on ice, place in a 37 °C water bath and make sure that the culture has reached 37 °C again prior to infection with eluted phage. Incubate the mix of *E. coli* culture and phage eluate for 45 min in a water bath at 37 °C without shaking.
2. Check the remaining uninfected *E. coli* culture for phage contamination by plating 50 μL on a small LB/Cam agar plate and 50 μL on a small LB/Kan agar plate. Incubate the agar plates overnight at 37 °C (*see* **Note 14**).
3. Determine the phage selection titer (output titer) of each selection via plating of serial dilutions. For this purpose, remove 200 μL of infected cells (after incubation in a water bath) into a separate tube prior to centrifugation. For a better workflow, serial dilution should be performed later (after **step 4**). Centrifuge the remaining infected *E. coli* culture for 5 min at 4,000 × *g* and 4 °C.
4. Discard the supernatant and resuspend each pellet in 600 μL 2-YT medium. Plate the cell suspension evenly on 2–3 large LB/Cam agar plates using sterile glass beads or other devices applicable for bacteria plating (*see* **Notes 15** and **16**). Incubate the large LB/Cam plates overnight at 30 °C. If the colonies are still very small on the following morning, continue incubation at 37 °C for 1–2 h.

5. Determine the output titer of each selection. Use four small LB/Cam agar plates per selection to check the phage selection titer. Perform a serial dilution by plating 100 μL of infected cells (from **step 3**) undiluted, diluted 10^{-1}, 10^{-2}, and 10^{-3}. Incubate the agar plates overnight at 37 °C. Clean working bench with 3% RBS solution and 80% EtOH.
6. (Next day) The output titer of each selection round is calculated from the colony forming units (cfu) on the phage selection output agar plates prepared in **step 5** the day before. Determine output titer according to the following equation (*see* **Notes 17** and **18**):

V_1 = Volume of infected cells [mL]

V_2 = Volume of undiluted cells plated [mL]

$$\text{Output } 1st\,/\,2nd\,/\,3rd \text{ round} = \frac{\text{counted colonies} \times V_1}{V_2}$$

3.1.6 Recovery of Panning Output

1. Scrape off the bacteria (which contain the selected phagemids) from each large selection agar plate with 3 mL freezing medium #1 using a new, sterile Drygalski spatula. Collect the bacterial suspensions belonging to the same selection in a 13 mL snap cap tube. Take care to prepare homogeneous bacteria suspensions.
2. Prepare 2–3 aliquots of each selection pool in 2 mL micro tubes and keep on ice.
3. After the last panning round, continue with preparation of polyclonal DNA as described in Subheading 3.2.
4. After earlier panning rounds, inoculate 10 mL growth medium for each antigen selection in a 50 mL Falcon tube with 10–50 μL bacterial suspension prepared in **step 1** resulting in an $OD_{600nm} < 0.3$.
5. Incubate the cultures for 30–90 min at 37 °C in a shaker at 250 rpm until an OD_{600nm} of 0.5 is reached.
6. Store the micro tubes with the remaining bacteria suspension at −80 °C.

3.1.7 Phage Production and Precipitation

1. Transfer 5 mL of the *E. coli* culture containing the selected phagemids to a 13 mL disposable snap cap tube and add a proper amount of helper phage, equivalent to at least 4*E*+10 total units of phage per 5 mL bacterial culture (*see* **Note 19**). Incubate for 30 min in a water bath at 37 °C without shaking and then for 30 min at 37 °C shaking at 250 rpm.
2. To check for helper phage infection, prepare 1:1000 dilutions of the selection cultures before and after infection. Plate 30 μL

of the dilutions each on a LB/Kan agar plate and incubate overnight at 37 °C. It is not necessary to count the colonies, but a significant increase in the number of colonies after helper phage infection should be visible. In general, it takes ~48 h for the infected bacteria to grow on LB/Kan agar plates.

3. Spin down the bacteria at 4,000 × *g* for 5 min at 4 °C and discard the helper phage containing supernatant. Add 1 mL induction medium #1 to the pellet containing the infected bacteria in the snap cap tube and carefully resuspend the bacteria.
4. Transfer the culture to 19 mL induction medium #1 in a shake flask or 50 mL falcon tube (= final volume of 20 mL). Ensure proper circulation of oxygen (e.g., by not closing but only loosely attaching the lid of the Falcon tube with tape) and incubate overnight at 22 °C, shaking at 250 rpm (for 18–20 h) (*see* **Note 20**).
5. (Next day) Spin down the bacteria from the overnight culture for 10 min at 4,000 × *g* at 4 °C using a 50 mL disposable tube. Transfer the supernatant containing the Fab presenting phage to a new 50 mL falcon tube and discard the bacterial pellet.
6. The cleared phage-containing supernatant should be purified and concentrated by a precipitation step as described next:
   (a) Add 5 mL chilled PEG/NaCl solution (precooled to −20 °C) to the phage-containing supernatant, mix and incubate for at least 30 min on ice (*see* **Note 21**).
   (b) Spin down the precipitated phage for a minimum of 30 min at 13,000 × *g* and 4 °C. Discard the supernatant and tap the tube on a stack of paper towels to remove residual PEG/NaCl. Allow to stand upside down on the paper towels for about 5 min (*see* **Note 22**).
   (c) Resuspend the phage pellet in 500–1000 μL sterile 1× PBS and transfer the phage suspension to a sterile 1.5 mL low binding reaction tube.
   (d) Rotate the phage suspension for at least 30 min at 4 °C to completely dissolve the phage.
   (e) Pellet residual bacterial debris at max. speed in a tabletop centrifuge for 2 min at room temperature and transfer the supernatant into a new sterile 2 mL low binding reaction tube.
   (f) Phages can be stored at 4 °C up to 2–3 weeks. For long term storage, we recommend a storage at −80 °C adding glycerol to a final concentration of 20% glycerol.

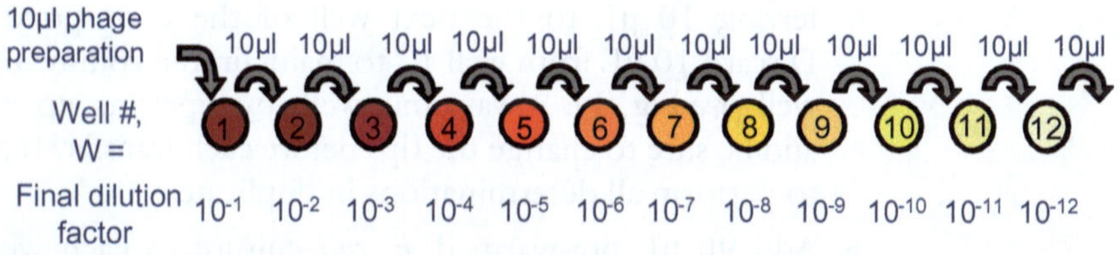

Fig. 1 Serial dilution for spot titration

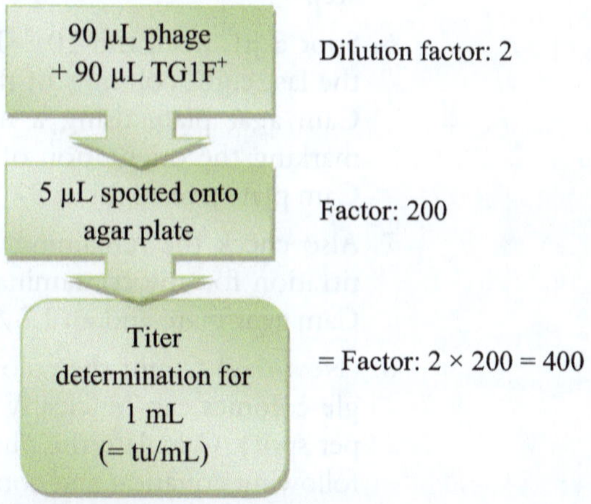

Fig. 2 Explanation of phage titer calculation

7. Determine the input phage titer for the next round of panning by spot titration (described below, *see* Figs. 1 and 2). Use a total phage input of at least 1*E*+11 phages for the next panning round (*see* **Notes 23** and **24**).
8. The number of total panning rounds may be chosen individually depending on the size and quality of the applied library. We recommend three rounds of panning in order to ensure sufficient selectivity.

3.1.8 Spot Titration (Determination of Input Titer for 2nd and 3rd Panning Round)

1. Dry a large LB/Cam agar plate by pre-incubation of plate with open lid for 2 h at 37 °C in an incubator. The use of dry plates is essential.
2. In a phage-free working space inoculate 100 mL 2-YT medium with *E. coli* (e.g., TG1) culture. The *E. coli* will be used for infection with the selected phage (*see* **Note 3**). Incubate the culture at 37 °C shaking at 250 rpm, until an OD_{600nm} of 0.6–0.8 is reached. Keep the *E. coli* on ice, until they are required for phage infection.
3. Add 90 µL 2-YT medium to wells A1–H12 of a 96-well round bottom microtiter plate using a multichannel pipette.
4. Add 10 µL of the phage preparation to a well of the first column (e.g., well A1) and perform a serial 1:10 dilution by trans-

ferring 10 μL to the next well of the corresponding row. Discard 10 μL from well 12 to maintain the volume of 90 μL/well (*see* Fig. 1). Always mix wells by pipetting up and down and be sure to change the tips before each transfer. It is advised to perform all determinations in duplicates.

5. Add 90 μL pre-warmed *E. coli* culture to each well of the microtiter plate containing the serial phage dilution. Incubate for 30 min at 37 °C in an incubator. Keep *E. coli* culture for **step** 7.
6. Spot 5 μL from the 10^{-5}–10^{-12} dilutions (i.e., dilutions from the last eight columns of the 96-well plate) onto a large LB/Cam agar plate using a multichannel pipette. Take care of marking the orientation of the plate. Incubate the large LB/Cam plate overnight at 37 °C.
7. Also check the remaining uninfected *E. coli* culture used for titration for any contamination by plating 50 μL on an LB/Cam agar plate and an LB/Kan agar plate, respectively.
8. (Next day) Count the colonies from those spots in which single colonies can be clearly identified (= less than 30 colonies per spot). Calculate the phage titer (tu/mL) according to the following equation and continue with the next panning round (*see* Fig. 2):

$$\text{Phage titer}\,[tu\,/\,mL] = 400 \times \text{Number of colonies} \times 10^{W}$$

W = Column number of the well from which the corresponding spots originate (e.g., column number 1, 2, …)

The total phage amount is calculated according to the following equation:

$$\text{Total phage amount}\,[tu] = V(\text{Phage preparation})[mL] \times \text{Phage titer}\,[tu\,/\,mL]$$

The total phage input for next panning round is calculated according to the following equation:

$$\text{Total phage input } for \text{ next selection round}\,[tu] = V(\text{Phage preparation used } for \text{ selection})[mL] \times \text{Phage titer}[tu\,/\,mL]$$

3.2 Subcloning of Selected Phage Pools into Expression Vector

Even though an initial screening is feasible with pIII fusion protein [19], we highly recommend subcloning the output of the last panning round into an expression vector system in which Fabs are produced as soluble proteins without being fused to pIII. This way, single positive hits are already identified in a format being appropriate for purification or further in-depth characterization. In the vector system used herein the Fab encoding region is flanked by unique restriction sites which enables subcloning into an expression vector by means of *Xba*I and *Eco*RI restriction enzymes. Other digestion strategies may be necessary when working with different phagemid vector systems.

1. Prepare samples as described in Subheading 3.1.6, **steps 1** and **2**.
2. Determine the OD_{600nm} of a 1:200 diluted bacterial suspension of each panning strategy. Calculate which volume equals the amount of bacteria contained in 1 mL of an OD_{600nm} 24 suspension according to the equation below and use this volume for a DNA preparation. Store the rest of the aliquots at −80 °C (*see* **Note 25**).

$$V\,(\text{bacterial suspension})\,[\mu L] = \frac{OD_{600nm}\ \text{desired} \times 1000\ \mu L}{OD_{600nm}\ (\text{bacterial suspension}) \times \text{dilution factor}}$$

3. Wash the bacteria twice to remove the glycerol: Add 2-YT medium for a final volume of 2 mL, pellet the bacteria by centrifugation for 8 min at 4,000 × *g* and 4 °C and discard the glycerol containing supernatant. Repeat washing with 2-YT medium a second time. Glycerol has to be removed as it may impair DNA preparation efficiency.
4. Perform two plasmid DNA preparations per selection using a suitable DNA preparation kit according to instructor's manual (*see* **Note 26**). Determine the plasmid DNA concentration, e.g., by absorbance measurement at 260 nm.
5. Digest 5 μg of the polyclonal DNA in a final volume of 200 μL with appropriate amounts of restriction enzymes *Eco*RI and *Xba*I according to Table 1.

Table 1
Digestion setup and ligation setup

Digestion	Ligation		
	Ligation	Vector background ctrl.	Insert background ctrl.
5 μg DNA 20 μL of 10× CutSmart buffer	4 μL of 5× ligase buffer	4 μL of 5× ligase buffer	4 μL of 5× ligase buffer
50 U Xba I	1 μL of vector DNA (50 ng)	1 μL of vector DNA (50 ng)	–
50 U EcoRI-HF	Insert DNA (5× molar excess to vector)	–	Insert DNA (5× molar excess to vector)
	1 μL T4 DNA ligase	1 μL T4 DNA ligase	1 μL T4 DNA ligase
H_2O ad 200 μL	H_2O ad 20 μL	H_2O ad 20 μL	H_2O ad 20 μL

6. Also digest 5 μg of the expression vector DNA in a similar manner. It is advisable to use restriction enzymes from a supplier allowing a double digest. Digest for 2–3 h at 37 °C.
7. After the digestion incubation time, inactivate the restriction enzymes in the sample for 20 min at 65 or 80 °C depending on the restriction enzymes used.
8. (Next day) Mix the digested samples with 20 μL 10× gel loading buffer and run with an appropriate DNA ladder on a preparative 1% agarose gel at about 10 V/cm gel distance (~2–3 h).
9. Cut out the bands of the digested Fab-encoding insert DNA and the expression vector DNA. Extract DNAs from the agarose gel slice using an appropriate gel extraction kit and elute with 50 μL of nuclease-free H_2O or with an appropriate elution buffer supplied with the purification kit.
10. Quantify the eluted DNA samples on a 1% analytical agarose gel by loading a mix of 5 μL DNA, 4 μL H_2O, and 1 μL 10× gel loading buffer. Also load an appropriate amount of a relevant DNA ladder on the gel to estimate the amount of expression vector DNA and Fab-encoding insert DNA. Separate the DNA by electrophoresis at 10 V/cm gel distance.
11. The digested Fab-encoding fragment is then ligated to the prepared expression vector fragment. For this, calculate ~5× molar excess of insert compared to vector and set up 20 μL ligations according to Table 1. Include a negative control ligation without insert DNA to check the vector background. A high number of resulting clones in this control ligation may indicate the presence of partially digested vector DNA or of truncated plasmids. Ligate also insert without vector DNA to check for the presence of truncated plasmids.
12. Mix the calculated volume of sterile distilled water, vector and DNA insert in a sterile 1.5 mL low binding tube, heat to 56 °C for 10 min and cool to room temperature. Add temperature sensitive ligase buffer and ligase to the samples and incubate the ligation mix for at least 2 h at room temperature (or overnight at 16 °C). Heat the reaction to 65 °C for 10 min to inactivate the ligase.
13. (Next day) For preparation of the transformation, desalt the sample by precipitation (*see* **Note 27**):
    (a) Fill the ligation samples to 50 μL with sterile H_2O, add 500 μL 2-butanol and 1 μL glycogen, and incubate for at least 5 min at room temperature on a rotator. Spin down the precipitated DNA for 30 min at maximum speed and 4 °C in a tabletop centrifuge (at least 20,000 × *g*).
    (b) Carefully remove the supernatant (the pellet of precipitated DNA might be hardly visible). Add 500 μL of 70%

ethanol precooled to −20 °C and spin for 15 min at maximum speed and 4 °C.

(c) Carefully remove the ethanol and air-dry the DNA for about 15 min. Resuspend the precipitated DNA samples in 10 μL sterile H_2O and incubate on a shaker for about 30 min at 37 °C to ensure that the DNA is completely dissolved.

(d) Dry large LB/Cam agar plates.

14. Transform electro-competent *E. coli* cells (e.g., DH5α or TG1F-) cells with 5 μL desalted ligation sample. Perform transformation as follows (*see* **Notes 28** and **29**):

(a) Thaw competent cells on ice and precool electroporation cuvettes at −20 °C. For each transformation sample, prewarm 950 μL SOB medium in a sterile 1.5 mL tube. Mix 5 μL of the ligated DNA sample with 45 μL competent cells and transfer the mixture into a chilled electroporation cuvette. Store the rest of the ligation at −20 °C.

(b) Dry the cuvette with a paper towel and perform electroporation as suggested by the manufacturer. Immediately after electroporation, transfer the electroporated cells into pre-warmed SOB medium and incubate for 1 h at 37 °C shaking at ~750 rpm.

(c) Centrifuge the bacterial suspension for 5 min at 1500 × *g*. Discard the supernatant and resuspend the pellet in 300–450 μL SOB medium.

15. Plate an appropriate volume of the transformation sample onto large LB/Cam agar plates to obtain single colonies (*see* **Note 30**). Incubate all plates overnight at 30 °C.

16. A retransformation of polyclonal DNA after **step 14** is strongly recommended to avoid mixed plasmids. For this, omit **step 15** and plate the bacterial suspension onto two large LB/Cam agar plates. Incubate all plates overnight at 30 °C. The next day, scrape the bacteria with 3–5 mL 2-YT medium using a new, sterile Drygalski spatula. Collect the bacterial suspensions belonging to the same transformation sample in a 13 mL disposable snap cap tube, determine the OD (*see* **step 2** in Subheading 3.2) and perform new polyclonal plasmid DNA preparations using a suitable DNA preparation kit according to instructor's manual.

17. Determine the plasmid DNA concentration, e.g., by absorbance measurements at 260 nm. Dilute the DNA to 50 pg/μL and perform a transformation of 50 pg DNA as described in **steps 14** and **15**.

3.3 Preparation of Selection Plates for Subsequent Primary Screening

1. (Next day) For each panning, prepare one flat-bottom 384-well microtiter plate with 60 μL growth medium per well. Inoculate each well with a single colony picked from the LB/Cam agar plates prepared (see **step 15** or **17** in Subheading 3.2), e.g., using a sterile toothpick. Seal the inoculated microtiter plates with gas-permeable foil, put the lid on top and shake overnight at 400 rpm and 37 °C (*see* **Note 31**). These plates will be the master plates.
2. (Next day) Prepare one flat bottom 384-well microtiter plate per master plate with 50 μL induction medium #2 per well. These plates will be used for cell screening ('expression plates'). Carefully remove the gas-permeable seals from the master plates and inoculate each well of the expression plate with 5 μL from the corresponding well of the master plate. Add 20 μL of freezing medium #2 to each well of the master plates, seal plates with aluminum foil and store master plates at −80 °C (*see* **Notes 32–34**).
3. Place a lid on the inoculated expression plates and shake at 37 °C and 400 rpm until the cultures become slightly turbid (~2–4 h) with an OD_{600nm} of ~0.5. Seal the expression plates with gas-permeable foil, put a lid on top and shake overnight at 400 rpm and 22 °C (*see* **Note 31**).
4. (Next day) Add 15 μL of 2× lysis buffer to each well of the expression plate and shake for 60 min at 400 rpm and 22 °C to lyse the bacteria.
5. For a subsequent cell screening, block the Fab-containing *E. coli* lysates by adding 15 μL of blocking buffer (final ~2.5% FBS) to each well and shake the expression plates for at least 30 min at 400 rpm and 22 °C.
6. Freeze expression plates (BEL lysates) at −80 °C for screening as described in the next chapter (*see* **Note 35**).

3.4 Primary Multiplex Cell Screening

Upon several rounds of selection of specific antibody candidates, it is highly recommended to perform a primary cell screening to evaluate binding of antibodies to multiple target cells (e.g., presenting antigen derived from different species or presenting antigen of one species on different cell lines) and/or to unwanted cell lines (e.g., non-target expressing cells). Such screening typically includes multiple rounds of hit-picking and repeated production of Fab-containing lysates. Using multiplexed screening approaches reduces necessary steps in the workflow, thus saving time and costs.

In multiplexing, different cell populations can be distinguished by pre-labeling the cells with distinct amounts of a fluorochrome, establishing a unique signature of fluorescence intensity for each population ("fluorescent barcoding") (*see* **Note 36**). The color coded cells are then physically combined and mixed together with antibodies (e.g., Fab-containing lysates, *see* Subheading 3.3) to be tested (*see* **Note 37**). After data acquisition, the cells from each sample can be identified and individually evaluated for antibody binding. For multiplexing approaches, we recommend to use a fluorophore (e.g., Calcein) for pre-staining of cells in combination with a secondary reagent (e.g., Alexa Fluor 647) for detection of bound antibodies with fluorescence emission which avoids the need for compensation.

All volumes indicated in this section refer to T175 cell culture flasks.

3.4.1 Fluorescent Barcoding of Cells for Multiplex Flow Cytometry

1. Grow a sufficient amount of cells (i.e., antigen-expressing and target negative cells). If necessary, add agents which stimulate antigen expression. Discard cultivation medium and wash cells once with 10 mL PBS. If required, detach cells (e.g., with 3 mL versene) for ~5–8 min at 37 °C. Monitor detachment under microscope. Resuspend cells in growth medium. Count cell suspensions and adjust cell numbers to 4*E*+06 cells/mL in growth medium.
2. For staining with Calcein, add an equal volume of dye-solution (2× concentrated) in growth medium to cells and incubate at 37 °C and 5% CO_2 for the required staining time (*see* **Note 38**). Spin down cells at 200 × *g* at 4 °C for 5 min, discard supernatant and wash with an adequate volume of FACS buffer. Repeat washing step twice.
3. Resuspend cells in the original volume in FACS buffer (4*E*+06 cells/mL). Cells are now stained and can be used for multiplexed analysis. Combine individual cell lines and proceed with flow cytometry (*see* **Notes 39** and **40**).

3.4.2 Flow Cytometric Staining Procedure in 384-well Format

1. For screening in 384-well format, transfer 15 μL of cell suspension/well in 384-well V-bottom plates. Generally, cell numbers of around 500–1000 cells/μL are ideal after mixing with BEL lysates. Depending on the number of cell lines to be tested in parallel, cell numbers might be adapted accordingly (*see* **Notes 40** and **41**).
2. Before using BEL lysates for FACS analysis, centrifuge plates for 5 min at 3,000 × *g* to pellet remaining cellular debris and clearing the BEL lysate supernatant (*see* **step 6** in Subheading 3.3) (*see* **Note 39**). Add 15 μL of pre-cleared and blocked BEL lysates to the respective wells on screening plates already containing cells; ensure that you use only supernatants (prepared as described in Subheading 3.3). Transfer *E. coli* lysate of an irrelevant Fab or 15 μL of FACS buffer to the negative control wells O24 and P24. Transfer 15 μL of FACS buffer to the negative control wells M24 and N24 and *E. coli* lysates of an irrelevant Fab to the negative control wells K24 and L24 (*see*

Table 2

Controls for flow cytometry screening

Well	Purpose	Antigen	Primary antibody	Secondary antibody
I24 J24	Pos. ctrl	Antigen of interest(cell mix)	BEL of reference antibody or purified reference antibody	Anti-reference antibody Alexa Fluor 647
K24 L24	Neg. ctrl	Antigen of interest(cell mix)	BEL of irrelevant antibody	FACS buffer
M24 N24	Neg. ctrl	Antigen of interest(cell mix)	FACS buffer	FACS buffer
O24 P24	Background	Antigen of interest(cell mix)	BEL of irrelevant antibody or FACS buffer	Anti-human Fab Alexa Fluor 647

Notes 32 and **34**, Table 2). Incubate cells with Fab-containing lysate in the dark for 60 min at 4 °C, gently shaking.

3. Add 50–70 μL of FACS buffer. Spin plates at 200 × *g* at 4 °C for 5 min. Aspirate supernatants and leave around 15–20 μL per well. Vortex plates shortly to resuspend cells.
4. Repeat washing as described in **step 3** twice using 80 μL of FACS buffer. Add 15 μL of Fab specific secondary antibody (Alexa Fluor 647-labeled) diluted as proposed by the manufacturer in FACS buffer. Incubate for 30–60 min at 4 °C in the dark, gently shaking.
5. Repeat washing as described in **step 3** twice using 80 μL of FACS buffer. Add 10–15 μL FACS buffer to adjust to the original volume (*see* **Note 42**). Perform measurement at flow cytometer.
6. Check signals of controls in row 24 (Table 2):
   (a) Wells I and J: positive control for preparation of *E. coli* lysates and/or flow cytometry procedure.
   (b) Wells K and L: background of BEL lysate on screening antigen.
   (c) Wells M and N: background of screening antigen.
   (d) Wells O and P: background of Fab detection antibody on screening antigen.
7. Determine the hit rate of the primary screenings. After data acquisition, the cell lines from each sample can be identified according to their fluorescence signature and individually evaluated for antibody binding. Positive signals should be at least

5× higher than the background signal (*see* **Note 43**). Positive hits can be selected from the respective position on the master plate and subjected to sequencing for CDR uniqueness, subsequent large scale expression and in-depth functional characterization.

4 Notes

1. This protocol describes a panning procedure with an antibody Fab-fragment based library. However, other library types such as scFv-based or scaffold-based libraries may be applied as well.
2. Phage might stick to the surface of tubes and may therefore be lost. This effect can be reduced by using siliconized, low binding tubes.
3. Ensure that an *E. coli* F+ strain harboring an F episome is used. The F pilus is essential to mediate the adsorption of the M13 phage to the bacterial cell.
4. Of utmost importance is an appropriate quality check of the target cells. Also, a high level of antigen expression on the cell surface should be confirmed before starting a panning. This can be done with suitable control antibodies via flow cytometry or cell-based ELISA.
5. If an (even very low) endogenous expression of the antigen on the adsorption cell line cannot be excluded, we highly recommend performing a differential whole cell panning instead, where post- or pre-adsorption steps are not necessary. Alternatively, residual endogenous expression of the target antigen can be eliminated by a si/shRNA knock-down approach.
6. We generally recommend the post-adsorption approach. At this stage the total number of phage is significantly lower which might improve adsorption efficiency.
7. Pre-experiments using a target-specific reference antibody may help to identify the optimal method for detachment of adherent cells expressing the target antigen.
8. If soluble agents, such as ligands or metal ions, are required for proper conformation of the target antigen, add them to all buffers used e.g., for blocking, washing, during pre- or post-adsorption steps and during primary screening.
9. We recommend keeping live cells cold during the entire panning process, including incubation at 4 °C and washing with ice-cold buffer.

10. Alternatively, fixed cells may be used for panning. In contrast to live cells, fixed cells may be incubated at room temperature during the panning procedure. We recommend fixing cells with 4% paraformaldehyde (PFA) in 1× PBS for 15 min on ice. Fixation, however, may affect antigen integrity.
11. The centrifugation speed depends on the size and robustness of the cells. It should be high enough to efficiently pellet the cells from the suspension and sufficiently low to avoid cell damage by mechanical stress. 200 × *g* is suitable for cell lines such as HEK293.
12. The pH of a glycine/Tris mixture (1000 μL glycine and 25 μL Tris base) should be checked for neutral pH, e.g., by pH indicator strips prior to post-adsorption.
13. Elution with DTT, which is an alternative elution method for disulfide-linked Fab phage complexes as it is the case for the Ylanthia library, is not recommended since most cell lines are sensitive to DTT and lysis of cells may occur in the buffers containing DTT.
14. Within the experimental setup described, the used library phage harbor a chloramphenicol resistance, while the helper phage confers the kanamycin resistance gene. If other antibiotic resistances are used, the agar plates should be adapted accordingly.
15. We recommend to use three large LB/Cam agar plates per selection after the first panning round and two large LB/Cam agar plates per selection for each subsequent panning round to ensure proper growth of all *E. coli* cells harboring a phagemid.
16. Do not use more than ~20 glass beads per large LB/Cam agar plate, as increased amount of glass beads may decrease the panning output.
17. The washing stringency of subsequent panning rounds can be adjusted according to the output titer of the previous selection round: in general, output titers should range within 1*E*+04 and 1*E*+08 cfu. For output titers >1*E*+08 cfu, number and duration of individual washing steps can be adjusted in the subsequent selection round for a more stringent selection. Keep in mind that certain cell lines may not tolerate increased washing steps.
18. Example: Number of colonies: 50.

Volume plated: 100 μL of 1:100 diluted infected TG1 cells (equals 1 μL = 0.001 mL of undiluted cells).

$$\text{Output } 1^{st} \;/\; 2^{nd} \;/\; 3^{rd} \text{ round} = \frac{50 \times 30\, mL}{0.001\, mL} = 1.5E + 06\, cfu.$$

19. For phage production, the *E. coli* TG1F+ carrying the selected phagemids are infected with helper phage which supply all the proteins required for complete assembly of functional phage.
20. The induction medium must not contain glucose which would inhibit the IPTG-induced Fab expression.
21. Usually, clouds of precipitating phage become visible after the incubation step on ice.
22. PEG/NaCl is toxic to *E. coli* and should therefore be thoroughly removed from the phage preparation.
23. Alternatively, A_{268nm} measurement can be used to determine titers of phage preparations. Determine the titer of the phage preparation according to the following equation (an $A_{268nm} = 1$ corresponds to a phage titer of 5*E*+12 tu/mL):

Phage titer [tu/mL] = A_{268nm} × (dilution factor of phage preparation) × 5*E*+12 tu/mL.

Using this method, the total phage titer is determined, while, in contrast, the titer determination via spot titration only covers infective phage. For this reason, A_{268nm} measurements tend to yield up to 10- to 100-fold higher titers than spot titration. The input phage titer for the next round of panning should therefore be 100,000× the size of the output titer of the previous selection round, in order to cover the whole range of diversity.

24. The expected phage titer lies in the range of 1*E*+12 and 1*E*+13 tu/mL.
25. Example: The 1:200 diluted bacterial suspension has an OD_{600nm} of 0.5.

$$V\,(\text{bacterial suspension})\,[\mu L] = \frac{24 \times 1000\,\mu L}{0.5 \times 200} = 240\,\mu L.$$

Use 240 μL of the undiluted bacterial suspension for each DNA preparation.

26. When using a DNA preparation kit, try to use a kit which enables the removal of endA nuclease to avoid the degradation of DNA (endA nuclease might be contained in plasmid preparations from TG1 cells).
27. Alternatively, desalting by dialysis is possible.
28. Alternatively, a heat shock transformation of chemically competent cells can be performed. In this case, the desalting of the ligation can be omitted.
29. DH5α *E. coli* cells do not require IPTG to induce expression from the lac promoter (even though the strain expresses the Lac repressor) as the copy number of most plasmids exceeds the repressor number in the cells. If obtaining the maximal

levels of expression is required, add IPTG to a final concentration of 1 mM.

30. It is recommended to plate different volumes between 50 and 500 μL to provide enough single colonies for picking the next day.
31. Covering the plates with a lid during overnight incubation is important to reduce evaporation.
32. For 384-well screening, inoculate wells I24 and J24 with a control clone, which will serve as positive control. For negative controls, inoculate the medium in wells K24, L24, O24, and P24 with an irrelevant Fab, which does not bind to the screening antigen, or leave wells O24 and P24 empty.
33. Do not store master plates at 4 °C. Avoid multiple freezing and thawing of master plates, as this will result in a lower number of growing clones after ~3–4 cycles.
34. Ideally, a positive control antibody directed against the target positive cell line(s) is available in a format which allows its preparation in BEL lysate. Alternatively, a control antibody can be spiked into an *E. coli* BEL lysate which does not contain a Fab.
35. Remember to culture sufficient cells for the subsequent high throughput flow cytometric screening as described in Subheading 3.4.
36. Individual cell lines behave differently and therefore it is recommended to evaluate staining conditions for each cell line. It is advised to stain individual cell lines using multiple dye concentrations: Testing a range of concentrations starting at 0.1 μM is recommended. Decent fluorescent staining was still observed at concentrations as low as 0.00025 μM. Typical staining times range from a few minutes up to around 60 min. Measure fluorescence signals in the appropriate channel of a flow cytometer for individual cell lines; make a graphic overlay of histograms to find staining conditions that allow separation of distinct cell lines (*see* Fig. 3).
37. Before using the identified staining conditions in a screening, verify that the cell lines can all be individually identified after being physically mixed together. It has been observed that residual dye from high fluorescence cells samples might stain low intensity/unstained cells. This can lead to a shift of intensity and to the overlap of signals from distinct cell lines, making it impossible to separate such samples from each other. As a result, always use as little dye as necessary.
38. Alternatively, cells may be stained with CellTracker Green. Avoid amine- and thiol-containing buffers and always use serum free medium for preparing dye solutions with

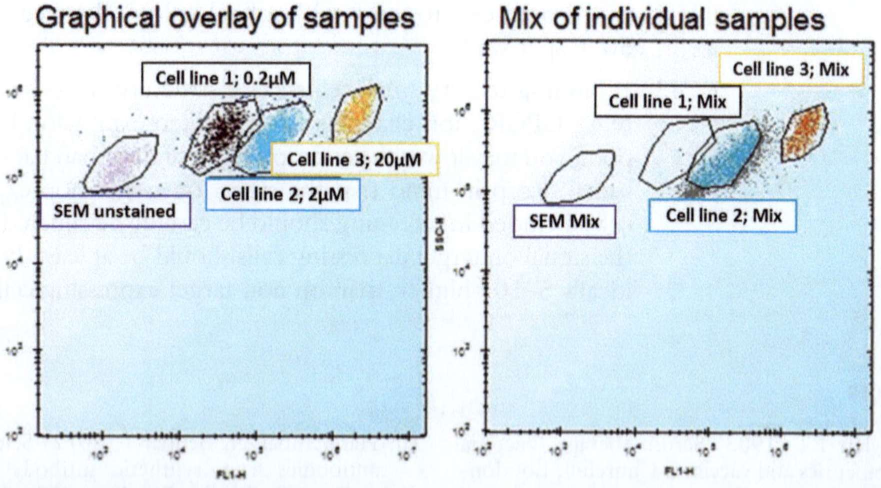

Fig. 3 Labeling of cells with high dye concentrations causes unstained populations to show substantial fluorescence after mixing

CellTracker Green. The optimal concentration of the probe for staining varies depending upon the cell line and application. To maintain normal cellular physiology and reduce potential staining artifacts, keep the final dye concentration as low as possible. For staining, add an equal volume of 2× dye-solution in serum-free medium to cells and incubate at 37 °C and 5% CO_2 for 30 min. Spin down cells, discard supernatants and add the same volume of cell-specific culture medium as used during staining. Incubate for another 30 min at 37 °C and 5% CO_2. Wash cells with an adequate volume of FACS buffer. Cells can now be used for primary screening.

39. In contrast to CellTracker Green, Calcein staining is not stable and analysis should be conducted in a timeframe of ~6 h to avoid shifting of cell populations towards lower signals and reduced differentiation of distinct cell lines.
40. Cell counts of around 250 and higher give robust data. Ideally, around 1000 events/population should be analyzed. Consequently, it is also possible to evaluate antibody binding to multiple cell lines in parallel without overloading the system; i.e., for multiplexing of four cell lines, it is not necessary to increase cell numbers of individual cell lines fourfold; however, slightly increasing cell numbers in such cases might be beneficial.
41. An additional Fab expression check of all picked clones is recommended for verification of Fab expression in *E. coli* lysates using e.g., an Fd specific anti-human IgG as coating material.
42. To ensure that a sufficient volume/well (and therefore a sufficient amount of cells) is analyzed during the measurement,

we recommend to work with a final volume between 30 and 50 μL per well.

43. If binding to very small cell surface structures is to be assessed (e.g., GPCRs, ion channels) or if antigen expression levels are poor, you may lower the criterion to 2× higher than background signal. Keep in mind that the signal on target non-expressing cells included in screening should be carefully evaluated as well: the signal on target expressing cells should be at least 2× higher, ideally 5–10× higher, than on non-target expressing cells.

References

1. Hewlett RT (1903) Serum therapy: bacterial therapeutics and vaccines. Churchill, London
2. Smith GP (1985) Filamentous fusion phage: novel expression vectors that display cloned antigens on the virion surface. Science 228(4705):1315–1317
3. McCafferty J, Griffiths AD, Winter G, Chiswell DJ (1990) Phage antibodies: filamentous phage displaying antibody variable domains. Nature 348(6301):552–554
4. Wang XX, Shusta EV (2005) The use of scFv-displaying yeast in mammalian cell surface selections. J Immunol Methods 304(1–2):30–42. doi:10.1016/j.jim.2005.05.006
5. Amstutz P, Forrer P, Zahnd C, Pluckthun A (2001) In vitro display technologies: novel developments and applications. Curr Opin Biotechnol 12(4):400–405
6. Fuchs P, Breitling F, Dubel S, Seehaus T, Little M (1991) Targeting recombinant antibodies to the surface of *Escherichia coli*: fusion to a peptidoglycan associated lipoprotein. Nat Biotechnol 9(12):1369–1372. doi:10.1038/nbt1291-1369
7. Michelfelder S, Lee M, deLima-Hahn E, Wilmes T, Kaul F, Müller O, Kleinschmidt JA, Trepel M (2007) Vectors selected from adeno-associated viral display peptide libraries for leukemia cell-targeted cytotoxic gene therapy. Exp Hematol 35(12):1766–1776. doi:10.1016/j.exphem.2007.07.018
8. He M, Taussig MJ (1997) Antibody-ribosome-mRNA (ARM) complexes as efficient selection particles for in vitro display and evolution of antibody combining sites. Nucleic Acids Res 25(24):5132–5134
9. Bradbury AR, Sidhu S, Dubel S, McCafferty J (2011) Beyond natural antibodies: the power of in vitro display technologies. Nat Biotechnol 29(3):245–254. doi:10.1038/nbt.1791
10. Harel Inbar N, Benhar I (2012) Selection of antibodies from synthetic antibody libraries. Arch Biochem Biophys 526(2):87–98. doi:10.1016/j.abb.2011.12.028
11. Miersch S, Sidhu SS (2012) Synthetic antibodies: concepts, potential and practical considerations. Methods 57(4):486–498. doi:10.1016/j.ymeth.2012.06.012
12. Vaughan TJ, Williams AJ, Pritchard K, Osbourn JK, Pope AR, Earnshaw JC, McCafferty J, Hodits RA, Wilton J, Johnson KS (1996) Human antibodies with sub-nanomolar affinities isolated from a large non-immunized phage display library. Nat Biotechnol 14(3):309–314. doi:10.1038/nbt0396-309
13. Tiller T, Schuster I, Deppe D, Siegers K, Strohner R, Herrmann T, Berenguer M, Poujol D, Stehle J, Stark Y, Hessling M, Daubert D, Felderer K, Kaden S, Kolln J, Enzelberger M, Urlinger S (2013) A fully synthetic human Fab antibody library based on fixed VH/VL framework pairings with favorable biophysical properties. MAbs 5(3):445–470
14. Rothe C, Urlinger S, Lohning C, Prassler J, Stark Y, Jager U, Hubner B, Bardroff M, Pradel I, Boss M, Bittlingmaier R, Bataa T, Frisch C, Brocks B, Honegger A, Urban M (2008) The human combinatorial antibody library HuCAL GOLD combines diversification of all six CDRs according to the natural immune system with a novel display method for efficient selection of high-affinity antibodies. J Mol Biol 376(4):1182–1200. doi:10.1016/j.jmb.2007.12.018
15. Knappik A, Ge L, Honegger A, Pack P, Fischer M, Wellnhofer G, Hoess A, Wolle J, Pluckthun A, Virnekas B (2000) Fully synthetic human combinatorial antibody libraries (HuCAL) based on modular consensus frameworks and CDRs randomized with trinucleotides. J Mol

Biol 296(1):57–86. doi:10.1006/jmbi.1999.3444

16. van den Brulle J, Fischer M, Langmann T, Horn G, Waldmann T, Arnold S, Fuhrmann M, Schatz O, O'Connell T, O'Connell D, Auckenthaler A, Schwer H (2008) A novel solid phase technology for high-throughput gene synthesis. Biotechniques 45(3):340–343
17. Chun E, Thompson AA, Liu W, Roth CB, Griffith MT, Katritch V, Kunken J, Xu F, Cherezov V, Hanson MA, Stevens RC (2012) Fusion partner toolchest for the stabilization and crystallization of G protein-coupled receptors. Structure 20(6):967–976. doi:10.1016/j.str.2012.04.010
18. Robertson N, Jazayeri A, Errey J, Baig A, Hurrell E, Zhukov A, Langmead CJ, Weir M, Marshall FH (2011) The properties of thermostabilised G protein-coupled receptors (StaRs) and their use in drug discovery. Neuropharmacology 60(1):36–44. doi:10.1016/j.neuropharm.2010.07.001
19. Burton DR, Scott JK, Silverman GJ (2001) Phage Display Cold Spring Harb Press, Cold Spring Harbor, NY, USA

Chapter 6

Generating Conformation and Complex-Specific Synthetic Antibodies

Marcin Paduch and Anthony A. Kossiakoff

Abstract

Phage display is commonly used to identify and isolate binders from large combinatorial libraries. Here we present phage selection protocols enabling generation of synthetic antibodies capable of recognizing multiprotein complexes and conformational states. The procedure describes stages of the experiment design, optimization, and screening, as well as provides the framework for building downstream assays with an end goal of isolating bioactive antibodies for future therapeutic use. The methods described are also applicable to screening directly on cells and can be ported to other in vitro directed evolution systems utilizing non-immunoglobulin scaffolds.

Key words Antibody phage display, Synthetic antibody, Conformation specific

1 Introduction

Since its introduction, phage display has become a powerful method for isolating binders from large libraries that routinely contain billions of unique variants (~10^{10}) [1]. There have been many groups that have contributed to its capabilities making it now recognized as a powerful in vitro alternative to widely used hybridoma methods [2, 3].

To date, a number of antibody-based phage display libraries have been developed and successfully used for generating functional binders, utilizing several types of selection strategies, to a variety of types of target antigens including soluble proteins, membrane proteins and protein complexes. In particular, several synthetic antibody libraries based on "reduced genetic code" concepts were developed allowing for a larger number of positions in the CDR loops to be diversified with no loss in binding capability [4–6]. Unlike antibody libraries obtained from natural sources that include a considerable level of potentially unwanted sequence diversity in the frameworks [7, 8], synthetic libraries either utilize a single scaffold [4, 5] or defined set of scaffolds [6] allowing for

Thomas Tiller (ed.), *Synthetic Antibodies: Methods and Protocols*, Methods in Molecular Biology, vol. 1575,
DOI 10.1007/978-1-4939-6857-2_6,

precise optimization of loop composition and framework leading to the advantage of making expression and stability much more predictable [9, 10].

One of many advantages of an in vitro selection process is the ability to design selection strategies capable of generating antibodies with customized functions without the limitations of the mammalian immune system like immune tolerance in B cell populations [11], presence of immunodominant epitopes [12] or challenges related to humanization [13]. Further, exquisite control of the selection conditions makes it possible to obtain synthetic antibodies that recognize a specific conformational state [14–18], regio-selective antibodies targeting specific regions of the target protein [19, 20], recognizing multiprotein complexes [21, 22] as well as capture and stabilize transient protein–protein interactions [18]. Protein engineering tools used for generation of synthetic antibodies are a mainstay of biotherapeutic development [23, 24] where customized functions are commonly used to elicit defined responses.

Applications in which synthetic antibodies can be used range from western blots, ELISA and related plate based immunoassays to immunostaining, immunoprecipitation and chromatin immunoprecipitation (ChIP) [10]. Nonetheless, the unique advantage lies in their ability to perform advanced functions based on custom designed properties where dynamic properties of the system can be studied [14, 17, 25]. This has led to the use of synthetic antibodies targeting conformations and multiprotein complexes as effective crystallization chaperones contributing the most of crystal contacts in the lattice and aiding structure solution through molecular replacement [14, 16–19, 21, 22].

Here we describe a simple workflow to isolate antibodies that are selective to conformations and protein complexes. The methods described should be augmented with knowledge of library production, antigen engineering, and assay development for downstream validation of generated antibodies, which are beyond the scope of this chapter.

2 Materials

1. EZ-Link NHS-SS-PEG$_4$-Biotin (NHS ester of biotin with an ethyl-1,3-dithioproprionate and tetraethyleneglycol spacer) (Pierce).
2. EZ-Link Biotin-HPDP (Pyridyldithiol-activated biotin with a hexylenediamine spacer) (Pierce).
3. BirA enzyme, 3.8 mg/mL (Avidity).
4. 10× Biomix-A, 0.5 M bicine buffer, pH 8.3.
5. 10× Biomix-B, 100 mM ATP, 100 mM magnesium acetate, 500 μM d-biotin.

6. Dithiothreitol (DTT).
7. HRV 3C protease, 1 mg/mL.
8. TEV protease, 1 mg/mL.
9. Sera-Mag SpeedBeeds blocked streptavidin-coated paramagnetic particles (GE Healthcare).
10. Sera-Mag SpeedBeads neutravidin-coated paramagnetic particles (GE Healthcare).
11. KingFisher Flex magnetic bead processor with 96-deep well magnet head (Thermo Scientific).
12. KingFisher 96 DW plates.
13. KingFisher 96 DW tip combs.
14. EL406 plate washer (BioTek).
15. Microplate, 96-well, PP, V-bottom (Greiner).
16. Microplate, 96-deep well MasterBlock, 2.4 mL (Greiner).
17. Microplate, 24-Square Deep Well Plate, 10 mL (Axygen).
18. Microplate, 384-well, PS, F-bottom, high binding (Greiner).
19. Microplate, 384-well, PP, V-bottom (Greiner).
20. 2×YT medium, 10 g yeast extract, 16 g tryptone, 5 g NaCl, add water to 1 L and adjust pH to 7.0, autoclave.
21. TB medium, 24 g yeast extract, 12 g tryptone, 2.31 g KH_2PO_4, 12.54 g K_2HPO_4, 5 g glycerol, add water to 1 L and adjust pH to 7.0, autoclave.
22. LB Agar plates with 100 μg/mL ampicillin.
23. Autoinduction mix, 50% glycerol, 0.3% glucose, 1.25% lactose, 80 mM $MgSO_4$.
24. Antifoam 204.
25. Tetracycline, 2 mg/mL in 50% EtOH, store at −20 °C.
26. Ampicillin, 100 mg/mL in water, store at −20 °C.
27. Chloramphenicol, 15 mg/mL in 100% EtOH, store at −20 °C.
28. d-Biotin, 5 mM in DMSO, store at 4 °C.
29. *E. coli* XL1-Blue cells, *recA1 endA1 gyrA96 thi-1 hsdR17 supE44 relA1 lac* [*F′ proAB lacIqZΔM15 Tn10 (TetR)*] (Agilent).
30. OverExpress C43(DE3) cells, *F- ompT hsdSB (rB- mB-) gal dcm* (DE3) (Lucigen) transformed with pTUM4 vector [26].
31. M13 KO7 helper phage.
32. PEG/NaCl, 20% PEG 8000, 2.5 M NaCl, filter sterilized.
33. PBS, filter sterilized, store at 4 °C.
34. PBS-BSA, PBS, 0.2% BSA, filter sterilized, store at 4 °C.

35. PBST-BSA, PBS, 0.05% Tween, 0.2% BSA, filter sterilized, store at 4 °C.
36. HRP/anti-M13 antibody conjugate (GE Healthcare).
37. NeutrAvidin (Pierce).
38. TMB substrate (Pierce).
39. Omega E-Z 96 Fastfilter Plasmid Kit (Omega Bio-Tek).
40. HiTrap MabSelect SuRe 5 mL (GE Healthcare).
41. Resource S 1 mL (GE Healthcare).
42. Lysis buffer, 50 mM Tris, 500 mM NaCl, 0.01 mg/mL DNAse, 0.5 mM $MgCl_2$, 2.5 μg/mL leupeptin, 1% Triton X-100, pH 8.0.
43. Equilibration buffer, 50 mM Tris, 500 mM NaCl, pH 8.0.
44. Elution buffer, 0.1 M acetic acid.
45. Regeneration solution, 0.1 M NaOH.
46. IEX equilibration buffer, 50 mM sodium acetate pH 5.0.
47. IEX elution buffer, 50 mM sodium acetate, 2 M NaCl, pH 5.0.

3 Methods

3.1 Experiment Design

Synthetic antibodies that recognize discrete conformations or protein complexes preferentially bind to given state of the target effectively shifting the equilibrium and stabilizing interacting components Fig. 1 [27]. Antibodies with affinities in the nM range can contribute significant free energy to protein complex

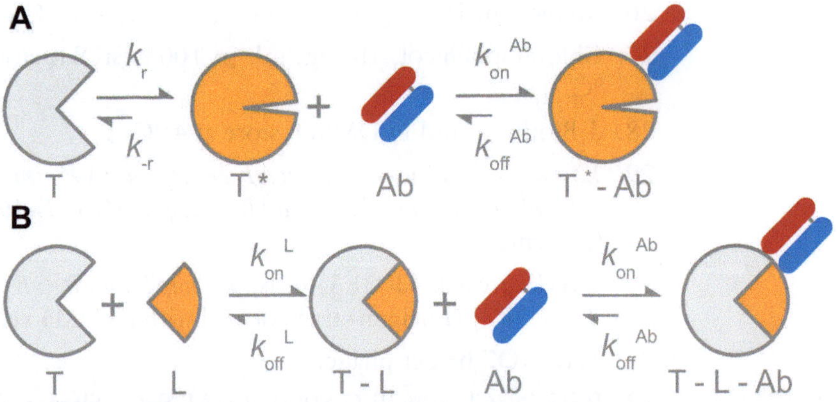

Fig. 1 Synthetic antibodies recognizing (**a**) conformations and (**b**) complexes. Figure presents two simplest modes of binding where in saturating concentrations of synthetic antibody [Ab] k_{on}^{Ab} drives formation of the complex effectively shifting the equilibrium $k_{obs1} = k_{on}^{Ab}[Ab]$, $k_{obs2} = k_r$, $k_{obs2} = k_{on}^{L}$

formation. Naturally more complicated modes of binding have been observed [28], therefore, all sources of conformational heterogeneity have to be taken into account when designing a selection strategy, for instance, in situations where binding of a partner protein can only occur in the cofactor bound state [18]. The ability to separate the conformational states and intrinsic stability of investigated complexes dictate the choice of selection techniques. Extensive characterization of target proteins prior to performing selection is preferred and strongly correlates with selection success. If mutants, construct variants, small molecule drugs, or cofactors can be used as stabilizers of a given conformational state (Fig.1a stable conformation T*) or can aid complex formation (Fig.1b stable complex T-L), the selection strategy can sometimes be simplified eliminating the need for extensive screening of obtained phage pools. Of course, if such stabilization approaches are used, confirmation is required that observed states are indeed physiologically relevant.

3.1.1 Subtraction and Competition Selection Schemes

Synthetic antibodies specific to conformations and complexes can be obtained from two different types of selection: positive or negative. Since subtraction and competition schemes have different strengths, the choice of type of selection is largely dictated by the properties of the system under study and the desired outcome.

The concept of subtractive selection is presented in Fig. 2a where negative and positive selections are performed sequentially in two steps. In step 1 the library is "pre-cleared" by incubating the phage library with the undesirable protein component or conformational state immobilized on solid support (magnetic beads or plastic plate). In this step unwanted binders are captured and effectively subtracted from the library. This pre-cleared library is used for positive selection in step 2 with the primary selection target.

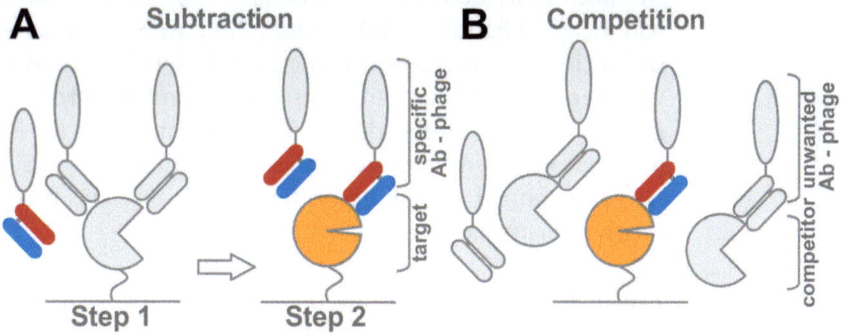

Fig. 2 Synthetic antibody selection strategies. (**a**) Subtractive selection—competitor (gray sector shape "open") in step 1 (negative selection) is added to the beads, library is pre-cleared and added in step 2 (positive selection) to the new beads containing target of interest (orange sector shape "closed"), (**b**) competition selection—competitor is added in excess in solution along with captured biotinylated target

In contrast, in the competitive selection strategy, negative and positive selections are performed in a single step (Fig. 2b). The competitor is added in large excess to the mixture of biotinylated target and the phage library. This effectively removes all binders to the competitor that might also react with the desired selection target. The biotinylated target can then be pulled down using magnetic beads, capturing only target-specific phages while untagged competitor and any of the phages bound to it are washed away. In our experience, due to smaller number of steps required, the competition selection scheme usually produces phage pools of higher diversity and is preferred over subtraction strategy.

Competition selection is better suited for targets with clear separation of conformational states and cases when high affinity cofactors can be used. In cases when ligand is binding weakly to the target subtraction strategy is usually a better choice. Antibodies that bind exclusively to the macromolecular complex (Fig. 2b) are ideally obtained in competition selection. In this particular scenario, target (T) is immobilized on a magnetic bead and saturating concentration of ligand (L) is added. This guarantees that the complex is formed even if the target–ligand affinity is low and antibodies recognizing the free ligand are not captured. In subsequent rounds of selection, this strategy can be further expanded by replacing the target with ligand on the bead surface while using an excess of solution target competitor. This ensures that antibodies recognizing both uncomplexed ligand (L) or target (T) are not inadvertently enriched in the phage pools.

3.1.2 Synthetic Antibody Selection and Screening Workflow

The sections below describe the principal steps required for successful generation of synthetic antibodies by phage display. First, the target preparation and quality control are discussed in Subheading 3.2 with particular focus on ensuring homogeneity, stability and effective bead capture and release. Next, the solution phase phage selection process is described in Subheading 3.3. Methods described in this section are ideally semi-automated using robotic bead handlers. Subheading 3.4 covers initial screening and specificity assessment of obtained binders with a focus on simple, yet effective tools that can be used to distinguish promising clones from poor ones. Subheading 3.5 presents methods for bench scale production of synthetic antibodies and outlines biophysical characterization aimed at ranking of the individual clones based on their relative performance.

3.2 Target Protein Preparation

Effective antigen optimization is pivotal to the success of any synthetic antibody screening campaign. Challenges like engineering solubility or modification of surface charge, conformational flexibility or presence of post translational modifications have to be addressed in the process. There are several protein systems that are especially difficult.

1. Multiprotein complexes and proteins that weakly bind to substrates or DNA pose multiple challenges. Since usually not all components constituting the large multiprotein complex will be stably binding, such systems can be reduced to binary stable interactions and studied as such. Alternatively, constructs can be optimized to stabilize multicomponent complex formation [21, 22]. In cases of DNA–protein complexes, nonspecific binding is usually attributed to electrostatics and has to be controlled by modification of the chemical milieu. Mild cross-linkers can be used, but usually such modification of the target results in fewer binders and lower enrichments.
2. Proteins with high isoelectric points (pI > 10) usually yield fewer specific binders because they can bind nonspecifically to the negatively charged phage coat [10]. In select cases, shielding or neutralizing surface charges can prove effective [29].
3. Presence of immunodominant epitopes is hard to predict based on knowledge of antigen sequence or even 3D structure. It is common that proteins have antigenic hot-spots frequently coinciding with natural binding sites. This in turn can be useful for generation of antibodies for biological studies. If the goal of the selection is to avoid a hot-spot, that surface epitope can be blocked by introduction of mutations in the hot spot region or masking the epitope with a known synthetic antibody that binds to undesirable surface of the protein.
4. Proteins with a high conformational flexibility are inherently more difficult to select on due to high entropic cost of synthetic antibody binding. To effectively increase chances of obtaining synthetic antibodies to such proteins conditions has to be devised in which chemical composition of the selection buffer is optimized in a way that limits conformational sampling by the protein of interest. This can be usually accomplished by adjusting pH, introducing ions or adding small molecule ligands or partner proteins. In the protocol described in Subheading 3.3.2 we often use the osmolyte trimethylamine N-oxide (TMAO), as a protein stabilizer.
5. Post translational modifications (PTMs) of proteins have properties that can prove very useful or detrimental to selection success. There are cases where the majority of protein surface is modified making in highly heterogeneous and difficult to select on [30]. Antibodies targeting PTMs often prove very selective, but difficult to characterize. Often glycans can be removed with PNGase, streamlining the selection process but later validation on fully modified protein is required. A similar logic applies to phosphorylation where phosphomimics can be used followed by validation on the naturally occurring protein form.

The antigens used in selection must be of high quality. Homogeneity and monodispersity are probably two most important parameters to optimize. It is advantageous to check if antigens are functional. This can be based on tests such as an enzymatic assay, binding to partner proteins or cell assays. Also, since the selection process takes usually 4 days antigens should be appropriately stored in conditions that will minimize degradation or aggregation. Denatured and/or aggregated protein samples are the principal problem for why synthetic antibody generation fails.

The protocol described in Subheading 3.3.2 relies on a specific capture and release step that utilizes a biotin tag to maximize solution capture of phage displaying the desired synthetic antibodies. High affinity binding of the tag is essential in this case since it must be able to withstand the rigorous washing steps that are necessary for removing nonspecific binders present in the library. In the simplest implementation, a protein can be chemically modified through amine moieties or exposed Cys residues with biotin-derivatized reagents with linkers that can be cleaved by reducing agents (Subheading 3.2.1). Chemical biotinylation is the only choice for proteins that are obtained from natural sources due to lack of recombinant protein or the presence of post translational modifications hard to reproduce in the recombinant systems. Therefore, if the protein sample is stable and readily available, this method can be used as the first approach. Biotinylated protein samples have to be rigorously tested for the level of biotinylation in pull-down assay described in Subheading 3.2.3.

An inevitable drawback of chemical biotinylation is that eukaryotic proteins are frequently not stable enough to withstand chemical modification of their surface. In some cases biotinylation can impair protein function by affecting its conformation and stability. Therefore, an approach utilizing BirA and in vivo or in vitro biotinylation (Subheading 3.2.2) of an Avi-tag fused to the protein of interest should be used first [10, 31]. This method is particularly well suited for high throughput applications and provides multiple advantages over chemical biotinylation. In an Avi-tag selection system, specific release of the protein can be achieved by introduction of either TEV or HRV 3C protease cut sites in the tag sequence. Positioning of the tag (N or C terminus of the protein) and linker length might have to be optimized. Also for multimeric proteins enzymatic cutting conditions have to be optimized by evaluating multiple target–magnetic bead ratios and cut times. Enzymatic elution cannot be used if protein of interest aggregates in the cleavage buffer and effectively is not eluted from the beads (Subheading 3.2.3).

3.2.1 Chemical Biotinylation

1. Dialyze or desalt 5 nmol of protein of interest to the buffer of choice. The starting protein concentration should be in 1–100 μM range (*see* **Note 1**).
2. Dissolve biotinylation reagent EZ-Link NHS-SS-PEG$_4$-Biotin for amine labeling or EZ-Link Biotin-HPDP for Cys labeling in dry DMSO or buffer at 10 mg/mL (*see* **Note 2**).
3. Mixed dissolved biotinylation reagent with the protein to obtain 2:1–20:1 molar ratio. Adjust the ratio based on molecular weight (Lys and free Cys content) and efficiency (*see* **Note 3**).
4. After 60 min of incubation, quench the reaction by adding TRIS buffer to a final concentration of 10 mM, pH 7.5–8 (this step is not required for EZ-Link biotin-HPDP reagent) (*see* **Note 4**).
5. Transfer the reaction to the dialysis tubing and dialyze against a buffer of choice overnight, preferably for 24 h (*see* **Note 5**).
6. Determine the protein concentration by the method of your choice and proceed to pull-down test using streptavidin-coated beads (Subheading 3.2.3) (*see* **Note 6**).

3.2.2 In Vitro Biotinylation

1. Combine Biomix-A and Biomix-B solutions with 5 nmol of Avi-tagged protein at 50 μM concentration or higher and 3 μg of BirA enzyme (*see* **Note 7**).
2. Incubate the reaction mixture for 1 h at 37 °C.
3. Desalt the reaction mixture on a gel filtration column or preferably affinity-purify the protein (*see* **Note 8**).
4. Determine the protein concentration and proceed to the pull-down test using streptavidin-coated beads (Subheading 3.2.3).

3.2.3 Pull-Down Test to Determine the Extent of Biotinylation and Efficiency of Antigen Release

1. Wash 5 μL of Sera-Mag SpeedBeeds blocked streptavidin particles three times with 10 μL of binding buffer (*see* **Note 9**).
2. For a protein of up to ~150 kDa mix 3 μg of biotinylated protein with 5 μL of streptavidin magnetic beads suspension.
3. Incubate for 15 min and collect the supernatant (unbound fraction). Keep the total volume below 10 μL so the whole sample can be loaded on the SDS-PAGE gel (*see* **Note 10**).
4. Incubate for 15 min and recover supernatant, use magnet to collect the beads at the side of the tube.
5. Wash the beads once with 10 μL of binding buffer.
6. Add 10 μL elution buffer-binding buffer supplemented with 100 mM fresh DTT, 5 μg of TEV protease or 3 μg of HRV 3C protease (*see* **Note 11**) mix gently with a pipette tip and allow to incubate for 15 min in room temperature. Collect eluate in a separate tube.

7. Wash the beads once with 10 μL of binding buffer and keep the beads.
8. Run SDS-PAGE gel with all collected fractions (unbound fraction, washes, elution and beads after elution) and determine percentage of the protein captured and released from the beads (*see* **Note 12**).
9. Biotinylated protein can be aliquoted in tubes and frozen at −80 °C or stored at 4 °C.

3.3 Phage Library Selection

Library selection is performed based on the chosen selection strategy (Subheading 3.1.1) and executed as described in Subheading 3.3.2. In fact, both of the selection types described can be combined in the process based on the desired outcome and limitations of the system studied. This particular implementation of phage selection can be easily adapted to a high-throughput format where many targets and conditions are evaluated in parallel allowing for direct comparisons of selection strategies. Screening of multiple targets or their variants can be performed easily often leading to outcomes not accessible to lower throughput methodologies.

High throughput processes rely on a KingFisher magnetic bead handler that allows automation of most phage selection steps (capture, washing and release) (Subheading 3.3.2). Recovered phages are propagated in *E. coli* and the selection process can be repeated with an increased level of selective pressure. For instance, in standard selections, the most critical selection pressure is introduced by lowering concentration of the target and the phage particles used each round. We generally gradually reduce the target concentration from 100 to 10 nM. Selection conditions are supplemented with cofactors when required and up to 10 μM competitors are included throughout the selection process for added stringency. Progress of the selection is monitored by phage titering to check recovery of phage particles, as well as propagation and enrichment levels.

3.3.1 Phage Library Preparation

In our selections, we routinely use Fab fragment libraries built on a single antibody framework and a diversity of >10^{10} unique clones where both light chain and heavy chain are randomized [4, 5, 32]. Detailed library construction and transformation protocols have been described previously [33] and can be used for the preparation of phage library samples used in methods described in this chapter. The selection procedures outlined below, should also work for other antibody-display formats based on alternative library designs. It has to be noted that quality of the library strongly affects selection outcomes and is pivotal for success of the antibody generation process.

3.3.2 Competitive and Subtractive Phage Display Selection

The library selection is conducted utilizing a chosen selection strategy (Subheading 3.1.1) and appropriate targets (Subheading 3.2). The process combines manual and automated steps (indicated in the text) and takes several days depending on the number of selection rounds required. In successful selections, enrichment can be observed as early as round 2 while the best performing clones are usually obtained after three rounds of screening. Extending the selection beyond four rounds of screening usually drastically limits diversity of the obtained phage pools and provides no advantage over selection redesign. In such cases using alternative antigens, competitors or phage libraries should be considered.

(A) Round 1 of selection:

1. In a 250 mL shaker flask inoculate 10 mL of 2×YT medium supplemented with tetracycline 20 μg/mL with 200 μL of overnight starting culture of XL1-Blue cells and grow at 37 °C shaking intensely for 3 h or until OD_{600} reaches 0.5 (*see* **Note 13**).
2. In a 50 mL conical tube combine 16 mL PBS, 3.5 mL PEG/NaCl, 1 mL of 2×10^{12} cfu/mL synthetic antibody library phage stock in 50% glycerol and place on ice for 30 min. Spin down at 8000 × *g* for 20 min, discard supernatant immediately, flash spin the tube and aspirate remaining supernatant. Resuspend the pellet in 1 mL of PBST-BSA (*see* **Note 14**).
   (a) Competitive selection: add competitors preferably at 1 μM concentration and incubate for 15 min before adding to the "Protein binding plate".
   (b) Subtractive selection: add 100 μL of 1 μM biotinylated competitor to 10 μL of washed Sera-Mag Streptavidin paramagnetic particles, wash once with one volume of PBST-BSA, place on magnet for 1 min, remove the supernatant and add the resuspended phage library. Incubate for 15 min on a rotator. After the incubation is complete magnet down the beads and carefully collect the phage library and add it to the "Protein binding plate".
3. Prepare KingFisher 96 DW selection plates (*see* **Note 15**).
   (a) Protein binding plate: 100 μL of 1 μM biotinylated target protein, and 5 μL of washed Sera-Mag Streptavidin paramagnetic particles.
   (b) Biotin block plate: 999 μL of PBST-BSA and 1 μL of 5 mM biotin.
   (c) Beads wash plate: 1000 μL of PBST-BSA.
   (d) Library plate: add 1000 μL of resuspended phage to row A.
   (e) Wash 1 plate: 1000 μL of PBST-BSA.
   (f) Wash 2 plate: 1000 μL of PBST-BSA.
   (g) Bead release plate: 500 μL of PBST-BSA.

(h) Tip load plate: place KingFisher Flex tip comb in an empty plate.

4. Run automated KingFisher protocol:

    (a) Load the tip comb from the "Tip load plate".

    (b) Mix the beads in the "Protein binding plate" at slow speed for 15 min. Collect the beads for 30 s five times.

    (c) Release the beads into the "Biotin block plate" and mix at a medium speed for 1 min. Collect the beads for 30 s five times.

    (d) Release the beads into the "Beads wash plate" and mix at a medium speed for 1 min. Pause and pool contents of eight wells in the column into a single well in row A of 96-well plate. Collect the beads for 30 s five times.

    (e) Release the beads into the "Library plate" and mix at a medium speed for 1 h in room temperature. Collect the beads 30 s five times.

    (f) Release the beads into the "Wash 1 plate" and mix at a fast speed for 1 min. Collect the beads 30 s five times.

    (g) Release the beads into the "Wash 2 plate" and mix at a fast speed for 1 min. Collect the beads 30 s five times.

    (h) Release the beads into the "Bead release plate" and mix at a fast speed for 1 min do not collect the beads.

    (i) Drop the comb back into the "Tip load plate".

5. Completely transfer 500 μL of beads from the "Bead release plate" to 5 mL of the log phase XL1-Blue cells (freshly grown in **step 1**) in a 250 mL baffled flask and incubate 20 min (*see* **Note 16**). Add 30 mL of 2×YT medium prewarmed to 30 °C, supplemented with ampicillin 100 μg/mL and 10^{10}/mL M13 KO7 helper phage. Shake at 260 rpm in 30 °C for 20 h.

6. Transfer the phage cultures to conical tubes and spin down the cells at 8000 × *g*, 4 °C for 10 min. Collect the supernatant to a fresh tube with 6 mL of chilled PEG/NaCl solution. Gently mix and incubate on ice for 30 min. Spin down at 8000 × *g*, 4 °C for 10 min and discard the supernatant immediately. Flash spin to remove the remaining supernatant. Resuspend the phage in 0.5 mL of PBST-BSA (supplemented with compounds or competitors). Spin again at 10,000 × *g*, 4 °C for 10 min and transfer the supernatant to a new tube. Prepared phage will be used as an input for round 2 of selection (*see* **Note 17**).

    (a) Competitive selection: add competitor at 1 μM concentration and incubate for 15 min before adding to the "Protein binding plate".

(b) Subtractive selection: add 100 μL of 1 μM biotinylated competitor to 10 μL of washed Sera-Mag streptavidin paramagnetic particles, wash once with one volume of PBST-BSA, place on magnet for 1 min, remove the supernatant and add resuspended phage. Incubate for 15 min on a rotator. After the incubation is complete separate the beads with a magnet and carefully collect the phage library to be added to the "Protein binding plate".

(B) Round 2 to Round 4 of selection (*see* **Note 18**):

1. Prepare the log phase cells as described in Subheading 3.3.2, A **step 1**.
2. Prepare KingFisher 96 DW selection plates.
   (a) Protein binding plate: 5 μL of 1 μM biotinylated target for round 2 and 1 μL for rounds 3 and 4, add 20 μL of phage solution for round 2 (prepared in Subheading 3.3.2, A **step 6**) and 10 μL for round 3 and 4 (prepared in Subheading 3.3.2, B **step 8**). Bring the total volume of the well to 100 μL with PBST-BSA.
   (b) Bead plate: add 1.6 μL of Sera-Mag SpeedBeads neutravidin-coated paramagnetic particles for round 2 and 0.3 μL for round 3 and 4, add PBST-BSA to the final volume of 100 μL.
   (c) Biotin block plate: 99 μL of PBST-BSA and 1 μL of 5 mM biotin.
   (d) Wash 1 plate: 100 μL of PBST-BSA.
   (e) Wash 2 plate: 100 μL of PBST-BSA.
   (f) Wash 3 plate: 100 μL of PBST-BSA.
   (g) Elution plate: 100 μL of PBST-BSA containing 100 mM DTT or 5 μg of TEV protease or 3 μg of HRV 3C protease (*see* **Note 19**).
   (h) Tip load plate: place KingFisher Flex tip comb in an empty plate.
3. Run automated KingFisher protocol:
   (a) Load the tip comb from the "Tip load plate".
   (b) Mix the beads in the "Protein binding plate" at a medium speed for 15 min.
   (c) Premix the beads at a fast speed in the "Bead plate", collect for 10 s three times and release to the "Protein binding plate", mix for 15 min at a fast speed and collect for 10 s three times.
   (d) Release the beads into the "Beads plate", mix the beads at a fast speed for 30 s and collect the beads for 10 s three times.

(e) Release the beads into the "Biotin block plate", mix at a fast speed for 30 s and collect the beads for 10 s three times.

(f) Release the beads into the "Wash 1 plate", mix at a fast speed for 30 s and collect the beads for 10 s three times.

(g) Release the beads into the "Wash 2 plate", mix at a fast speed for 30 s and collect the beads for 10 s three times.

(h) Release the beads into the "Wash 3 plate", mix at a fast speed for 30 s and collect the beads for 10 s three times.

(i) Release the beads into the "Elution plate", mix at fast a speed for 15 min and collect the beads for 10 s three times.

(j) Release the beads into the "Biotin block plate".

(k) Drop the comb back into the "Tip load plate".

7. Infect *E. coli* XL1-Blue log phase cells and make serial dilutions to determine the enrichment ratios.

   (a) Dispense 90 μL of the log phase XL1-Blue *E. coli* cells (freshly grown in **step 1**) into rows A and E of a Greiner 96-well microplate, fill the remaining wells with 90 μL of 2×YT medium. Serial dilutions will be prepared in top and bottom half of the plate so 12 targets can be titered on a single 96-well plate.

   (b) Transfer 10 μL of eluted phage from "Elution plate" to rows A and E maintaining original order of "+" and "−" wells from the "Protein binding plate". Agitate slowly in room temperature for 20 min and make 3 tenfold serial dilutions (rows B, C, D, and F, G, H) each time transferring 10 μL of material and mixing well (*see* **Note 20**).

   (c) Plate 10 μL of each of the serial dilutions for wells "+" and wells "−" in sections of the single ampicillin agar plate, do not discard the serial dilution plate.

   (d) Transfer contents of the wells "+" in rows A and E to the individual wells of 24-deep well block filled with 3 mL 2×YT medium supplemented with ampicillin 100 μg/mL and 10^{10}/mL M13 KO7 helper phage. Seal the plate with a breathable film and shake at 260 rpm in 30 °C for 20 h.

8. Spin down the cells in 24-deep well blocks at 4000 × *g*, 4 °C for 20 min. Transfer the supernatants to the fresh plate filled with 600 μL of chilled PEG/NaCl solution. Gently mix and incubate on ice for 30 min. Spin down at 4000 × *g*, 4 °C for 1 h and very gently discard the supernatant. Flash spin the plates and aspirate the remaining supernatant, avoid aspirating the pellet. Resuspend the phage in 100 μL of buffer PBST-BSA (supplemented with compounds or competitors) and transfer to a fresh 96-well Greiner microplate. Spin again at 4000 × *g*, 4 °C for 20 min and transfer the supernatant to a new plate.

(a) Competitive selection: add competitor at 1 μM concentration and incubate for 15 min before transferring to the "Protein binding plate".

(b) Subtractive selection: add 100 μL of 1 μM biotinylated competitor to 10 μL of washed Sera-Mag streptavidin paramagnetic particles, wash once with one volume of PBST-BSA, place on magnet for 1 min, remove the supernatant and add the resuspended phage. Incubate for 15 min on the rotator. After the incubation is complete separate the beads with a magnet and carefully collect the phage library to be added to the "Protein binding plate".

9. Examine the agar plates with phage titers after the overnight growth. In case of a lack of enrichment perform another round of selection (go to Subheading 3.3.2, B **step 1** using phage from **step 8**). If a significant enrichment ratio is observed (more than 2, *see* **Note 21**) plate the remaining medium from the individual wells on the saved serial dilution plate on agar plates and incubate at 37 °C until colonies reach ~0.5 mm in size. Proceed to clone picking (Subheading 3.4.1, **step 1**) and ELISA testing.

3.4 Primary Screening

A successful selection experiment usually results in a robust enrichment with low background binding. The typical enrichment after round 3 usually ranges between 2- and 100-fold, but it has to be noted that specific binders can still be obtained even if little to no enrichment is observed. The objective of primary screening is to assess both the affinity and specificity of the selected clones (Subheading 3.4.1). These clones are then sequenced to determine the diversity of the obtained phage pool (Subheading 3.4.2). In this step only sequence unique binders are chosen and data are correlated with specificity information obtained from ELISA.

3.4.1 Competitive Phage ELISA

Binders obtained in phage selection are first tested in competitive phage ELISA assay. The ELISA will be performed using all compounds, competitors and target variants utilized in selection process. With automation, usually 384 synthetic antibody clones can be managed easily. Reducing this number, below 96 clones tested, can lead to a loss in quality of the final pool of characterized binders. This assay is based on the standard single point competitive phage ELISA format and is performed with 20 nM competitor. Incubation and wash steps are optimized to capture clones with K_D values of 20 nM or lower with low k_{off} values. Clones for sequencing (Subheading 3.4.2) are chosen based on several parameters derived from ELISA data: (a) overall strength of ELISA signal and (b) competition ratio–fraction of the signal retained when protein coating the ELISA plate competes for phage binding with non-biotinylated protein added in solution. While the signal correlates to the overall affinity and display level of synthetic antibody on a phage, the competitive ratio indicates specificity of the binding and relative off-rate of binding. This assay becomes even more powerful if variants of the same protein, alternate conformations, compound bound states etc. can be tested side by side along with unrelated protein (e.g., BSA) on the same 384-well plate. Ideally 96 clones can be tested in 384-well format (4 × 96) where each quadrant can contain different variant of the target or condition tested. Specificity can be also assessed by inclusion of cell lysate in the binding conditions or blood serum. Note, however, competitive phage ELISA should be used only as a guideline for ranking the clones and is not a replacement for comprehensive biophysical analysis of obtained binders (Subheading 3.5.2).

1. Pick 96 single colonies from agar plates (prepared in Subheading 3.3.2, B **step 9**) into a 96-deep well block filled with 600 μL 2×YT supplemented with ampicillin 100 μg/mL and 10^{10}/mL M13 KO7 helper phage. Seal the plate with breathable film. Grow for 20 h, shaking at 280 rpm in 30 °C.
2. Coat two 384-well ELISA plates per target with 20 μL 2 μg/mL NeutrAvidin dissolved in PBS buffer. Incubate for 1 h in 37 °C or 16 h at 4 °C.
3. Discard coating solution and add 60 μL of blocking buffer PBS-BSA and incubate for 1 h at 37 °C or 16 h at 4 °C.
4. Spin down the phage growth plate from **step 1**.
5. Prepare two phage dilution plates as follows:

   (a) Dilution A: add 11 μL of phage supernatant to 13 μL of PBST-BSA.

   (b) Dilution B: add 11 μL of phage supernatant to 13 μL of 50 nM target protein and incubate for 15 min (*see* **Note 22**).
6. Wash blocked ELISA plates (prepared in **step 3**) three times with PBST on EL406 plate washer and add 13 μL of 50 nM biotinylated target protein.
7. Incubate for 15 min and add 10 μL of 23 μM biotin in PBST (10 μM final concentration) to block residual binding sites on the plate and wash three times with PBST.
8. Apply diluted phage transferring 20 μL from dilution A and B plates (**step 5**) to the coated ELISA plates (prepared in **step 6**).
9. Shake for 15 min at room temperature and wash three times.
10. Add 20 μL of 5000× diluted HRP/anti-M13 antibody and incubate for 30 min with shaking.
11. Wash three times and add 20 μL of TMB substrate.

12. Incubate from 5 to 15 min with shaking and stop the reaction by adding 20 μL of 1 M H_3PO_4.
13. Read the absorbance at 450 nm in a plate reader (*see* **Note 23**).
14. Plot a signal of dilution A as a function of the ratio of dilution B to A signals.
15. For further analysis pick the clones with the highest signal from dilution A and lowest ratio of dilution B to A signals (*see* **Note 24**).

3.4.2 Clone Sequencing and Sequence Analysis

Successful clones picked based on ELISA data (Subheading 3.4.1) should be analyzed by DNA sequencing to determine their CDR loop composition and their uniqueness. Usually decoding the CDR loop H3 sequence is sufficient for this purpose since this loop is the most diverse and should contain unique sequence signatures. Nevertheless, sequence comparison of all CDR loops is advisable [5]. Also, it is a good practice to analyze the entire open reading frame of the synthetic antibody sequence checking the scaffold for mutations. For big datasets spanning many targets, it is advisable to use a relational database or dedicated LIMS system to store antibody sequences. This should allow for broad comparisons and facilitate pruning of the nonunique binders originating from closely related targets and multiple selection experiments.

1. Prepare log phase XL1-Blue *E. coli* cells as described in Subheading 3.3.2, A **step 1**.
2. Dispense 10 μL of the log phase XL1-Blue *E. coli* cells into each well of a 96-deep well block.
3. Cherry pick successful phage clones from the cultures used for the ELISA experiment (Subheading 3.4.1, **step 4**) by transferring 1 μL of supernatant to 96-deep well block with XL1-Blue *E. coli* cells.
4. Add 900 μL of TB medium to the plate and shake overnight at 37 °C, 260 rpm.
5. Isolate plasmid DNA using Omega E-Z 96 Fastfilter Plasmid Kit according to the manufacturer's protocol. Elute the dsDNA with 100 μL of sterile water adjusted to pH 8.5.
6. Sequence open reading frames of both heavy chain and light chain.
7. Analyze sequencing data using a multiple sequence alignment tool such as MAFFT [34]. Compare identity of the randomized CDR loops and check entire antibody scaffold for mutations (*see* **Note 25**).

3.5 Secondary Screening

Downstream characterization of the synthetic antibodies usually begins with conversion to the soluble protein expression format (Subheading 3.5.1). While there are several different cloning vectors that can be used for this purpose [31], expression usually requires optimization to boost efficiency [35]. The most streamlined cloning methods generally rely on site specific recombination [36, 37] and can be easily adopted for high-throughput environments.

3.5.1 Large Scale Synthetic Antibody Expression and Purification

Since comprehensive secondary screening frequently requires mg quantities of binders, the synthetic antibodies are expressed at 1 L. scale using Ptac promoter vectors like pSFV4 [31]. The procedure relies on heat lysis of the bacterial cells and employs dual chromatography using first Protein A affinity, followed by ion exchange. The method yields from 5 to 20 mg of >95% pure synthetic antibody and scales well up to 6 L of media. Synthetic antibodies can be safely stored in a frozen form and their quality can be assessed when needed.

1. Transform 10 μL of C43 chemocompetent cells (*see* **Note 26**) with synthetic antibody plasmid DNA and directly inoculate starter culture of 20 mL 2×YT medium supplemented with ampicillin 100 μg/mL and chloramphenicol 15 μg/mL. Grow at 37 °C, 220 rpm for 16 h.
2. Use 10 mL of the overnight culture to inoculate 1 L. TB medium supplemented with ampicillin 100 μg/mL, 16 mL of autoinduction mix and 1 mL Antifoam 204. Grow for 20 h in a baffled Fernbach flask at 25 °C, 220 rpm.
3. Pellet the cells at 5000 × *g*, 15 min, 4 °C and discard the supernatant. Do not freeze the pellet.
4. Resuspend the pellet in a lysis buffer. Use 5 mL of buffer per 1 g of the pellet. Place on a rotator for 1 h and make sure that the cells are fully resuspended.
5. Heat the whole cell suspension to 65 °C in a water bath for 30 min and immediately chill on ice for 15 min (*see* **Note 27**).
6. Clear the lysate by spinning it down at 16000 × *g* for 15 min at 4 °C. Collect the supernatant and filter through a 0.22 μm filter before loading it onto the purification column.
7. Regenerate HiTrap MabSelect SuRe 5 mL with 1.5 column volumes of the regeneration solution at 3 mL/min.
8. Equilibrate the column with 2 volumes of equilibration buffer.
9. Load the supernatant onto the column at 5 mL/min and wash with 8 column volumes of the equilibration buffer or wait until absorbance 280 nm reaches a baseline (*see* **Note 28**).
10. Elute a protein with 4 column volumes of elution buffer at 3 mL/min, collect fractions.
11. Regenerate the column with 1.5 column volumes of the regeneration solution and equilibrate with 2 column volumes of the equilibration buffer for short term storage.

12. Recharge the Resource S 1 mL with 2 column volumes of the IEX elution buffer and equilibrate with 8 column volumes of the IEX equilibration buffer at 3 mL/min.
13. Load fractions containing synthetic antibody from **step 10** onto the column at 3 mL/min.
14. Wash column with 1 column volume of the IEX equilibration buffer at 3 mL/min or until absorbance 280 nm reaches a baseline.
15. Elute protein with a 0–50% linear gradient of 25 column volumes of IEX elution buffer at 3 mL/min collecting 1 mL fractions.
16. Regenerate the column with 2 column volumes of the regeneration solution and equilibrate with the IEX equilibration buffer for short term storage.
17. Run SDS-PAGE gel on all fractions of interest, pool and dialyze fractions containing pure synthetic antibody to the buffer of choice.

3.5.2 Biophysical Characterization

A number of techniques can be employed to characterize binders. The minimal set should include affinity, specificity, stability, and solubility assessment, as described below.

Surface plasmon resonance (SPR) is probably the best known method to determine binding kinetics of macromolecules. SPR can yield accurate binding kinetics and affinity data, as well as being useful for epitope binning [30, 38, 39]. Its major limitation is that it requires high quality soluble antigens. However, since the phage display experiments likewise have that requirement, the same set of antigens can be used for both selection and SPR. Affinities of synthetic antibodies to different conformations, variants of proteins or compound bound states, as used in selection, can be easily obtained and compared. While affinity data can be obtained readily for simple 1:1 Langmuir model, binding involving more complex kinetics is harder to analyze. In such cases equilibrium analysis can be used, but only if steady state of interaction can be reached [40].

Fluorescence-based thermal shift information obtained using differential scanning fluorimetry (DSF) [41] is an established method for identification of hits from small molecule libraries [42]. Similarly, it provides useful information regarding stability of both target proteins and synthetic antibodies as well as quantifies stabilization effects. In a majority of cases, synthetic antibody binding stabilizes the proteins to which they bind and this effect can be assessed by concentration dependent shifts in the thermal stability of the resulting complexes [43]. The main advantage of the assay is that is requires a miniscule amount of protein compared to methods like differential scanning calorimetry or chemical and thermal unfolding monitored by circular dichroism. The assay is performed in microplate format using readily available RT-PCR instrument [41, 44].

Size exclusion chromatography can be performed in cases where information about sample homogeneity and size distribution of molecules is required. An additional boost in detection capabilities can be provided by either differential static light scattering (DSLS) or multi-angle light scattering (MALS) [45]. The MALS method is very useful to monitor synthetic antibody complex formation prior to crystallization and for characterization of large multicomponent complexes in cryo-electron microscopy.

Solubility and nonspecific binding of synthetic antibodies can be assessed with multiple methods. A particularly useful and straightforward method is a cross-interaction chromatography (CIC) [46]. In this method antibodies with poor solubility can be quickly identified since their retention times inversely correlate with solubility. The assay takes only minutes to perform and simple FPLC coupled with autosampler can screen many antibodies and conditions in relatively short time.

Second useful method for assessment of self-association and aggregation is affinity-capture self-interaction nanoparticle spectroscopy (AC-SINS) [47]. The method is based on detection of wavelengths of maximum absorbance upon aggregation of gold nanoparticles coated with studied antibody. This method can be readily adapted to high throughput environments, but should be considered complementary to CIC and not a substitute since types of interactions detected by both methods are different.

4 Notes

1. A non-amine buffer of pH 7–8 should be used for biotinylation reactions using NHS ester; salt, detergents, or stabilizing cofactors or compounds can be added; do not add reducing agents. Total amount of protein used in phage display and primary testing can be as little as 5 nmol nonetheless losses can occur due to precipitation during labeling and extra protein might be necessary for troubleshooting.
2. Avoid exposure of reagent powder to humidity, and prevent condensation of water on the reagent bottle, you can safely store the reagent in DMSO for about 2 weeks without noticeable drop in activity, or conversely reagent dissolved in water has to be used immediately.
3. Ideally not more than three biotins per 20 kDa molecule should be attached in amine labeling reaction. For biotin-HPDP Cys labeling lower ratios of biotinylation reagent: protein can be used. Protein should be in fully reduced state and desalted from excess reducing agents immediately before the reaction. Keep the reaction at room temperature and occa-

sionally mix it by gently flipping the tube. Do not place on ice since biotinylation reagents are often insoluble in low temperatures.

4. Incubation times longer than 60 min or excessive vortexing is not recommended and can result in protein precipitation.
5. Ion exchange and affinity chromatography are great alternatives to dialysis and can be used to completely remove excess reagent speeding up the process. Gel filtration chromatography can be also used for desalting nevertheless in select cases it might not allow for efficient removal of excess biotinylation reagent which can comigrate with protein during separation.
6. Efficiency of biotinylation can be also tested by mass spectrometry, HPLC chromatography or HABA test (Pierce).
7. Avoid buffers with high detergent or salt concentration, if possible use Biomix A buffer and supplemented it with cofactors if required by the protein of interest. You can speed up the reaction by keeping protein concentration high (concentrations lower than 10 μM will require extending the reaction time).
8. The goal of this step is to remove free unincorporated biotin. Removal of BirA is not required (it will not interfere with the selection process) but it is going to affect protein concentration measurement. If possible, purify the protein of interest by affinity or gel filtration chromatography.
9. Pick buffer composition based on the type of elution method (TEV, HRV 3C, DTT) and requirements of the protein. The selection buffer used in Subheading 3.3.1 will be additionally supplemented with 0.2% BSA, detergent and protein stabilizers which should be avoided in the steps where enzymatic elution will be performed.
10. For small proteins higher amount might be required to properly visualize them on the gel. For membrane proteins use up to 50 μL of beads and 5 μg of protein.
11. Choice of the elution condition depends on the type of biotinylation and protease site used in the tag. In cases when TEV or HRV 3C protease will not fully cut the protein off the magnetic beads, 0.1 M glycine pH 2.7 can be used to elute bound phages.
12. Usually combined 80% pull-down efficiency and elution from magnetic beads should be obtained. Lower efficiencies make the solution capture selection difficult and should be avoided. Negative results are usually indication of incomplete desalting of the reaction mixture.
13. Culture should be grown in parallel to the steps below so the cells can be used for phage infection immediately after the

selection step is complete (Subheading 3.3.2, A **step 5** and B **step** 7). The starting culture should be prepared the night before the experiment from a single bacterial colony inoculated into 2 mL of 2×YT medium supplemented with tetracycline 20 μg/mL. It is a good practice to setup additional control cultures with antibiotics commonly used in the lab, in particular kanamycin and ampicillin. It is imperative that the cell cultures are prepared fresh and not contaminated with library phagemid or helper phage used in the selection process.

14. The amount of phage library specified is sufficient for multiplexed sorting of eight targets. Multiplexing maximizes efficiency of the selection process. Note that only compatible targets can be combined during round 1. Targets are immobilized separately on the beads (**step 4**a–d) and pooled together column-wise into the single well in row A of the 96-well plate. In **step 4**e pooled beads loaded with eight antigens are incubated with a library placed in a single well in row A. In the KingFisher Flex up to 48 targets can be selected on at the same time (48 wells are used for negative controls). In a competition experiment, soluble competitor interacts with phages bearing binders selective for the competitor before they are transferred to the well in "Protein binding plate" containing target of interest. The incubation time should be sufficient to capture both low and high affinity binders and effectively make them unavailable for binding to the selection target. In a subtractive selection, phages specific for the competitor are separated from the library in a separate incubation step. Phage pre-clearing as outlined on Fig. 2a should be performed immediately before adding the phage to the "Protein binding plate." Do not carry over any beads from the pre-clearing step. The PBST-BSA buffer should be supplemented with compounds that are going to be used in screening or replaced with selection buffer of choice containing BSA and detergent.

15. Selection is performed column-wise in batches of eight targets. Volumes are listed as volumes per well of the 96-well plate. Compounds or partner proteins with a low affinity should be included in all wells. If selection utilizes small molecule compounds their water solubility and stability should be considered. The use of organic solvent is not recommended and DMSO up to 5% can be tolerated. Competitors should be added in concentration 100 times higher than the target molecule.

16. During the infection agitate the culture slowly in a room temperature to allow complete infection and minimize shearing forces. In the described method phage is propagated in a single flask for eight individual targets, therefore the same phage sample will be used for those eight targets on round 2.

17. Phage purification steps can be automated and 10,000 × *g* spin step can be substituted with 50 °C heating to kill remaining bacterial cells. Phage clearing steps can be also automated on the KingFisher magnetic bead handler.
18. All remaining rounds of selection follow the same protocol and only the amount of protein and phage concentration are modified to control the stringency of selection. If Avi-tagged constructs are used without enzymatic release by HRV 3C or TEV protease it might be necessary to pre-clear the library using biotin blocked Sera-Mag Streptavidin paramagnetic particles to lower the nonspecific background in selection. Each target requires two wells on all selection plates listed below: one for the target selection "+" and the second one for the negative control where no target protein is added "–". Place the wells "+" and "–" next to each other in selection plates to speed up the phage pipetting and titering. Compounds should be added to the "Protein binding plate" at concentration at least ten times higher than expected K_D. Low affinity compounds should be also added in all wash plates. Adding protein competitor in all wells can increase the stringency in competition selection.
19. Always use freshly prepared DTT solution. It is expected that TEV or HRV 3C will release all of the target protein with bound phage from the paramagnetic particles within 15 min. Alternatively if Avi-tag construct is used without proteolytic release add 100 mM Glycine-HCl pH 2.1 to the elution plate. That solution needs to be neutralized after the selection is complete with 1 M HEPES pH 7.5 before performing phage infection.
20. Change tips between the serial dilution steps to prevent nonspecific carryover of cells on the outside of the tip. Before performing the titration, prewarm the agar plates that will be used.
21. The progress of phage display selection can be characterized by an enrichment parameter defined as the ratio of the phage count recovered from the target containing condition compared to the count of the background binding phages in the negative control. The background binding can be diminished by both optimizing the target and switching to the beads with a minimal background binding. Binders specific to neutravidin or streptavidin can be eliminated in the library clearing step. Enrichment after round 3 varies in the range between 2- and 100-fold and 10- and 1000 after round 4. Low or no enrichment and especially low recovery in competition experiment indicate poor selection performance and lower the chances of obtaining synthetic antibodies for the desired conformation or complex. One probable cause is a small difference in a confor-

mation between a target and competitor being used. In such a case perform selection with a lower stringency or switch to the subtraction selection scheme with increased phage input. For challenging targets, final concentration of 10 nM may be too low. Target concentrations can be adjusted starting from round 1 and gradually reduced from round to round, e.g., 500 nM (round 1)–100 nM (round 4).

22. Each of the quadrants of the 384-well plate can be occupied by a different target variant, e.g., quadrant 1: antigen used in selection, quadrant 2: competitor used in selection, quadrant 3: small molecule bound form of the target, quadrant 4: well left blank for background binding assessment. The advantage of such arrangement is that all those conditions are tested side by side on the single plate generating more consistent data. Dilution plates from **step 5**a and **step 6** must have corresponding order of antigens in quadrants of 384-well plates. Competitor concentration can be adjusted from 10 to 100 nM. Phage can be more diluted, up to 50 fold—the goal is to obtain a moderate signal from dilution A.
23. Alternatively, plates can be read in kinetic mode at 650 nm without adding 1 M H_3PO_4 in **step 12**.
24. In competitive ELISA out of 96 tested clones usually 25–80% are positive. If comparisons are made to multiple forms of a target (*see* **Note 22**) and hit picking is performed based on multiple criteria yields from successful selection will be lower 10–50%. In case of problematic targets low affinity binders can be obtained (high B to A signal ratio). In such case it is advisable to perform another selection with optimized targets or proceed with affinity maturation of obtained binders.
25. Primary screening can be substituted or supplemented with deep sequencing where binding and nonbinding phage pools can be sequenced and comprehensively queried [48]. Clone specificity can be inferred from bioinformatic analysis of the obtained data. In cases when a pool of sequence unique binders is low selection can be redone with lower stringency parameters or epitope masking experiments can be used. Alternatively, phage clones can be screened right after round 2 to query pool with potentially higher diversity.
26. Always start from fresh transformation; transform late during the day so the culture will not overgrow. It is not necessary to plate and pick single colony for expression. For protocols describing the preparation and transformation of chemically competent cells, refer to Maniatis et al. [49].
27. Make sure that the lysis temperature is reached, especially if a larger volume is used.

28. For most reproducible results it is advised to perform all liquid chromatography steps on a FPLC system. It is beneficial to use high performance columns for best separation results.

29. The eluted solution can be directly loaded onto ion exchange column (**step 12**) without neutralization or desalting step. The synthetic antibody is typically more than 95% pure following Protein A purification, and for most applications, it is sufficient. If further purification is required proceed to **step 13**.

Acknowledgments

We thank Vincent Lu, Daniel King, and Svitlana Usatyuk for contributing to the development of protocols. This work was supported by National Institutes of Health Grants: U54-GM087519, U54-HG006436.

References

1. Hoogenboom HR (2005) Selecting and screening recombinant antibody libraries. Nat Biotechnol 23:1105–1116. doi:10.1038/nbt1126
2. Bradbury A, Velappan N, Verzillo V et al (2003) Antibodies in proteomics I: generating antibodies. Trends Biotechnol 21:275–281. doi:10.1016/S0167-7799(03)00112-4
3. Bradbury A, Velappan N, Verzillo V et al (2003) Antibodies in proteomics II: screening, high-throughput characterization and downstream applications. Trends Biotechnol 21:312–317. doi:10.1016/S0167-7799(03)00117-3
4. Miller KR, Koide A, Leung B et al (2012) T Cell receptor-like recognition of tumor in vivo by synthetic antibody fragment. PLoS One 7:e43746. doi:10.1371/journal.pone.0043746
5. Persson H, Ye W, Wernimont A et al (2012) CDR-H3 diversity is not required for antigen recognition by synthetic antibodies. J Mol Biol. doi:10.1016/j.jmb.2012.11.037
6. Prassler J, Thiel S, Pracht C et al (2011) HuCAL PLATINUM, a synthetic fab library optimized for sequence diversity and superior performance in mammalian expression systems. J Mol Biol 413:261–278. doi:10.1016/j.jmb.2011.08.012
7. Johnson G, Wu TT (2001) Kabat database and its applications: future directions. Nucleic Acids Res 29:205–206
8. Wu TT, Johnson G, Kabat EA (1993) Length distribution of CDRH3 in antibodies. Protein Struct Funct Genet 16:1–7
9. Tiller T, Schuster I, Deppe D et al (2013) A fully synthetic human Fab antibody library based on fixed VH/VL framework pairings with favorable biophysical properties. mAbs 5:445–470. doi:10.4161/mabs.24218
10. Hornsby M, Paduch M, Miersch S et al (2015) A high through-put platform for recombinant antibodies to folded proteins. Mol Cell Proteomics 14:2833–2847. doi:10.1074/mcp.O115.052209
11. Zhou H, Wang Y, Wang W et al (2009) Generation of monoclonal antibodies against highly conserved antigens. PLoS One 4:e6087. doi:10.1371/journal.pone.0006087
12. Scheerlinck JP, DeLeys R, Saman E et al (1993) Redistribution of a murine humoral immune response following removal of an immunodominant B cell epitope from a recombinant fusion protein. Mol Immunol 30:733–739
13. Saldanha JW (2009) Humanization of recombinant antibodies. Cambridge University Press, New York, NY
14. Li Q, Wanderling S, Paduch M et al (2014) Structural mechanism of voltage-dependent gating in an isolated voltage-sensing domain. Nat Struct Mol Biol 21:244–252. doi:10.1038/nsmb.2768
15. Gao J, Sidhu SS, Wells JA (2009) Two-state selection of conformation-specific antibodies.

Proc Natl Acad Sci U S A 106:3071–3076. doi:10.1073/pnas.0812952106

16. Rizk SS, Paduch M, Heithaus JH et al (2011) Allosteric control of ligand-binding affinity using engineered conformation-specific effector proteins. Nat Struct Mol Biol 18:437–442. doi:10.1038/nsmb.2002
17. Shukla AK, Manglik A, Kruse AC et al (2013) Structure of active β-arrestin-1 bound to a G-protein-coupled receptor phosphopeptide. Nature 497:137–141. doi:10.1038/nature12120
18. Mateja A, Paduch M, Chang H-Y et al (2015) Structure of the Get3 targeting factor in complex with its membrane protein cargo. Science 347:1152–1155. doi:10.1126/science.1261671
19. Uysal S, Vásquez V, Tereshko V et al (2009) Crystal structure of full-length KcsA in its closed conformation. Proc Natl Acad Sci U S A 106:6644–6649. doi:10.1073/pnas.0810663106
20. Welch BD, Paduch M, Leser GP et al (2014) Probing the functions of the paramyxovirus glycoproteins F and HN with a panel of synthetic antibodies. J Virol 88:11713–11725. doi:10.1128/JVI.01707-14
21. Stuwe T, Bley CJ, Thierbach K et al (2015) Architecture of the fungal nuclear pore inner ring complex. Science 350:56–64. doi:10.1126/science.aac9176
22. Stuwe T, Correia AR, Lin DH et al (2015) Architecture of the nuclear pore complex coat. Science 347:1148–1152. doi:10.1126/science.aaa4136
23. Wittrup KD, Verdine GL (2012) Protein engineering for therapeutics. Academic, New York, NY
24. Sidhu SS (2005) Phage display in biotechnology and drug discovery. CRC Press, Boca Raton, FL
25. Brawley CM, Uysal S, Kossiakoff AA, Rock RS (2010) Characterization of engineered actin binding proteins that control filament assembly and structure. PLoS One 5:e13960. doi:10.1371/journal.pone.0013960
26. Schlapschy M, Grimm S, Skerra A (2006) A system for concomitant overexpression of four periplasmic folding catalysts to improve secretory protein production in *Escherichia coli*. Protein Eng Des Sel 19:385–390. doi:10.1093/protein/gzl018
27. Vogt AD, Pozzi N, Chen Z, Di Cera E (2014) Essential role of conformational selection in ligand binding. Biophys Chem 186:13–21. doi:10.1016/j.bpc.2013.09.003
28. James LC, Roversi P, Tawfik DS (2003) Antibody multispecificity mediated by conformational diversity. Science 299:1362–1367. doi:10.1126/science.1079731
29. Lamboy JA, Arter JA, Knopp KA et al (2009) Phage wrapping with cationic polymers eliminates nonspecific binding between M13 phage and high pI target proteins. J Am Chem Soc 131:16454–16460. doi:10.1021/ja9050873
30. Zhang X, Hoey RJ, Lin G et al (2012) Identification of a tetratricopeptide repeat-like domain in the nicastrin subunit of γ-secretase using synthetic antibodies. Proc Natl Acad Sci U S A 109:8534–8539. doi:10.1073/pnas.1202691109
31. Zhong N, Loppnau P, Seitova A et al (2015) Optimizing production of antigens and Fabs in the context of generating recombinant antibodies to human proteins. PLoS One 10:e0139695. doi:10.1371/journal.pone.0139695
32. Fellouse FA, Esaki K, Birtalan S et al (2007) High-throughput generation of synthetic antibodies from highly functional minimalist phage-displayed libraries. J Mol Biol 373:924–940 S0022-2836(07)01062-5
33. Tonikian R, Zhang Y, Boone C, Sidhu SS (2007) Identifying specificity profiles for peptide recognition modules from phage-displayed peptide libraries. Nat Protoc 2:1368–1386. doi:10.1038/nprot.2007.151
34. Katoh K, Asimenos G, Toh H (2009) Multiple alignment of DNA sequences with MAFFT. Methods Mol Biol Clifton NJ 537:39–64. doi:10.1007/978-1-59745-251-9_3
35. Robinson M-P, Ke N, Lobstein J et al (2015) Efficient expression of full-length antibodies in the cytoplasm of engineered bacteria. Nat Commun. doi:10.1038/ncomms9072
36. Zhang Y, Werling U, Edelmann W (2012) SLiCE: a novel bacterial cell extract-based DNA cloning method. Nucleic Acids Res 40:e55. doi:10.1093/nar/gkr1288
37. Gibson DG, Young L, Chuang R-Y et al (2009) Enzymatic assembly of DNA molecules up to several hundred kilobases. Nat Methods 6:343–345. doi:10.1038/nmeth.1318
38. Nieba L, Nieba-Axmann SE, Persson A et al (1997) BIACORE analysis of histidine-tagged proteins using a chelating NTA sensor chip. Anal Biochem 252:217–228. doi:10.1006/abio.1997.2326
39. Rich RL, Myszka DG (2007) Higher-throughput, label-free, real-time molecular interaction analysis. Anal Biochem 361:1–6. doi:10.1016/j.ab.2006.10.040

40. Rich RL, Myszka DG (2010) Grading the commercial optical biosensor literature-class of 2008: "The Mighty Binders". J Mol Recognit 23:1–64. doi:10.1002/jmr.1004
41. Niesen FH, Berglund H, Vedadi M (2007) The use of differential scanning fluorimetry to detect ligand interactions that promote protein stability. Nat Protoc 2:2212–2221. doi:10.1038/nprot.2007.321
42. Lo M-C, Aulabaugh A, Jin G et al (2004) Evaluation of fluorescence-based thermal shift assays for hit identification in drug discovery. Anal Biochem 332:153–159. doi:10.1016/j.ab.2004.04.031
43. Cimmperman P, Baranauskiene L, Jachimoviciūte S et al (2008) A quantitative model of thermal stabilization and destabilization of proteins by ligands. Biophys J 95:3222–3231. doi:10.1529/biophysj.108.134973
44. Vedadi M, Niesen FH, Allali-Hassani A et al (2006) Chemical screening methods to identify ligands that promote protein stability, protein crystallization, and structure determination. Proc Natl Acad Sci U S A 103:15835–15840. doi:10.1073/pnas.0605224103
45. Nettleship JE, Brown J, Groves MR, Geerlof A (2008) Methods for protein characterization by mass spectrometry, thermal shift (ThermoFluor) assay, and multiangle or static light scattering. Methods Mol Biol Clifton NJ 426: 299–318. doi:10.1007/978-1-60327-058-8_19
46. Jacobs SA, Wu S-J, Feng Y et al (2009) Cross-interaction chromatography: a rapid method to identify highly soluble monoclonal antibody candidates. Pharm Res 27:65–71. doi:10.1007/s11095-009-0007-z
47. Liu Y, Caffry I, Wu J et al (2014) High-throughput screening for developability during early-stage antibody discovery using self-interaction nanoparticle spectroscopy. mAbs 6:483–492. doi:10.4161/mabs.27431
48. Matochko WL, Chu K, Jin B et al (2012) Deep sequencing analysis of phage libraries using Illumina platform. Methods 58:47–55. doi:10.1016/j.ymeth.2012.07.006
49. Sambrook J, Fritsch EF, Maniatis T (1989) Molecular cloning: a laboratory manual, 2nd edn. Cold Spring Harbor Laboratory Press, Cold Spring Harbor, NY

Chapter 7

High-Throughput IgG Conversion of Phage Displayed Fab Antibody Fragments by AmplYFast

Andrea Sterner and Carolin Zehetmeier

Abstract

Phage display of antibody libraries is an invaluable strategy in antibody discovery. Many synthetic antibody library formats utilize monovalent antibody binding fragments (Fab), displayed on filamentous phage and expressed in *Escherichia coli* for selection and screening procedures, respectively. For most therapeutic applications, however, the final antibody candidate favors a bivalent immunoglobulin G (IgG) format, due to its particular effector function, half-life, and avidity.

Here, we present an optimized subcloning method, termed AmplYFast, for the fast and convenient conversion of phage-displayed monovalent Fab fragments into full-length IgG or immunoglobulins of any other isotype. By using biotinylated primers, unique mammalian expression vectors, and multi-well plates, AmplYFast combines the rapid amplification, digestion, and ligation of recombinant Ig heavy and light chain sequences in an easy-to-operate high-throughput manner. Thus, AmplYFast improves quality and efficiency in DNA cloning and significantly minimizes timelines to antibody lead identification.

Key words Immunoglobulin, Phage display, IgG, Fab, DNA cloning, Biotinylated primer, Recombinant antibody

1 Introduction

Phage display of antibody libraries is an important in vitro technology for the discovery of monoclonal antibodies with therapeutic or diagnostic potential or value in basic research [1, 2]. Several antibody libraries with varying size and sources are described [3–7], mainly comprising single-chain variable fragments (scFv) or Fab-based phage display libraries, which predominantly attempt to mirror the diversity of natural human antibody repertoires [8, 9]. As a prime example, the fully synthetic human Fab phage display antibody library Ylanthia® covers a broad range of diverse CDR structures with biophysically optimal VH/VL pairings [10]. Within iterative selection rounds (referred to as "panning"), in which the Fab is displayed at the surface of the phage and presented to the antigen of interest ("target"), target-specific Fab-displaying phage are getting enriched. Screening for specific target binding ("hits")

Thomas Tiller (ed.), *Synthetic Antibodies: Methods and Protocols*, Methods in Molecular Biology, vol. 1575,
DOI 10.1007/978-1-4939-6857-2_7, © Springer Science+Business Media LLC 2017

can be conveniently performed in high throughput by using soluble Fab molecules expressed in *E. coli* [11]. However, for comprehensive functional testing and in-depth biophysical characterization of the antibody, especially when the desired final format is an IgG, reformatting of identified Fab hits into IgG, followed by IgG expression and purification, becomes indispensable.

In course of this, Ylanthia comprises three major vector systems, each with different features. The display phagemid vector pYPdis10 enables the Fab to be presented at the surface of the phage covalently linked via a cysteine disulfide bridge to the phage coat protein pIII. The bacterial expression vector, pYBex10, is required for Fab expression and screening of crude *E. coli* lysates and is complemented by a mammalian expression vector, pYMex10, for full-length IgG production. While the subcloning of Ig heavy and light chain sequences from pYPdis10 into the bacterial expression vector pYBex10 system is a straightforward one-step cloning procedure (Fig. 1a), the transfer of the Fab antibody sequences into the mammalian pYMex10 vector requires in standard protocols two cloning steps. This is due to the essential exchange of the prokaryotic expression cassette (present in pYPdis10 and pYBex10) with a cassette containing a cytomegalovirus (CMV) promotor and

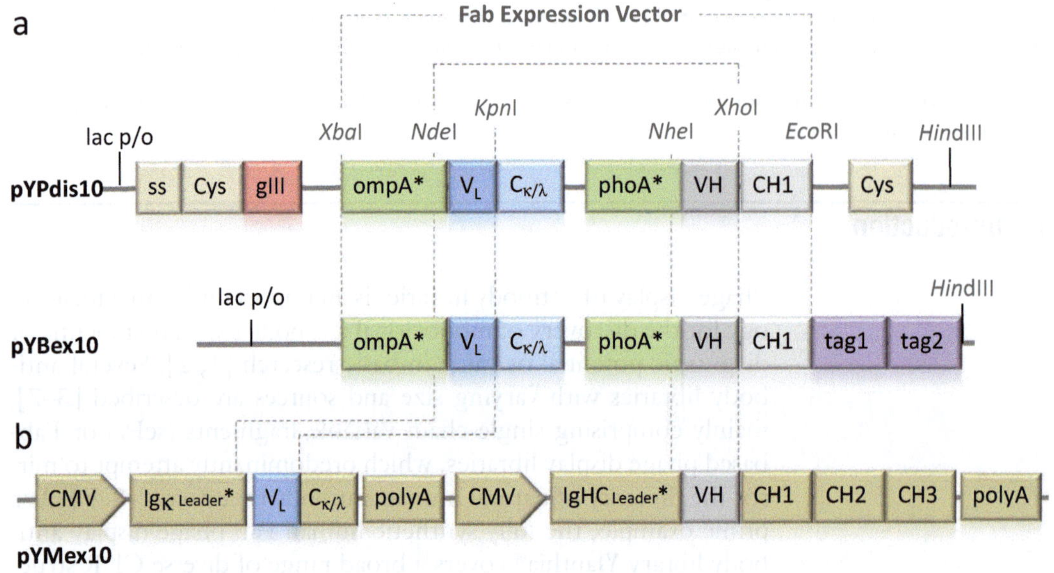

Fig. 1 (**a**) Modular design and compatible restriction sites allow easy subcloning between pYPdis10 Fab display and pYBex10 Fab bacterial expression vectors via *XbaI* and *EcoRI*. (**b**) Conversion between Fab and full-length IgG format is accomplished by transferring the antibody sequence via *NdeI* and *XhoI*, followed by the introduction of a mammalian expression cassette with a polyA and a CMV-driven promotor sequence via *KpnI* and *NheI*. ompA*, phoA* are modified leader sequences. VL and VH indicate the variable regions, Cκ/λ and CH1, CH2, CH3 depict the constant region domains of the antibody's light and heavy chain, respectively. Cys: cysteine, *lac* p/o: lac promoter/operator sequence, gIII: gene III encoding for phage coat protein pIII. Schematic distances and sizes are not to scale

a polyA sequence region suited for mammalian expression, in addition to the transfer of the antibody variable region segments from the bacterial to the mammalian vector system (Fig. 1b).

Classical cloning methods to switch antibody formats from Fab to IgG utilize established standard technologies based on size-dependent electrophoretic separation methods [12]. These usually require high amounts of both, donor and acceptor vector DNA for restriction digest and subsequent DNA purification steps via laborious preparative agarose gel electrophoresis and gel extraction. Especially, the preparative gel electrophoresis procedure is very inconvenient and time consuming when dealing with high numbers of samples. Furthermore, purification of DNA from agarose gels is very often accompanied by significant loss of DNA yields, making an upscaling of the input DNA material absolutely inevitable.

AmplYFast circumvents these downsides and represents a rapid method for convenient and efficient parallel cloning of many antibody V region encoding donor DNA inserts (e.g., derived from pYPdis10, pYBex10) into an acceptor vector backbone (e.g., from the pYMex10 vector series). The following description is linked to the Ylanthia® vector system [10]; however, the protocol can easily be adapted, if modified accordingly by appropriate primer design, to any other vector system. The AmplYFast procedure is subdivided into two parts. Each part comprises a PCR applying one biotinylated and one unmodified primer, a purification step with streptavidin-coupled beads, a restriction digest, and a ligation step. In the first part of AmplYFast, the "Fab vector backbone" and variable antibody sequences are amplified and the bacterial expression cassette is replaced by a cassette suitable for mammalian expression. The second part of the method includes the transfer of the antibody sequence in conjunction with the mammalian expression cassette to the acceptor vector backbone which defines the final antibody format. By the use of biotinylated primers and streptavidin-coupled magnetic beads, the need for agarose gel purification steps becomes obsolete. Furthermore, most steps of the AmplYFast procedure can be automated, allowing a fast reformatting of hundreds of different antibody sequences from the Fab into IgG format in just a few days (Fig. 2).

2 Materials

Prepare all solutions using DNase- and RNase-free ultrapure water. Prepare and store all reagents on ice or at 4 °C (unless indicated otherwise).

2.1 Equipment

1. Thermomixer (e.g., Thermomixer comfort, Eppendorf AG, Hamburg, Germany).
2. PCR Cycler (e.g., Mastercycler pro, Eppendorf AG, Hamburg, Germany).

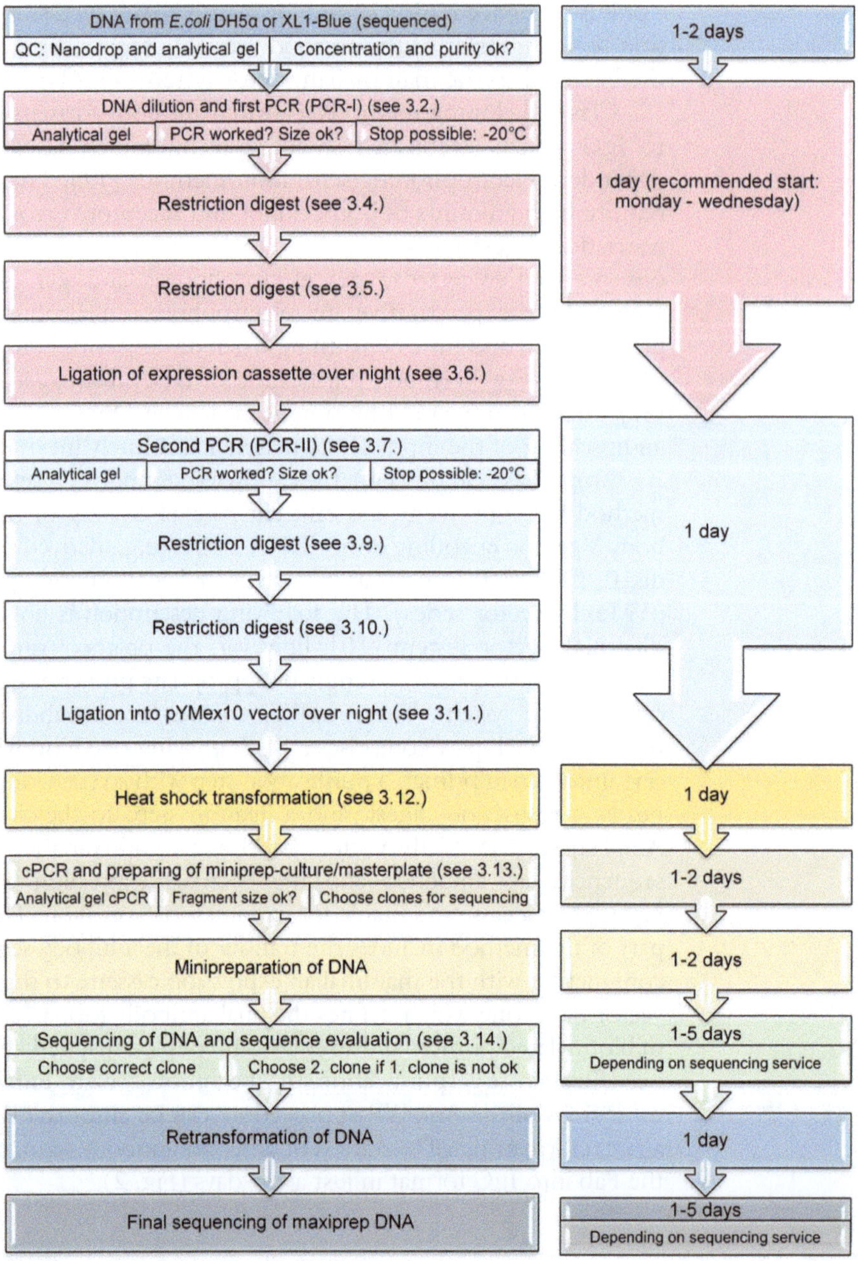

Fig. 2 Flow diagram of the AmplYFast procedure. The individual steps are referenced to the respective chapter sections

3. Magnetic Separator (e.g., NucleoMag SEP for 96-well MTP, Machery-Nagel Inc., Bethlehem, PA, USA).
4. Centrifuge (e.g., Sigma 6 K15, Sigma Laborzentrifugen GmbH, Osterode am Harz, Germany).

Table 1

Primer sequences for PCR amplification and sequencing of the final vector construct are indicated

Primer	Name	Sequence
A	AmplY_NheI_fw	5′- GAT AAG CAT GCG TAG GAG AAA - 3′
B	AmplY_KpnI_k_rv	5′- Biotin - GCC GCT TTT CAG CTG TTC ATC - 3′
C	AmplY_KpnI_l_rv	5′- Biotin - GGC TTG CAG TTC TTC GCT - 3′
D	AmplY_NdeI_fw	5′- Biotin - GTG GCC CTG GCA GGC TTT G - 3′
E	AmplY_XhoI_rv	5′- CCA GCG GAA ACA CGC TC - 3′
F	AmplY_cPCR_fw	5′- TAA TAC GAC TCA CTA TAG GG - 3′
G	AmplY_cPCR_rv	5′- GGG AAG TAG TCC TTG ACC AG - 3′
H	AmplY_kYMin_fw	5′- GTG TTG AGC GGG ACT CTG - 3′
I	AmplY_lYMin_fw	5′- CGT TTA AAC GTC GAG GAA TTC GG - 3′
J	AmplY_YMin_rv	5′- GCC ACC AGC AGC AGA AAG AA - 3′
K	AmplY_VH_rv	5′- GGA AAG TAG TCC TTC ACG A -3′

The given sequences correspond to the Ylanthia vector series; however, primer can be designed accordingly to match any other expression vector system. Restriction sites, given in the primer name, are for orientation purpose only, as the primer sequences here do not contain the respective restriction sites

2.2 Primer

See Table 1.

2.3 Reagents and Components

1. *E. coli* chemically competent cells (e.g., DH5α, Invitrogen, San Diego, CA, USA, e.g., XL1-Blue, Agilent, Santa Clara, CA, USA).
2. Mammalian expression cassettes (*see* **Note 1**).
   (a) pYMin_Igkappa (20 ng/μL, digested with *Kpn*I, *Nhe*I).
   (b) pYMin_Iglambda (20 ng/μL, digested with *Kpn*I, *Nhe*I).
3. Mammalian IgG expression acceptor vector backbone (*see* **Note 2**).
   (a) pYMex10_h_IgG1 (20 ng/μL, digested with *Nde*I, *Xho*I).
   (b) pYMex10_h_IgG2 (20 ng/μL, digested with *Nde*I, *Xho*I).
   (c) pYMex10_h_IgG3 (20 ng/μL, digested with *Nde*I, *Xho*I).
   (d) pYMex10_h_IgG4 (20 ng/μL, digested with *Nde*I, *Xho*I).
4. LB agar plates containing 100 μg/mL ampicillin and 1% glucose (LB/Amp plates).
5. LB agar plates containing 34 μg/mL chloramphenicol and 1% glucose (LB/Cam plates).
6. Washing Buffer: 5 mM Tris, 0.5 M NaCl, 0.25 M EDTA.

7. T4 DNA-Ligase (1 U/µL) (e.g., Invitrogen, San Diego, CA, USA).
8. dNTP mix (2.5 mM per dNTP) (e.g., Invitrogen, San Diego, CA, USA).
9. High Fidelity DNA-Polymerase (2 U/µL) (e.g., Phusion® High-Fidelity Polymerase, New England Biolabs, Ipswich, MA, USA).
10. DNA-Polymerase Reaction Buffer (e.g., Buffer GC (5×) (supplied with Phusion Kit, New England Biolabs, Ipswich, MA, USA).
11. *Kpn*I-HF (20 U/µL) (e.g., New England Biolabs, Ipswich, MA, USA).
12. *Nhe*I-HF (20 U/µL) (e.g., New England Biolabs, Ipswich, MA, USA).
13. *Xho*I (20 U/µL) (e.g., New England Biolabs, Ipswich, MA, USA).
14. *Nde*I (20 U/µL) (e.g., New England Biolabs, Ipswich, MA, USA).
15. Appropriate RE-Buffer (e.g., CutSmart® Buffer supplied with enzymes from New England Biolabs, Ipswich, MA, USA).
16. ATP (10 mM) (e.g., New England Biolabs, Ipswich, MA, USA).
17. Thermosensitive Alkaline Phosphatase (1 U/µL, e.g., FastAP Thermo Scientific, Rockford, IL, USA).
18. PCR Cover Strips (e.g., Sarstedt, Nümbrecht, Germany).
19. 96-well round-bottom Micro Well Plates (e.g., Nunc, Thermo Scientific, Rockford, IL, USA).
20. Low binding reaction tubes, 1.5 mL (e.g., LoBind Tubes, Eppendorf AG, Hamburg, Germany).
21. Adhesive cover foil (e.g., BZO, Biozym Scientific GmbH, Hessisch Oldendorf, Germany).
22. Streptavidin-coupled magnetic beads (e.g., Dynabeads® M-280, Invitrogen, Thermo Fisher Scientific, MA, USA).
23. DNA Preparation Kit (e.g., QIAprep® Spin Miniprep Kit, Qiagen, Hilden, Germany).
24. Taq-DNA-Polymerase (e.g., GoTaq® G2 Green Mastermix (Promega Corporation, WI, USA).
25. DNA Ladder (e.g., GeneRuler™ DNA Ladder Mix, Thermo Scientific, Rockford, IL).
26. 96-well PCR Plates (e.g., Abgene, Thermo Scientific, Rockford, IL, USA).
27. Deep Well Plates, 2.0 mL (e.g., U96 Deep Well Plates, Nunc, Thermo Scientific, Rockford, IL, USA).

28. Gas-permeable Membrane (e.g., Thermo Scientific, Rockford, IL, USA).
29. SOB medium: 2% (w/v) tryptone, 0.5% yeast extract, 10 mM NaCl, 2.5 mM KCl, 10 mM $MgCl_2$, 10 mM $MgSO_4$, in H_2O.
30. H_2O (sterilized, PCR grade).

2.4 The Ylanthia Vector System

2.4.1 Phage Fab Display Vector pYPdis10

In the tri-cistronic phagemid Fab display vector pYPdis10 the variable region (V) light and heavy chain sequences of the Fab are preceded by modified prokaryotic *ompA** and *phoA** leader sequences, respectively. Upstream of the V region sequence is the gIII sequence encoding for the phage coat protein pIII. N-terminally (for pIII) and C-terminally (for Fd chain) introduced cysteine residues are the basis for the formation of disulfide-linked Fd-pIII complexes during phage assembly and the subsequent incorporation of the complete Fab-pIII complex into the phage particle.

2.4.2 Bacterial Fab Expression Vector pYBex10

In the bacterial Fab expression vector pYBex10, the variable light and heavy chain sequences of the Fab are preceded by modified prokaryotic *ompA** and *phoA** leader sequences, respectively. Upstream of the V region is an Isopropyl β-D-1-thiogalactopyranoside (IPTG) and lactose inducible promotor. The sequence section between the *Kpn*I and *Nhe*I restriction sites consists of the light chain constant region sequence, and the modified *phoA** leader sequence that precedes the V region of the heavy chain (Fig. 3).

2.4.3 Mammalian IgG Expression Vector: pYMex10

In the mammalian IgG expression vector pYMex10, the V region light and heavy chain sequences of the IgG are encoded on one plasmid (Fig. 4). They are preceded by the natural Vkappa leader and a consensus VH leader sequence, respectively. Each chain is translated together with its N-terminal leader sequence from a separate transcript. Transcription is driven by two cytomegalovirus (CMV) promotor sites. The acceptor vector backbone contains the constant region of the heavy chain, the polyA sequence of human growth hormone (hGH), the AmpR resistance gene, and a CMV promotor site (Fig. 5a). The eukaryotic expression cassette between *Kpn*I and *Nhe*I is comprised of the constant light chain region, a Simian-Virus 40 (SV40) (for pYMin_Igkappa) or an hGH (for pYMin_Iglambda) polyA region, one CMV promotor site, and the leader sequence for the heavy chain (Fig. 5b).

3 Methods

3.1 Transformation of E. coli with Plasmid DNA

To ensure a high quality of template DNA material for the individual PCR samples and to prevent the use of mixed plasmids, it is recommended to re-transform molecular biology optimized nuclease deficient *E. coli* cells (e.g., DH5a or XL1-Blue) with a very

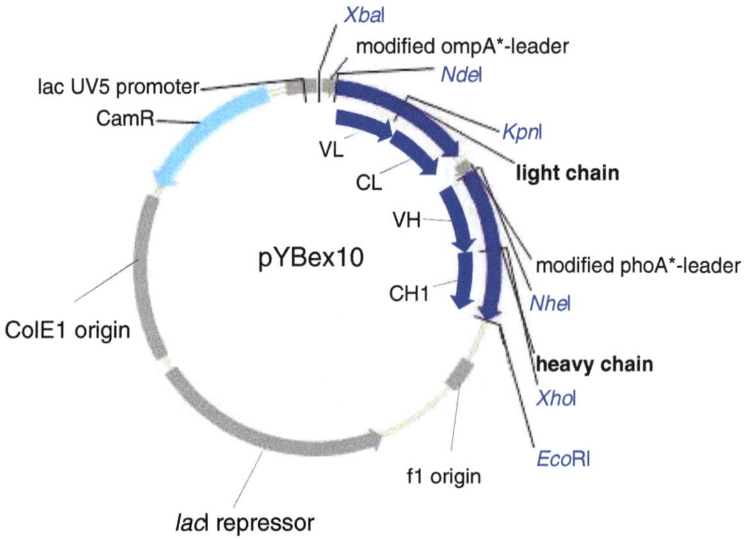

Fig. 3 Schematic map of the pYBex10 bacterial Fab expression vector. The pYBex10 plasmid carries a chloramphenicol resistance gene (camR) and an origin for double-stranded (ColE1 ori) DNA replication. *XbaI*, *NdeI*, *KpnI*, *NheI*, *XhoI*, and *EcoRI* indicate conserved restriction sites for cloning or exchange of domains. ompA* and phoA* are modified leader sequences

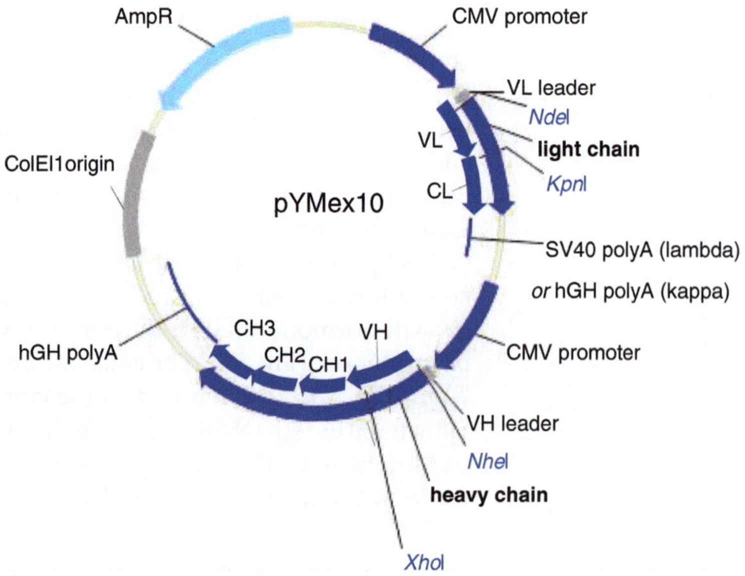

Fig. 4 Schematic map of the pYMex10 mammalian IgG expression vector. The pYMex10 plasmid carries an ampicillin resistance gene (ampR) and an origin for double-stranded (ColE1 ori) DNA replication. *NdeI*, *KpnI*, *NheI*, and *XhoI* indicate conserved restriction sites for cloning or exchange of domains

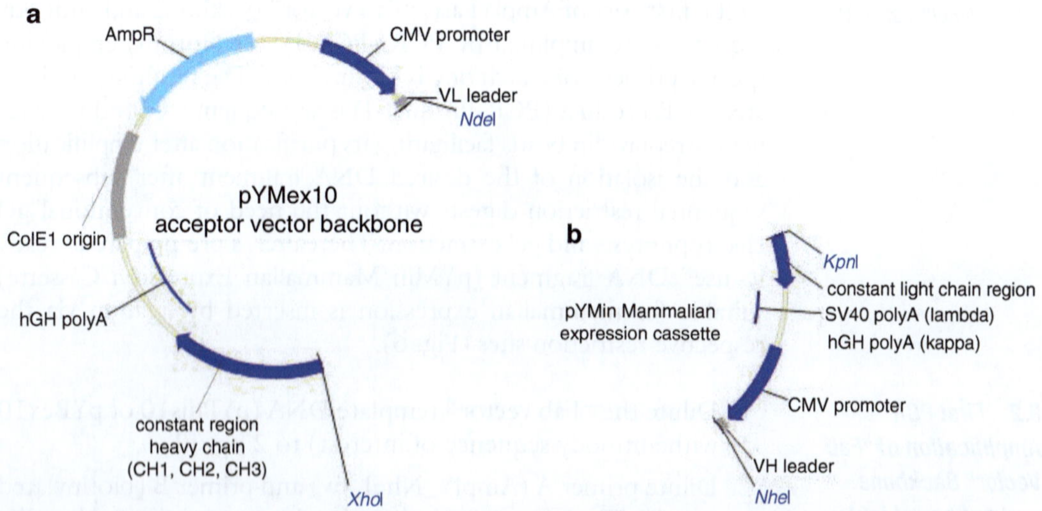

Fig. 5 (**a**) The pYMex10 acceptor vector backbone and (**b**) pYMin mammalian expression cassette insert

dilute solution (about 100 pg/transformation) of the starting sample DNA (e.g., pYPdis10 or pYBex10 "Fab vector") (*see* **Note 3**).

1. Thaw chemically competent *E. coli* DH5α (or *E. coli* XL1-Blue) cells on ice.
2. Transform chemically competent *E. coli* cells with 100 pg of DNA by adding 5 μL of DNA (20 pg/μL) directly to 45 μL of *E. coli* cells.
3. Gently mix and incubate on ice for 30 min. Heatshock the cells for 45 s in a 42 °C water bath, and then place them on ice for 2 min.
4. Add 500 μL of pre-warmed bacterial growth SOB medium and incubate for 1 h at 37 °C under agitation (600 rpm) using a thermomixer. Finally, spread the transformed bacteria onto LB agar plates containing 34 μg/mL chloramphenicol and 1% glucose.
5. Incubate the plates overnight at 37 °C in a static incubator.
6. Pick a single colony from the agar plate using a sterile pipette tip and inoculate a culture of 5 mL LB medium containing 34 μg/mL of the selective antibiotic chloramphenicol. Incubate overnight at 37 °C with vigorous shaking.
7. Perform plasmid DNA purification (e.g., according to the protocol provided with the QIAprep® Spin Miniprep Kit).
8. Determine the DNA concentration and verify DNA quality by spectrometric measurement at A260 and A280 nm (e.g., by using a Nanodrop Device, Thermo Fisher Scientific) (*see* **Note 4**).

3.1.1 AmplYFast PART 1

In the first part of AmplYFast, "Fab vector" backbone and antibody sequences are amplified by PCR (PCR-I) (*see* **Note 5**) employing specific primers, of which one is biotinylated. The resulting biotinylated PCR product (PCR-Product-I) is subsequently bound to magnetic streptavidin beads facilitating its purification after amplification and the isolation of the desired DNA fragment after subsequent sequential restriction digests without the need of conventional gel electrophoresis and gel extraction. Thereafter, a pre-prepared "ready to use" DNA fragment (pYMin Mammalian Expression Cassette) suitable for mammalian expression is inserted by ligation via the respective restriction sites (Fig. 6).

3.2 First PCR Amplification of "Fab Vector" Backbone and Antibody Light and Heavy Chain Sequences

1. Dilute the "Fab vector" template DNA (pYPdis10 or pYBex10 with antibody sequence of interest) to 2 ng/μL.
2. Dilute primer A (AmplY_NheI_fw) and primer B (biotinylated AmplY_KpnI_k_rv) for kappa samples and primer A (AmplY_NheI_fw) and primer C (biotinylated AmplY_KpnI_l_rv) for lambda samples with fresh, sterile 1× TE buffer to 20 pmol/μL (*see* **Note 6**). A complete list of primers is found in Table 1.

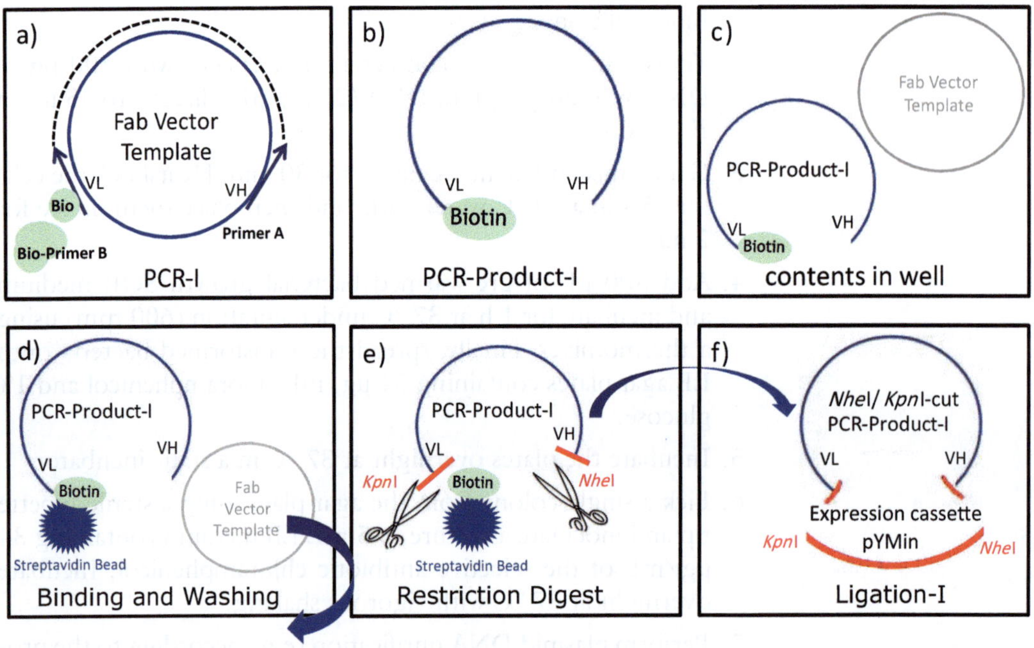

Fig. 6 Schematic outline of AmplYFast PART 1. (**a**) The antibody and "Fab vector" backbone sequences are getting amplified by PCR employing specific primers, of which one is biotinylated. The resulting biotinylated PCR product (PCR-Product-I) (**b**, **c**) is subsequently bound to magnetic streptavidin beads (**d**) facilitating its purification after amplification and restriction digest (**e**) without the need of conventional gel electrophoresis. Thereafter, the "ready to use" (pYMin) cassette suitable for mammalian expression is inserted by ligation via the respective restriction sites (Ligation-Product-I) (**f**)

3. Thoroughly mix the following components and put the Master Mix on ice (*see* **Note 7**). Per reaction use 5 μL Buffer GC (5×), 2 μL dNTPs (2.5 mM), 0.5 μL forward primer (20 pmol/μL), 0.5 μL reverse primer (20 pmol/μL), 0.25 μL Phusion® High-Fidelity DNA Polymerase (2 U/μL) (*see* **Note 8**) and 15.75 μL H_2O. Adjust the volumes according to the number of processed samples. For each sample, add 24 μL of the Master Mix into the well of a 96-well PCR plate (*see* **Note 9**).
4. Add 1 μL of the respective template DNA dilution (2 ng/μL) per well and seal the wells with PCR cover strips.
5. Perform the PCR using a PCR thermocycler applying the following thermal cycling conditions: 2 min at 98 °C; 15 s at 98 °C, 15 s at 60 °C, and 1 min 30 s at 72 °C, 20 cycles; and 5 min at 72 °C. After the completion of the cycles, immediately cool down to 4 °C (*see* **Note 10**).
6. Prepare a 1.2% agarose gel for analytical inspection of the samples. Load 2 μL of the PCR products and 3 μL of DNA ladder (*see* **Note 11**).
7. After electrophoretic separation, visualize the PCR products in the gel on a UV table and confirm their sizes using the DNA ladder. The expected PCR product sizes are ~4800 bp for pYPdis10_Igkappa or pYBex10_Igkappa templates and ~4700 bp for pYPdis10_Iglambda or pYBex10_Iglambda (Fig. 7).

3.3 Binding of Biotinylated First PCR Product (PCR-Product-I) to Streptavidin-Coupled Magnetic Beads

1. Pipette 50 μL streptavidin-coupled magnetic beads per sample into a low binding plate or tube (*see* **Note 12**).
2. Wash the beads by adding 80 μL Washing Buffer to the beads, mix gently until beads are resuspended, and capture beads via magnetic separator for 1 min until the supernatant is clear (*see* **Note 13**). Discard the Washing Buffer. Repeat this procedure three times (*see* **Note 14**).

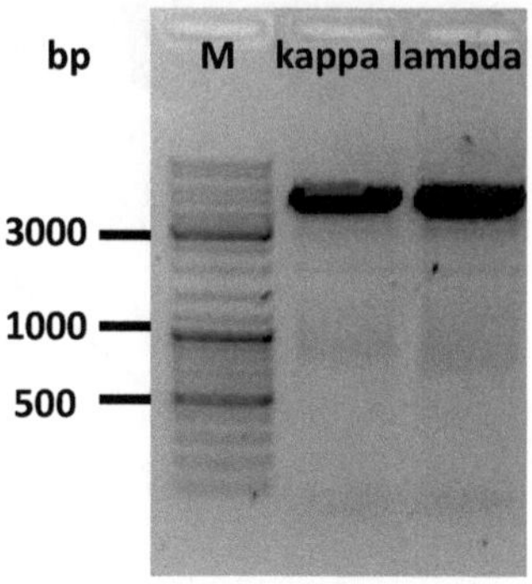

Fig. 7 Agarose gel of PCR products (2 μL of 25 μL) from the first amplification of kappa and lambda clones. As marker (M) the GeneRuler™ DNA Ladder Mix is shown. The expected PCR product size is ~ 4800 bp for kappa and ~ 4700 bp for lambda templates

3. After washing, resuspend the beads by adding 75 μL Washing Buffer per sample (*see* **Note 15**).
4. Per sample add 23 μL of biotinylated PCR product (PCR-Product-I) to the well containing 75 μL of washed bead suspension (~ 100 μL total volume) and seal the wells tightly with adhesive foil.
5. Place the plate into a thermomixer and set the temperature to 25 °C. Close the lid of the thermomixer and incubate the plate for 30 min at 1000 rpm (*see* **Note 16**).
6. Set the plate on the magnetic separator (Fig. 8) to capture the beads for 1 min. Gently lift the foil and carefully aspirate and discard the supernatant with a multichannel pipette.
7. Wash the beads three times with 80 μL Washing Buffer (*see* Subheading 3.3, **step 2**). After the last washing step, keep the washed beads with bound biotinylated PCR product (PCR-Product-I) in Washing Buffer until further use.

3.4 Restriction Digest with *Nhe*I

1. For the *Nhe*I reaction mix of the streptavidin bead-purified biotinylated PCR product (PCR-Product-I), combine the following components in the given order and mix via snipping the tube. Per reaction use 88 μL H_2O, 10 μL 10× CutSmart Buffer, 1 μL *Nhe*I-HF (20 U/μL), 1 μL FastAP (1 U/μL) (*see* **Note 17**).
2. Set the temperature of a thermomixer to 37 °C.
3. Carefully discard the Washing Buffer from the wells and add 100 μL of the *Nhe*I reaction mix to the biotinylated PCR product (PCR-Product-I) captured by the beads.
4. Seal the wells tightly with adhesive foil and incubate the plate for 60 min at 1000 rpm in a thermomixer.

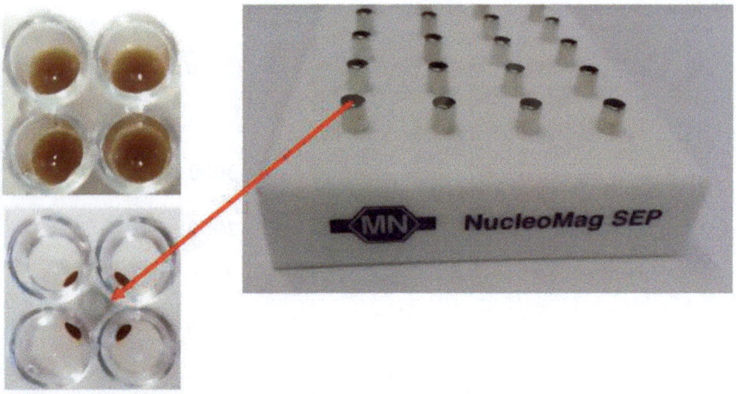

Fig. 8 Magnetic separator adapted for 96-well microtiter plates (e.g., NucleoMag® SEP, Machery-Nagel) is a convenient magnetic device designed for simple and rapid separation of magnetic beads. Beads as freely dispersed in solution (upper panel) and beads attracted to the magnet (bottom panel)

5. After the incubation, centrifuge the plate briefly at 2000 × *g*, to collect any condensate that might have accumulated underneath the cover foil during incubation. Collect the beads by using the magnetic separator for 5 min (Fig. 8) and discard the supernatant.
6. Wash the beads three times with 80 μL Washing Buffer and discard supernatant before adding the *Kpn*I digestion.

3.5 Restriction Digest with KpnI

1. For the *Kpn*I reaction mix prepare the following components in the given order and mix via snipping the tube. Per reaction use 88 μL H_2O, 10 μL 10× CutSmart Buffer, 1 μL *Kpn*I-HF (20 U/μL), 1 μL FastAP (1 U/μL).
2. Pipette 100 μL *Kpn*I reaction mix to the beads and seal the wells tightly with adhesive foil. Incubate for 60 min at 37 °C with 1000 rpm in a thermomixer. After the *Kpn*I digest, the desired KpnI-NheI DNA-fragment comprising the "Fab vector" backbone and antibody sequences to be used for ligation is contained in the supernatant (*see* **Note 18**).

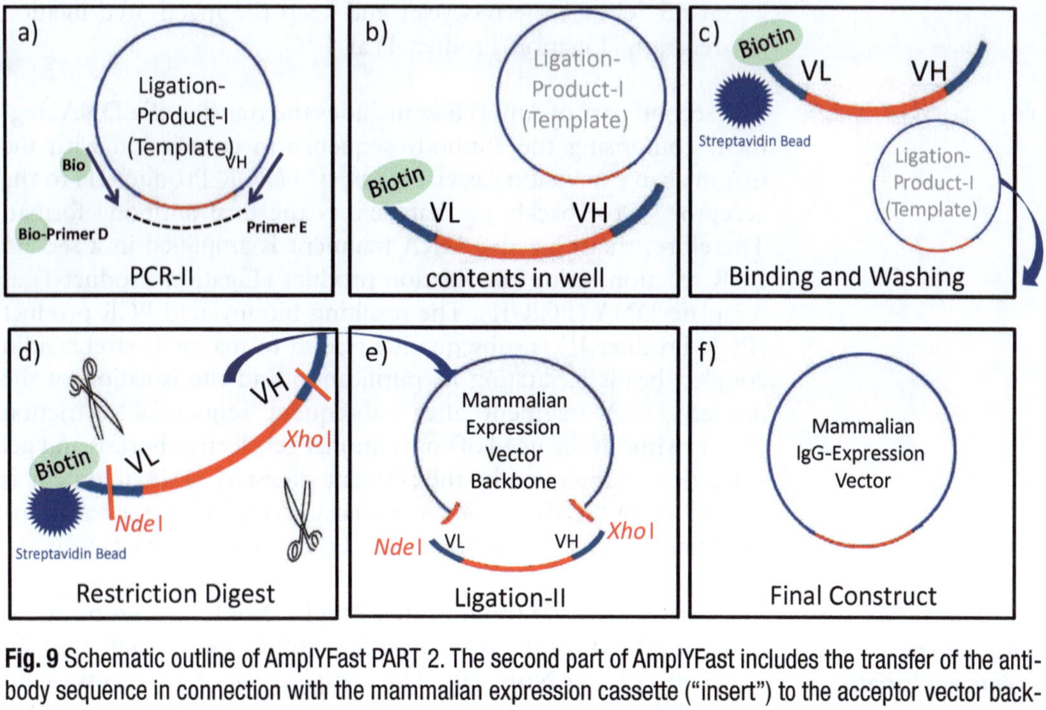

Fig. 9 Schematic outline of AmplYFast PART 2. The second part of AmplYFast includes the transfer of the antibody sequence in connection with the mammalian expression cassette ("insert") to the acceptor vector backbone which defines the final antibody format. First, the insert is amplified by PCR utilizing one biotinylated and one unmodified primer (**a**) and the Ligation-Product-I as templated DNA (PCR-Product-II). The biotinylated PCR-Product-II, comprising the variable antibody sequences of the individual starting clones and the constant sequences originating from the "ready to use" pYMin expression cassette (**b**), is captured on magnetic streptavidin beads (**c**) for convenient purification after digest (**d**) with the respective restriction enzymes. (e) Subsequently, the digested PCR-Product-II is ligated into the "ready to use" acceptor vector backbone to obtain the final vector construct (**f**)

3. Centrifuge the plate briefly at 2000 × g, to remove any condensate on the foil. Let the beads bind to the magnetic separator for 5 min (Fig. 8) and transfer the supernatant (~ 100 μL) into a fresh 96-well PCR plate.
4. Constrain the *Kpn*I enzyme activity by placing this plate for 20 min at 80 °C in a thermocycler.

3.6 Insertion of Mammalian Expression Cassette into Vector Backbone by Ligation

1. Prepare the ligation mix in the given order and mix thoroughly. Per reaction use 0.5 μL "ready to use" pYMin expression cassette (20 ng/μL; either pYMin_Igkappa or pYMin_Iglambda *Kpn*I/*Nhe*I digested), 1 μL freshly thawed ATP (10 mM), and 1 μL T4 DNA-Ligase (1 U/μL).
2. Dispense 2.5 μL of the ligation mix per well into a fresh 96-well PCR plate and add 7.5 μL of the *Nhe*I/*Kpn*I digested PCR-Product-I contained in the supernatant (*see* **Note 19**).
3. Seal the 96-well PCR plate with PCR cover strips and incubate the ligation samples (10 μL total volume per sample) at 16 °C overnight (12–14 h) in a thermocycler.
4. After the incubation, inactivate the T4 DNA Ligase for 20 min at 65 °C in a thermocycler and keep the inactivated ligation reaction (Ligation-Product-I) at 4 °C.

3.6.1 AmplYFast PART 2

The second part of AmplYFast includes the transfer of a DNA fragment comprising the antibody sequence in connection with the mammalian expression cassette ("insert") (PCR-Product-II) to the acceptor vector backbone that defines the final antibody format. Therefore, the respective DNA fragment is amplified in a second PCR reaction, using the ligation product (Ligation-Product-I) as template DNA (PCR-II). The resulting biotinylated PCR product (PCR-Product-II) is subsequently bound to magnetic streptavidin coupled beads facilitating its purification and the isolation of the desired DNA fragment after subsequent sequential restriction digests without the need of conventional gel electrophoresis and gel extraction. Subsequently, the relevant digested DNA fragment is ligated into a "ready to use" mammalian acceptor vector backbone to obtain the final vector construct (IgG vector construct) (Fig. 9).

3.7 Second PCR Amplification of Antibody Variable Region Sequence and Eukaryotic Expression Cassette

1. Dilute primer D (biotinylated AmplY_NdeI_fw) and primer E (AmplY_XhoI_rv) with fresh and sterile 1× TE buffer to 20 pmol/μL (*see* **Note 20**). The complete list of primers is found in Table 1.
2. Pipette the following components in the given order and mix thoroughly. Per reaction use 5 μL Buffer GC (5×), 2 μL dNTPs (2.5 mM), 0.5 μL forward primer (20 pmol/μL), 0.5 μL reverse primer (20 pmol/μL), 0.25 μL Phusion Polymerase (2 U/μL), and 13.75 μL H_2O.

3. Dispense 22 μL of the mix per well into a 96-well PCR plate.
4. As template for the second PCR reaction, use 3 μL of Ligation—Product-I, add it to the 22 μL PCR mix, and store the residual 7 μL at 4 °C as backup.
5. Seal the PCR plate with PCR cover strips and perform the PCR using a PCR thermocycler applying the following thermal cycling conditions: 2 min at 98 °C; 15 s at 98 °C, 15 s at 60 °C, and 42 s at 72 °C, 20 cycles; and 5 min at 72 °C. After completion of the cycles, immediately cool down to 4 °C.
6. Prepare a 1.2% agarose gel for analytical inspection of the PCR products. Load 2 μL of the PCR products and 3 μL of DNA ladder.
7. After electrophoretic separation, analyze the PCR products on a UV table and estimate their sizes using the DNA ladder. The expected PCR product size for pYMin_Igkappa ~2100 bp and for pYMin_Iglambda is ~2500 bp (Fig. 10).

3.8 Binding of Biotinylated Second PCR Product (PCR-Product-II) to Streptavidin-Coupled Magnetic Beads

The binding procedure is as described in Subheading 3.3.

1. Use 50 μL streptavidin-coupled magnetic beads per sample.
2. Wash the beads by adding 80 μL of Washing Buffer to the beads, mix gently until beads are resuspended, and capture beads via magnetic separator for 1 min until the supernatant is clear. Discard the Washing Buffer. Repeat this procedure three times.
3. After washing, resuspend the beads by adding 75 μL Washing Buffer per sample and dispense 75 μL of bead solution into the wells of a new 96-well plate.
4. Add 23 μL of the PCR sample to the bead solution and close the wells tightly with adhesive foil.

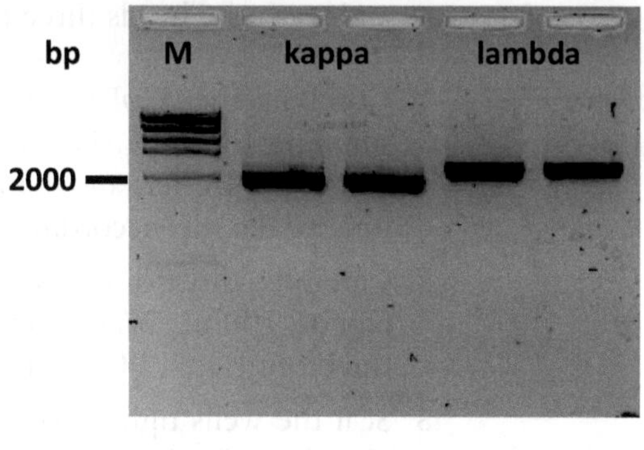

Fig. 10 Agarose gel of PCR products from the second amplification of kappa and lambda clones. As marker (M) the High Mass DNA ladder (Life Technologies) is shown. The expected PCR product size for kappa inserts is ~ 2100 bp and for lambda templates is ~ 2500 bp

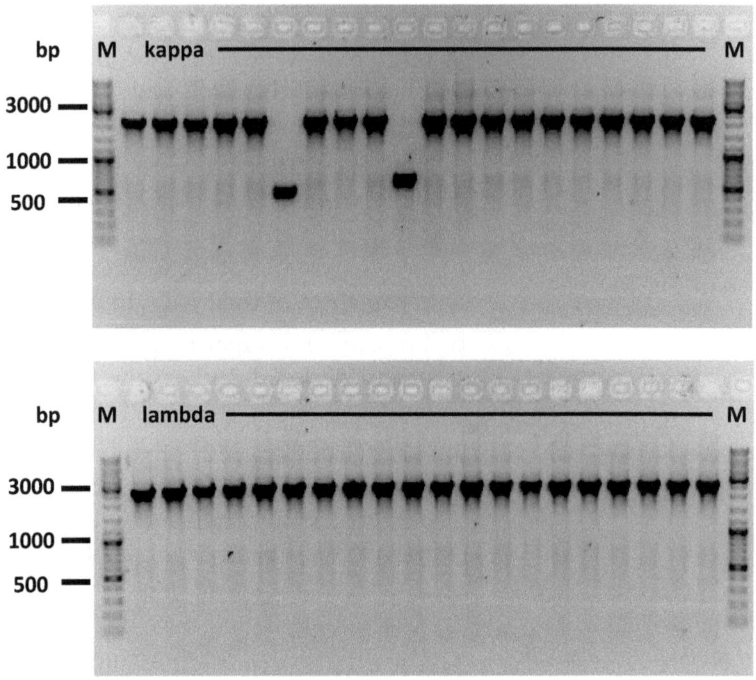

Fig. 11 Representative 1.5% agarose gel picture showing colony PCR products of kappa (top) and lambda (bottom) amplifications. As marker (M) the GeneRuler™ DNA Ladder Mix is shown. Expected PCR product sizes are ~ 2200 bp for kappa and ~ 2600 bp for lambda templates

5. Set the temperature of the thermomixer to 25 °C and place the plate into the thermomixer. Close the lid and incubate for 30 min at 1000 rpm.
6. After the incubation, place the 96-well plate on the magnetic separator to capture the beads and then detach carefully the foil. Discard the supernatant by using a multichannel pipette and wash the beads three times with 80 μL Washing Buffer.

3.9 Restriction Digest with *Xho*I

1. Combine the *Xho*I digest in the following order and mix by snipping the tube. Per reaction use 89 μL H_2O, 10 μL 10× CutSmart Buffer, and 1 μL *Xho*I (20 U/μL). Adjust the volume of the mix according to the number of samples.
2. Carefully discard the Washing Buffer from each well and pipette 100 μL *Xho*I reaction mix into each well that contains the biotinylated PCR-Product-II coupled to the beads.
3. Seal the wells tightly with adhesive foil and place the 96-well plate into the thermomixer. Incubate for 60 min at 37 °C with 1000 rpm.
4. After the incubation, centrifuge the plate briefly at 2000 × *g* to remove potential condensate under the foil, bind the beads to

the magnetic tray, and discard the supernatant of the wells with a multichannel pipette. Wash the beads three times with 80 μL Washing Buffer and discard the supernatant just before adding the next digestion mix.

3.10 Restriction Digest with NdeI

1. Prepare the *Nde*I digestion in the following order and mix by snipping the tube. Per reaction use 89 μL H_2O, 10 μL 10× CutSmart Buffer and 1 μL *Nde*I (20 U/μL). Adjust the volume of the mix according to the number of samples.
2. Pipette 100 μL of the *Nde*I digestion mix in each well on the beads and seal the wells tightly with adhesive foil. Incubate for 60 min at 37 °C with 1000 rpm. After the digestion with *Nde*I the desired *Xho*I-*Nde*I digested DNA fragment (PCR-Product-II cut with *Xho*I-*Nde*I) comprising the antibody sequences together with the inserted mammalian expression cassette to be used for subsequent ligation is contained in the supernatant.
3. After incubation, briefly centrifuge the plate to remove potential condensate under the foil. Let us bind the beads to the magnetic tray for 1 min (Fig. 8) and transfer 100 μL supernatant into a new 96-well PCR plate.
4. Inactivate the *Nde*I enzyme by placing this new plate for 20 min at 80 °C in a thermocycler.

3.11 Ligation into pYMex10 IgG Acceptor Vector Backbone

1. As (mammalian) acceptor vector backbone for the subsequent ligation, thaw the "ready to use" *Nde*I/*XhoI* digested pYMex10_IgG expression vector fragment.
2. Prepare the ligation mix in the following order and mix thoroughly. Per reaction use 1 μL digested "ready to use" IgG acceptor vector backbone (20 ng/μL), 1 μL freshly thawed ATP (10 mM) (*see* **Note 21**), and 1 μL T4 DNA-Ligase (1 U/μL).
3. Dispense 3 μL of the ligation mix into a fresh 96-well PCR plate and add 7 μL *Xho*I/*Nde*I digested PCR-Product-II DNA fragment (from Subheading 3.10, **step 2**) to the ligation mix (*see* **Note 22**).
4. Seal the 96-well plate with PCR cover strips and incubate the ligation reactions at 16 °C overnight (12–14 h) in a thermocycler.
5. After the incubation inactivate the T4 DNA Ligase for 20 min at 65 °C in the thermocycler and keep the ligation sample after inactivation at 4 °C.

3.12 Transformation of *E. coli* DH5α Cells

1. Fill a 96 deep well plate (DWP) with 900 μL SOB medium per well and put the plate into the thermomixer at 37 °C to prewarm the medium (*see* **Note 23**).

2. Prepare a bucket with ice for the chemical competent *E. coli* DH5α cells and thaw the cells on ice for about 15 min. Adjust the temperature of a water bath to 42 °C. Dry a respective number of small LB/Amp agar plates upside down in a 37 °C incubator.
3. Add 10 μL of the inactivated ligation sample to 100 μL chemical competent *E. coli* DH5α cells per well in a 96-well PCR plate and incubate the plate on ice for 30 min.
4. Heat-shock the ice-cooled samples in a water bath at 42 °C for exactly 45 s. Then immediately put the cells on ice for 2 min.
5. Transfer the cells into the pre-warmed SOB medium to a final volume of about 1 mL. Seal the DWP with a gas-permeable foil and incubate the cells for 1 h at 37 °C with 500 rpm in the thermomixer.
6. After incubation, plate 100 μL of the transformed *E. coli* DH5α sample onto a dry small LB/Amp agar plate. Store the residual 900 μL of the transformation sample as backup at 4 °C, if replating becomes necessary.
7. Incubate the LB/Amp plates at 37 °C overnight (16 h).

3.12.1 AmplYFast QC

To identify correct colonies, an "insert check" colony PCR (cPCR) is performed. Plasmid DNA is extracted from selected correct clones, followed by DNA sequencing of the antibody's VH and VL regions to confirm their identity to the initial Fab clone sequence.

3.13 "Insert Check" Colony PCR

1. The next day, check the LB/Amp agar plates for colony growth (*see* **Note 24**).
2. Prepare the cPCR Master Mix on ice (*see* **Note 25**). Per sample use 11.5 μL H_2O, 12.5 μL 2×GoTaq® Green master mix, 0.5 μL forward primer F (AmplY_cPCR_fw, 20 pmol/μL), 0.5 μL reverse primer G (AmplY_cPCR_rv, 20 pmol/μL). Add 25 μL of the cPCR Master Mix per well to a 96-well PCR plate (*see* **Note 26**).
3. As template material for each cPCR reaction pick a single colony from a LB/Amp plate with a sterile toothpick. Transfer the toothpick into one well with cPCR Master Mix and rotate the toothpick 2–3 times.
4. Transfer the toothpick from the cPCR Master Mix to inoculate corresponding wells of a 96 DWP prefilled with 1000 μL LB/Amp medium per well. Leave the toothpick for the duration of the picking procedure in the medium.
5. After completing one row of the 96-well plate, close the row with a PCR cover strip and continue with the second row (depending on the number of samples). Once the picking procedure has been completed, remove the toothpicks from the wells of the DPW, seal the plate containing the culture medium

inoculated with the picked colonies with a gas-permeable membrane, and incubate overnight (16 h) at 37 °C while shaking at 800 rpm.

6. Perform the cPCR in a thermocycler with the following thermal cycling conditions: 10 min at 95 °C; 30 s at 95 °C, 30 s at 55 °C, and 160 s at 72 °C, 30 cycles; and 8 min at 72 °C. After the completion of the cycles, immediately cool down to 4 °C.
7. Prepare a 1.2% agarose gel for analytical inspection of the cPCR samples. After the cPCR run, load 3 μL of the cPCR products and 3 μL of DNA ladder.
8. After electrophoretic separation, analyze the cPCR products on a UV table and estimate their sizes using the DNA ladder. Expected cPCR product sizes are ~2200 bp for pYMex10_IgG_kappa and ~2600 bp for pYMex10_IgG_lambda templates (Fig. 11).
9. On the next day, use the grown cultures from the DWP for plasmid purification of one clone per sample with the expected cPCR product size (e.g., DNA preparation according to the protocol provided with the QIAprep® Spin Miniprep Kit).

3.14 Sequencing of Individual IgG Expression Vector DNA

To confirm the open-reading frame of the full-length antibody coding sequence and to verify the correctness of the variable region sequence, sequencing of the final vector DNA is recommended. Sequencing primers H - K are listed in Table 1.

1. Submit ~1 μg (30–50 ng/μL, 30 μL) of sample DNA to a sequencing service provider and request Sanger sequencing with the respective primers (*see* **Note 27**).
2. Analyze the sequencing results by aligning the obtained sequences to in silico cloned expression vector constructs. Confirm that the full-length Ig light and heavy chain sequences were inserted in frame with the respective leader and constant region sequences and that no mutations were introduced during the reformatting process in comparison to the VH and VL template sequences.
3. Once the correctness of the final plasmid DNA sequences has been confirmed, retransform the DNA (*as described under* Subheading 3.1), purify large amounts of ultrapure, transfection grade plasmid DNA (e.g., by using Qiagen plasmid maxi kit to get DNA of good quality for transfection).

4 Notes

1. It is recommended to prepare the pYMin mammalian expression cassettes via double *Nhe*I/*Kpn*I restriction digest and agarose gel purification on a large-scale basis. Store the

digested and linearized "ready to use" cassettes at −20 °C until they are needed.

2. It is recommended to prepare the respective pYMex10 acceptor vector backbone via double *Nde*I/*Xho*I restriction digest and agarose gel purification on a large-scale basis. Store the digested and linearized "ready to use" acceptor vector backbone at −20 °C until it is needed. The pYMex10 vector series comprises all human IgG isotypes.
3. For all molecular biology procedures, an effective outcome strongly depends on the quality of the starting DNA material. PCRs, in general, are prone to fail, for example, if the template DNA is contaminated with other DNA species or inappropriate DNA concentration. In case of mixed DNA as a starting template, the PCR product consists of heterogeneous amplified DNA species [13].
4. Results of DNA concentrations determined via spectrometric measurement could be affected by buffers, salts, and other interference factors [14]. If the 260/280 nm ratio is out of range for pure DNA (generally accepted ratio for DNA is about 1.8), it is recommended to confirm the DNA concentration by semiquantitative analytical agarose gel electrophoresis. Hence, 50–100 ng DNA (as determined by spectroscopy) is mixed with loading dye and subjected to a 1.2% agarose-gel, together with an appropriate DNA reference marker (e.g., GeneRuler® DNA ladder, Thermo Fisher Scientific). For the determination of the DNA concentration an analytical software (e.g., Quantity One, Bio-Rad) can be used.
5. For all PCR steps use an exclusive PCR working environment, consisting of a PCR workstation (PCR hood), sterile DNAse and RNase-free filter tips, and pipettes to avoid cross-contamination [15]. At this PCR work bench, do not work with DNA templates to prevent contamination. Routinely, wipe the PCR area with DNA and DNAse removal agents (e.g., DNAaway, Thermo Fisher Scientific) and/or use UV light to inactivate potential DNA contaminants.
6. As the binding site of the reverse primer is in the light chain constant (CL) region, which is either of the Ig kappa or Ig lambda type, care has to be taken to choose the appropriate primer matching the respective CL region.
7. Keep all PCR reagents, polymerases, and PCR mixes on ice or in a cooling container while preparing the PCR. Do not use excess of biotinylated primers in the PCR mixes since residual, non-incorporated biotinylated oligonucleotides will bind to the streptavidin beads and might interfere with subsequent purification and reaction steps.

8. Use of a high-fidelity DNA polymerase with proof reading feature significantly reduces the insertion of mutations during the PCR.
9. For each pair of primers include a negative control (i.e., sample without DNA template).
10. Immediately after adding the template DNA to the PCR mix start the PCR run, to prevent degradation of primers by the proof reading feature of the Phusion polymerase. Avoid long-term storage of the final PCR product at 4 °C or RT to prevent DNA degradation.
11. Alternatively to controlling all PCR products on an agarose gel (e.g., when sample size exceeds 100), one can include for each primer pair 1–2 random samples of pYPdis10_Igkappa or pYBex10_Igkappa and 1–2 samples of pYPdis10_Iglambda or pYBex10_Iglambda DNA as surrogate positive controls.
12. All binding steps of biotinylated PCR products to streptavidin-coupled magnetic beads are recommended to be performed in an untreated polystyrene 96-well plate or low binding tubes.
13. Attraction of the magnetic beads to the magnetic pins of the separation plate depends on the magnetic strength of the pins, the selected separation plate, distance of the separation plate from the magnetic pins, and the volume to be processed. Check the individual times for complete attraction of the beads to the magnetic pins before starting the washing procedure.
14. Washing of beads in plates can be facilitated by using a multichannel pipette.
15. While processing a large number of samples it is convenient to wash all beads for the samples together and dispense 75 μL of the washed beads into the wells before adding the biotinylated PCR products to the wells.
16. After incubation it is possible that condensate has accumulated underneath the adhesive cover foil. To prevent spill-over between samples while detaching the foil, centrifuge the plate briefly before detaching the adhesive seal.
17. Use a low binding reaction tube for preparing the restriction digest mix.
18. At this step pay close attention not to discard the supernatant, as the second restriction enzyme digest releases the desired DNA fragment into the solution while the biotinylated 5′ end of the PCR product stays bound to the beads (and can thereby be captured from the solution).
19. Keep the residual volume of the supernatant containing the *Kpn*I-*Nhe*I DNA fragment comprising the "Fab vector" backbone and antibody sequences stored at −20 °C as backup.

20. For the second PCR amplification, one set of primers (one biotinylated primer and one non-biotinylated primer) is used for both pYMin_Igkappa and pYMin_Iglambda samples.
21. Make sure that ATP is added to the ligation reaction. The ATP concentration is very crucial for the successful ligation by T4 DNA ligase. Therefore, we recommend using aliquots of T4 DNA ligase buffer, which have been properly stored at −20 °C and not thawed repeatedly.
22. Keep the residual volume of the supernatant containing the *Nde*I/*XhoI* DNA fragment comprising the mammalian pYMex10_IgG expression acceptor vector backbone fragment with the target insert sequence stored at −20 °C as backup.
23. The Deep Well Plate is used for cultivation of *E. coli* after the transformation.
24. Typically, 5–50 colonies are expected. If there are no colonies on the plate, it is possible to plate the whole transformation approach which was kept at 4 °C onto one plate. To check for IgG clones with correct inserts, it is recommended to screen 2–3 colonies by "insert check" cPCR.
25. Prepare the cPCR Master Mix with 10–15% volume excess.
26. Prepare a negative (without DNA template) and a positive (with DNA template) control.
27. For optimal sequencing results (chromatogram reading lengths >800 bp) use purified DNA at the appropriate concentration (usually ~ 50 ng/μL, depending on sequencing provider).

Acknowledgment

The authors like to thank Stephanie Patzelt and Christoph Erkel for their support and scientific discussions, as well as Thomas Tiller and Katja Siegers for comments on the manuscript.

References

1. Bradbury ARM, Marks JD (2004) Antibodies from phage antibody libraries. J Immunol Methods 290(1–2):29–49. doi:10.1016/j.jim.2004.04.007
2. Hoogenboom HR (2002) Overview of antibody phage-display technology and its applications. Methods Mol Biol 178:1–37
3. Weber M, Bujak E, Putelli A et al (2014) A highly functional synthetic phage display library containing over 40 billion human antibody clones. PLoS One 9(6):e100000. doi:10.1371/journal.pone.0100000
4. Rothe C, Urlinger S, Löhning C et al (2008) The human combinatorial antibody library HuCAL GOLD combines diversification of all six CDRs according to the natural immune system with a novel display method for efficient selection of high-affinity antibodies. J Mol Biol 376(4):1182–1200. doi:10.1016/j.jmb.2007.12.018
5. Prassler J, Thiel S, Pracht C et al (2011) HuCAL PLATINUM, a synthetic Fab library optimized for sequence diversity and superior performance in mammalian expression sys-

tems. J Mol Biol 413(1):261–278. doi:10.1016/j.jmb.2011.08.012

6. Li K, Zettlitz KA, Lipianskaya J et al (2015) A fully human scFv phage display library for rapid antibody fragment reformatting. Protein Eng Des Sel 28(10):307–316. doi:10.1093/protein/gzv024
7. Hoet RM, Cohen EH, Kent RB et al (2005) Generation of high-affinity human antibodies by combining donor-derived and synthetic complementarity-determining-region diversity. Nat Biotechnol 23(3):344–348. doi:10.1038/nbt1067
8. Miersch S, Sidhu SS (2012) Synthetic antibodies: concepts, potential and practical considerations. Methods 57(4):486–498. doi:10.1016/j.ymeth.2012.06.012
9. Ponsel D, Neugebauer J, Ladetzki-Baehs K et al (2011) High affinity, developability and functional size: the holy grail of combinatorial antibody library generation. Molecules 16(12):3675–3700. doi:10.3390/molecules16053675
10. Tiller T, Schuster I, Deppe D et al (2014) A fully synthetic human Fab antibody library based on fixed VH/VL framework pairings with favorable biophysical properties. MAbs 5(3):445–470. doi:10.4161/mabs.24218
11. Krebs B, Rauchenberger R, Reiffert S et al (2001) High-throughput generation and engineering of recombinant human antibodies. J Immunol Methods 254(1–2):67–84
12. Slater GW, Kist TB, Ren H et al (1998) Recent developments in DNA electrophoretic separations. Electrophoresis 19(10):1525–1541. doi:10.1002/elps.1150191003
13. Kalle E, Gulevich A, Rensing C (2013) External and semi-internal controls for PCR amplification of homologous sequences in mixed templates. J Microbiol Methods 95(2):285–294. doi:10.1016/j.mimet.2013.09.014
14. Gallagher SR, Desjardins PR (2001) Quantitation of DNA and RNA with Absorption and Fluorescence Spectroscopy. In: Haines JL, Korf BR, Morton CC et al (eds) Current protocols in human genetics. John Wiley & Sons, Inc., Hoboken, NJ
15. Dieffenbach CW, Dveksler GS (1993) Setting up a PCR laboratory. Genome Res 3(2):S2–S7. doi:10.1101/gr.3.2.S2

Chapter 8

Utilization of Selenocysteine for Site-Specific Antibody Conjugation

Xiuling Li and Christoph Rader

Abstract

Site-specific conjugation methods are becoming increasingly important in building next-generation antibody-drug conjugates. We have developed a site-specific conjugation technology based on monoclonal antibodies with engineered selenocysteine (Sec) residues, named selenomabs. Here, we provide protocols for the engineering, expression, and purification of selenomabs in single-chain variable fragment (scFv)-Fc format. Methods for selective conjugation of selenomabs to selenol-reactive compounds and analytical characterization of selenomab conjugates are also included.

Key words Antibody-drug conjugates, Site-specific conjugation, Selenomab, Selenocysteine

1 Introduction

Antibody-drug conjugates (ADCs) as an emerging class of cancer therapy have attracted extensive interest by biotech and pharma companies in the past 5 years, inspired by the profound clinical benefit of two FDA-approved ADCs, brentuximab vedotin (Adcetris®) and ado-trastuzumab emtansine (Kadcyla®) as well as encouraging data from numerous ongoing clinical trials [1–3]. The concept of ADCs is using monoclonal antibodies (mAbs) to deliver highly potent cytotoxic agents specifically to tumor cells, thereby improving therapeutic indices and minimizing the side effects of chemotherapy [4, 5]. Traditionally, the drugs are randomly conjugated to an antibody via lysine side chain amines (ado-trastuzumab emtansine) [6] or interchain cysteine thiol groups (brentuximab vedotin) [7]. These conventional ADCs are heterogeneous mixtures of multiple species with varying drug-to-antibody ratio (DAR) and different conjugation sites. Each of these species has distinct pharmacokinetic (PK) and pharmacodynamic (PD) properties [8, 9]. To overcome this heterogeneity, site-specific conjugation methods have been developed to produce

Thomas Tiller (ed.), *Synthetic Antibodies: Methods and Protocols*, Methods in Molecular Biology, vol. 1575,
DOI 10.1007/978-1-4939-6857-2_8, © Springer Science+Business Media LLC 2017

homogenous ADCs with improved stability, PK and PD properties, and therapeutic index [10, 11].

We have developed a site-specific conjugation technology based on mAbs with engineered selenocysteine (Sec) residues, which we named selenomabs [12–15]. Sec, the 21st natural amino acid, is structurally similar to cysteine, but with a selenium instead of a sulfur atom in its side chain. In eukaryotes, Sec is encoded by the stop codon UGA and its incorporation requires the presence of a Sec insertion sequence (SECIS) in the 3′ untranslated region (UTR) of the mRNA [16]. Despite little primary sequence conservation, most SECIS elements have a common stem loop structure with the consensus sequence AUGA—(A)AA—GA. Nonetheless, the distance between the UGA codon and the SECIS element is highly variable in different selenoprotein mRNAs [17]. The Sec incorporation machinery plays an essential role in mammalian cells by governing the expression of ~25 natural selenoproteins and can be harnessed for the expression of recombinant selenoproteins [18].

The selenol (SeH) group of Sec (p*K*a 5.2) is more nucleophilic than the thiol (SH) group of Cys (p*K*a 8.3). Taking advantage of their different chemical reactivities, we showed that under mildly acidic (pH 5.2) and reducing (0.1 mM dithiothreitol (DTT)) conditions small molecules having electrophilic functionalities can be selectively conjugated to engineered Sec residues without modifying Cys residues or any other amino acids or carbohydrates in selenomabs [12]. Indeed, corresponding thiomabs with engineered Cys residues are not conjugated under these conditions [12, 19]. Consequently, selenomab conjugations avoid the extra step of reformation of disulfide bridges required for thiomab conjugation [10].

Here, we provide a protocol for the cloning, expression, and purification of selenomabs in scFv-Fc format with a His_6 tag. We incorporate Sec either at the C-terminus or the C_H3 loop of the Fc fragment respectively, and refer to the resulting selenomabs as *scFv-Fc-Sec-His*$_6$ and *scFv-Fc(Ser396Sec)-His*$_6$. To increase the Sec incorporation efficiency, we tried several SECIS elements from the 3′UTR of human or other eukaryotic selenoprotein mRNAs. The SECIS element from human thioredoxin reductase 1 (TXNRD1), which contains a C-terminal Sec, was used to produce scFv-Fc-Sec-His_6 in human embryonic kidney (HEK) 293 c18 cells (Fig. 1a). The non-canonical SECIS elements (GGGA) of protozoan parasites and its AUGA mutant were reported to mediate highly efficient Sec insertion in mammalian cells [20]. We used an AUGA mutant of the GGGA-type *Toxoplasma gondii* SelT 3′UTR to guide the insertion of Sec at the C_H3 loop (Fig. 1b). Using this protocol and our previously published protocol for the generation of scFv-Fc [21], we reformatted the anti-HER2 mAb trastuzumab from IgG1 to scFv-Fc-Sec-His_6 (Fig. 2a) and scFv-Fc(Ser396Sec)-His_6 (Fig. 2b).

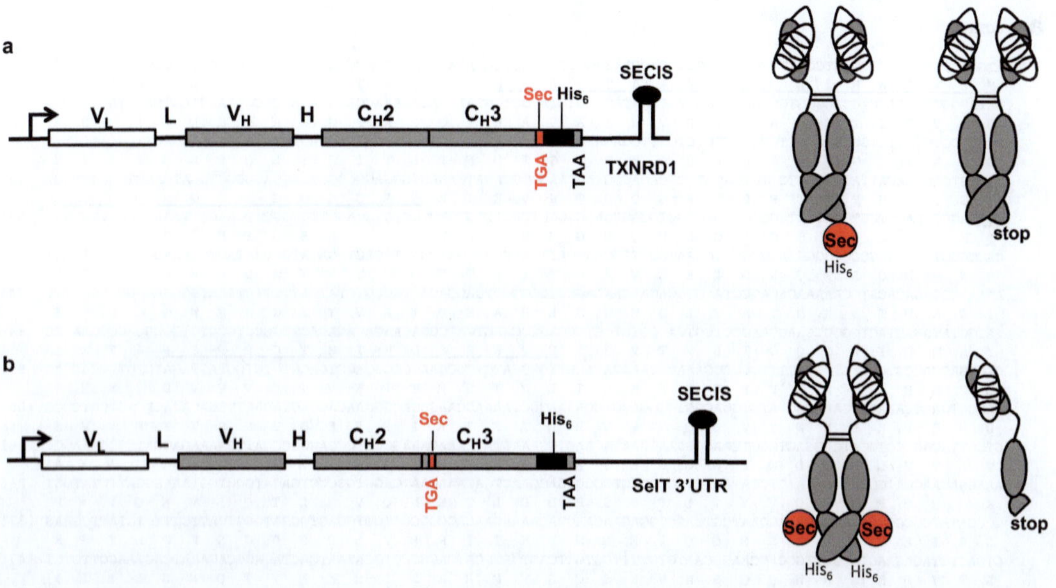

Fig. 1 Expression cassette of (**a**) pCEP4/scFv-Fc-Sec-His_6 and (**b**) pCEP4/scFv-Fc(Ser396Sec)-His_6. The variable domain of the light chain (V_L) is shown in *white*. The variable domain (V_H), second constant domain (C_H2), and third constant domain (C_H3) of the heavy chain are shown in *gray*. Linker (L) and hinge (H) are depicted by *black lines*. The TGA codon for Sec is in *red* and the TAA stop codon is shown in *gray*. A partial TXNRD1 3′UTR containing the SECIS element was used for expression of pCEP4/scFv-Fc-Sec-His_6. Approximately 20% of the product contains Sec with the remainder being truncated stop protein. The 3′UTR of *Toxoplasma gondii* SelT was used for expression of pCEP4/scFv-Fc(Ser396Sec)-His_6. The truncated byproduct protein cannot homodimerize

After tandem Protein A and immobilized metal ion affinity chromatography (IMAC), purified trastuzumab scFv-Fc-Sec-His_6 was recovered with a yield of up to 4 mg/L. By co-transfection with SECIS-binding protein 2 (SBP2) followed by one-step Protein G purification (Fig. 3), the yield of trastuzumab scFv-Fc(Ser396Sec)-His_6 can reach 3 mg/L. This protocol also encompasses the selective conjugation of selenomabs to compounds with selenol-reactive groups (Fig. 4). We recently compared a series of suitable electrophiles, including maleimide, iodoacetamide, methylsulfone oxadiazole, and allenamide derivatives [22]. Additionally, we also describe methods for analyzing the conjugates, such as hydrophobic interaction of chromatography (HIC) (Fig. 5) and electrospray ionization high resolution mass spectrometry (ESI-HRMS). Engineered selenomabs have been successfully used to conjugate biotin, fluorophores, and cytotoxic drugs for mAb-based research and applications. This protocol provides a detailed description of the rapid and simple production, conjugation, and analysis of selenomab conjugates.

a

```
GCTAGC  6

ATGGACTGGACCTGGCGAATCCTTTTCCTGGTCGCCGCAGCAACAGGGGCCCACTCGGATATCCAGATGACCCAGTCCCCGAGCTCCCTGTCCGCCTCTGTGGGC  111
 M  D  W  T  W  R  I  L  F  L  V  A  A  A  T  G  A  H  S  D  I  Q  M  T  Q  S  P  S  S  L  S  A  S  V  G
GATAGGGTCACCATCACCTGCCGTGCCAGTCAGGATGTGAATACTGCTGTAGCCTGGTATCAACAGAAACCAGGAAAAGCTCCGAAACTACTGATTTACTCGGCA  216
 D  R  V  T  I  T  C  R  A  S  Q  D  V  N  T  A  V  A  W  Y  Q  Q  K  P  G  K  A  P  K  L  L  I  Y  S  A
TCCTTCCTCTACTCTGGAGTCCCTTCTCGCTTCTCTGGATCCAGATCTGGGACGGATTTCACTCTGACCATCAGCAGTCTGCAGCCGGAAGACTTCGCAACTTAT  321
 S  F  L  Y  S  G  V  P  S  R  F  S  G  S  R  S  G  T  D  F  T  L  T  I  S  S  L  Q  P  E  D  F  A  T  Y
TACTGTCAGCAACATTATACTACTCCTCCCACGTTCGGACAGGGTACCAAGGTGGAGATCAAAGGGGGAGGCGGGTCAGGAGGGGGTGGATCAGGTGGCGGTGGA  426
 Y  C  Q  Q  H  Y  T  T  P  P  T  F  G  Q  G  T  K  V  E  I  K  G  G  G  G  S  G  G  G  G  S  G  G  G  G
TCGGAGGTTCAGCTGGTGGAGTCTGGCGGTGGCCTGGTGCAGCCAGGGGGCTCACTCCGTTTGTCCTGTGCAGCTTCTGGCTTCAACATTAAAGACACCTATATA  531
 S  E  V  Q  L  V  E  S  G  G  G  L  V  Q  P  G  G  S  L  R  L  S  C  A  A  S  G  F  N  I  K  D  T  Y  I
CACTGGGTGCGTCAGGCCCCGGGTAAGGGCCTGGAATGGGTTGCAAGGATTTATCCTACGAATGGTTATACTAGATATGCCGATAGCGTCAAGGGCCGTTTCACT  636
 H  W  V  R  Q  A  P  G  K  G  L  E  W  V  A  R  I  Y  P  T  N  G  Y  T  R  Y  A  D  S  V  K  G  R  F  T
ATAAGCGCAGACACATCCAAAAACACAGCCTACCTGCAGATGAACAGCCTGCGTGCTGAGGACACTGCCGTCTATTATTGTTCTAGATGGGGAGGGGACGGCTTC  741
 I  S  A  D  T  S  K  N  T  A  Y  L  Q  M  N  S  L  R  A  E  D  T  A  V  Y  Y  C  S  R  W  G  G  D  G  F
TATGCTATGGACTACTGGGGTCAAGGAACCCTGGTCACCGTCTCCTCGGAGCCCAAGTCATCGGACAAGACACACACTTGCCCTCCCTGTCCCGCACCGGAATTG  846
 Y  A  M  D  Y  W  G  Q  G  T  L  V  T  V  S  S  E  P  K  S  S  D  K  T  H  T  C  P  P  C  P  A  P  E  L
CTGGGAGGACCCTCCGTGTTTCTCTTTCCCCCGAAACCCAAGGATACCTTGATGATCTCGCGAACCCCGGAGGTGACGTGTGTGGTAGTAGATGTGTCCCACGAG  951
 L  G  G  P  S  V  F  L  F  P  P  K  P  K  D  T  L  M  I  S  R  T  P  E  V  T  C  V  V  V  D  V  S  H  E
GACCCTGAAGTAAAGTTCAACTGGTATGTGGACGGTGTGGAGGTCCATAACGCGAAAACCAAGCCACGGGAAGAGCAGTACAATTCGACATACAGAGTCGTGTCG 1056
 D  P  E  V  K  F  N  W  Y  V  D  G  V  E  V  H  N  A  K  T  K  P  R  E  E  Q  Y  N  S  T  Y  R  V  V  S
GTCCTTACAGTCCTCCATCAGGACTGGCTTAACGGGAAGGAATACAAGTGCAAAGTATCAAACAAAGCCCTCCCAGCCCCGATTGAAAAGACGATCTCGAAAGCC 1161
 V  L  T  V  L  H  Q  D  W  L  N  G  K  E  Y  K  C  K  V  S  N  K  A  L  P  A  P  I  E  K  T  I  S  K  A
AAAGGACAACCACGCGAGCCGCAAGTCTATACTCTGCCGCCGTCCCGCGACGAACTCACAAAGAATCAGGTGTCCCTCACATGCCTGGTAAAAGGATTCTATCCT 1266
 K  G  Q  P  R  E  P  Q  V  Y  T  L  P  P  S  R  D  E  L  T  K  N  Q  V  S  L  T  C  L  V  K  G  F  Y  P
TCCGATATCGCGGTCGAATGGGAAAGCAATGGTCAGCCGGAGAACAACTACAAAACAACACCGCCGGTGTTGGACTCCGATGGGTCATTCTTTCTCTATTCGAAG 1371
 S  D  I  A  V  E  W  E  S  N  G  Q  P  E  N  N  Y  K  T  T  P  P  V  L  D  S  D  G  S  F  F  L  Y  S  K
CTGACTGTAGACAAGTCCCGCTGGCAGCAAGGCAATGTGTTTTCGTGCTCAGTGATGCATGAGGCCCTGCATAATCACTACACGCAGAAATCACTAAGCTTGTCT 1476
 L  T  V  D  K  S  R  W  Q  Q  G  N  V  F  S  C  S  V  M  H  E  A  L  H  N  H  Y  T  Q  K  S  L  S  L  S
CCGGGTGCTTGACATCACCATCACCATCACTAA                                                                          1509
 P  G  A  U  H  H  H  H  H  H  *
GCCCCAGTGGATGCTGTTGCCAAGACTGCAAACCACTGGCTCGTTTCCGTGCCCAAATCCAAGGCGAAGTTTTCTAGAGGGTTCTTGGGCTCTTGGCACCTGCGT 1614
GTCCTGTGCTTACCACCGCCCAAGGCCCCCTTGGATCTCTTGGATAGGAGTTGGTGAATAGAAGGCAGGCAGCATCACACTGGGGTCACTGACAGACTTGAAGCT 1719
GACATTTGGCAGGGCATCGAAGGGATGCATCCATGAAGTCACCAGTCTCAAGCCCATGTGGTAGGCGGTGATGGAACAACTGTCAAATCAGTTTTAGCATGACCT 1823
TTCCTTGTGGA                                                                                                1834

CTCGAG 1840
```

b

```
TGAGATATCGCGGTCGAATGGGAAAGCAATGGTCAGCCGGAGAACAACTACAAAACAACACCGCCGGTGTTGGACTCCGATGGGTCATTCTTTCTCTATTCGAAG 1371
 U  D  I  A  V  E  W  E  S  N  G  Q  P  E  N  N  Y  K  T  T  P  P  V  L  D  S  D  G  S  F  F  L  Y  S  K
CTGACTGTAGACAAGTCCCGCTGGCAGCAAGGCAATGTGTTTTCGTGCTCAGTGATGCATGAGGCCCTGCATAATCACTACACGCAGAAATCACTAAGCTTGTCT 1476
 L  T  V  D  K  S  R  W  Q  Q  G  N  V  F  S  C  S  V  M  H  E  A  L  H  N  H  Y  T  Q  K  S  L  S  L  S
CCGGGTGCTCATCACCATCACCATCACTAA                                                                             1503
 P  G  A  H  H  H  H  H  H  *
AGCGTGGTACCAGAGGAGGGAAATAGTTGGTTAAATATGGACATAGAGGACGGAGGAAGCTGCGGATTTTTGAGGCGGGGAGCGGTGTCAGTTTGGACGGTGGAG 1608
TTGTGTGGAGAGAGAAGGCGGCAGGAAGTACGCAGGCACAGAAACTGCCGGAGAAGGCGGAAGAGCGAGTCGCGAGTGGATTTCGTGGAGCGTCACGGTTGTGTA 1713
GCTGAGGTCGGACGGTTTCATGCGGGTCGCGGGACGAGGTATGTACGAAAAATGTGGAAGTGGTAGTCCGGCGATTCCAATGCCAGCGGCTTGAGACTTTCTGTA 1818
GATCCACCGGAAGACGGGTATGGTTTATCACCTCGGATAACGCTGCGAatGAGGATGCTGGCAGAAACCTCTCCATTCGAGGCAGCTGGCATCTGATAGTTGGCT 1923
TTTCTGTGTTGAAGATCGTATCCGCCTCTTGTGATCTACTGACAGAATGGCGCACACTTTTGTACCGTGGCAAGTTGTGTCTTTTGGTGAGAAAAGGCAAAGAAA 2028
GAACAACCAAGACTTGAAAAGGACAGTTGGTCGTCGTAACTGTCAAGCCTACGGTGCGTTTCAGGT                                         2095

CTCGAG 2101
```

Fig. 2 DNA and amino acid sequence of trastuzumab scFv-Fc-Sec-His$_6$ and scFv-Fc(Ser396Sec)-His$_6$. (**a**) Using custom DNA synthesis, the variable domains of humanized anti-human HER2 mAb trastuzumab linked with a $(G_4S)_3$ linker were fused to the hinge-C_H2-C_H3 portion of human IgG1 followed by a C-terminal Sec, His$_6$ tag, TAA stop codon and partial of TXNRD1 3′UTR. The DNA sequence of trastuzumab was the sequence of humAb4D5–8 [24]. Flanking sequences with *Nhe*I and *Xho*I restriction sites at the 5′ and 3′ ends, respectively, are shown in *bold italics.* An internal *Hind*III restriction site is shown in *bold. Underlined* amino acid sequences indicate signal peptide, $(G_4S)_3$ linker, hinge, and His$_6$ tag. The *underlined* DNA sequence marks the SECIS element. Note that the amino acid sequence of the hinge (EPKSSDKTHTCPPCP) contains a C-to-S mutation compared to the wild-type human IgG1 hinge (EPKSCDKTHTCPPCP) to remove the cysteine that forms a disulfide bridge with C_L in the IgG1 molecule. (**b**) Sec was incorporated at the C_H3 loop region replacing the original Ser396 residue. The TAA stop codon is followed by the *Toxoplasma gondii* SelT 3′UTR. The *underlined* DNA sequence marks the SECIS element. Its wild-type GGGA sequence is mutated to atGA. Note that neither 3′UTR contains its polyA site. It is replaced with the SV40 polyA site encoded by pCEP4 downstream of *Xho*I

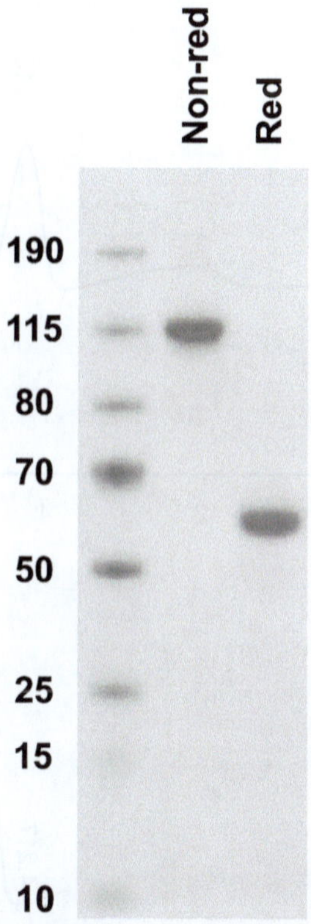

Fig. 3 SDS-PAGE analysis of trastuzumab scFv-Fc(Ser396Sec)-His_6 purified by Protein G affinity chromatography. For each lane, 3 μg of protein in NuPAGE LDS Sample Buffer with (*Red*) or without (*Non-red*) NuPAGE Sample Reducing Reagent was separated by electrophoresis on a 1.5-mm, 10-well NuPAGE Novex 4–12% Bis-Tris Gel using NuPAGE MES SDS Running Buffer followed by staining with SimplyBlue SafeStain based on Coomassie Brilliant Blue G-250. A molecular weight standard (PageRuler Plus Prestained Protein Ladder; Thermo Scientific) is given on the *left* in kDa. No truncated stop protein was detectable

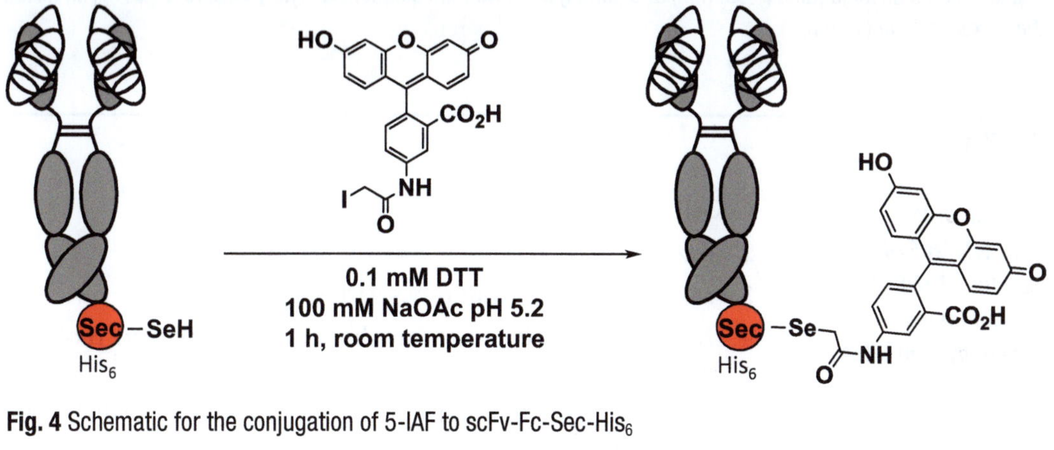

Fig. 4 Schematic for the conjugation of 5-IAF to scFv-Fc-Sec-His_6

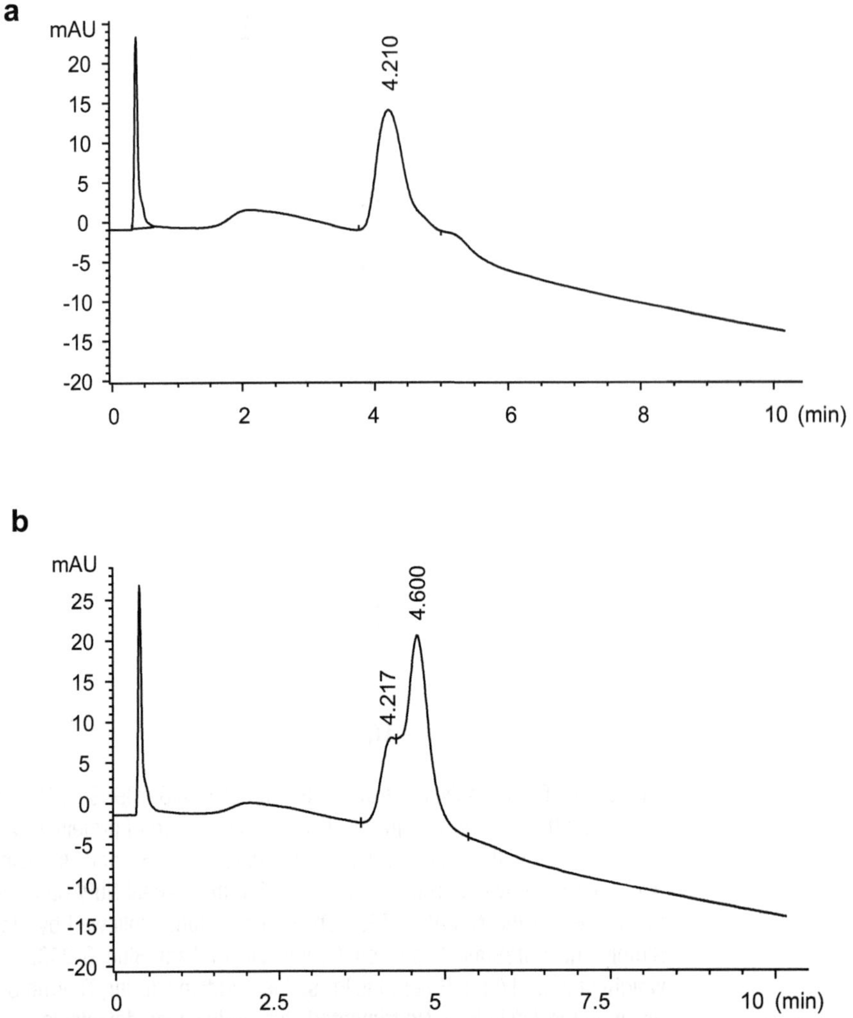

Fig. 5 Analytical characterization of selenomab conjugates by HIC-HPLC. (**a**) HIC-HPLC of unconjugated trastuzumab scFv-Fc-Sec-His$_6$, retention time is ~4.2 min. (**b**) HIC-HPLC of trastuzumab scFv-Fc-Sec-His$_6$/5-IAF. The peaks indicate unconjugated (~4.2 min) and conjugated (4.6 min) selenomab. The y axes refer to A_{280} (in mAU), the *x* axes to time (in min)

2 Materials

2.1 Cloning of pCEP4/scFv-Fc-Sec-His$_6$-TXNRD1 SECIS Plasmid via Restriction Sites

1. Cell spreader, sterile.
2. Razor blades.
3. 14-mL round-bottom tubes with snap caps.
4. *Nhe*I-HF and 10× CutSmart buffer (New England Biolabs), store at −20 °C.
5. *Xho*I and 10× CutSmart buffer (New England Biolabs), store at −20 °C.

6. QIAprep Spin Miniprep Kit (Qiagen).
7. Millipore ultrapure water. Sterilize by filtration through a 0.22-μm filter and store at room temperature.
8. E.Z.N.A. Gel Extraction Kit (Omega Bio-tek).
9. NanoDrop (Thermo Scientific).
10. LB medium: dissolve 25 g LB broth powder (Teknova) in 1 L water, autoclave, and store at room temperature.
11. One Shot TOP10 Chemically Competent *E. coli* (Life Technologies), store at −80 °C.
12. Glycerol stock of *E. coli* strain TOP10 containing a custom synthesized scFv-Fc-Sec-His$_6$-TXNRD1 SECIS encoding DNA fragment (e.g., DNA2.0, Menlo Park, CA) in a plasmid with ampicillin (or kanamycin) resistance gene (Fig. 2). Store at −80 °C.
13. SOC medium (Life Technologies), store at room temperature.
14. 100 μg/μL carbenicillin: dissolve 1 g carbenicillin disodium in 10 mL water, sterilize by filtration through a 0.22-μm filter, and store in 1-mL aliquots at −20 °C.
15. 50% (v/v) glycerol: dilute 1 vol 100% (v/v) glycerol with 1 vol water, autoclave, and store at room temperature.
16. ExpressLink T4 DNA ligase and 5× T4 DNA ligase reaction buffer (Life Technologies), store at −20 °C.
17. DNA sequencing primers:
    pCEP forward (5′-AGCAGAGCTCGTTTAGTGAACCG-3′).
    CH2 reverse (5′-CGACTCTGTATGTCGAATTG-3′).
    EBV reverse (5′-GTGGTTTGTCCAAACTCATC-3′).

2.2 Cloning of pCEP4/scFv-Fc(Ser396Sec)-His$_6$-SelT 3′ UTR Plasmid

1. *Pfu Turbo* DNA polymerase and 10× Cloned *Pfu* DNA polymerase reaction buffer (Agilent Technologies), store at −20 °C.
2. Site-directed mutagenesis primers:
   Forward (5′-GGTAAAAGGATTCTATCCT**TGA**GATATCGCGGTCGAATGG-3′).
   Reverse (5′-CCATTCGACCGCGATATC**TCA**AGGATAGAATCCTTTTACC-3′).
3. *Dpn*I (10 U/μL) (New England Biolabs).
4. XL1-Blue Chemically Competent *E. coli* (Agilent Technologies), store at −80 °C.
5. DNA sequencing primers:
   CMV forward (5′-CGCAAATGGGCGGTAGGCGTG-3′).
   CH2 reverse (5′-CGACTCTGTATGTCGAATTG-3′).
   BGH reverse (5′-TAGAAGGCACAGTCGAGG-3′).
6. Phusion High-Fidelity (HF) DNA polymerase (20 U/μL) and 5× Phusion HF buffer (New England Biolabs), store at −20 °C.

7. SelT 3′UTR gBlocks (Integrated DNA Technologies): overlap with Fc fragment starting upstream of *Hind*III site (AAGCTT), remove TGA codon, and keep the six His codons (*see* Fig. 2b).
8. SelT 3′UTR PCR primers:
   Forward (5′-tcactAAGCTTGTCTCCG-3′)
   Reverse (5′-agtcCTCGAGACCTGAAACGC-3′)
9. Other materials listed in Subheading 2.1.

2.3 Production of scFv-Fc-Sec-His$_6$ in Mammalian Cells

1. pCEP4/scFv-Fc-Sec-His$_6$-TXNRD1 SECIS plasmid.
2. Medium for cell growth: High glucose DMEM supplemented with 10% (v/v) fetal bovine serum (FBS), 100 μg/mL streptomycin, and 100 U/mL penicillin (all from Life Technologies), store at 4 °C.
3. Medium for antibody production: High glucose DMEM supplemented with 100 μg/mL streptomycin, 100 U/mL penicillin, and 1 μM sodium selenite (Sigma-Aldrich), store at 4 °C.
4. 293 c18 cells, a derivative of HEK293 cells expressing EBNA1 (American Type Culture Collection; CRL-10852).
5. PEI transfection reagent: dissolve 100 mg polyethylenimine (MW~25,000 Da) (Polysciences) in 100 mL water, adjust pH to 7.0. Aliquot and store at −20 °C.
6. 500-mL 0.45-μm Stericup Filter Unit (Millipore).
7. HiTrap Protein A HP columns (GE Healthcare), store at 4 °C.
8. Peristaltic P-1 pump (GE Healthcare).
9. PBS (Life Technologies).
10. Protein A or G elution buffer: 0.5 M ice-cold acetic acid.
11. Neutralization buffer: 1 M Tris–HCl pH 8.0.
12. Column storage buffer: 20% (v/v) ethanol.
13. HisTrap column washing buffer: 25 mM imidazole, 0.5 M NaCl in PBS, pH 7.4. HisTrap column elution buffer: 0.5 M imidazole, 0.5 M NaCl in PBS, pH 7.4.
14. PD-10 Desalting Columns (GE Healthcare).
15. 15-mL Amicon Ultra Centrifugal Filter Devices with a 30-kDa MWCO membrane (EMD Millipore).
16. UV photometer (Eppendorf).
17. Disposable cuvettes (Eppendorf).
18. NuPAGE Novex 4–12% Bis-Tris Gel 1.5 mm, ten well (Life Technologies), store at 4 °C.
19. NuPAGE MES SDS Buffer Kit with MES SDS Running Buffer, Sample Reducing Agent, Antioxidant, and LDS Sample Buffer (Life Technologies).
20. SimplyBlue SafeStain based on Coomassie Brilliant Blue G-250 (Life Technologies).

2.4 Production of scFv-Fc(Ser396Sec)-His$_6$ in Mammalian Cells

1. pCEP4/scFv-Fc(Ser396Sec)-His$_6$-SelT 3′UTR plasmid.
2. pCEP4/SBP2 plasmid.
3. HiTrap Protein A or G HP columns (GE Healthcare).
4. All other materials listed in Subheading 2.3.

2.5 Conjugation to Selenol-Reactive Compounds

1. Conjugation buffer: 100 mM NaOAc buffer, pH 5.2.
2. 100 mM DTT in water, made freshly.
3. 50 mM 5-iodoacetamide-fluorescein (5-IAF) (Marker Gene Technologies) in DMSO, store at −80 °C.
4. 10 mM methylsulfone phenyloxadiazole fluorescein [23] in DMSO, store at −80 °C.
5. NAP-10 columns (GE Healthcare).
6. PBS (Life Technologies).
7. 4-mL Amicon Ultra Centrifugal Filter Devices with a 30-kDa MWCO membrane (EMD Millipore).

2.6 Drug-to-Antibody Ratio (DAR) Analysis

1. TSKgel Butyl-NPR column (4.6 mm ID × 3.5 cm, 2.5 μm, Tosoh Bioscience) in conjunction with an HPLC instrument, such as Agilent 1100 series HPLC system.
2. Mobile phase A buffer: aqueous solution of 1.5 M ammonium sulfate, 25 mM sodium phosphate (pH 7.0).
3. Mobile phase B buffer: 75% (v/v) aqueous solution of 25 mM sodium phosphate (pH 7.0) and 25% (v/v) isopropanol.
4. PNGase F (500 U/μL) (New England Biolabs), store at −20 °C.
5. Ab SpinTrap: Prepacked Protein G Sepharose High Performance columns (GE Healthcare).
6. Elution buffer: 0.5 M acetic acid.
7. Neutralizing buffer: 1 M Tris–HCl pH 8.0.
8. NAP-10 columns (GE Healthcare).
9. PBS (Life Technologies).
10. 0.5-mL Amicon Ultra Centrifugal Filter Devices with a 30-kDa MWCO membrane (EMD Millipore).

3 Methods

3.1 Cloning of pCEP4/scFv-Fc-Sec-His$_6$-TXNRD1 SECIS Plasmid via Restriction Sites

1. From frozen glycerol stocks, streak out [1] *E. coli* strain TOP10 containing a custom synthesized scFv-Fc-Sec-His$_6$-TXNRD1 SECIS encoding DNA fragment with flanking *Nhe*I and *Xho*I restriction sites (Fig. 2) in a plasmid with an ampicillin resistance gene and [2] *E. coli* strain TOP10 containing mammalian expression vector pCEP4 (Invitrogen) on LB-agar carbenicillin (100 μg/mL) plates and incubate at 37 °C for 16 h (*see* **Note 1**).

2. Pick up two single colonies from each plate to inoculate 5 mL LB medium containing 100 μg/mL carbenicillin in four 14-mL round-bottom tubes with snap caps, and grow culture at 37 °C and 250 rpm for 12–14 h.
3. Take 500 μL of each culture and add 500 μL of 50% (v/v) glycerol in a 2-mL screw top tube or cryovial, gently mix, and store at −80 °C temporally.
4. Isolate the plasmid DNA from the rest of the culture using the QIAprep Spin Miniprep Kit and measure the plasmid concentration using a NanoDrop instrument.
5. Set up a 50-μL *Nhe*I/*Xho*I double digestion reaction containing 1× CutSmart buffer, 1 μL *Nhe*I-HF (20 U/μL), 1 μL *Xho*I (20 U/μL), 2 μg scFv-Fc-Sec-His_6-TXNRD1 SECIS plasmid, or 2 μg pCEP4. Incubate at 37 °C for 4 h (*see* **Note 2**).
6. Separate the double cut scFv-Fc-Sec-His_6-TXNRD1 SECIS encoding DNA fragment (~1.8 kb) and pCEP4 vector (~10 kb) by electrophoresis on a 1% (w/v) agarose gel in 1× TAE buffer. Purify the two DNA fragments by using the E.Z.N.A. Gel Extraction Kit (*see* **Note 3**).
7. Set up a 10-μL ligation reaction containing 1× T4 DNA ligase buffer, 50 ng double cut and excised pCEP4, 50 ng double cut and excised scFv-Fc-Sec-His_6-TXNRD1 SECIS, and 0.5 μL ExpressLink T4 DNA ligase (5 U/μL). Incubate at room temperature for 10 min.
8. Add 1 μL of the above reaction sample to 25 μL One Shot TOP10 Chemically Competent *E. coli* and incubate the cells on ice for 30 min, followed by 30 s at 42 °C, and then place back on ice for 2 min. Add 200 μL of SOC medium to the reaction tube and shake at 37 °C and 225 rpm for 1 h (*see* **Note 4**).
9. Spread the 200 μL on an LB-agar carbenicillin plate and incubate for 16 h at 37 °C.
10. Randomly pick three colonies from the plate and inoculate in 5 mL LB medium with 100 μg/mL carbenicillin in three 14-mL round-bottom tubes with snap caps, and grow culture at 37 °C and 250 rpm for 12–14 h.
11. Isolate the plasmid DNA with the QIAprep Spin Miniprep Kit using 30 μL sterile Millipore ultrapure water for elution. Measure the concentration of each plasmid using the NanoDrop instrument.
12. Set up three 30-μL *Nhe*I/*Xho*I double digestion reactions containing 1× CutSmart buffer, 0.5 μL *Nhe*I-HF (20 U/μL), 0.5 μL *Xho*I (20 U/μL), and 1 μg plasmid from above preparations. Incubate at 37 °C for 2 h.

13. Directly add 5 μL 6× gel loading dye solution, and separate by electrophoresis on a 1% (w/v) agarose gel in 1× TAE buffer. Clones that have the scFv-Fc-Sec-His_6-TXNRD1 SECIS encoding DNA fragment reveal two bands of ~10 and ~1.8 kb. Further confirm the insert by DNA sequencing using primers pCEP forward (upstream from *Nhe*I), CH2 reverse (internal), and EBV reverse (downstream from *Xho*I).
14. Store the confirmed pCEP4/scFv-Fc-Sec-His_6-TXNRD1 SECIS plasmid at −20 °C and the bacterial glycerol stocks at −80 °C.

3.2 Cloning of pCEP4/scFv-Fc(Ser396Sec)-His_6-SelT 3′ UTR Plasmid

1. *Site-directed mutagenesis.* Set up 50-μL PCR reaction containing 1× Cloned *Pfu* DNA polymerase reaction buffer, 0.8 mM total dNTP mixture, 0.2 μM of each forward and reverse mutagenic primers, 10 ng pcDNA3.1/scFv-Fc-Sec-His_6-TXNRD1 SECIS plasmid, and 1 μL *Pfu Turbo* DNA polymerase (2.5 U/μL) (*see* **Note 5**).
2. Run the PCR using the following program: 2 min at 95 °C for denaturation, followed by 16 cycles of 30 s denaturation at 95 °C, 60 s annealing at 55 °C, and 8 min of extension at 68 °C.
3. Add 1 μL *Dpn*I (10 U/μL) directly to the PCR sample and incubate at 37 °C for 1 h (*see* **Note 6**).
4. Add 1 μL of the *Dpn*I-treated sample to 50 μL XL1-Blue Chemically Competent *E. coli* and incubate the cells on ice for 30 min, followed by 30 s at 42 °C, and then place back on ice for 2 min. Add 200 μL of SOC medium to the reaction tube and shake at 37 °C and 225 rpm for 1 h.
5. Spread the 200 μL on an LB-agar carbenicillin plate and incubate for 16 h at 37 °C.
6. Randomly pick six colonies from the plate and inoculate in 5 mL LB medium with 100 μg/mL carbenicillin in six 14-mL round-bottom tubes with snap caps, and grow culture at 37 °C and 250 rpm for 12–14 h.
7. Take 500 μL of each culture and add 500 μL of 50% (v/v) glycerol in a 2-mL screw top tube or cryovial, gently mix, and store at −80 °C temporally.
8. Isolate the plasmid DNA with the QIAprep Spin Miniprep Kit using 50 μL water for elution. Measure the concentration of each plasmid using the NanoDrop instrument. The desired selenocysteine mutation and any unwanted nonspecific mutations can be identified by DNA sequencing using primers CMV forward (upstream from *Nhe*I), CH2 reverse (internal), and BGH reverse (downstream from *Xho*I).
9. Pool confirmed pcDNA3.1/scFv-Fc(Ser396Sec)-His_6-TXNRD1 SECIS plasmids and store at −20 °C.

10. Set up a 50-μL PCR reaction containing 1× Phusion HF buffer, 0.8 mM total dNTP mixture, 0.2 μM of each forward and reverse SelT 3′UTR primers, 10 ng SelT 3′UTR gBlocks, and 1 μL Phusion HF DNA polymerase (2 U/μL).
11. Run the PCR using the following program: 2 min at 95 °C for denaturation, followed by 25 cycles of 30 s denaturation at 95 °C, 20 s annealing at 55 °C, and 15 s of extension at 72 °C.
12. Identify the PCR product (~650 bp) by electrophoresis on a 1% (w/v) agarose gel in 1× TAE buffer. Purify the DNA fragment by E.Z.N.A. Gel Extraction Kit.
13. Set up a 50-μL *Hind*III/*Xho*I double digestion reaction containing 1× CutSmart buffer, 1 μL *Hind*III-HF (20 U/μL), 1 μL *Xho*I (20 U/μL), 2 μg pcDNA3.1/scFv-Fc(Ser396Sec)-His_6-TXNRD1 SECIS or 2 μg SelT 3′UTR PCR product. Incubate at 37 °C for 4 h.
14. Separate the double cut pcDNA3.1/scFv-Fc(Ser396Sec) fragment (~6.2 kb) and SelT 3′UTR PCR product (~650 bp) by electrophoresis on 1% (w/v) agarose gel in 1× TAE buffer. Purify two DNA fragments using the E.Z.N.A. Gel Extraction Kit.
15. Set up a 10 μL ligation reaction containing 1× T4 DNA ligase buffer, 50 ng double cut and excised pcDNA3.1/scFv-Fc(Ser396Sec), 50 ng double cut and excised SelT 3′UTR fragment, and 0.5 μL ExpressLink T4 DNA ligase (5 U/μL). Incubate at room temperature for 10 min.
16. Add 1 μL of above reaction sample to 25 μL One Shot TOP10 Chemically Competent *E. coli* and incubate the cells on ice for 30 min, followed by 30 s at 42 °C, and then place back on ice for 2 min. Add 200 μL of SOC medium to the reaction tube and shake at 37 °C and 225 rpm for 1 h.
17. Repeat **steps 5–8**.
18. Pool confirmed pcDNA3.1/scFv-Fc(Ser396Sec)-His_6-SelT 3′UTR plasmid and store at −20 °C.
19. *Cloning of pCEP4/scFv-Fc(Ser396Sec)-His_6-SelT 3′UTR plasmid* via *restriction sites.* Repeat **steps 5–13** in Subheading 3.1 using pcDNA3.1/scFv-Fc(Ser396Sec)-His_6-SelT 3′UTR plasmid instead of scFv-Fc-Sec-His_6-TXNRD1 SECIS plasmid.
20. Store confirmed pCEP4/scFv-Fc(Ser396Sec)-His_6-SelT 3′UTR plasmid at −20 °C and bacterial glycerol stocks at −80 °C.

3.3 Production of Selenomab scFv-Fc-Sec-His_6 in Mammalian Cells

1. *Plasmid maxi-preparation.* Scrape the frozen bacterial glycerol stock containing pCEP4/scFv-Fc-Sec-His_6-TXNRD1 SECIS plasmid using a sterile pipette tip, inoculate 500 mL LB medium with 100 μg/mL carbenicillin in a 2-L Erlenmeyer flask, and shake at 37 °C and 250 rpm for 14–16 h. Isolate the

plasmid DNA with a QIAGEN HiSpeed Plasmid Maxi Kit following the manufacturer's instructions (*see* **Note 7**).

2. *Transient transfection of 293 c18 cells.* The day before transfection, prepare ten 140-cm^2 cell culture dishes and seed 1.0×10^7 293 c18 cells into each dish with 20 mL medium for cell growth (*see* Subheading 2.3). Grow cells overnight in a cell culture incubator with humidified atmosphere containing 5% CO_2 at 37 °C (*see* **Note 8**).
3. Dilute 10 × 30 μg (Σ300 μg) pCEP4/scFv-Fc-Sec-His$_6$-TXNRD1 SECIS plasmid into a total volume of 14.4 mL of Opti-MEM medium in a 50-mL centrifuge tube. Subsequently add 10 × 60 μL (Σ600 μL) PEI transfection reagent directly to the diluted plasmid, mix gently by swirling, and incubate at room temperature for 15 min (*see* **Note 9**).
4. Carefully add 1.5 mL transfection complex dropwise to each dish of cells and mix the medium gently.
5. Incubate the cells in the CO_2 incubator at 37 °C for 8–16 h.
6. Aspirate growth medium, wash gently with 10 mL PBS once, and add 20 mL medium for antibody production (*see* Subheading 2.3) to each dish and incubate the cells in the CO_2 incubator at 37 °C for 72 h.
7. Collect the medium, store at 4 °C, add 20 mL fresh medium for antibody production, and incubate cells for another 72 h.
8. Repeat **step** 7.
9. Collect the third supernatant. The total volume of the three harvested supernatants is ~600 mL.
10. *Purification of total antibody by Protein A affinity chromatography.* Spin the pooled harvests at 10,000 × *g* for 20 min, pour the supernatant in a 500-mL 0.45-μm Stericup Filter Unit, and filter under vacuum.
11. Load the ~600-mL filtrate directly on a PBS pre-equilibrated prepacked 1-mL HiTrap Protein A HP column using a peristaltic P-1 pump at 1 mL/min at 4 °C (*see* **Note 10**).
12. Wash the column with 50 mL ice-cold PBS at 1 mL/min.
13. Prepare ten 1.5-mL microcentrifuge tubes containing 0.65 mL 1 M Tris–HCl (pH 8.0) and place them under the column to collect the elution at 0.5 mL/tube. Elute the column with 5 mL 0.5 M ice-cold acetic acid at a flow rate of 1 mL/min.
14. Regenerate the column with 10 mL water followed by 10 mL 20% (v/v) ethanol, store at 4 °C.
15. Identify the fractions that contain the eluted protein by measuring the absorbance at 280 nm with a disposable cuvette in a UV photometer.

16. *Selenomab purification by HisTrap column.* Pool the fractions that contain the eluted protein in a 50-mL centrifuge tube, add 0.05 vol of HisTrap column elution buffer, mix well, and load on a washing buffer pre-equilibrated prepacked 1-mL HisTrap column using a peristaltic P-1 pump at 0.5 mL/min at 4 °C.
17. Place a 15-mL Amicon Ultra Centrifugal Filter Device with a 30-kDa MWCO membrane under the HisTrap column to collect the flow-through which contains the truncated scFv-Fc without Sec and His_6 tag (scFv-Fc-stop). Place on ice.
18. Wash column with washing buffer at 1 mL/min for 10 min.
19. Place a second 15-mL Amicon Ultra Centrifugal Filter Device with a 30-kDa MWCO membrane under the HisTrap column.
20. Elute the column with 10 mL elution buffer.
21. Concentrate both flow-through (scFv-Fc-stop) and eluate (scFv-Fc-Sec-His_6) to a final volume of ~2.5 mL by centrifuging 20 min at 3000 × *g* at 4 °C in a refrigerated benchtop centrifuge with swinging bucket rotor.
22. Prepare two PD-10 Desalting Columns. Pour off the column storage solution and equilibrate columns with 25 mL 100 mM NaOAc (pH 5.2) (*see* **Note 11**).
23. Add concentrated HisTrap column flow-through and eluate to two separate PD-10 columns. Let the samples enter the packed bed completely. Discard the flow-through (*see* **Note 12**).
24. Elute the column with 3.5 mL 100 mM NaOAc (pH 5.2).
25. Measure the concentration of the scFv-Fc-stop and scFv-Fc-Sec proteins by measuring their absorbance at 280 nm. Use this absorbance to calculate the protein concentration based on the estimation that 1 mg/mL of an immunoglobulin protein gives an absorbance of 1.4. Dilute or concentrate the protein to a suitable storage concentration (typically 0.5–1 mg/mL), aliquot, and store at −80 °C.
26. Analyze the purity and integrity of the scFv-Fc-Sec-His_6 protein on an SDS-PAGE gel under reducing and nonreducing conditions following staining with Coomassie Brilliant Blue G-250.

3.4 Production of Selenomab scFv-Fc(Ser396Sec)-His_6 in Mammalian Cells

1. Maxi-preparation of plasmid pCEP4/scFv-Fc(Ser396Sec)-His_6-SelT 3′UTR and pCEP4/SPB2, as described in Subheading 3.3, **step 1** (*see* **Note 13**).
2. Prepare 293 c18 cells for transfection as described in Subheading 3.3, **step 2**.
3. Dilute 10 × 20 μg (Σ200 μg) pCEP4/scFv-Fc(Ser396Sec)-His_6-SelT 3′UTR plasmid and 10 × 10 μg (Σ100 μg) pCEP4/SPB2 plasmid into a total volume of 14.4 mL of Opti-MEM

medium in a 50-mL centrifuge tube. Subsequently add 10 × 60 μL (Σ600 μL) PEI transfection reagent directly to the diluted plasmid, mix gently by swirling, and incubate at room temperature for 15 min.

4. Repeat **steps 4–15** in Subheading 3.3 using a Protein G HP column (*see* **Note 14**).
5. Combine the fractions in a 15 mL-Amicon Ultra Centrifugal Filter Device (30 kDa) and concentrate the sample to a final volume of ~500 μL by centrifuging 20 min at 3000 × *g* at 4 °C in a refrigerated benchtop centrifuge with swinging bucket rotor. If necessary, continue the centrifugation until the volume of the concentrate is down to ~500 μL. Discard the filtrate and add 15 mL 100 mM NaOAc (pH 5.2) to the concentrate. Repeat the centrifugation as before until the volume of the concentrate is again down to ~500 μL. Repeat this step one more time (*see* **Note 15**).
6. Measure the concentration of the scFv-Fc(Ser396Sec)-His$_6$ protein by measuring its absorbance at 280 nm. Calculate the protein concentration based on the estimation that 1 mg/mL of an immunoglobulin protein gives an absorbance of 1.4. Dilute or concentrate the protein to a suitable storage concentration (typically 0.5–1 mg/mL), aliquot, and store at −80 °C.
7. Analyze the purity and integrity of the scFv-Fc(Ser396Sec)-His$_6$ protein on an SDS-PAGE gel under reducing and nonreducing conditions following staining with Coomassie Brilliant Blue G-250 (Fig. 3).

3.5 Conjugation to Selenol-Reactive Compounds

1. Add 1 μL of 100 mM DTT to 1 mL 4 μM (~0.5 mg) selenomab prepared in Subheadings 3.3 and 3.4 in conjugation buffer and incubate at room temperature for 20 min (*see* **Note 16**).
2. Dilute 5-IAF or methylsulfone phenyloxadiazole fluorescein, or a corresponding selenol-reactive drug derivative to 2 mM in DMSO. Add 20 μL of each solution to the reaction mixture and incubate at room temperature for 1 h (*see* **Note 17**).
3. Wash and equilibrate a NAP-10 column with conjugation buffer, add the reaction mixture to the column, and let the sample enter the packed bed completely.
4. Elute the column with 1.5 mL conjugation buffer.
5. Concentrate and buffer exchange the sample into PBS using a 4-mL Amicon Ultra Centrifugal Filter Device with a 30-kDa MWCO membrane. Store the conjugate at −80 °C.

3.6 Drug-to-Antibody Ratio (DAR) Analysis

1. *Hydrophobic interaction chromatography (HIC)*: Dilute samples (2 mg/mL) 1:1 in mobile phase A buffer and load 10 μL on the column.

2. Elute at a flow rate of 0.8 mL/min with a gradient of 0% B to 100% B over 12 min for scFv-Fc-Sec-drug conjugates and 30 min for scFv-Fc(Ser396Sec)-drug conjugates. Monitor the protein peak using A_{280} absorbance (*see* **Note 18**).
3. Use ChemStation software (Agilent) to resolve and quantify selenomab conjugate species and DAR.
4. *Electrospray ionization high resolution mass spectrometry (ESI-HRMS)*: De-glycosylate the conjugated or unconjugated selenomab by adding 5 μL PNGase F to 200 μg sample in PBS and incubate at 37 °C for 18 h.
5. Add the sample to an Ab SpinTrap column (Protein G Sepharose) pre-washed with PBS and incubate for 4 min while gently mixing.
6. Centrifuge for 30 s at 100 × *g*.
7. Wash the column with 600 μL PBS three times.
8. Add 400 μL elution buffer and mix by inversion.
9. Place the column in a 2-mL microcentrifuge tube containing 30 μL neutralizing buffer. Centrifuge for 30 s at 100 × *g* and collect the eluate.
10. Repeat **step 9** and collect the second eluate.
11. Combine the two eluates and load on a NAP-10 Column pre-equilibrated with 0.1% (v/v) formic acid.
12. Elute the column with 1.5 mL 0.1% (v/v) formic acid and concentrate the sample to 2 mg/mL using a 0.5-mL Amicon Ultra Centrifugal Filter Device with a 30-kDa MWCO membrane (*see* **Note 19**).
13. Add 20 μL of the sample to 180 μL 30% (v/v) acetonitrile/0.1% (v/v) formic acid and immediately inject at 3 μL/min into a Q Exactive Hybrid Quadrupole-Orbitrap mass spectrometer.
14. Acquire MS data using a continuous survey scan (1600–3000 Th) at a resolution of 17,500 or 35,000 (at 200 *m/z*) with a maximum injection time of 50–100 ms and a 1–5 × 10^5 ion count target.
15. Perform deconvolution manually by averaging the molecular masses obtained for all detectable peaks in the spectra (*see* **Note 20**).

4 Notes

1. Use *Kpn*I instead of *Nhe*I if there is no internal *Kpn*I site within the DNA fragment. *Kpn*I will give a higher ligation efficiency than *Nhe*I. Synthesis of three gBlock fragments (up to 750 bp in length for one fragment) and running overlap

extension PCR to fuse the fragments is an alternative to custom synthesis of the full length gene.

2. Double digestion condition can be found on this website: www.neb.com/tools-and-resources/interactive-tools/double-digest-finder.
3. Elute the DNA with 15 μL water to achieve higher concentrations but lower yield.
4. Use 50 μL or more if using chemically competent *E. coli* made in house.
5. scFv-Fc-Sec-His$_6$-TXNRD1 SECIS was cloned into vector pcDNA3.1 using *Nhe*I and *Xho*I sites following our cloning protocol in Subheading 3.1. The smaller pcDNA3.1/scFv-Fc-Sec-His$_6$-TXNRD1 SECIS plasmid is preferred over the larger pCEP4/scFv-Fc-Sec-His$_6$-TXNRD1 SECIS plasmid as template plasmid for site-directed mutagenesis.
6. *Dpn*I digests the template plasmid. Incubation for 1 h is sufficient.
7. Using 15 mL buffer P1, 15 mL buffer P2, and 15 mL buffer P3 for 500 mL culture. After adding buffer P3, mix thoroughly, then centrifuge the mixture at 3000 × *g* or higher speed for 5 min before pouring into the barrel of the QIAfilter Cartridge. pCEP4 is a low copy vector, and elution of the DNA with 1 mL water or elution buffer typically yields a plasmid concentration of 1 μg/μL. If the plasmid concentration is lower than 0.5 μg/μL, revive the transformed bacteria by streaking out the frozen glycerol stocks on LB-agar carbenicillin (100 μg/mL) plates and incubate at 37 °C for 14–16 h. Pick a well-isolated single colony and proceed as described with inoculating 500 mL LB medium with 100 μg/mL carbenicillin and plasmid maxi-preparation.
8. We recommend 293 c18 cells but have also used other strains of HEK 293 cells successfully. 293 c18 cells adhere better to plates than other strains of HEK 293 cells.
9. The amount of PEI used for each plate must not exceed 60 μL due to its cytotoxicity. We noted a dramatic loss of PEI's transfection efficiency during storage of opened bottles. By contrast, the PEI solution can be stored at −20 °C for up to 1 year without significant loss of transfection efficiency. PEI stock solution (5 mg/mL) can be made by dissolving all PEI powder (2 g) into 0.2 M HCl (400 mL) and storing at −20 °C. Use Polysciences' PEI (MW ~25,000 Da) with catalog number 23966. Other sizes of PEI do not work as efficiently as this one.
10. The binding capacity of a 1-mL HiTrap Protein A HP column is 20 mg IgG1. Thus, the loading volume depends on the expected yield of antibody.

11. The equilibration buffer is the final storage buffer for the selenomabs prior to conjugation. Depending on their variable domains, some selenomabs may aggregate and precipitate at pH 5.2. PBS or TBS are alternative equilibration buffers.
12. If the sample volume is less than 2.5 mL, add equilibration buffer to adjust the volume to 2.5 mL after the sample has entered the packed bed completely.
13. SBP2 cDNA was purchased from OriGene Technologies (SC112316) and cloned into pCEP4 via *Kpn*I/*Xho*I following our cloning protocol in Subheading 3.1. We found that co-transfection of SBP2 gave a 50% yield increase of trastuzumab scFv-Fc(Ser396Sec)-His$_6$ when using the SelT SECIS element. No yield increase was found when co-transfecting SBP2 with trastuzumab scFv-Fc(Ser396Sec)-His$_6$ using the TXNRD1 SECIS element.
14. The truncated trastuzumab scFv-Fc(Ser396stop) protein can bind Protein A because it contains a human variable heavy chain domain that belongs to the V_H3 subfamily. However, it does not bind Protein G. Thus, one-step Protein G purification is sufficient to get pure trastuzumab scFv-Fc(Ser396Sec)-His$_6$. Other scFv-Fc(Ser396stop) proteins without V_H3 subfamily sequences cannot bind to Protein A, enabling one-step Protein A purification.
15. A PD-10 desalting column also can be used to concentrate and buffer exchange the purified selenomab.
16. To demonstrate selective conjugation to the selenomab, use the purified scFv-Fc-stop protein as negative control. The final concentration of DTT can be up to 4 mM for a reaction containing 0.5 mg/mL of selenomab. As DTT is not stable in solution, use freshly made DTT for each conjugation reaction. DTT cannot be replaced with TCEP because it inhibits iodoacetamide-selenol conjugation.
17. Iodoacetamide derivatives are sensitive to light and should be stored in the dark. Use freshly prepared solution if the conjugation efficiency is low. The final DMSO concentration in the reaction mixture can be up to 5% (v/v).
18. Chromatographic separation of selenomab conjugate species by HIC is feasible for compounds that add substantial hydrophobicity to the protein. We found that the conjugation site also has great impact on the hydrophobicity of the selenomab conjugate. For example, an auristatin drug derivative added less hydrophobicity when conjugated to scFv-Fc(Ser396Sec) compared to scFv-Fc-Sec. Increasing the elution time may improve separation.

19. The membrane of this Amicon Ultra Centrifugal Filter Device will bind ~50 μg protein.
20. The unconjugated Sec at the C-terminus is converted to dehydroalanine during ESI-HRMS. Therefore, the mass of unconjugated selenomab is calculated as containing dehydroalanine instead of Sec.

References

1. Chari RV, Miller ML, Widdison WC (2014) Antibody-drug conjugates: an emerging concept in cancer therapy. Angew Chem Int Ed Engl 53:3796–3827
2. de Goeij BE, Lambert JM (2016) New developments for antibody-drug conjugate-based therapeutic approaches. Curr Opin Immunol 40:14–23
3. Sassoon I, Blanc V (2013) Antibody-drug conjugate (ADC) clinical pipeline: a review. Methods Mol Biol 1045:1–27
4. Adair JR, Howard PW, Hartley JA, Williams DG, Chester KA (2012) Antibody-drug conjugates—a perfect synergy. Expert Opn Biol Ther 12:1191–1206
5. Sievers EL, Senter PD (2013) Antibody-drug conjugates in cancer therapy. Annu Rev Med 64:15–29
6. Lewis Phillips GD, Li G, Dugger DL, Crocker LM, Parsons KL, Mai E, Blattler WA, Lambert JM, Chari RV, Lutz RJ, Wong WL, Jacobson FS, Koeppen H, Schwall RH, Kenkare-Mitra SR, Spencer SD, Sliwkowski MX (2008) Targeting HER2-positive breast cancer with trastuzumab-DM1, an antibody-cytotoxic drug conjugate. Cancer Res 68:9280–9290
7. Doronina SO, Toki BE, Torgov MY, Mendelsohn BA, Cerveny CG, Chace DF, DeBlanc RL, Gearing RP, Bovee TD, Siegall CB, Francisco JA, Wahl AF, Meyer DL, Senter PD (2003) Development of potent monoclonal antibody auristatin conjugates for cancer therapy. Nat Biotechnol 21:778–784
8. Hamblett KJ, Senter PD, Chace DF, Sun MM, Lenox J, Cerveny CG, Kissler KM, Bernhardt SX, Kopcha AK, Zabinski RF, Meyer DL, Francisco JA (2004) Effects of drug loading on the antitumor activity of a monoclonal antibody drug conjugate. Clin Cancer Res 10:7063–7070
9. Boswell CA, Mundo EE, Zhang C, Bumbaca D, Valle NR, Kozak KR, Fourie A, Chuh J, Koppada N, Saad O, Gill H, Shen BQ, Rubinfeld B, Tibbitts J, Kaur S, Theil FP, Fielder PJ, Khawli LA, Lin K (2011) Impact of drug conjugation on pharmacokinetics and tissue distribution of anti-STEAP1 antibody-drug conjugates in rats. Bioconjug Chem 22:1994–2004
10. Junutula JR, Raab H, Clark S, Bhakta S, Leipold DD, Weir S, Chen Y, Simpson M, Tsai SP, Dennis MS, Lu Y, Meng YG, Ng C, Yang J, Lee CC, Duenas E, Gorrell J, Katta V, Kim A, McDorman K, Flagella K, Venook R, Ross S, Spencer SD, Lee Wong W, Lowman HB, Vandlen R, Sliwkowski MX, Scheller RH, Polakis P, Mallet W (2008) Site-specific conjugation of a cytotoxic drug to an antibody improves the therapeutic index. Nat Biotechnol 26:925–932
11. Panowksi S, Bhakta S, Raab H, Polakis P, Junutula JR (2014) Site-specific antibody drug conjugates for cancer therapy. MAbs 6:34–45
12. Hofer T, Thomas JD, Burke TR Jr, Rader C (2008) An engineered selenocysteine defines a unique class of antibody derivatives. Proc Natl Acad Sci U S A 105:12451–12456
13. Hofer T, Skeffington LR, Chapman CM, Rader C (2009) Molecularly defined antibody conjugation through a selenocysteine interface. Biochemistry 48:12047–12057
14. Cui H, Thomas JD, Burke TR Jr, Rader C (2012) Chemically programmed bispecific antibodies that recruit and activate T cells. J Biol Chem 287:28206–28214
15. Vire B, Skarzynski M, Thomas JD, Nelson CG, David A, Aue G, Burke TR Jr, Rader C, Wiestner A (2014) Harnessing the fcmu receptor for potent and selective cytotoxic therapy of chronic lymphocytic leukemia. Cancer Res 74:7510–7520
16. Hatfield DL, Gladyshev VN (2002) How selenium has altered our understanding of the genetic code. Mol Cell Biol 22:3565–3576
17. Kryukov GV, Castellano S, Novoselov SV, Lobanov AV, Zehtab O, Guigo R, Gladyshev

VN (2003) Characterization of mammalian selenoproteomes. Science 300:1439–1443

18. Labunskyy VM, Hatfield DL, Gladyshev VN (2014) Selenoproteins: molecular pathways and physiological roles. Physiol Rev 94:739–777
19. Li X, Patterson JT, Sarkar M, Pedzisa L, Kodadek T, Roush WR, Rader C (2015) Site-specific dual antibody conjugation via engineered cysteine and selenocysteine residues. Bioconjug Chem 26:2243–2248
20. Novoselov SV, Lobanov AV, Hua D, Kasaikina MV, Hatfield DL, Gladyshev VN (2007) A highly efficient form of the selenocysteine insertion sequence element in protozoan parasites and its use in mammalian cells. Proc Natl Acad Sci U S A 104:7857–7862
21. Yang J, Rader C (2012) Cloning, expression, and purification of monoclonal antibodies in scFv-Fc format. Methods Mol Biol 901: 209–232
22. Pedzisa L, Li X, Rader C, Roush WR (2016) Assessment of reagents for selenocysteine conjugation and the stability of selenocysteine adducts. Org Biomol Chem 14:5141–5147
23. Patterson JT, Asano S, Li X, Rader C, Barbas CF III. (2014) Improving the serum stability of site-specific antibody conjugates with sulfone linkers. Bioconjug Chem 25:1402–1407
24. Carter P, Presta L, Gorman CM, Ridgway JB, Henner D, Wong WL, Rowland AM, Kotts C, Carver ME, Shepard HM (1992) Humanization of an anti-p185HER2 antibody for human cancer therapy. Proc Natl Acad Sci U S A 89:4285–4289

Chapter 9

Solubility Characterization and Imaging of Intrabodies Using GFP-Fusions

Emilie Renaud, Pierre Martineau, and Laurence Guglielmi

Abstract

Ectopically expressed intracellular recombinant antibodies, or intrabodies, are powerful tools to visualize proteins and study their function in fixed or living cells. However, many intrabodies are insoluble and aggregate in the reducing environment of the cytosol. To solve this problem, we describe an approach based on GFP-tagged intrabodies. In this protocol, the GFP is used both as a folding-reporter to select correctly folded intrabodies and as a fluorescent tag to localize the scFv and its associated antigen in eukaryotic cells. Starting from a scFv gene cloned in a retroviral vector, we describe retrovirus production, cell line transduction, and soluble intrabody characterization by microscopy and FACS analysis.

Key words Intrabody, scFv, GFP, Folding, Aggregation, Degradation

1 Introduction

Intrabodies are recombinant antibody fragments expressed in cells. They have been mainly used to study protein function and localization within live cells. Indeed, an intrabody can modulate the activity of its cognate antigen and thus induce a phenotypic change that will give clues about the protein function within the cell [1, 2]. As such, intrabodies are analogous to RNAi but at the protein level. In addition, when fused to the green fluorescent protein (GFP), intrabodies become powerful probes to visualize intracellular targets in live cells [3, 4]. Of interest, intrabodies maintain the high specificity and affinity of the antibody molecule for its antigen, and are thus also able to target a particular protein conformation or posttranslational modification [4].

The most widely used format of intrabodies is the single-chain variable fragment (scFv) since it is the smallest antibody fragment (28 kDa) still containing the full antigen-binding site. ScFvs are

The original version of this chapter was revised. The erratum to this chapter is available at: DOI 10.1007/978-1-4939-6857-2_27

Thomas Tiller (ed.), *Synthetic Antibodies: Methods and Protocols*, Methods in Molecular Biology, vol. 1575, DOI 10.1007/978-1-4939-6857-2_9, © Springer Science+Business Media LLC 2017

made up by joining with a flexible linker the variable domains of the heavy (VH) and the light chain (VL), and are thus encoded by a single gene, contrary to regular antibodies or Fab fragments. This facilitates the cloning of the gene in different vectors for intrabody expression and also avoids dissociation of the two chains within the cell. Such scFvs are now routinely selected by screening antibody libraries using phage-display in *E. coli*. Isolated scFvs are then produced in *E. coli* periplasm or cytoplasm for further in vitro characterization and finally tested, as intrabodies, in an appropriate mammalian cell model.

Despite advances in the development of optimized libraries based on hyperstable frameworks [5–10], the intracellular expression and activity of many scFvs is limited, making the approach still challenging. The reducing environment that pertains in the cytosol and the nucleus of mammalian cells prevents the formation of the two conserved disulphide bridges of the scFv, resulting in a decrease in its stability and folding efficiency. Consequently, only a small proportion of intrabodies is soluble and functional in the cell cytoplasm and, in many cases, intrabodies form insoluble aggregates or are quickly degraded by the proteolytic machinery of the cell.

We describe here an approach to easily identify the most soluble and stable intrabodies by using the GFP as a folding-reporter, as originally proposed in *E. coli* [11]. This allows the visualization of scFv-GFP fusion proteins in live cells, an easy evaluation of intrabody steady-state expression levels, a sensitive detection of protein aggregates, and even in some cases a first evaluation of the intrabody activity in the cell. The approach relies on the fact that during protein synthesis there is a kinetic competition between folding, aggregation, and proteolytic degradation. During protein folding, aggregation-prone folding intermediates will have different fates depending on the protein expression level. If over-expressed in the cytoplasm in mammalian cells, the degradation machinery will be overwhelmed and scFv-GFP will accumulate as insoluble and strongly fluorescent aggregates. Conversely, when expressed at low level, the intrabody and its companion GFP will be efficiently degraded, resulting in weakly fluorescent cells (Fig. 1) [12]. Therefore, an aggregation-prone scFv fused to the GFP can be distinguished by FACS from a soluble one, provided it is expressed at a sufficiently low level to avoid the formation of fluorescent aggregates: the former will give a weak fluorescent signal, whereas expression of the latter will result in strongly fluorescent cells.

To obtain a sufficiently low expression level, we propose to transfer scFv genes by retroviral infection in mammalian cell lines. Using this approach, only a few scFv-GFP genes enter each cell, while transient transfection would result in the transfer of thousands of copies in a single cell. In addition, the intrabody will be stably and evenly expressed by all the cells.

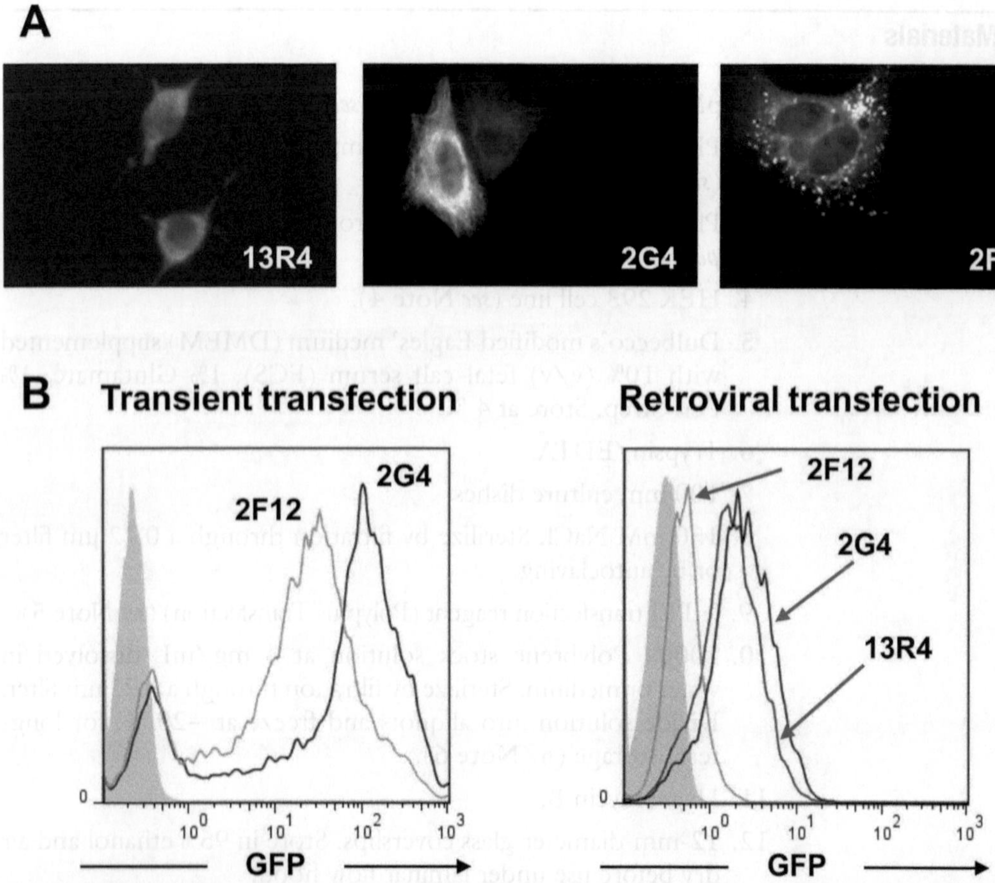

Fig. 1 Cytoplasmic expression of scFvs in HeLa cells. The three scFvs were selected from the same single-framework library [12] and differ only in their CDR3 sequences. Despite their high homology, each scFv has different properties when expressed as an intrabody in the cell cytoplasm and nucleus. Panel A shows the expression pattern of three intrabody-GFP fusions expressed at high levels in HeLa cell cytoplasm (transient transfection using a pCDNA3-derived plasmid). Intrabody 13R4 is highly soluble and does not recognize any cellular protein (13R4); 2G4 is also highly soluble and recognizes alpha-tubulin; 2F12 is aggregation-prone and forms large fluorescent aggregates. Panel B shows FACS analysis of HeLa cells expressing GFP-tagged intrabodies. In the *left panel*, cells were transiently transfected with pCMV-EGFP plasmid containing the indicated scFv, as in panel A. Using this approach, the GFP fluorescence signal does not permit to distinguish soluble and insoluble intrabodies since aggregated proteins are also fluorescent. In the *right panel*, HeLa cells were infected with retrovirus containing 2F12-GFP, 13R4-GFP, and 2G4-GFP fusions. The low expression level of aggregation-prone 2F12 intrabody leads to its degradation by the proteasome and therefore to a low GFP fluorescence signal

In this chapter, we describe procedures to obtain a low intracellular expression of scFv in eukaryotic cells using retroviral transfection. We detail protocols for retrovirus production (Subheading 3.1) and cell line transduction (Subheading 3.2), and then for the monitoring of intrabody expression by microscopy (Subheading 3.3) and FACS (Subheading 3.4) analysis.

2 Materials

1. pMSCV-EGFP plasmid DNA (*see* **Notes 1** and **2**).
2. Plasmid DNA encoding an amphotropic envelop (VSV-G) (*see* **Notes 1** and **3**).
3. Plasmid DNA encoding the retroviral packaging genes *gag* and *pol* (*see* **Note 1**).
4. HEK 293 cell line (*see* **Note 4**).
5. Dulbecco's modified Eagles' medium (DMEM) supplemented with 10% (v/v) fetal calf serum (FCS), 1% Glutamate, 1% Pen-Strep. Store at 4 °C.
6. Trypsin/EDTA.
7. 100 mm culture dishes.
8. 150 mM NaCl. Sterilize by filtration through a 0.22 μm filter or by autoclaving.
9. jetPEI transfection reagent (Polyplus Transfection) (*see* **Note 5**).
10. 1000× Polybrene stock solution at 4 mg/mL dissolved in water or medium. Sterilize by filtration through a 0.22 μm filter. Divide solution into aliquots and freeze at −20 °C for long-term storage (*see* **Note 6**).
11. Hygromycin B.
12. 12-mm-diameter glass coverslips. Store in 95% ethanol and air dry before use under laminar flow hood.
13. Phosphate-buffered saline (PBS): dissolve 8 g NaCl, 0.2 g KCl, 0.61 g Na_2HPO_4, and 0.2 g KH_2HPO_4 in 800 mL distilled water. Adjust volume to 1 L with distilled water. Sterilize by autoclaving. The pH should be about 7.4 and does not need to be adjusted.
14. 2% (w/v) Paraformaldehyde (PFA) diluted in PBS (*see* **Notes 7** and **8**).
15. 0.01% (v/v) Triton X-100 diluted in PBS. Prepare the day of use using a 10% stock solution (*see* **Note 9**).
16. Hoechst 33342 dye solution at 500 μg/mL diluted in water (*see* **Note 10**).
17. Prolong (Invitrogen) (*see* **Note 11**).

3 Methods

3.1 Retrovirus Production

The following protocol assumes that the sequence of the scFv gene has been cloned in a pMSCV retroviral expression vector [7] in frame with the gene encoding the enhanced GFP (Fig. 2). We describe here a protocol of retrovirus production by transient transfection

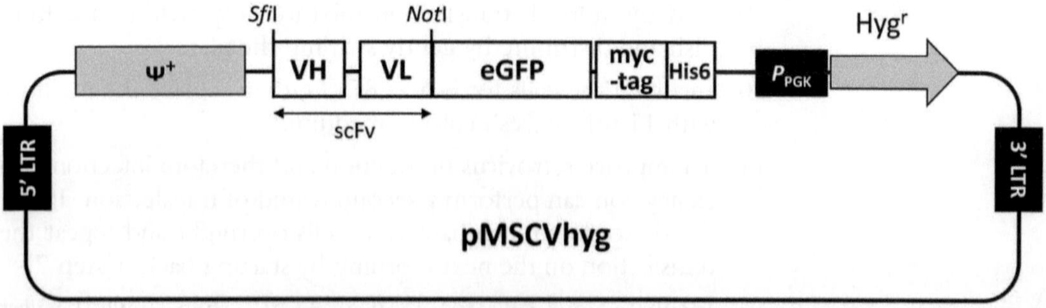

Fig. 2 Retroviral pMSCVhygSN-EGFP plasmid. The pMSCVhygSN-EGFP plasmid is derived from the pMSCVhyg plasmid (Clontech) by insertion of the eGFP, a myc-tag and a his-tag sequence successively cloned in frame. The vector contains: (1) the retroviral packaging signal, Ψ^{+}, which promotes high-titer virus production, (2) the hygromycin resistance gene, (3) the long terminal repeat (LTR) promoter that drives high level and constitutive expression of the scFv gene. *SfiI* and *NotI* unique cloning sites are compatible with most phage-display vectors allowing direct cloning of the selected scFvs

in the human HEK 293 cell line (*see* **Note 12**). Cells are grown in DMEM supplemented with 10% heat-inactivated FCS at 37 °C in a 5% CO_2 humidified atmosphere.

1. Aspirate and discard the culture medium and rinse the cells with trypsin/EDTA.
2. Add enough trypsin/EDTA to cover the cell monolayer and incubate at 37 °C for approximately 3 min until cells become round and start to float (*see* **Note 13**).
3. Neutralize trypsin with a twofold excess of supplemented culture medium.
4. Centrifuge the cells at 300 × *g* and resuspend the pellet in fresh medium.
5. Count the cells and plate 2 × 10^6 cells in a 100 mm diameter culture dish (*see* **Notes 14** and **15**).
6. Incubate the cells at 37 °C and let them grow until the next day.
7. On the day of transfection, replace the culture medium with 5 mL of fresh DMEM with 10% FCS.
8. For each dish of cells that has to be transfected, set up two tubes. In the first tube, dilute 5 μg of the three DNA plasmids into 250 μL of 150 mM NaCl as follows: 1 μg of plasmid DNA encoding *gag* and *pol* genes, 1 μg of plasmid DNA encoding the envelope, and 3 μg of pMSCV plasmid DNA containing the scFv-GFP fusion. In the second tube, mix 15 μL of jetPEI solution with 250 μL of 150 mM NaCl solution. Vortex briefly both tubes then add 250 μL of the jetPEI solution (second tube) to the first tube containing the 250 μL of DNA solution. Vortex 15 s, and incubate for 20 min at room temperature (*see* **Note 14**).

9. Add the 500 μL transfection mixture drop-wise to a culture dish and distribute by gently swirling the plate.
10. Incubate the cells for 5 h at 37 °C, then replace the medium with 11 mL of fresh culture medium.
11. To improve retrovirus production and therefore infection efficiency you can perform a second round of transfection. If you decide to do so, incubate your cells overnight and repeat the transfection on the next morning by starting back at **step 7**.
12. Incubate your cells for 48 h at 37 °C, then slowly harvest supernatant containing retroviral particles, paying attention not to detach 293 cells.
13. Filter the supernatant through a 0.4 μm sterile filter. The supernatant is now ready to use for the infection of the recipient cell line (*see* **Notes 16** and **17**).

3.2 Cell Line Transduction and Selection for Stable Integration

The following method describes transduction of adherent cell lines, such as HeLa or MCF7.

1. The day prior to infection, plate 1 million of your target cells in a 100 mm dish and incubate them at 37 °C (*see* **Note 18**).
2. The next day, remove medium, then add 5 mL of fresh medium, 1 mL of viral supernatant (*see* **Note 19**) and Polybrene at 4 μg/mL final concentration.
3. Incubate overnight at 37 °C.
4. Change the medium of the infected cells and grow cells for an additional 24 h before replacing the culture medium with fresh medium supplemented with hygromycin B (Invitrogen) as selecting agent (*see* **Notes 20–22**).

3.3 Microscopy Analysis

1. Using forceps, deposit one 12-mm-diameter glass coverslip per well in a 24-well plate.
2. Plate 25,000 cells/well.
3. Twenty-four to 48 h later, wash the cells once with PBS.
4. Cover the cells with 300 μL PBS-2% PFA to fix them. Incubate for 10 min at room temperature (*see* **Note 23**).
5. Wash the cells three times with PBS.
6. Cover the cells with 300 μL of PBS-0.01% Triton X-100 to permeabilize the cells.
7. Incubate 5 min at room temperature.
8. Wash the cells three times with PBS.
9. To visualize DNA, add Hoechst dye to a final concentration of 5 μg/mL in PBS.
10. Incubate 5 min at room temperature in the dark (*see* **Note 24**).

11. Wash the cells three times with PBS.
12. Place 12 μL of Prolong mounting medium on the surface of a clean glass microscope slide. Recover a coverslip with a forceps and place it cell-side down onto the Prolong medium. Leave the slide overnight at room temperature before examining the cells under a fluorescent microscope (*see* **Note 25**).

3.4 FACS Analysis of Intrabody Expression Level

1. Harvest the transduced and non-transduced (negative control) cells, as described above. Count and transfer 5×10^5 cells per sample in a FACS tube. Pellet the cells by centrifugation for 5 min at $300 \times g$.
2. Discard supernatant and resuspend the pellet in 0.5 mL of PBS, and analyze immediately at a suitable flow cytometer equipped with the appropriate laser and filter settings or store the cells at 4 °C in the dark.

4 Notes

1. All the vectors used in this protocol were purchased from Clontech. Other manufacturers such as Cell Biolabs also provide retrovirus expression systems.
2. The pMSCV vector is used for the retroviral expression of a scFv of interest under the control of an LTR constitutive promoter. Figure 2 describes the pMSCVhygSN-EGFP plasmid for the expression of the scFv as an N-terminal fusion with the enhanced GFP (eGFP), a c-myc, and a His6 tag [7].
3. Depending on the cell line you want to transduce, choose the appropriate retroviral envelope (ecotropic, amphotropic, pantropic). The VSV-G envelope is used for pantropic virus production that allows all mammalian species cell lines transduction.
4. Stably transfected 293 packaging cell lines expressing Gag and Pol proteins are available at Clontech. They also provide several retroviral packaging cell lines each expressing, in addition to Gag and Pol, a different envelope to make ecotropic, amphotropic, or pantropic retrovirus.
5. The HEK 293 cell line is easy to transfect. You may choose other transfecting reagents such as FuGENE (Promega) or LipofectAMINE (Invitrogen). Refer to the manufacturer's instructions for experimental conditions. The 293 cells can also be efficiently transfected using calcium phosphate.
6. Frozen aliquots are stable for up to 1 year. The working stock can be stored at 4 °C for up to 2 weeks.
7. PFA is highly toxic by inhalation. Manipulate under a chemical hood.

8. Concentrated (16% or 32%) stable PFA aqueous solution is available. You can also prepare the PBS-2% PFA solution: Dissolve 2 g of PFA powder in 90 mL of water heated to 60 °C, add 100 μL of a 2 M NaOH solution, stir until the solution becomes clear, complete with 10 mL of 10× PBS, adjust the pH to 7.4, aliquot and store at −20 °C.
9. Undiluted Triton X-100 is a viscous fluid. Prepare a 10% stock solution by adding 1 mL of Triton X-100 to 9 mL of distilled water. Mix at room temperature on a rotating wheel until the detergent is completely dissolved. Store at 4 °C.
10. Prepare a Hoechst 33342 dye solution by dissolving the powder in water to a final concentration of 50 mg/mL. Do not dilute the dye in PBS as it will precipitate at this concentration. Store aliquots at −20 °C.
11. Alternatively, you can prepare Mowiol coverslip mounting solution: add 6 g of glycerol and 2.4 g of Mowiol 4-88 to 6 mL of distilled water and leave at room temperature for 2 h. Add 12 mL of 0.2 M Tris (pH 8.5) and incubate at ~55 °C until Mowiol has dissolved. Clarify by centrifugation at 5000 × *g* for 15 min. Store in 1 mL aliquots at −20 °C.
12. Retrovirus manipulation requires Biosafety Level 2 (BL2) or BL3 physical containment. Check with your local authority before starting your experiment.
13. HEK 293 cells attach weakly to the culture flask or dishes. Cells detach easily by just flushing the dishes. Here, the trypsinization step is important to get single cell suspension and reproducible experiments.
14. Prepare as many dishes as necessary, knowing that, in this protocol, around 10 mL of supernatant containing retrovirus is produced per 100 mm dish, and 1 mL of supernatant will permit to transduce one million of cells. To scale up, you can also transfect 293 cells in 150 mm dishes.
15. To improve attachment of 293 cells to the plastic, you can coat the culture dishes with poly-D-lysine. Cover the 100 mm dishes with a sterile 5 μg/mL poly-D-Lysine solution diluted in distilled water. Incubate the plates for 2 h at room temperature. Wash three times with PBS before plating your cells.
16. This step is important to get rid of the 293 cells present in the supernatant.
17. For a better efficiency, perform infection immediately with fresh supernatant containing retroviral particles. VSV-G-pseudotyped retrovirus are however stable at −80 °C and aliquots can be kept frozen and used for at least 3 months.
18. Retrovirus only infects dividing cells. Thus, make sure your cells grow well during the infection by adjusting the number of seeded cells.

19. In this protocol, we do not concentrate retroviral particles before infection. All cell lines tested so far in the group (e.g., HeLa, MCF7, HCT116, SW48, DLD-1, HepG2) are efficiently transduced using these experimental conditions. However, in this experiment, you will not control the number of insertion events per cell. For making libraries you may need to determine the viral titer to use a precise MOI during the infection. Using the procedure described here, we routinely obtain a viral titer of about 1–5 × 10^7 cfu/mL. Many protocols are available to measure your viral titer, for instance in the retroviral gene transfer and expression user manual of Clontech.
20. You should start to see the scFv-GFP fusion protein expression 6–24 h after cell transduction. If you wish to improve transduction efficiency by increasing the gene copy number per cell, you can perform a second infection 24 h after the first one.
21. Because each cell line has its own sensitivity to a given antibiotic, you have to determine, before the transduction step, the optimal concentration of hygromycin B necessary to kill the uninfected cells. For most cell lines, the optimal concentration will lie between 50 and 1000 μg/mL. For instance, optimal concentration is 100 μg/mL for HCT116, and 500 μg/mL for HeLa and MCF7. When performing the retroviral transduction, do not forget to provide a control dish of uninfected target cells as a positive control of the antibiotic selection efficacy. Hygromycin B-induced killing is slow and may take 10–14 days.
22. To guarantee a stable expression of the scFv-GFP protein and avoid silencing, always maintain the antibiotic selective pressure in your cell cultures.
23. Paraformaldehyde cross-links proteins and permits detection of soluble proteins. Alternatively, you can use methanol that fixes the cells by dehydration. Methanol precipitates proteins and permeabilizes the cells at the same time, so that soluble proteins from the cytoplasm may be extracted. Use therefore methanol fixation for nuclear and cytoskeletal protein detection. Immerse coverslips in ice-cold absolute methanol (about 1 mL), then incubate for 10 min at −20 °C. Gradually rehydrate the cells by adding PBS in aliquot of about 500 μL to the methanol covering your coverslips. Finally wash cells once in PBS.
24. To prepare a 5 μg/mL solution of Hoechst in PBS, perform a 100-fold intermediate dilution of the stock solution in water. This will avoid precipitation of the dye in PBS.
25. According to the manufacturer's instructions, long-term storage is possible by keeping your slide in the dark at 4 or −20 °C.

References

1. Lobato MN, Rabbitts TH (2003) Intracellular antibodies and challenges facing their use as therapeutic agents. Trends Mol Med 9:390–396
2. Visintin M, Meli GA, Cannistraci I, Cattaneo A (2004) Intracellular antibodies for proteomics. J Immunol Methods 290(1-2):135–153
3. Rinaldi AS, Freund G, Desplancq D, Sibler AP, Baltzinger M, Rochel N, Mély Y, Didier P, Weiss E (2013) The use of fluorescent intrabodies to detect endogenous gankyrin in living cancer cells. Exp Cell Res 319(6):838–849
4. Cassimeris L, Guglielmi L, Denis V, Larroque C, Martineau P (2013) Specific In Vivo Labeling of Tyrosinated alpha-Tubulin and Measurement of Microtubule Dynamics Using a GFP Tagged, Cytoplasmically Expressed Recombinant Antibody. PLoS One 8:e59812
5. Philibert P, Stoessel A, Wang W, Sibler A, Bec N, Larroque C, Saven JG, Courtete J, Weiss E, Martineau P (2007) A focused antibody library for selecting scFvs expressed at high levels in the cytoplasm. BMC Biotechnol 7:81
6. Robin G, Sato Y, Desplancq D, Rochel N, Weiss E, Martineau P (2014) Restricted diversity of antigen binding residues of antibodies revealed by computational alanine scanning of 227 antibody-antigen complexes. J Mol Biol 426:3729–3743
7. Freund G, Desplancq D, Stoessel A, Weinsanto R, Sibler AP, Robin G, Martineau P, Didier P, Wagner J, Weiss E (2014) Generation of an intrabody-based reagent suitable for imaging endogenous proliferating cell nuclear antigen in living cancer cells. J Mol Recognit 27: 549–558
8. Auf der Maur A, Zahnd C, Fischer F, Spinelli S, Honegger A, Cambillau C, Escher D, Pluckthun A, Barberis A (2002) Direct in vivo screening of intrabody libraries constructed on a highly stable single-chain framework. J Biol Chem 277:45075–45085
9. Tanaka T, Chung GTY, Forster A, Lobato MN, Rabbitts TH (2003) De novo production of diverse intracellular antibody libraries. Nucleic Acids Res 31:e23
10. Villani ME, Carli MD, Donini M, Traversini G, Lico C, Franconi R, Benvenuto E, Desiderio A (2008) Validation of a stable recombinant antibodies repertoire for the direct selection of functional intracellular reagents. J Immunol Methods 329:11–20
11. Waldo GS, Standish BM, Berendzen J, Terwilliger TC (1999) Rapid protein-folding assay using green fluorescent protein. Nat Biotechnol 17(7):691–695
12. Guglielmi L, Denis V, Vezzio-Vie N, Bec N, Dariavach P, Larroque C, Martineau P (2011) Selection for intrabody solubility in mammalian cells using GFP fusions. Protein Eng Des Sel 24:873–881

Chapter 10

Antibody Validation by Immunoprecipitation Followed by Mass Spectrometry Analysis

Helena Persson, Charlotta Preger, Edyta Marcon, Johan Lengqvist, and Susanne Gräslund

Abstract

We describe a mass spectrometry-based approach for validation of antibody specificity. This method allows validation of antibodies or antibody fragments, against their endogenous targets. It can assess if the antibody is able to bind to its native antigen in cell lysates among thousands of other proteins, DNA, RNA, and other cellular components. In addition, it identifies other proteins the antibody is able to immunoprecipitate allowing for the assessment of antibody specificity and selectivity. This method is easily scalable, adaptable to different cell lines and conditions and has been shown to be reproducible between multiple laboratories.

Key words Immunoprecipitation, Antibody fragments, Mass spectrometry, Antibody validation

1 Introduction

When generating antibodies researchers are, in most cases, dependent on the use of recombinant proteins or synthetic peptides as antigens. This holds for antibodies produced by both in vitro selection and immunization-based methods. Also, most binding characterization assays are commonly performed using recombinant antigens. Nevertheless, the intended end-use of these antibodies, either as research tools in the laboratory or as diagnostic or therapeutic tools in the clinic, relies on the interaction with endogenous antigen as found in its natural environment.

This presents several challenges. Recombinant antigens are usually truncated versions of their endogenous counterparts that may or may not be folded correctly. As a consequence, the required epitope might not be accessible to the antibody in the native antigen. In addition, in its natural environment the antigen is surrounded by thousands of other proteins, DNA and RNA molecules, and other cellular components that may interfere with the epitope presentation and antibody-antigen binding.

Thomas Tiller (ed.), *Synthetic Antibodies: Methods and Protocols*, Methods in Molecular Biology, vol. 1575,
DOI 10.1007/978-1-4939-6857-2_10,

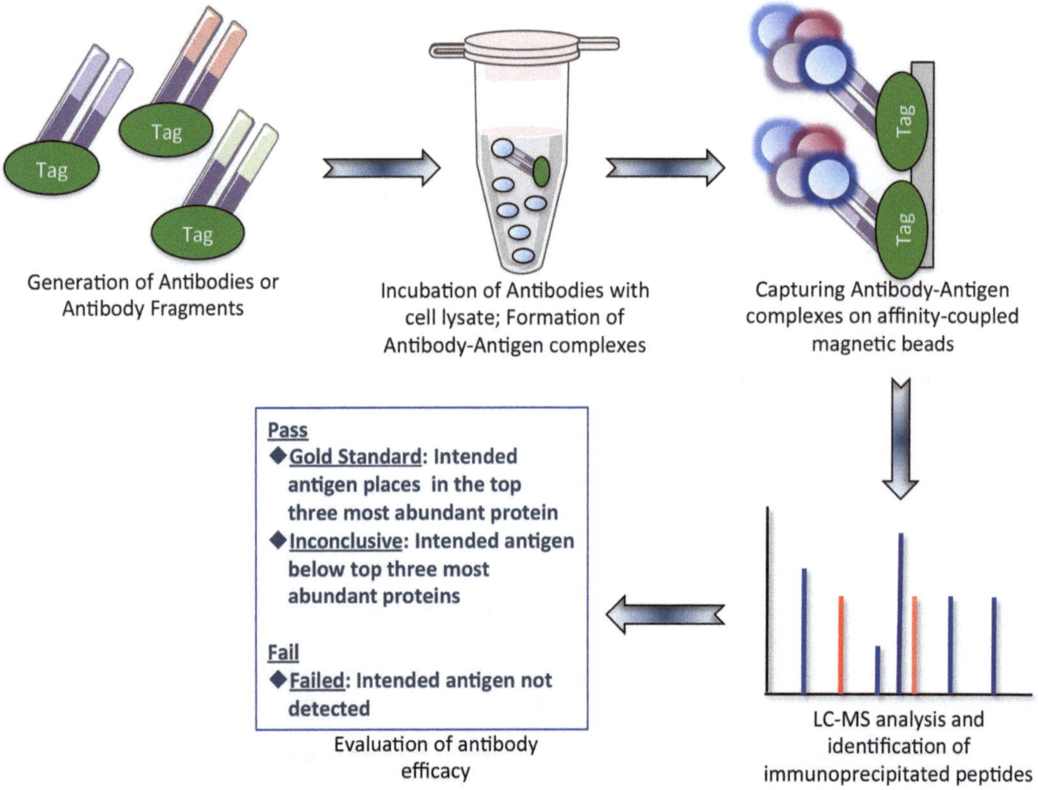

Fig. 1 Mass spectrometry-based validation of antibodies. After incubation of antibodies or antibody fragments and whole cell lysates, antibody-antigen complexes are captured on an affinity matrix appropriate for antibody isotype or tag present on the antibody fragment. After several washes, antibody-antigen complexes are released from the matrix, trypsin digested and subjected to mass spectrometry analysis where proteins bound to the antibody are identified. After background removal, the remaining proteins are ranked based on abundance and the antibody is evaluated based on the relative abundance of its cognate antigen within all the identified proteins. Parts of the figure were modified from the power point image bank (www.servier.com/powerpoint-image-bank)

Here, we describe the use of immunoprecipitation (IP) followed by mass spectrometry (MS) readout for validation of antibody binding to its cognate endogenous antigen as found in its natural environment. The basic principle of the method is outlined in Fig. 1. The antibody or antibody fragment that is to be validated is incubated with a cell-lysate comprising the endogenously expressed cognate antigen. The antibody-antigen complex is then captured on affinity-coupled magnetic beads. The use of a magnet allows efficient enrichment of captured protein through washing and removal of unbound background components. Finally, the antibody-antigen complexes are eluted from the beads, trypsin digested and subjected to liquid chromatography (LC)-MS analysis, whereby the proteins bound to the antibody of interest are detected. The identity and abundance of each protein bound are determined based on the number of peptides recovered and can be used for antibody evaluation in terms of specificity and selectivity (Fig. 1).

To date, we have used the described IP-MS workflow to validate the specificity of thousands of unique antibodies, both full-length IgG and different antibody fragment formats [1]. Although the method has been developed for the analysis of antibody reagents, it can also be used to validate affinity reagents of other scaffold types. The method described here is attractive for several reasons: (1) it is easily adaptable to different cell lines (*see* **Note 4**); (2) it can be used across laboratories with little variation [1]; and (3) it can be automated, making it a fairly high-throughput method. It has been shown that antibodies validated through this method are very likely to also perform well in other applications where the aim is to detect a natively folded antigen, such as immunofluorescence (IF) and chromatin immunoprecipitation (ChIP), making IP-MS the method of choice for general antibody validation. The method can be used to compare antibodies between manufacturers and when the data is put into the public domain it can increase reproducibility of data within the scientific community. An additional advantage of this method, although not described in detail here, is that it also allows the identification of interacting protein partners and can provide functional predictions of antibody's target (Fig. 2) [3].

2 Materials

2.1 Cell Culture

1. FreeStyle 293-F cells.
2. FreeStyle 293 Expression medium.
3. Dulbecco's Phosphate Buffer Saline (DPBS), without calcium and magnesium.

2.2 Immunoprecipitation

1. Magnetic beads: For IPs using biotinylated antibody fragments, streptavidin beads are used, for IPs with FLAG-tagged antibody fragments, FLAG M2 magnetic beads are used, and for IPs using full-length IgG protein A/G beads are used (*see* **Note 1**).
2. Magnetic separator for microcentrifuge tubes.
3. Benzonase nuclease.
4. Protease inhibitor.
5. Stock solutions of 2 M Tris–HCl (pH 7.9), 5 M NaCl and 10% NP-40.
6. AFC Buffer (high salt): 10 mM Tris–HCl, pH 7.9, 420 mM NaCl, 0.1% NP-40. Prepare a 10× solution by mixing 2 mL 2 M Tris–HCl pH 7.9, 33.6 mL 5 M NaCl, 4 mL 10% NP-40,

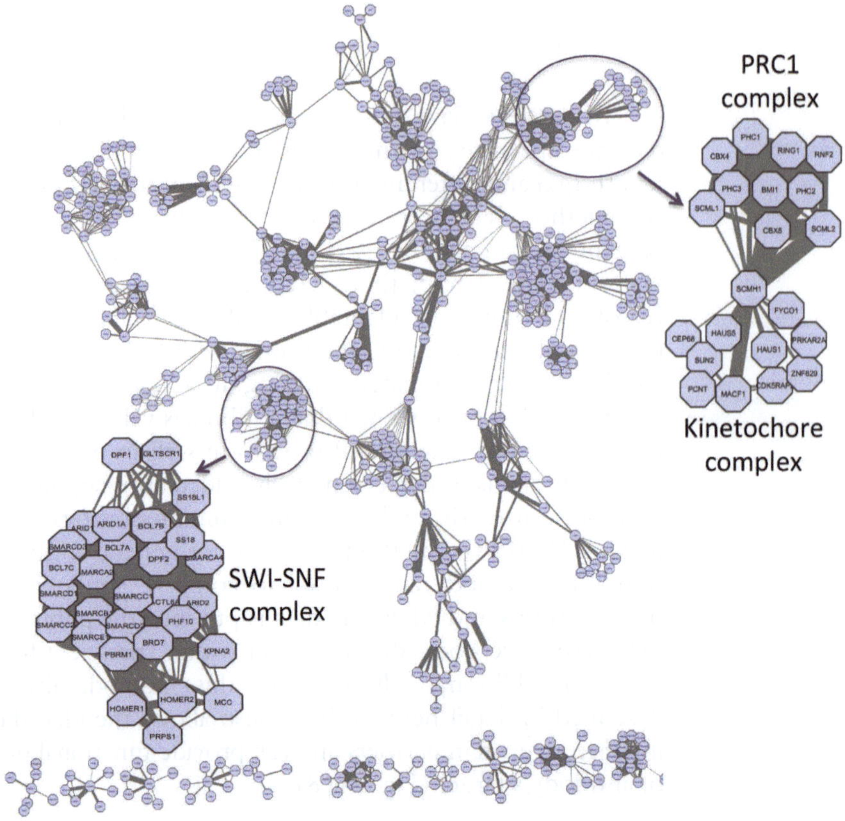

Fig. 2 Mass spectrometry-based validation approach can be used to identify an antigen's interacting proteins and to generate network diagrams. The network that is shown here was generated using Cytoscape [2]. The data came from a recent published paper where antibodies against 98 different chromatin related proteins were validated [3]. The edges represent the strength of the interaction based on the modified hypergeometric spectral counts (HGS) score [4]. The wider the edge, the stronger the interaction score. Many different complexes can be distinguished and two (SWI-SNF complex and PRC complex) are shown as larger diagrams. There are also connections between many of the complexes indicating that some proteins are part of multiple complexes. These kinds of analyses of protein-protein interactions and network diagrams can lead to predictions as to the function of novel, yet uncharacterized proteins.

and 0.4 mL water. The day of the experiment, dilute the buffer to a 1× solution (10 mM Tris–HCl, pH 7.9, 420 mM NaCl, 0.1% NP-40) and add protease inhibitor.

7. AFC Buffer (low salt): 10 mM Tris–HCl, pH 7.9, 100 mM NaCl, 0.1% NP-40. Prepare a 10× solution by mixing 2 mL 2 M Tris–HCl pH 7.9, 8 mL 5 M NaCl, 4 mL 10% NP-40, and 26 mL water. The day of the experiment, dilute the buffer to a 1× solution (10 mM Tris–HCl, pH 7.9, 100 mM NaCl, 0.1% NP-40) and add protease inhibitor.

8. AFC Buffer (low salt) without detergent: 10 mM Tris–HCl, pH 7.9, 100 mM NaCl. Prepare a 10× solution by mixing 2 mL 2 M Tris–HCl pH 7.9, 8 mL 5 M NaCl, 4 mL, and 30 mL

water. The day of the experiment, dilute the buffer to a 1× solution (10 mM Tris–HCl, pH 7.9, 100 mM NaCl) and add protease inhibitor.

9. Elution buffer: 0.5 M ammonium hydroxide.

2.3 Trypsin Digestion, Desalting, and LC-MS Analysis

1. Stock solutions of 1 M ammonium bicarbonate and 1 M Tris (2-carboxyethyl) phosphine hydrochloride (TCEP-HCl) in water.
2. Working solutions of 50 mM ammonium bicarbonate and 100 mM TCEP-HCl (*see* **Note 2**).
3. 500 mM iodoacetamide solution (*see* **Note 3**).
4. Mass spectrometry/sequencing grade trypsin.
5. ZipTip (Millipore).
6. Wetting and Equilibration solution (ZipTip): 70% acetonitrile with 0.1% formic acid.
7. Washing solution (ZipTip): 0.1% formic acid in water.
8. LC-MS sample plate or injection vials.
9. HiPPR™ Detergent Removal Spin Column Kit (ThermoFisher Scientific).
10. 2 mL microcentrifuge collection tubes.

2.4 LC-MS Data Acquisition

1. NanoLC-MS system (e.g., Q Exactive, Thermo Scientific).
2. NanoLC-MS analytical column (75 μm × 50 cm, P/N ES803, Thermo Scientific).
3. LC-solvents: A: 2% acetonitrile, 0.1% formic acid, B: 90% acetonitrile, 0.1% formic acid.

2.5 Analysis of IP-MS Data

1. Download and install the MaxQuant software package (version MaxQuant 1.5.3.30 or higher). Download and instructions can be found here (http://www.coxdocs.org/doku.php?id=maxquant:start).

3 Methods

Use non-autoclaved tubes and pipette tips, in all method sections except for "Subheading 3.1," as autoclaving might release chemicals from the plastics that may interfere with MS analysis.

3.1 Cell Culture

In this protocol we describe the use of HEK293 suspension cells (*see* **Note 4**), but it is also possible to use adherent HEK293 cells (*see* **Note 5**).

1. Split FreeStyle 293-F cells to 0.4×10^6 mL^{-1} and culture in 300 mL FreeStyle 293 Expression medium in 1 L Erlenmeyer

Flasks at 37 °C, 70% relative humidity, 5% CO_2, 125 rpm in a shaking incubator. Culture cells for 72 h until reaching a cell density of about 2.5×10^6 cells/mL.

2. Harvest 600 mL cells (2.5×10^6/mL) by centrifuging at 600 × *g*, 4 °C for 15 min. Subsequently, resuspend pellet in a total volume of 150 mL cold DPBS.
3. Aliquot the cells to 5 mL and transfer to 15 mL tubes before centrifuging at 600 × *g*, 4 °C for 15 min. Discard the supernatant and store the cell pellet (50×10^6/tube) in −80 °C freezer until used in IP-MS.
4. 50×10^6 cells are enough for ten captures, when resuspending pellet in 5 mL lysis buffer and using 0.5 mL per sample.

3.2 Preparing Cell Lysate

Carry out all procedures on ice or at 4 °C unless indicated otherwise.

1. Thaw frozen HEK293 cell pellet (50×10^6 cells) with 5 mL 1× AFC Buffer (high salt).
2. Freeze-thaw for three cycles by transferring the sample between ethanol/dry ice and 37 °C water bath. To prevent the samples from reaching above 4 °C, mix frequently.
3. Sonicate 12 times 1 s on/2 s off with 50% amplitude.
4. Add benzonase nuclease (to a final concentration of 25 units/mL) to remove DNA and RNA, and incubate for 30 min at 4 °C.
5. Centrifuge at 13,000 rpm for 30 min in a microfuge at 4 °C to remove cell debris. Discard the pellet and retain supernatant.
6. Measure the protein concentration with absorbance at 280 nm using 1 Abs = 1 mg/mL. The protein concentration usually is ~5 mg/mL.

3.3 Antibody Binding

Carry out all procedures on ice or at 4 °C unless indicated otherwise.

1. Use 500 μL clarified lysate per IP and add ~2 μg antibody or antibody fragment. Incubate the samples overnight at 4 °C with gentle rotation (*see* **Note 6**).
2. Prewash 20 μL magnetic beads per sample (*see* **Note** 7) with 1 mL high salt AFC buffer. After last wash, resuspend the beads with 20 μL low salt AFC buffer per sample. Add 20 μL beads to the lysate with antibodies, and incubate for 2 h at 4 °C with gentle rotation (*see* **Note 8**).
3. Wash the beads three times for 5 min with 1 mL low salt AFC buffer following two times for 5 min with 1 mL low salt AFC buffer without detergent.
4. Elute the immunoprecipitated proteins with 50 μL 0.5 M ammonium hydroxide four times (a total volume of 200 μL). Flash freeze in liquid nitrogen and store samples at −80 °C until trypsin digestion.

3.4 Trypsin Digestion

Use high purity MQ-grade, HPLC grade water, or equivalent in all the following steps of the protocol (tryptic digestion, ZipTip cleanup, and detergent removal).

1. Evaporate samples to dryness using a rotary evaporator.
2. Resuspend by adding 44 μL 50 mM ammonium bicarbonate (*see* **Notes 9** and **10**).
3. Add 1 μL 100 mM TCEP-HCl solution and incubate at 37 °C for 1 h on a shaker (500 rpm).
4. Cool samples to room temperature and add 1 μL 500 mM iodoacetamide solution.
5. Incubate at room temperature without light for 45 min on a shaker.
6. Add 1 μg of trypsin and incubate at 37 °C overnight on a shaker.
7. Continue with the detergent-removal protocol or continue with **step 8** below (if not including the detergent removal step).
8. Add 2 μL acetic acid to stop the digestion reaction.
9. Store at 4 °C until performing the ZipTip cleanup.

3.5 Detergent Removal Step

1. For a sample volume of 50 μL, add 100 μL of detergent removal resin slurry (50% slurry) to an empty spin column.
2. Remove the bottom cap and place the spin column in a 1.5 mL microcentrifuge tube, then spin for 1 min at 1500 × *g* to remove the storage buffer.
3. Add 50 μL equilibration buffer to the spin column, spin 1 min at 1500 × *g*, and discard flow though.
4. Repeat **step 3** twice.
5. Fit the bottom cap back on to the spin column to prevent draining the sample, then add the sample to the equilibrated detergent removal column (*see* **Note 11**), and incubate for 10 min at room temperature.
6. Remove the bottom cap and place the spin column into a clean collection tube.
7. Spin for 2 min 1500 × *g* to collect the sample, then continue with the ZipTip desalting protocol.

3.6 Reversed Phase (ZipTip) Desalting for LC-MS Analysis

1. If the sample has been processed using the detergent removal protocol, start by acidifying the samples by adding 2 μL acetic acid.
2. Aspirate 10 μL Wetting and Equilibration solution into ZipTip and dispense to waste (*see* **Note 12**).
3. Repeat **step 2** twice.

4. Aspirate 10 μL of Washing solution and dispense to waste.
5. Repeat **step 4** twice.
6. Aspirate the sample and dispense back into the same (sample) tube. Pipette up and down like this for a minimum of 20 times to ensure peptide binding to the tip.
7. Aspirate 10 μL of Washing solution and dispense to waste.
8. Repeat **step** 7 once.
9. Aspirate 10 μL Wetting and Equilibration solution and dispense into a collection (LC-MS sample) plate or vial (*see* **Note 13**).
10. Repeat **step 9** three times until total eluate volume is 40 μL.

Evaporate samples to dryness using a rotary evaporator. Store the plate frozen until LC-MS analysis.

3.7 LC-MS Data Collection

1. Standard reverse phase micro-capillary column and analytical columns can be used. Our previous analysis [1] indicated that the data is very similar between different laboratories when the same antibodies and the same lysate are used with slightly different machines and columns employed for the analysis. However, separation time is important and should be no less than 90 min at a flow rate of 250–300 nL/min to recover maximum number of peptides according to our experience. Eluted peptides are sprayed directly into an Orbitrap Velos or a Q Exactive mass spectrometer (Thermo Scientific) (*see* **Note 14**).
2. Reconstitute the sample in 15 μL of 3% acetonitrile, 0.1% formic acid.
3. Acquire data using a Q Exactive mass spectrometer coupled with a Dionex RSLC nanoLC-system (both from Thermo Scientific). Use the following LC-gradient (running at 250 nL/min), starting at 3% B-buffer, 5 min (3% B), 8 min (3% B), 95 min (40% B).
4. Set the MS data acquisition parameters as follows for the Q Exactive-instrument: MS-scan; (resolution 70 K, AGC 1e6, max injection time 100 ms), MS/MS scans; (resolution 17 K, AGC 2e5, max injection time 500 ms); Normalized-collision energy (NCE): 30; Dynamic precursor exclusion 30 s.

3.8 Protein Identification and Relative Quantification from IP-MS Data (for Thermo, Sciex, Agilent, or Bruker Instrument Data)

1. Download and install MaxQuant [5] from http://www.coxdocs.org/doku.php?id=maxquant:common:download_and_installation. Please note that you will also need to install.NET Framework 4.5 and MSFileReader.
2. Familiarize yourself with running the software as described here: http://www.coxdocs.org/doku.php?id=maxquant:manual:beginner.

3. Download the proper protein database file (in .FASTA-format) for your experiment. So-called reference proteome collections from Uniprot are a good starting point (http://www.uniprot.org/proteomes/).
4. Add the FASTA file to MaxQuant as described here: http://www.coxdocs.org/doku.php?id=maxquant:andromeda:configuration:protdatabases.
5. Load the raw files you want to analyze. MaxQuant will place the result files it creates in the same folder that you loaded the raw data files from (*see* **Note 15**).
6. Select the database (.FASTA file) that you want to use for the search.
7. Under the "Raw files" tab, select "No fractions."
8. To use the normalized spectral abundance factor (NSAF) [6] approach for protein quantification as described here, there are no other settings that you need to change from the default settings.
9. Start the search. Trouble shooting advice can be found here: http://www.coxdocs.org/doku.php?id=maxquant:viewer:troubleshooting.

3.9 Protein List Cleanup and Calculation of NSAF Values

1. When the search is done, locate the "proteinGroups.txt" file created by MaxQuant. It is located in the "txt" folder in the "combined" folder.
2. Open the list in Excel (*see* **Note 16**).
3. Remove all proteins (rows) that are flagged as "Reverse" and "Potential contaminant" (indicated by a "+"). These are hits in the reversed database (only used to determine the false-discovery rate in the identification step) and proteins that are commonly observed as contaminants in LC-MS analysis (e.g., human keratins).
4. Calculate the normalized spectrum abundance factor (NSAF) for each protein in each sample using the number of identified spectra (i.e., the number of spectral counts using the "MS/MS Count" value). The NSAF-value is calculated using the following formula:

$$\text{spectral counts} \,/ \sum \text{sample spectral counts} \times \bar{x}\ \text{sample spectral counts}_{(\text{batch or study})}$$

To correct for nonspecific binding, background proteins should be removed from the list. If many different and unrelated antibodies are included in the same batch, most proteins that appear in all immunocaptures can be considered background. The most basic form of background correction is to remove all proteins that are detected (i.e., have NSAF >0) in every sample.

If a negative control sample (i.e., with no antibody added) is included in the batch, the detected proteins in that sample can be used to define the nonspecific bead background. Further, existing published background lists may also be used to remove common nonspecific binders [1, 7]. However, one has to be aware that highly abundant proteins (and proteins present in multiple complexes) are often seen in many unrelated captures. Thus, it is important to examine the relative abundance of a particular target protein in all samples (*see* **Note 17**) (Fig. 3). For example, if a protein has 2–5 spectral counts in most (or all) captures but 15 in one of them, it is likely a true target for that particular antibody. In our experience, abundance of three times over background can generally be considered specific.

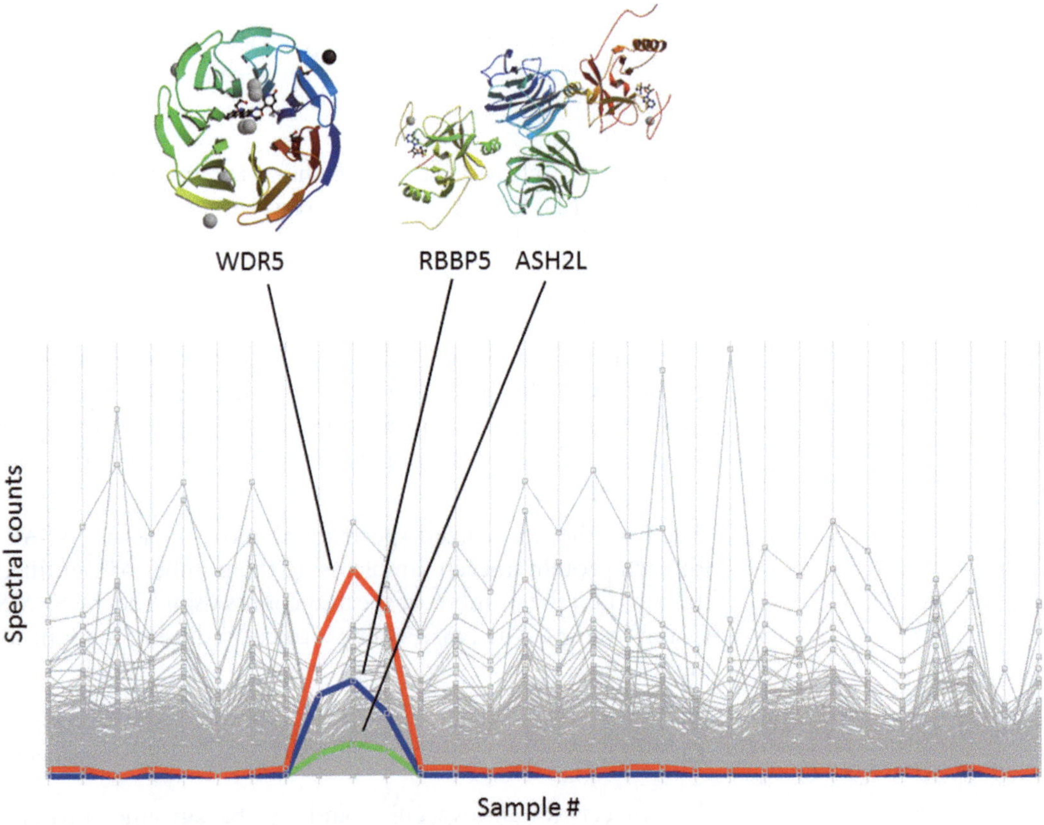

Fig. 3 Protein abundance (spectral counts) plotting and profile correlation can be used to discern target proteins from background and to detect interacting proteins. Using the Perseus software, proteins showing high correlation to the abundance profile of WDR5 (across a batch of IP-MS samples) were detected. The three proteins WDR5, RBBP5, and ASH2L constitute the core components of the H3K4 histone methyl transferase complex [8].

5. Sort the proteins detected in each sample based on the NSAF values (highest to lowest). Then check the rank of the target protein compared to the other proteins detected in that sample.
6. If the target protein is detected among the top three proteins (after background removal), the antibody can be considered to be of high quality [1].

4 Notes

1. Depending on the specific tag and isotype subclass of the antibody or antibody fragment other types of magnetic beads can be used.
2. As an alternative to 50 mM ammonium bicarbonate, 50 mM triethylammonium bicarbonate (TEAB) can be used.
3. Prepare iodoacetamide solution just prior to use and protect the solution from light, as iodoacetamide is light sensitive. Alternatively, aliquots of 500 mM iodoacetamide solution can be stored frozen, thawed, and used only once before discarding.
4. We have used HEK293 cells for their quick and easy reproduction and maintenance [9]. In addition, HEK293 cells express a large proportion of the human proteins. However, not all human proteins are expressed and those that are, differ greatly in abundance. Information about expression in various cell lines and human tissues can, for example, be obtained from www.proteomicsdb.org or www.proteinspire.org/MOPED. All of these parameters can affect antibody evaluation. Obviously, if the antibody target is not expressed in the cell line used, the antibody will register as a false negative. Similarly, if the cognate antigen is expressed weakly, the antibody may appear as a false negative. On the other hand, even a low affinity antibody can immunoprecipitate its highly abundant target.

   Looking at the abundance ratios (based on the counts of peptides detected in MS) between proteins bound to the antibody can lessen some of these problems as it helps determine antibody selectivity and presumably antibody efficacy, regardless of the levels of expression for the target protein. Nevertheless, it is important to choose the correct cell lines for antibody validation especially for targets that are not expressed in general cell lines.
5. Adherent cell preparation: Grow stable HEK293 cells to 80–90% confluency in 15 cm plates in DMEM media supplemented with 10% FBS and 1% ampicillin/streptomycin. One 15 cm plate typically generates 10^7–10^8 cells. Harvest five

plates at the time, and freeze cell pellets until needed. Five plates are usually enough for ten IPs.

6. The incubation time can be adjusted depending on the stability of the proteins, 2 h is sufficient for most antibodies.
7. To ensure capture of all antibody-antigen complexes make sure to use a high enough bead volume, taking the bead binding capacity as indicated by the manufacturer into consideration.
8. The incubation step and following washing steps can be automated using, e.g., KingFisher Flex for higher sample throughput.
9. Prepare 50 mM ammonium bicarbonate solution fresh each day.
10. Work in a clean environment to avoid polymer and keratin contamination. Hood is not necessary, but a clean working environment is essential.
11. It is important to add the sample directly to the resin, not to the tube wall.
12. Avoid introducing air into the ZipTip when pipetting up and down.
13. Depending on the injection format used in the nanoLC-MS step, samples may be eluted from the ZipTip directly to injection vials instead of to a plate.
14. Include analysis of at least one blank sample together with each sample batch to be able to subtract background present in the LC-MS system (instrument background) from the identified protein list.
15. If you start a second search from within the same folder, the first result files will be overwritten. To avoid this you can rename the "combined" folder that MaxQuant created to hold the result files from the previous search.
16. The Perseus software distributed together with MaxQuant can also be used instead of Excel.
17. A convenient and informative way of analyzing the data can be to plot the NSAF values for individual (target) proteins across all samples in a batch or in a study.

Acknowledgments

The Structural Genomics Consortium (SGC) is a registered charity (number 1097737) that receives funds from AbbVie, Bayer Pharma AG, Boehringer Ingelheim, Canada Foundation for Innovation, Eshelman Institute for Innovation, Genome Canada, Innovative Medicines Initiative (EU/EFPIA) [ULTRA-DD grant no.

115766], Janssen, Merck & Co., Novartis Pharma AG, Ontario Ministry of Economic Development and Innovation, Pfizer, São Paulo Research Foundation-FAPESP, Takeda, and the Wellcome Trust. We also gratefully acknowledge Leif Dahllund for assistance with cell culture work and Elena Ossipova for sample preparation before MS.

References

1. Marcon E, Jain H, Bhattacharya A, Guo H, Phanse S, Pu S et al (2015) Assessment of a method to characterize antibody selectivity and specificity for use in immunoprecipitation. Nat Methods 12:725–731
2. Shannon P, Markiel A, Ozier O, Baliga NS, Wang JT, Ramage D et al (2003) Cytoscape: a software environment for integrated models of biomolecular interaction networks. Genome Res 13:2498–2504
3. Marcon E, Ni Z, Pu S, Turinsky AL, Trimble SS, Olsen JB et al (2014) Human-chromatin-related protein interactions identify a demethylase complex required for chromosome segregation. Cell Rep 8:297–310
4. Guruharsha KG, Rual JF, Zhai B, Mintseris J, Vaidya P, Vaidya N et al (2011) A protein complex network of *Drosophila melanogaster*. Cell 147:690–703
5. Cox J, Mann M (2008) MaxQuant enables high peptide identification rates, individualized p.p.b.-range mass accuracies and proteome-wide protein quantification. Nat Biotechnol 26:1367–1372
6. Zybailov BL, Florens L, Washburn MP (2007) Quantitative shotgun proteomics using a protease with broad specificity and normalized spectral abundance factors. Mol Biosyst 3:354–360
7. Mellacheruvu D, Wright Z, Couzens AL, Lambert JP, St-Denis NA, Li T et al (2013) The CRAPome: a contaminant repository for affinity purification-mass spectrometry data. Nat Methods 10:730–736
8. Dou Y, Milne TA, Ruthenburg AJ, Lee S, Lee JW, Verdine GL et al (2006) Regulation of MLL1 H3K4 methyltransferase activity by its core components. Nat Struct Mol Biol 13:713–719
9. Thomas P, Smart TG (2005) HEK293 cell line: a vehicle for the expression of recombinant proteins. J Pharmacol Toxicol Methods 51:187–200

Chapter 11

Novel HPLC-Based Screening Method to Assess Developability of Antibody-Like Molecules

Neeraj Kohli and Melissa L. Geddie

Abstract

The discovery of antibodies that bind to targets with high affinity is now a routine exercise. However, it is still challenging to screen for candidates that, in addition to having excellent biological properties, also have optimal biophysical characteristics. Here, we describe a simple HPLC-based screening method to assess for developability factors earlier in the discovery process.

Key words Standup monolayer adsorption chromatography, Developability, Colloidal stability, Zenix column, Size exclusion chromatography

1 Introduction

In the last few decades, due to rapid advances in antibody technologies and development, monoclonal antibodies have emerged as the most promising and fastest growing class of biotherapeutics. They have transformed patient care in disease indications as diverse as oncology, ophthalmology, neurology, and inflammation [1–6]. The discovery of antibodies that bind to particular targets with high affinity is now a routine exercise and a variety of in vitro and in vivo approaches are currently available for this purpose [7–12].

Antibody discovery is typically accompanied by functional profiling that may include target binding assays, cell-based assays, and in vivo models. Traditionally, less attention was paid at this stage to other developability attributes such as stability, homogeneity, expression, solubility, and half-life. As a result, these challenges were identified much later, resulting in the immense investment of resources for establishing appropriate risk mitigation strategies. Sometimes, these challenges were irreducible, leading to the termination of promising programs. Therefore, over the years, companies have evolved their selection strategies, and currently, various developability assessment techniques are now employed to derisk

Thomas Tiller (ed.), *Synthetic Antibodies: Methods and Protocols*, Methods in Molecular Biology, vol. 1575,
DOI 10.1007/978-1-4939-6857-2_11, © Springer Science+Business Media LLC 2017

the molecules much earlier in the discovery process [13–17]. These strategies may include experimental approaches such as affinity capture self-interaction nanoparticle spectroscopy (AC-SINS) [18], cross-interaction chromatography [19], and baculovirus binding [20]. All of these early screening techniques have their own advantages and disadvantages, but one challenge common to all is reproducibility or batch-to-batch variation. Screening later in the lead identification process typically consists of short-term stability studies at various temperatures, freeze-thaw studies, and limited forced degradation studies [21, 22].

Prior to being formulated, the antibody must also go through several purification/chromatography steps, which are generally done over a range of pH and salt conditions. For commercial viability, the antibody is also expected to have a shelf-life of at least 2 years when formulated. Therefore, one critical factor for the successful development of any antibody program is high conformational and colloidal stability. At the discovery stage, the conformation stability of an antibody can easily be assessed in a high-throughput manner using techniques such as differential scanning fluorimetry (DSF) [23]; however, screening for colloidal instability is much more challenging. We recently published a novel screening technique [24], standup monolayer adsorption chromatography (SMAC), that can be used to assess colloidal stability of antibodies using only microgram quantities of protein. Briefly, we showed that for antibodies injected on a commercially available prepacked Zenix HPLC column, the retention times were inversely related to their colloidal stability. Antibodies prone to precipitation or aggregation were retained longer on the column and had broader peaks. One additional advantage of this technique is that the column we used is commercially available and produced under controlled conditions; therefore, there should be very little batch-to-batch variation.

2 Materials

2.1 Equipment

1. HPLC: Agilent 1200 HPLC with Diode Array Detector (DAD), Auto-sampler, with a quaternary or binary pumping system, or equivalent.
2. Computer software: Agilent Chemstation, or equivalent.
3. HPLC disposables: Low volume vials and caps as required.
4. pH Meter.
5. pH Electrode.
6. Analytical Column Zenix SEC300 (P/N 213300–4630).
7. Pipettors and tips.
8. Analytical Balance.

2.2 Chemicals

1. NaCl.
2. Sodium Phosphate Monobasic Monohydrate.
3. Sodium Phosphate Dibasic Heptahydrate.
4. Hydrochloric Acid.
5. Sodium Hydroxide.
6. Ethanol.
7. Molecular Weight Standards (e.g., Bio-Rad P/N 151-1901 or equivalent).

2.3 Preparation of Reagents

Prepare all solutions using ultrapure water (purity: 18 MΩcm or more) and analytical grade reagents. Prepare and store all reagents at room temperature (unless indicated otherwise). *Note*: Reagents may be scaled for appropriate volumes.

1. Mobile Phase Buffer: 150 mM Sodium Phosphate, pH 7.0. In a 2000 mL graduated cylinder, add approximately 1800 mL. Add 17.5 ± 0.1 g of Sodium Phosphate Monobasic Monohydrate and dissolve. Add 46.4 ± 0.1 g of Sodium Phosphate Dibasic Heptahydrate and dissolve. Check the pH of the buffer to ensure a pH of 7.0 ± 0.1 and adjust with HCl or NaOH as appropriate. Bring the volume up to 2000 mL and transfer to a 2 L bottle. Expiration is 1 month from the date of preparation.
2. Column storage solution: 50 mM Sodium Phosphate buffer with 0.02% sodium azide. When the column is not in use, it should be stored in this buffer.

2.4 Preparation of Controls, Reference Standard, and Samples

1. Molecular weight marker (MWM): Allow the Biorad Gel Filtration Standards (GFS) to come to room temperature and reconstitute with 0.5 mL of Milli-Q water. Dilute the reconstituted MWM solution 1:20 with Milli-Q water. Aliquot the diluted GFS by 50 μL into screw-capped polypropylene tubes and store at −80 °C for up to 1 year.
2. Reference Standard: Aliquot appropriate amounts of reference standard to have a column load of 30 μg plus 20 μL into a HPLC vial. We typically use an in-house clinical antibody as a reference standard. However, the user can use any available well-behaved antibody as reference standard.
3. Samples: Test antibodies can be expressed and purified according to any method of user's choice. Purified test antibodies need to be at a minimum concentration of 600 μg/mL. For screening purposes, we typically express antibodies (or antibody-like molecules) using Thermo Fisher's Expi293 system using manufacturer's recommended conditions (*see* **Note 1**). Obtain the sample concentration from A280 assay prior to running sample if applicable (*see* **Note 2**). Aliquot

appropriate amounts of samples to have a column load of 30 μg plus 20 μL into an HPLC vial.

Example: For an unknown sample with a concentration of 3.0 mg/mL, the calculation to determine the injection load would be as follows:

$$\frac{3.0mg}{mL} \times \frac{1mL}{1000mL} \times \frac{1000\mu g}{1mg} = \frac{3.0\mu g}{\mu L}$$

$$\frac{30\mu g}{3.0\mu g/\mu L} = 10\mu L$$

Therefore, the injection volume of the sample would be 10 μL, and the volume needed for the HPLC vial is 30 μL for a single injection of a sample. Calculate the load volume the same way for all molecules.

3 Methods

3.1 SMAC Assay

1. Connect the column to the column compartment, and place the appropriate line in the mobile phase.
   Open the purge valve and prime the lines.
2. Equilibrate column for 60 min at 0.350 mL/min flow rate with mobile phase buffer (*see* **Note 3**).
3. Write the running method using manufacturer's recommended procedure and verify that it has the following test parameters/conditions:
   - Mobile Phase A: 150 mM Sodium Phosphate pH 7.0.
   - Flow Rate: 0.35 mL/min.
   - Stoptime: 20 min.
   - Maximum Pressure: 125 bar.
   - Column Temperature: 20 °C.
   - Autosampler Temperature: 4 °C.
   - Detection Wavelength: 280 nm.
4. Load samples into the injection sequence. The reference standard should bracket the injections for each set of samples.
5. Start the method.

3.2 Data Analysis

3.2.1 System Suitability

1. Perform a system suitability assay to determine that the column is sufficient for the analysis of each molecule run in sequence.
2. Ensure that the molecular weight markers exhibit the characteristic elution pattern, as shown in Fig. 1, and that the following

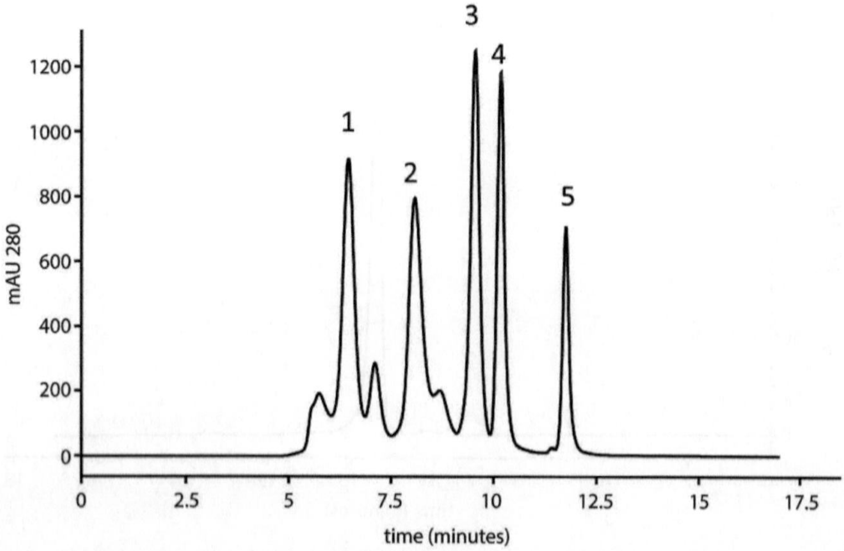

Fig. 1 Example chromatogram of the molecular weight marker

Table 1
System suitability: acceptable ranges

USP tailing	USP plate count	USP resolution between myoglobin (peak 4) and vitamin B12 (peak 5)
≤1.2	>10,000	≥3

criteria (Table 1) are met for the Vitamin B12 peak (which is the last eluting peak at ~12 min) (*see* **Note 4**).

3.2.2 Assay Acceptance

1. Verify that the chromatogram of the reference antibody exhibits the characteristic elution pattern, as shown in Fig. 2.

Determine that the percentage monomer of the reference standard falls within established control chart ranges. The difference (beginning minus end) for the retention time of the each reference standard monomer is ± 0.1 min. *See* the example chromatogram in Fig. 2.

3.2.3 Interpreting and Reporting Data

1. This assay is designed to screen antibodies for relative retention times.
2. Clones that are retained longer on the Zenix column have poorer colloidal stability in our experience.

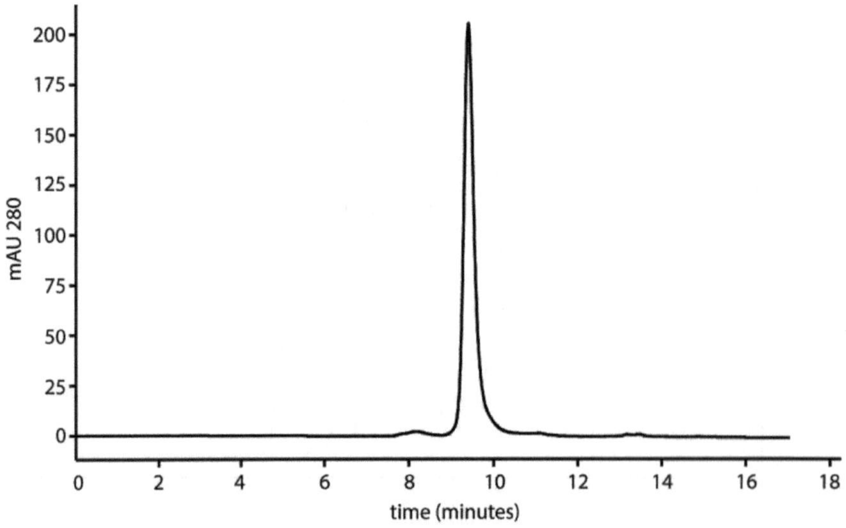

Fig. 2 Example chromatogram of reference antibody standard

Table 2
Retention times for antibodies on a Zenix SEC-300 column

Protein name	Retention time (min)
Reference antibody	9.4
Antibody A	11.9
Antibody B	9.8
Antibody C	>20
Antibody D	10.9
Antibody E	9.5

3. Absolute retention time may vary depending upon the HPLC system or length of tubings. Therefore, for this screening, we flag any antibody that has retention time more than 1 min off the reference standard.
4. For example, in our system, the reference antibody has a retention time of 9.4 min, so any antibody that has retention time greater than 10.4 min will be flagged.
5. *See* example Table 2. Antibodies A, C, and D will be flagged.
6. For an example of an antibody that was flagged, *see* Fig. 3. This shows a representative chromatogram of Antibody D. It was retained longer on the column and also had a broad elution profile.

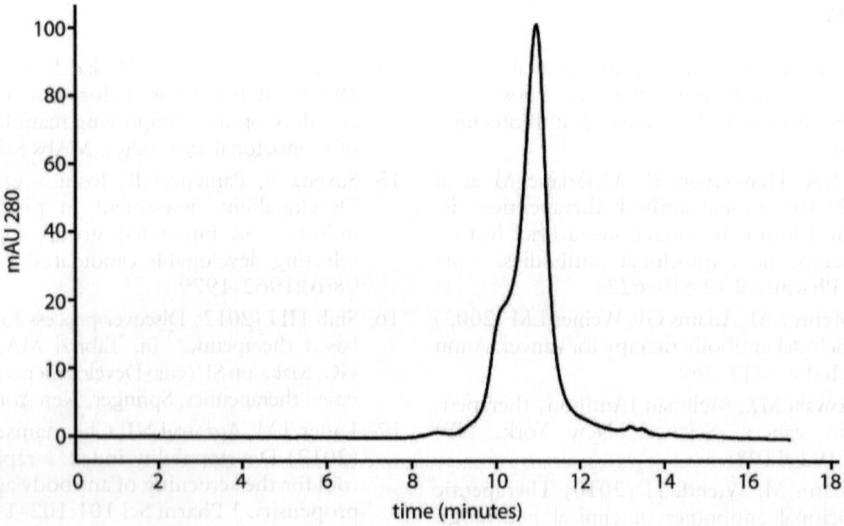

Fig. 3 Example chromatogram of antibody D

4 Notes

1. Ideally, all test antibodies should be dialyzed in 1× PBS before injection to prevent any buffer interference. However, in our experience if the antibody concentration is above 1000 μg/mL, there is very limited interference from the buffer, and no need for buffer exchange.
2. When measuring sample concentration using *A*280 values, make sure the *A*280/*A*260 ratio is around 0.57 ± 0.05. A280 values for dilute samples may be over-reported if the *A*280/*A*260 ratio is outside this range.
3. If, prior to equilibration, this column was also being used for another application that requires a different mobile phase, do not increase the flow rate of the mobile phase directly to 0.35 mL/min as the backpressure could exceed maximum recommended operating limit due to difference in viscosities of mobile phases. Therefore, when changing the mobile phase, prime with lines with new buffer, equilibrate the column at 0.1 mL/min for 10 column volumes and then re-equilibrate at 0.35 mL/min for 60 min.
4. We typically use United States Pharmacopeia (USP) methods for calculating the tailing factor and plate height. However, the user can use any other method of their choice for qualifying the column performance.

References

1. Bradbury ARM, Sidhu S, Dübel S et al (2011) Beyond natural antibodies: the power of in vitro display technologies. Nat Biotechnol 29:245–254
2. Buss NA, Henderson SJ, Mcfarlane M et al (2012) Monoclonal antibody therapeutics: history and future Introduction—a brief history of therapeutic monoclonal antibodies. Curr Opin Pharmacol 12:615–622
3. von Mehren M, Adams GP, Weiner LM (2003) Monoclonal antibody therapy for cancer. Annu Rev Med 54:343–369
4. Sliwkowski MX, Mellman IAntibody therapeutics in cancer. Science (New York, NY) 341:1192–1198
5. Buttmann M, Wiendl H (2010) Therapeutic monoclonal antibodies in clinical neurology. Nervenarzt 81:753–764 quiz 765–6
6. Chan AC, Carter PJ (2010) Therapeutic antibodies for autoimmunity and inflammation. Nat Rev Immunol 10:301–316
7. Chao G, Lau WL, Hackel BJ et al (2006) Isolating and engineering human antibodies using yeast surface display. Nat Protoc 1:755–768
8. Lee EC, Owen M (2012) The application of transgenic mice for therapeutic antibody discovery. Methods Mol Biol 901:137–148
9. Tiller T (2011) Single B cell antibody technologies. N Biotechnol 28(5):453–457
10. Bratkovič T (2010) Progress in phage display: evolution of the technique and its applications. Cell Mol Life Sci 67(5):749–767
11. Xu L, Kohli N, Rennard R et al (2013) Rapid optimization and prototyping for therapeutic antibody-like molecules. MAbs 5:237–254
12. Schaffitzel C, Hanes J, Jermutus L et al (1999) Ribosome display: an in vitro method for selection and evolution of antibodies from libraries. J Immunol Methods 231:119–135
13. Jarasch A, Koll H, Regula JT et al (2015) Developability assessment during the selection of novel therapeutic antibodies. J Pharm Sci 104(6):1885–1898
14. Yang X, Xu W, Dukleska S et al (2013) Developability studies before initiation of process development: improving manufacturability of monoclonal antibodies. MAbs 5:787–794
15. Saxena V, Panicucci R, Joshi Y et al (2009) Developability assessment in pharmaceutical industry: an integrated group approach for selecting developable candidates. J Pharm Sci 98(6):1962–1979
16. Shih HH (2012) Discover process for antibody-based therapeutics. In: Tabrizi MA, Bornstein GG, Klakamp SL (eds) Development of antibody-based therapeutics. Springer, New York, NY
17. Lauer TM, Agrawal NJ, Chennamsetty N et al (2012) Developability index: a rapid in silico tool for the screening of antibody aggregation propensity. J Pharm Sci 101:102–115
18. Liu Y, Caffry I, Wu J et al (2014) High-throughput screening for developability during early-stage antibody discovery using self-interaction nanoparticle spectroscopy. MAbs 6:483–492
19. Jacobs SA, Wu SJ, Feng Y et al (2010) Cross-interaction chromatography: a rapid method to identify highly soluble monoclonal antibody candidates. Pharm Res 27:65–71
20. Hotzel I, Theil FP, Bernstein LJ et al (2012) A strategy for risk mitigation of antibodies with fast clearance. MAbs 4:753–760
21. Blessy M, Patel RD, Prajapati PN et al (2014) Development of forced degradation and stability indicating studies of drugs—a review. J Pharm Anal 4(3):159–165
22. Hawe A, Wiggenhorn M, van de Weert M et al (2012) Forced degradation of therapeutic proteins. J Pharm Sci 101(3):895–913
23. Lavinder JJ, Hari SB, Sullivan BJ et al (2009) High-throughput thermal scanning: a general, rapid dye-binding thermal shift screen for protein engineering. J Am Chem Soc 131: 3794–3795
24. Kohli N, Jain N, Geddie ML et al (2015) A novel screening method to assess developability of antibody-like molecules. MAbs 0862:37–41

Chapter 12

Glycosylation Profiling of α/β T Cell Receptor Constant Domains Expressed in Mammalian Cells

Kai Zhang, Stephen J. Demarest, Xiufeng Wu, and Jonathan R. Fitchett

Abstract

Glycoprofiling recombinant proteins expressed and secreted from mammalian cells is key to understanding their interactions with glycoprotein receptors in vivo. Recently, recombinant T cell receptors (TCRs) are being considered as therapeutic moieties. Here we present a mass spectrometry based protocol with a "bottom up" approach to characterize glycosylation in recombinant fusion proteins with α/β TCR constant domains expressed in mammalian cells. The protocol focuses on using peptide mass mapping and mass spectrometry for *N*-linked glycan profiling, including analyses of site occupancy, glycan heterogeneity, and possible glycan compositions and structures.

Key words Glycoprofiling, Glycosylation, Glycopeptide, *N*-Linked, Mass spectrometry (MS), LC-MS/MS, ESI-MS, T Cell receptor (TCR)

1 Introduction

α/β T cell receptors (TCRs) are glycoproteins on the surface of T cells that recognize aberrantly expressed self-antigen peptides and non-self-antigen peptides displayed by major histocompatibility complex (MHC) proteins on the surfaces of target cells. This binding along with costimulation leads to T cell activation, proliferation, and other T cell modifications that may result in target cell killing. Glycosylation plays critical roles in the assembly, cell-membrane sub-compartmentalization, and activation of α/β TCRs [1]. Glycosylation promotes glycoprotein binding to ER chaperones calnexin and calreticulin that assist in proper folding [2]. α/β TCRs have generally slow folding β-sheet structures with complex disulfide bond arrangements that require chaperone assistance to reach their functional forms. The presence of branched glycans rich in Gal-GlcNAc (Gal: galactose; GlcNAc: N-acetylglucosamine) on α/β TCRs and CD8/CD4 leads to pentameric galectin 3 binding spacing the two receptors apart limiting costimulation and impacting their ability to assimilate within the critical TCR immune

Thomas Tiller (ed.), *Synthetic Antibodies: Methods and Protocols*, Methods in Molecular Biology, vol. 1575,
DOI 10.1007/978-1-4939-6857-2_12, © Springer Science+Business Media LLC 2017

synapse [3–5]. Thus, glycan occupancy and composition is critical for T cell activation and regulation.

We and others have focused recently on generating soluble α/β T cell receptors as putative therapeutic modalities for cancer and autoimmune therapy [6, 7]. Glycosylation occupancy and composition impacts the pharmacokinetics and pharmacodynamics of secreted soluble glycoproteins including soluble α/β TCRs. For example, glycan-mediated clearance through the endothelial asialoglycoprotein receptor (ASGPR) leads to hepatocytic accumulation of IgA and recombinant Factor VII and VIII and sequestration from serum [8–10]. ASGPR-mediated clearance of recombinant IgA superficially resembles what has been observed for recombinant α/β TCR constant domain fusions [6, 11]. Thus, understanding glycan occupancy and composition of α/β TCRs is key to understanding how they will express and behave as therapeutics.

Usually glycoprotein analysis falls into three categories: glycomics (analysis of released glycans), proteomics (analysis of non-glycosylated peptides from protein enzymatic digestion), and glycoproteomics (analysis of glycopeptides from protein enzymatic digestion) [12]. This protocol falls in glycoproteomics, which characterizes glycan structures in a site-specific manner with the main focus on *N*-linked glycosylation (*see* **Note 1**). Most *N*-linked glycans are covalently attached to the amide nitrogen of asparagine (Asn) side chains in peptide sequences with an Asn-X-Ser/Thr motif, where X represents any amino acid residue except proline. Occasionally *N*-glycans occur at Asn-X-Cys if the cysteine is in the reduced form. The glycan site occupancy in Asn-X-Ser/Thr motifs may be predicted by computational servers such as NGlycPred if a high resolution structure is available for the protein of interest [13]. The most common *N*-glycan linkage is GlcNAcβ1-Asn, which has a common core sequence consisting of two *N*-acetylglucosamine and three mannose residues $(Man)_3(GlcNAc)_2$. There are three general types of *N*-glycans: high mannose (also called oligomannose), complex, and hybrid [14]. Peptide N-glycosidase F (PNGase F) is widely used to release *N*-linked glycans [14], except for those with α1,3-linked core fucose, which is commonly found in plant glycoproteins and may be released by using PNGase A when no sialic acid is present [15].

O-linked glycans are usually covalently attached to the hydroxyl group of serine or threonine side chains. No specific consensus sequence has been identified for *O*-linked glycosylation sites, although there are a few publications describing computational prediction algorithms [16, 17]. The majority of *O*-linked glycans are of mucin-type (*O*-GalNAc), and other types such as *O*-Fucose, *O*-Glucose, *O*-GlcNAc, and *O*-Mannose are reported as well [18, 19]. Since there is no "universal" enzyme to remove all types of *O*-linked glycans, a series of enzymes may be used (*see* **Note 2**), with the caution that the deglycosylation might not be complete. Chemical release methods such as alkaline β-elimination [20, 21] and hydrazinolysis [22] are commonly used for further analysis on the released glycans, while the release with anhydrous trifluoromethanesulfonic acid (TFMS) may retain the primary structure of the protein [23]. The challenges with chemical release include the control of experimental conditions, the completeness of glycan removal, the way to maintain protein integrity, etc.

This chapter will focus on *N*-linked glycosylation profiling, and will consider *O*-linked glycans when they interfere with the analysis of *N*-linked glycosylation (*see* **Note 2**). In general, the same strategy applies to analysis for *O*-linked glycosylation, except for the challenges of completely removing the *O*-linked glycans.

Here we present a MS-based workflow that involves protein denaturation, reduction, alkylation and proteolytic digestion ("bottom up"), followed by reversed phase UPLC (Ultra Performance Liquid Chromatography) separation, and ESI (ElectroSpray Ionization) tandem mass spectrometry analysis (*see* **Note 3**). The protein digest with *N*-linked deglycosylation treatment is analyzed in parallel (*see* **Note 4**). Data analysis focuses on *N*-linked glycosylation profiling: identification of *N*-linked glycosylation site(s), occupancy of each site, heterogeneity, and possible glycan compositions and structures for each site. We apply the method to an IgG_TCR protein [6] to evaluate Cα/Cβ glycosylation during expression and secretion from HEK293 cells. Ultimately, the method could be used to characterize immunoprecipitated receptors from primary immune cells to help correlate glycan occupancy with phenotype.

2 Materials

2.1 Protein Materials

Antibody/TCR fusion proteins denoted 'IgG_TCRs' were constructed, expressed and purified as described previously [6]. These proteins comprise an antibody Vκ region fused to the TCRβ constant domain and an antibody VH-TCRα-IgG1-Fc fusion that assemble to form a structure resembling a native-like IgG-heterotetramer.

2.2 Reagents and Supplies

All the solvents, such as water and acetonitrile, should be in LC/MS grade. Use mass spectrometry grade reagents or the highest purity whenever possible. Filter prepared solutions through a 0.22 μm filter as needed.

1. Denaturing buffer: 6.4 M guanidine HCl, 0.1 M Tris, pH 7.4. Combine 100 μL of 1 M Trizma hydrochloride buffer, pH 7.4 with 900 μL 8.0 M guanidine HCl solution, and mix well.

2. Reducing agent: 0.1 M dithiothreitol (DTT) in water. Dissolve 7.7 mg DTT in 100 μL water to obtain 0.5 M DTT. Then combine 50 μL of 0.5 M DTT with 200 μL water to obtain 0.1 M DTT. Prepare fresh daily.
3. Alkylating agent: 0.2 M iodoacetamide in water. Dissolve 9.3 mg iodoacetamide in 100 μL water to obtain 0.5 M iodoacetamide. Then combine 80 μL of 0.5 M iodoacetamide with 120 μL water to obtain 0.2 M iodoacetamide, and store in dark (e.g., wrap it with aluminum foil). Prepare fresh daily, preferably right before usage.
4. Digestion buffer: 25 mM Trizma, 20 mM $CaCl_2$, pH 7.4. Add 25 mL of 1 M Trizma hydrochloride buffer pH 7.4 and 20 mL of 1 M $CaCl_2$. QS to 1 L with water, and mix well.
5. Desalting and buffer exchange to digestion buffer: Desalting column (e.g., Zeba spin desalting column 7K MWCO, or NAP-5 column) or dialysis cassettes (7K MWCO or 10K MWCO).
6. Trypsin solution: 0.1 μg/μL trypsin (Promega, Madison, WI, USA). Add 100 μL of water to 100 μg of trypsin (mass spectrometry grade), and mix well.
7. Peptide-*N*-glycosidase F (PNGase F) (Prozyme, Hayward, CA, USA), 2.5 U/mL or above.
8. Mobile phase A: 0.1% formic acid (Thermo Fisher Scientific, Waltham, MA USA) in water. Add 1 mL formic acid to 1 L water and mix well.
9. Mobile phase B: 0.1% formic acid in acetonitrile. Add 1 mL formic acid to 1 L acetonitrile and mix well.
10. Solvent for ESI calibration solution and lock mass solution: 0.1% formic acid in 50:50 (v:v) water:acetonitrile. Add 1 mL formic acid to 500 mL water and 500 mL acetonitrile, and mix well.
11. MS ESI calibration solution: 0.8 pmol/μL of [Glu^1]-fibrinopeptide B (Waters, Milford, MA, USA) in 0.1% formic acid in 50:50 (v:v) water:acetonitrile. Dissolve 0.1 mg of [Glu^1]-fibrinopeptide B in 1 mL of water and mix well, resulting in 63.6 pmol/μL stock solution. The stock solution is stored as 160-μL aliquots at −20 °C until usage. The working solution is prepared by mixing one 160 μL aliquot stock solution with 12.58 mL of 0.1% formic acid in 50:50 (v:v) water:acetonitrile.
12. Lock mass solution: 2 ng/μL Leucine Enkephalin (Waters, Milford, MA, USA) in 0.1% formic acid in 50:50 (v:v) water:acetonitrile. Add 7.5 mL of water to 3 mg of Leucine Enkephalin to obtain 400 ng/μL of Leucine Enkephalin stock solution. The stock solution is stored as 1-mL aliquots at −20

°C until usage. The working solution is prepared by mixing 1 mL of 400 ng/μL Leucine Enkephalin with 199 mL of 0.1% formic acid in 50:50 (v:v) water:acetonitrile.

13. Reversed Phase HPLC column (C18 or similar), preferably sub-2 μm particle size (e.g., Acquity BEH C18 or HSS T3 column from Waters, Milford, MA, USA).

2.3 Equipment

1. Chromatographic system: HPLC (High Performance Liquid Chromatography) or UPLC (Ultra Performance Liquid Chromatography), e.g., Acquity UPLC (Waters, Milford, MA, USA).
2. High resolution ESI mass spectrometer, e.g., Synapt G2 (Waters, Milford, MA, USA).
3. Vacuum concentrator (e.g., Labconco CentriVap concentrator).
4. pH meter.
5. Incubator.

3 Methods

3.1 Peptide Mapping Sample Preparation

1. Add 30 μg of protein to an Eppendorf tube, and dry to nearly completeness in a centrifugal vacuum concentrator without using heat (*see* **Note 5**). If the protein concentration is 10 μg/μL or higher, no need to concentrate.
2. Add 100 μL denaturing buffer (6.4 M guanidine HCl, 0.1 M Tris, pH 7.4) to the protein tube to reconstitute (*see* **Note 6**).
3. Add 11 μL of freshly prepared 0.1 M DTT to a final concentration of 10 mM DTT, and mix well. Incubate at 37 °C for 1 h.
4. Add 12 μL of freshly prepared 0.2 M iodoacetamide to a final concentration of 20 mM iodoacetamide, and mix well. Incubate at 25 °C in dark for 45 min (e.g., wrapped in aluminum foil).
5. Use desalting column or dialysis cassette (7K MWCO or 10K MWCO) to desalt and buffer exchange the protein sample with digestion buffer: 25 mM Trizma, 20 mM $CaCl_2$, pH 7.4.
6. Add 0.1 μg/μL of trypsin to the denatured, reduced, alkylated protein sample tube at the trypsin to protein ratio of 1:20 (w:w), and incubate at 37 °C overnight (16–18 h). A digest blank is prepared in the same way, except using digestion buffer alone without protein.
7. Split the protein digest into two aliquots. To one aliquot, 2 μL of 2.5 U/mL PNGase F is added. Incubate at 37 °C for an additional 1 h.

8. Add 1 μL of formic acid to each tube to quench the reaction.
9. The resulting two tryptic digests, with and without PNGase F treatment, are ready to be analyzed.
10. If a secondary enzyme is needed for complementary digestion (*see* **Note 2**), prepare the material in the similar way as in Subheadings 3.1, **steps 1–5** prior to digestion, where a proper digestion buffer for that secondary enzyme should be used.

3.2 LC/MS Setup and Data Acquisition

1. The peptides from digestions are to be analyzed using an electrospray ionization (ESI) tandem mass spectrometer coupled with reversed phase HPLC or UPLC. UPLC is preferred for better chromatographic resolution and faster analysis.
2. As an example, an Acquity UPLC (Waters, Milford, MA, USA) is used for chromatographic separation. A portion of each digested sample is injected into a reversed phase column (sub-2 μm particle size), for example, Acquity BEH C18 (1.7 μm, 2.1 × 100 mm) or HSS T3 (1.8 μm, 2.1 × 100 mm) set at 40 °C. The injection amount varies, depending on the sensitivity of the mass spectrometer. Mobile phase A is 0.1% (or 0.2%) formic acid in water, and mobile phase B is 0.1% (or 0.2%) formic acid in acetonitrile. As a starting point, a linear gradient from 1 to 35% mobile phase B over 60 min with the constant flow rate of 0.2 mL/min may be used to elute peptides. The solvent peaks are diverted to the waste for 3 min before the flow is switched to ESI source.
3. An ESI tandem mass spectrometer such as a Synapt G2 (Waters, Milford, MA, USA) is used, where positive ion mode and resolution mode are selected. Calibrate the mass spectrometer prior to use by using [Glu^1]-Fibrinopeptide B, and the mass accuracy with the external lock mass correction (by using Leucine Enkephalin) is 1 ppm or better. The ESI source capillary voltage is at 3 kV, and the source temperature is at 100 °C. A low cone voltage should be used to minimize carbohydrate fragmentation (*see* **Note 7**).
4. If the Synapt G2 is employed for mass spectrometry analysis, MS^E mode with the *m/z* range from 135 to 1950 in the continuum mode may be chosen for data acquisition, where three functions are collected. Function 1 acquires low energy precursor ion mass spectra, Function 2 acquires the elevated-energy fragmental ion spectra with trap collision energy ramping from 15 to 45 eV, and Function 3 is the lock mass channel. Function 1 generates MS data for identification and quantification, and Function 2 produces MSMS data for identification through data-independent acquisition (DIA). The lock mass reference (Function 3) is introduced through LockSpray for one acquisition every 30 s, which can be the

singly charged ion of Leucine Enkephalin at *m/z* of 556.2771, and/or the doubly charged ion of [Glu^1]-Fibrinopeptide B at *m/z* of 785.8426.

5. In general, MSMS data can be generated from data-independent acquisition (DIA) or data-dependent acquisition (DDA). For Synapt G2, besides elevated-energy function in MS^E mode, survey scan and MSMS modes are also available, allowing different energy settings (trap collision energy, transfer collision energy, and/or cone voltage) to produce fragmental ions.

3.3 Data Analysis

1. Predict *N*-linked glycosylation sites by searching for Asn-X-Ser/Thr consensus sequence, where X is any amino acid except proline.
2. Because samples are reduced and alkylated with iodoacetamide, the modification of Cysteine to Carbamidomethyl Cysteine (CAM), resulting in the monoisotopic mass increase of 57.0215 Da, should be considered as a fixed modification.
3. The MS and MSMS data are processed by using commercial software, such as BiopharmaLynx and ProteinLynx Global Server (PLGS) (Waters, Milford, MA, USA). Peptides are identified based on mass accuracy in MS data and fragmental ions in MSMS data. Typical mass accuracy used for peptide identification of masses from Synapt G2 (Waters, Milford, MA, USA) with lock mass correction is 3 ppm or better. For the glycopeptides with ionization suppression, especially when occurred at low abundances, the mass accuracy may not be as good (up to 6 ppm).
4. The peptide identifications from auto-processing software should be verified manually, especially for the critical gylcopeptides to insure there are no automated misassignments. The verification approaches include the comparison of enzymatic digests with and without PNGase F treatment, the manual check on mass accuracy, and the evaluation of fragmental ions of peptides and glycopeptides. These approaches are elaborated in Subheadings 3.3, **step 5**–7.
5. The chromatograms of the tryptic digests with and without PNGase F treatment are compared. The unique masses observed in the sample without PNGase F treatment are considered to be glycopeptides where *N*-linked glycosylation is involved. On the other hand, the unique masses observed in the sample with PNGase F treatment are most likely the deglycosylated peptides.
6. The comparison is illustrated in Fig. 1, which shows the expanded view of the Total Ion Chromatograms (TICs) for tryptic digests with and without PNGase F treatment, for the elution time window where HC Cα Asn185 is involved. The

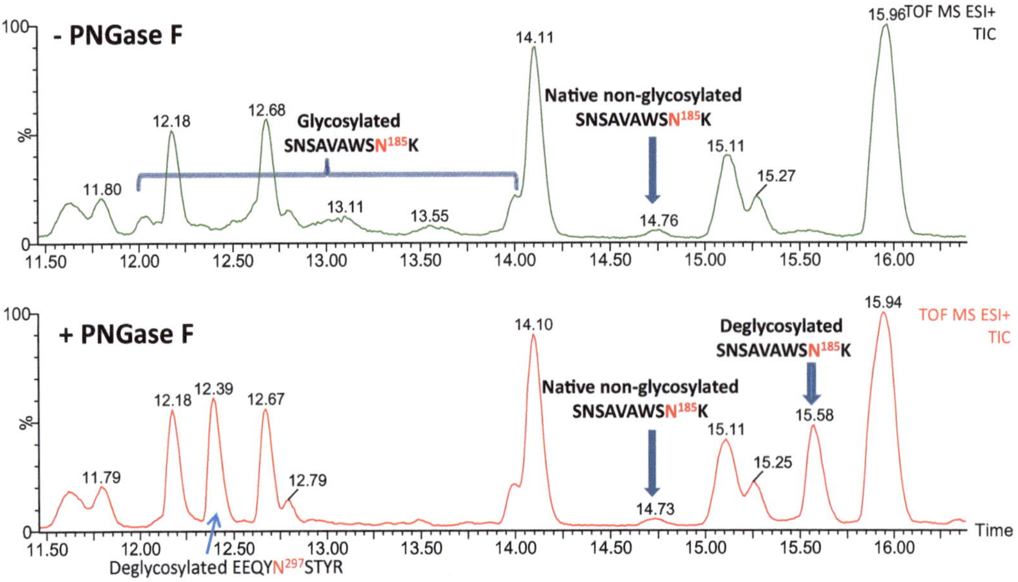

Fig. 1 Expanded view of Total Ion Chromatograms (TICs) for tryptic digests in the region where HC Cα Asn185 is involved. The peptide containing HC Cα Asn185 is observed as glycopeptides, deglycosylated peptide, and native non-glycosylated peptide. *Top panel*: digest in the absence of PNGase F treatment, where multiple glycopeptides spread between 11.9 and 14.1 min. *Bottom panel*: digest in the presence of PNGase F treatment, where the deglycosylated form (doubly charged ion with *m/z* of 532.77) is observed at 15.58 min

peptide containing HC Cα Asn185 is observed as glycopeptides, deglycosylated peptide, and native non-glycosylated peptide. Compared to non-glycosylated peptides, the intensities of glycopeptides are low due to the suppression of ionization, which is common. During *N*-linked deglycosylation, the occupied Asn is expected to be converted to Asp, resulting in the mass increase of 0.984 Da. Therefore, the mass difference can be used to differentiate the *N*-linked degylcosylated peptide from the native peptide.

7. The MSMS spectra from DIA or DDA for peptides involved in *N*-linked gycosylation may help with the identification. For example, for the native and modified peptides SNSAVAWSN^{185}K observed at 15.6 and 14.7 min in Fig. 1, the fragmental b and y ions are verified as shown in Fig. 2. The top panel shows the fragmental ions of a doubly charged ion, $m/z = 532.27$, for the low abundance native non-glycosylated peptide, and the bottom panel shows fragmental ions of a doubly charged ion, $m/z = 532.76$, for the *N*-linked deglycosylated (modified) peptide, where the mass increase of 0.984 Da in the precursor ion is due to the conversion of Asn185 to Asp upon deglycosylation.
8. Upon treatment with PNGase F, the conversion of Asn to Asp during *N*-linked deglycosylation leads to the mass increase of

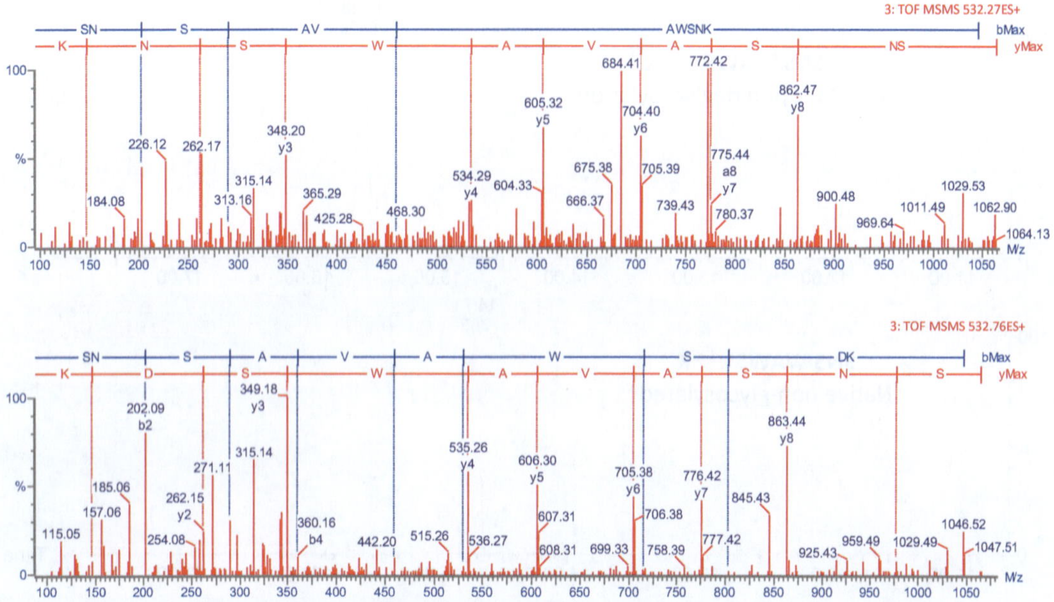

Fig. 2 MSMS spectra of peptide SNSAVAWSN[185]K. *Top panel*: fragmental ions of a doubly charged ion, *m/z* = 532.27, for the native non-glycosylated peptide (low abundance). *Bottom panel*: fragmental ions of a doubly charged ion, *m/z* = 532.76, for the *N*-linked deglycosylated (modified) peptide, where the mass increase of 0.984 Da in the precursor ion is due to the conversion of Asn185 to Asp upon deglycosylation

0.984 Da for the peptide containing Asn ("modified peptide"). The unoccupied Asn remains the same during deglycosylation, so the mass of the peptide involved remains the same ("native peptide"). Therefore occupancy of each predicted *N*-linked glycosylated site can be determined using deglycosylated sample data:

$$\text{Occupancy} = \frac{\text{Occupied}}{(\text{Occupied} + \text{unoccupied})} \times 100\% \quad (1)$$

or more specifically,

$$\text{Occupancy} = \frac{\text{modified peptide}}{(\text{modified peptide} + \text{native peptide})} \times 100\% \quad (2)$$

The amount of modified peptide and native peptide are calculated using either ion intensity or integrated area of the extracted ion of the peptide (*see* **Notes 8** and **9**).

9. An example of occupancy calculation is shown in Fig. 3, where the integrated areas of modified peptide (top panel) and native peptide (bottom panel) are used for calculation. In this case, the occupancy of HC Cα Asn185 is calculated by using Eq. 2, and found to be nearly fully *N*-linked glycosylated (97.3% occupancy).

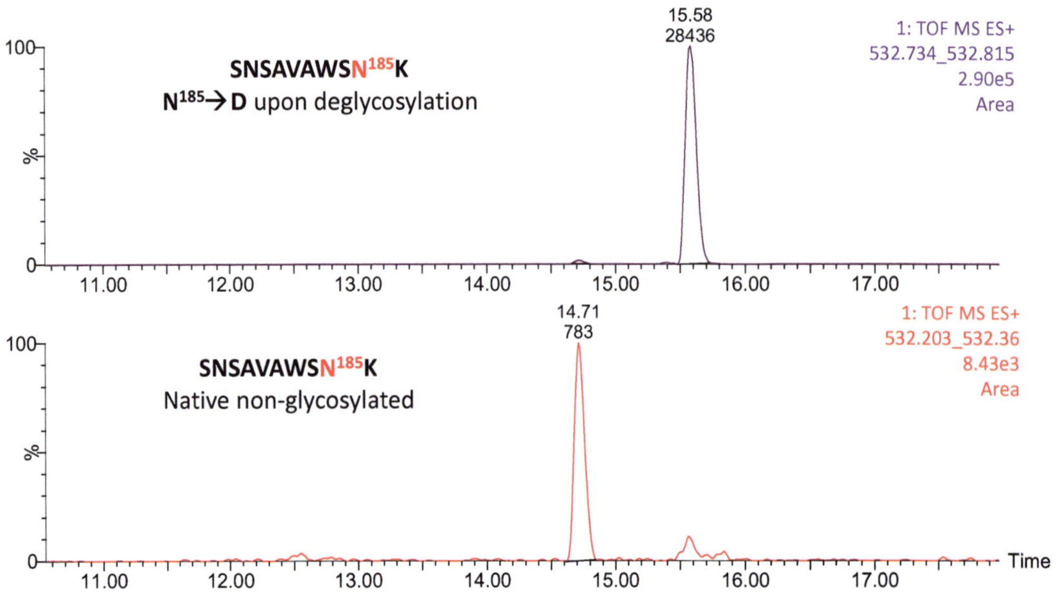

Fig. 3 Occupancy calculation for HC Cα Asn185 for peptide SNSAVAWSN^{185}K. *Top panel*: extracted ion chromatogram for the modified peptide (Asn^{185} is converted to Asp upon deglycosylation), where the retention time and integrated area are annotated on *top* of the largest peak, which the *Y*-axis is normalized to. *Bottom panel*: extracted ion chromatogram for the native peptide. The occupancy was calculated to be 97.3% using Eq. 2

10. The procedures above may be repeated for all the possible *N*-linked glycosylation sites to determine the site occupancy of each Asn.
11. Two complementary approaches can be used to locate/verify glycopeptides. First, glycopeptides should be unique in the sample without *N*-linked deglycosylation. If a mass is observed in both samples with and without PNGase F treatment at the same retention time, it does not belong to a glycopeptide. Second, search for oxonium fragment ions ("signatures") in MSMS data from Collision Induced Dissociation (CID), which is the most widely used fragmentation method. The CID fragmentation of a glycopeptide typically leads to preferential cleavage of glycosidic bonds [24]. Therefore, it generates ions at m/z of 366, 204, 292, and 657 for Hex-HexNAc, HexNAc, sialic acid (NeuAc), and HexNAc-Hex-NeuAc, respectively [25], where Hex is hexose, and HexNAc is N-acetylhexosamine, such as N-acetylglucosamine (GlcNAc) or N-acetylgalactosamine (GalNAc). In a recent study, signature ions at *m/z* of 366, 204, 292, and 657 in the MSMS spectra are used to identify glycopeptides, as shown in the left panel of Fig. 4.
12. When multiple glycosylation sites are present in the protein, the assignment of observed glycopeptides to the corresponding native peptide needs to be done with caution. Since the

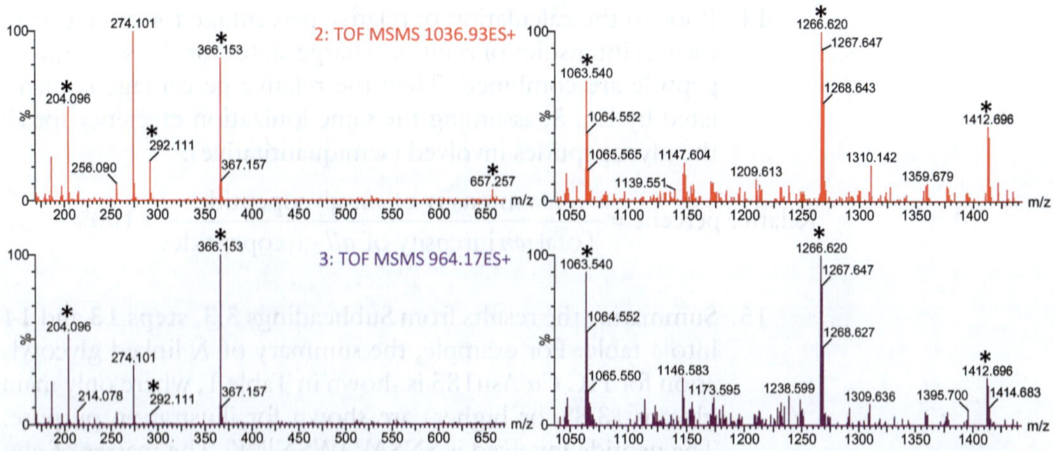

Fig. 4 MSMS spectra for glycopeptides of SNSAVAWSN[185]K. *Left panel*: Expanded view of signature ions at *m/z* of 366, 204, 292, and 657 (*asterisks*). *Right panel*: Expanded view of singly charged ions at *m/z* of 1063.5, 1266.6, and 1412.7 (*asterisks*) for native peptide, native peptide with HexNAc, and native peptide with (HexNAc+Hex), respectively

CID fragmentation always leads to preferential cleavage of glycosidic bonds, rather than the cleavage along the peptide backbone [24], it may generate the ions of native peptide. In the previous example, besides the signature ions shown in Fig. 4 (left panel), the fragmentation on glycopeptides of SNSAVAWSN[185]K also generates singly charged ions at *m/z* of 1063.5, 1266.6, and 1412.7 for native peptide, native peptide with HexNAc, and native peptide with (HexNAc+Hex), respectively, as shown in the right panel of Fig. 4. Since most likely the native peptides of different N-linked glycosylation sites have different masses, these fragmental ions will help assigning the glycopeptide in question to one of the possible native peptides.

13. At each glycosylation site, the masses of glycans are determined from the mass differences between the observed glycopetides and the native peptide. Depending on the scope of glycoprofiling, the cutoffs for non-assigned versus assigned glycopeptide peaks could be arbitrary (for example, 0.5%). In consideration of the common core $(Man)_3(GlcNAc)_2$, possible glycan compositions can be estimated by using GlycoMod [26], UniCarbKB [27], or other computational tools [28]. The selection of mass tolerance for searching possible compositions in computational tools depends on the ion intensity of the glycopeptide and the mass accuracy performance of the specific mass spectrometer. For example, for the first round of search, the mass tolerance may be set at 10 ppm with lock mass correction, and 0.1 Da without lock mass correction (*see* **Note 10**).

14. Prior to the calculation of relative percentage for each glycan, the ion intensities of multiple charge states for the same glycopeptide are combined. Then the relative percentage is calculated by Eq. 3, assuming the same ionization efficiency for all the glycopeptides involved (semiquantitative):

$$\text{Relative percent} = \frac{\textit{Ion}\,\text{intensity of}\,\textit{one}\,\text{glycopeptide}}{\text{Total}\,\textit{ion}\,\text{intensity of}\,\textit{all}\,\text{glycopeptides}} \times 100\% \quad (3)$$

15. Summarize the results from Subheadings 3.3, **steps 13** and **14** into a table. For example, the summary of *N*-linked glycosylation for HC Cα Asn185 is shown in Table 1, where only main glycans (3.8% or higher) are shown for illustration purpose. The peptide involved is SNSAVAWSN^{185}K. The masses of glycopeptides, possible glycan compositions, mass accuracy, and relative percentages are listed.
16. Construct the distribution profile for each glycosylated site, where *X*-axis is the mass of glycan in Dalton, and *Y*-axis is the relative percentage of intensity (*see* **Note 11**). The possible structures are illustrated for better visualization. For example, Fig. 5 shows *N*-linked glycosylation profile for HC Cα Asn185. Each glycan is shown as a vertical bar, and the height of bar reflects the relative abundance of that glycan. The possible

Table 1
Summary of *N*-linked glycosylation for HC Cα Asn185, where only main glycans (3.8% or higher) are shown

Measurement Mass of Glycopeptide (Da)	Possible Composition/Structure	Mass Error (ppm)	Relative Percent (%)
2278.9312	$(Hex)_2 + (Man)_3(GlcNAc)_2$	−0.5	6.7
3122.2458	$(Hex)_2(HexNAc)_2(Deoxyhexose)_1(NeuAc)_1 + (Man)_3(GlcNAc)_2$	0.4	5.9
3196.2769	$(Hex)_3(HexNAc)_3(Deoxyhexose)_1 + (Man)_3(GlcNAc)_2$	−1.3	3.8
3561.4106	$(Hex)_4(HexNAc)_4(Deoxyhexose)_1 + (Man)_3(GlcNAc)_2$	−0.8	7.4
3852.5027	$(Hex)_4(HexNAc)_4(Deoxyhexose)_1(NeuAc)_1 + (Man)_3(GlcNAc)_2$	−1.6	11.2
4055.5815	$(Hex)_4(HexNAc)_5(Deoxyhexose)_1(NeuAc)_1 + (Man)_3(GlcNAc)_2$	−1.6	4.7
4143.6079	$(Hex)_4(HexNAc)_4(Deoxyhexose)_1(NeuAc)_2 + (Man)_3(GlcNAc)_2$	0.9	14.0
4434.6987	$(Hex)_4(HexNAc)_4(Deoxyhexose)_1(NeuAc)_3 + (Man)_3(GlcNAc)_2$	−0.2	8.4

The peptide involved is SNSAVAWSN^{185}K

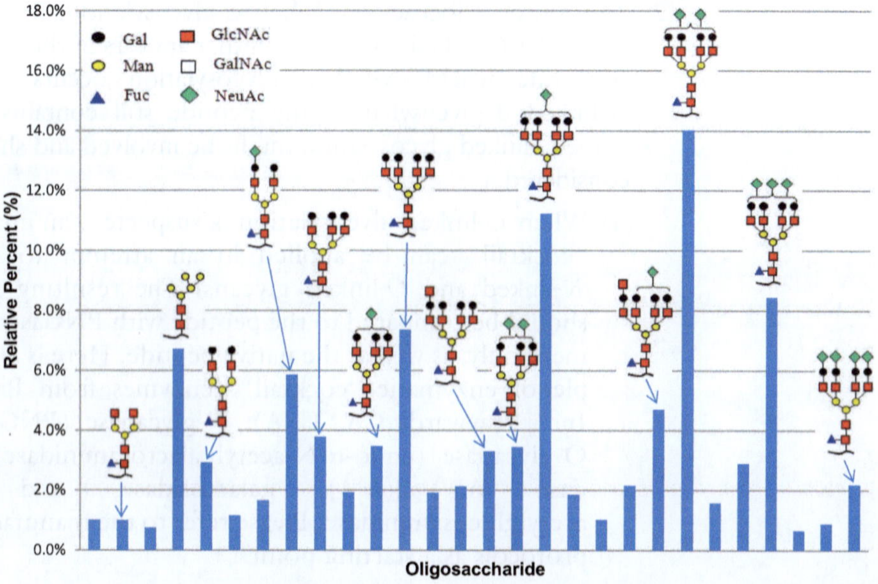

Fig. 5 *N*-Linked glycosylation profile for HC Cα Asn185. Each glycan is shown as a *vertical bar*, and the height of bar reflects the relative abundance of that glycan. The possible branch structures are estimated based on the masses of glycopeptides, as some are illustrated here. This glycosylation site is mainly occupied by bi-, tri-, and tetra-antennary complex type glycans that contain a fucose residue in the core region, while high mannose type is also observed

branch structures are estimated based on the masses of glycopeptides, as some are illustrated here. This glycosylation site is mainly occupied by bi-, tri-, and tetra-antennary complex type glycans that contain a fucose residue in the core region, while high mannose type is also observed. Note, there is a significant amount of terminal galactose moieties. Given that these galactose moieties are spread over the other *N*-linked glycosylation sites, the majority of protein likely contains terminal galactose in at some of its glycans. Free terminal galactose is known to associate with the glycan-binding ASGPR in the liver [8, 9]. The prevalence of terminal galactose on the IgG_TCR protein may explain the rapid serum clearance observed for the majority of the protein when administered into mice [6].

4 Notes

1. This protocol describes the bottom-up approach where the protein is denatured and digested to allow glycopeptides to be analyzed. The top-down approach might not be very informative due to the large size of protein and the numerous and complex glycans.

2. The protocol focuses on *N*-linked glycosylation. Be cautious about the *O*-linked glycosylation that appears in the same glycopeptide where *N*-linked glycosylation occurs. If after *N*-linked deglycosylation, the peptide still contains glycan mass, *O*-linked glycosylation might be involved and should be considered.
   (a) When *O*-linked glycosylation is suspected, an enzymatic "cocktail" can be applied in an attempt to remove *N*-linked and *O*-linked glycans. The resulting peptide should be compared to the peptide with PNGase F treatment only, as well as the native peptide. Here is an example of enzymatic "cocktail" (enzymes from ProZyme Inc., Hayward, CA, USA): N-glycanase (PNGase F), O-glycanase (endo-α-N-acetylgalactosaminidase), sialidase A, β(1–4) galactosidase, and β-N-acetylglucosaminidase. Please refer to the manufacturers' protocols as a starting point.
   (b) The probability of O-linked glycosylation sites might be predicted by computational tools [16].
   (c) If electron-transfer dissociation (ETD) or electron capture dissociation (ECD) is available for peptide backbone fragmentation, the exact sites of both N-linked and O-linked glycosylation could be defined [18].
   (d) Consider using another enzyme to separate the N-linked and O-linked glycosylation sites into different glycopeptides. Depending on the locations of glycosylation and the desired cleavage sites, other enzymes, such as Asp-N, Lys-C, chymotrypsin, Glu-C at pH 4, and Glu-C at pH 7 might be used. Please refer to the manufacturers' protocols for digestion conditions as a start point.
3. If the isolation of glycopeptides from non-glycopeptides is desired to enhance the ionization of glycopeptides, a few strategies are available for general glycopeptide enrichment, such as hydrophilic interaction LC (HILIC), affinity chromatography, and graphitized carbon LC [12, 25].
4. For *N*-linked glycosylation profiling, it is critical to analyze two aliquots in parallel, with and without PNGase F treatment, while keeping all other conditions identical. This helps identify the glycopeptides (unique in the native sample) and eliminate possible false positives of deamidated peptides from deglycosylation. For example, if a peptide with 0.984 Da mass increase appeared in both aliquots with similar ion intensity, most likely it is from the spontaneous deamidated Asn in the same peptide, rather than from the process of deglycosyaltion. The same strategy applies to *O*-linked glycosylation profiling, where the aliquots with and without enzymatic "cocktail" should be analyzed in parallel.

5. More protein material might be needed for peptide mapping, depending on the molecular weight of the glycoprotein, the amount of the glycosylation sites, the glycan heterogeneity, and the sensitivity of the mass spectrometer. The amount of N-glycanase in Subheading 3.1, **step 7** may need to be adjusted accordingly.
6. During the denaturing step, the typical final concentration of guanidine HCl is 6 M for effective denaturation.
7. "Soft" ionization should be used in the mass spectrometer to minimize glycan loss from glycopeptides [29]. A series of cone voltages in ESI should be evaluated to monitor the glycan loss while maintaining reasonable ionization for the glycopeptides.
8. If more than one peptide is liberated containing the Asn of interest (*N*-linked glycosylated) due to enzymatic mis-cleavage or incomplete cut, all the peptides should be included in the glycoprofiling calculation for that Asn.
9. For occupancy calculation in Subheading 3.3, **step 8**, be cautious about spontaneous deamidated Asn in the same peptide, which also appears as +0.984 Da from the expected native peptide, but should be excluded from the calculation for glycosylated peptide.
10. Please note the possible glycan compositions are derived from the accurate mass measurement of glycopeptides, considering the common core $(Man)_3(GlcNAc)_2$, structures observed in literature such as UniCarbKB [27], and the favorable sequential structures. This approach does not provide evidence to distinguish isomers (e.g., GlcNAc and GalNAc), nor evidence of the branch structures. The mass accuracy plays an important role here. During glycan composition search, the mass accuracy of 1 ppm or better provides high confidence in the search result, while a large mass tolerance (e.g., 0.1 Da) may result in more possibilities of the glycan compositions in the search result. When there are multiple possibilities from the search, MSMS data can help decipher the glycan composition, and the experimenter should consider the previously observed glycan structures and favorable sequential glycan structures.
11. This methodology provides possible glycan compositions and structures in a site-specific manner. The relative percentage of each glycan is semiquantitative (Fig. 5) due to differences in ionization efficiency for various glycopeptides.

Acknowledgments

The authors thank Bryan E. Jones and Wolfgang Glaesner for their managerial support, Flora Huang for the purification of TCR proteins, and Jayd Hanna and Benjamin Gutierrez for assistance with transient transfection and 293F cell culture.

References

1. Daniels MA, Hogquist KA, Jameson SC (2002) Sweet 'n' sour: the impact of differential glycosylation on T cell responses. Nat Immunol 3(10):903–910. doi:10.1038/ni1002-903
2. Deprez P, Gautschi M, Helenius A (2005) More than one glycan is needed for ER glucosidase II to allow entry of glycoproteins into the calnexin/calreticulin cycle. Mol Cell 19(2):183–195. doi:10.1016/j.molcel.2005.05.029
3. Demotte N, Stroobant V, Courtoy PJ, Van Der Smissen P, Colau D, Luescher IF, Hivroz C, Nicaise J, Squifflet JL, Mourad M, Godelaine D, Boon T, van der Bruggen P (2008) Restoring the association of the T cell receptor with CD8 reverses anergy in human tumor-infiltrating lymphocytes. Immunity 28(3):414–424. doi:10.1016/j.immuni.2008.01.011
4. Nabi IR, Shankar J, Dennis JW (2015) The galectin lattice at a glance. J Cell Sci 128(13):2213–2219. doi:10.1242/jcs.151159
5. Fujii H, Shinzaki S, Iijima H, Wakamatsu K, Iwamoto C, Sobajima T, Kuwahara R, Hiyama S, Hayashi Y, Takamatsu S, Uozumi N, Kamada Y, Tsujii M, Taniguchi N, Takehara T, Miyoshi E (2016) Core fucosylation on T cells, required for activation of T-cell receptor signaling and induction of colitis in mice. Is Increased in Patients With Inflammatory Bowel Disease. Gastroenterology. doi:10.1053/j.gastro.2016.03.002
6. Wu X, Sereno AJ, Huang F, Zhang K, Batt M, Fitchett JR, He D, Rick HL, Conner EM, Demarest SJ (2015) Protein design of IgG/TCR chimeras for the co-expression of Fab-like moieties within bispecific antibodies. MAbs 7(2):364–376. doi:10.1080/19420862.2015.1007826
7. Liddy N, Bossi G, Adams KJ, Lissina A, Mahon TM, Hassan NJ, Gavarret J, Bianchi FC, Pumphrey NJ, Ladell K, Gostick E, Sewell AK, Lissin NM, Harwood NE, Molloy PE, Li Y, Cameron BJ, Sami M, Baston EE, Todorov PT, Paston SJ, Dennis RE, Harper JV, Dunn SM, Ashfield R, Johnson A, McGrath Y, Plesa G, June CH, Kalos M, Price DA, Vuidepot A, Williams DD, Sutton DH, Jakobsen BK (2012) Monoclonal TCR-redirected tumor cell killing. Nat Med 18(6):980–987. doi:10.1038/nm.2764
8. Seested T, Nielsen HM, Christensen EI, Appa RS (2010) The unsialylated subpopulation of recombinant activated factor VII binds to the asialo-glycoprotein receptor (ASGPR) on primary rat hepatocytes. Thromb Haemost 104(6):1166–1173. doi:10.1160/TH10-06-0356
9. Rifai A, Fadden K, Morrison SL, Chintalacharuvu KR (2000) The N-glycans determine the differential blood clearance and hepatic uptake of human immunoglobulin (Ig) A1 and IgA2 isotypes. J Exp Med 191(12):2171–2182
10. Lenting PJ, Vans CJ, Denis CV (2007) Clearance mechanisms of von Willebrand factor and factor VIII. J Thromb Haemost 5(7):1353–1360. doi:10.1111/j.1538-7836.2007.02572.x
11. Lohse S, Meyer S, Meulenbroek LA, Jansen JH, Nederend M, Kretschmer A, Klausz K, Moginger U, Derer S, Rosner T, Kellner C, Schewe D, Sondermann P, Tiwari S, Kolarich D, Peipp M, Leusen JH, Valerius T (2016) An anti-EGFR IgA that displays improved pharmacokinetics and myeloid effector cell engagement in vivo. Cancer Res 76(2):403–417. doi:10.1158/0008-5472.CAN-15-1232
12. Kolli V, Schumacher KN, Dodds ED (2015) Engaging challenges in glycoproteomics: recent advances in MS-based glycopeptide analysis. Bioanalysis 7(1):113–131. doi:10.4155/bio.14.272
13. Chuang GY, Boyington JC, Joyce MG, Zhu J, Nabel GJ, Kwong PD, Georgiev I (2012) Computational prediction of N-linked glycosylation incorporating structural properties and patterns. Bioinformatics 28(17):2249–2255. doi:10.1093/bioinformatics/bts426
14. Stanley P, Schachter H, Taniguchi N (2009) N-Glycans. In: Varki A, Cummings RD, Esko JD et al (eds) Essentials of glycobiology, 2nd edn. Cold Spring Harbor Laboratory Press, Cold Spring Harbor, NY
15. Taga EM, Waheed A, Van Etten RL (1984) Structural and chemical characterization of a homogeneous peptide N-glycosidase from almond. Biochemistry 23(5):815–822
16. Steentoft C, Vakhrushev SY, Joshi HJ, Kong Y, Vester-Christensen MB, Schjoldager KT, Lavrsen K, Dabelsteen S, Pedersen NB, Marcos-Silva L, Gupta R, Bennett EP, Mandel U, Brunak S, Wandall HH, Levery SB, Clausen H (2013) Precision mapping of the human O-GalNAc glycoproteome through SimpleCell technology. EMBO J 32(10):1478–1488. doi:10.1038/emboj.2013.79
17. Nishikawa I, Nakajima Y, Ito M, Fukuchi S, Homma K, Nishikawa K (2010) Computational prediction of O-linked glycosylation sites that preferentially map on intrinsically disordered

regions of extracellular proteins. Int J Mol Sci 11(12):4991–5008. doi:10.3390/ijms11124991

18. North SJ, Hitchen PG, Haslam SM, Dell A (2009) Mass spectrometry in the analysis of N-linked and O-linked glycans. Curr Opin Struct Biol 19(5):498–506. doi:10.1016/j.sbi.2009.05.005
19. Brockhausen I, Schachter H, Stanley P (2009) O-GalNAc Glycans. In: Varki A, Cummings RD, Esko JD et al (eds) Essentials of glycobiology, 2nd edn. Cold Spring Harbor Laboratory Press, Cold Spring Harbor, NY
20. Kolarich D, Lepenies B, Seeberger PH (2012) Glycomics, glycoproteomics and the immune system. Curr Opin Chem Biol 16(1–2):214–220. doi:10.1016/j.cbpa.2011.12.006
21. Morelle W, Canis K, Chirat F, Faid V, Michalski JC (2006) The use of mass spectrometry for the proteomic analysis of glycosylation. Proteomics 6(14):3993–4015. doi:10.1002/pmic.200600129
22. Kuraya N, Hase S (1992) Release of O-linked sugar chains from glycoproteins with anhydrous hydrazine and pyridylamination of the sugar chains with improved reaction conditions. J Biochem 112(1):122–126
23. Edge AS (2003) Deglycosylation of glycoproteins with trifluoromethanesulphonic acid: elucidation of molecular structure and function. Biochem J 376(Pt 2):339–350. doi:10.1042/BJ20030673
24. Hogan JM, Pitteri SJ, Chrisman PA, McLuckey SA (2005) Complementary structural information from a tryptic N-linked glycopeptide via electron transfer ion/ion reactions and collision-induced dissociation. J Proteome Res 4(2):628–632. doi:10.1021/pr049770q
25. Alley WR Jr, Mann BF, Novotny MV (2013) High-sensitivity analytical approaches for the structural characterization of glycoproteins. Chem Rev 113(4):2668–2732. doi:10.1021/cr3003714
26. Cooper CA, Gasteiger E, Packer NH (2001) GlycoMod—a software tool for determining glycosylation compositions from mass spectrometric data. Proteomics 1(2):340–349. doi:10.1002/1615-9861(200102)1:2<340::AID-PROT340>3.0.CO;2-B
27. Campbell MP, Peterson R, Mariethoz J, Gasteiger E, Akune Y, Aoki-Kinoshita KF, Lisacek F, Packer NH (2014) UniCarbKB: building a knowledge platform for glycoproteomics. Nucleic Acids Res 42(Database issue):D215–D221. doi:10.1093/nar/gkt1128
28. Shubhakar A, Reiding KR, Gardner RA, Spencer DI, Fernandes DL, Wuhrer M (2015) High-throughput analysis and automation for glycomics studies. Chromatographia 78(5–6):321–333. doi:10.1007/s10337-014-2803-9
29. Ivancic MM, Gadgil HS, Halsall HB, Treuheit MJ (2010) LC/MS analysis of complex multiglycosylated human alpha(1)-acid glycoprotein as a model for developing identification and quantitation methods for intact glycopeptide analysis. Anal Biochem 400(1):25–32. doi:10.1016/j.ab.2010.01.026

Chapter 13

A Proximity-Based Assay for Identification of Ligand and Membrane Protein Interaction in Living Cells

Hongkai Zhang and Richard A. Lerner

Abstract

There is a need for a general reporter system to study the interactions between a ligand and membrane protein in living cells. Here, we demonstrate a method that allows identification of undiscovered ligand partners of a membrane protein in its natural milieu. This assay will be of importance especially for multiple pass transmembrane proteins such as ion channels and GPCRs.

Key words Proximity-based assay, Ligand, Membrane protein, Interaction, Reporter cell, Lentivirus

1 Introduction

A variety of two hybrid systems and protein fragment complementation (PCA) have been employed to study protein–protein interactions in living cells [1, 2]. Although these techniques are widely used to study cytoplasm protein interactions [3], membrane protein interactions [4], RNA–protein interactions [5], and even small molecule–protein interactions [6], there is no general-purpose reporter system to study the interactions between ligand and membrane protein in living cells.

We reason that one could take advantage of proximity effect, operating through effective molarity, to achieve this goal. The concept of effective molarity relates to the kinetic advantage of an intramolecular reaction relative to its intermolecular counterpart [7].

In the proximity-based assay, the ligand is displayed on cell surface by fusing to the N-terminus of platelet-derived growth factor receptor transmembrane domain (PDGFR-TM) and the Tobacco Etch Virus (TEV) protease is fused to the C-terminus of the PDGFR-TM. The membrane protein of interest is fused to a TEV substrate sequence and an artificial transcription factor GAL4-VP16 at its C-terminus. The interaction between the membrane tethered ligand and the membrane protein of interest causes

Thomas Tiller (ed.), *Synthetic Antibodies: Methods and Protocols*, Methods in Molecular Biology, vol. 1575,
DOI 10.1007/978-1-4939-6857-2_13, © Springer Science+Business Media LLC 2017

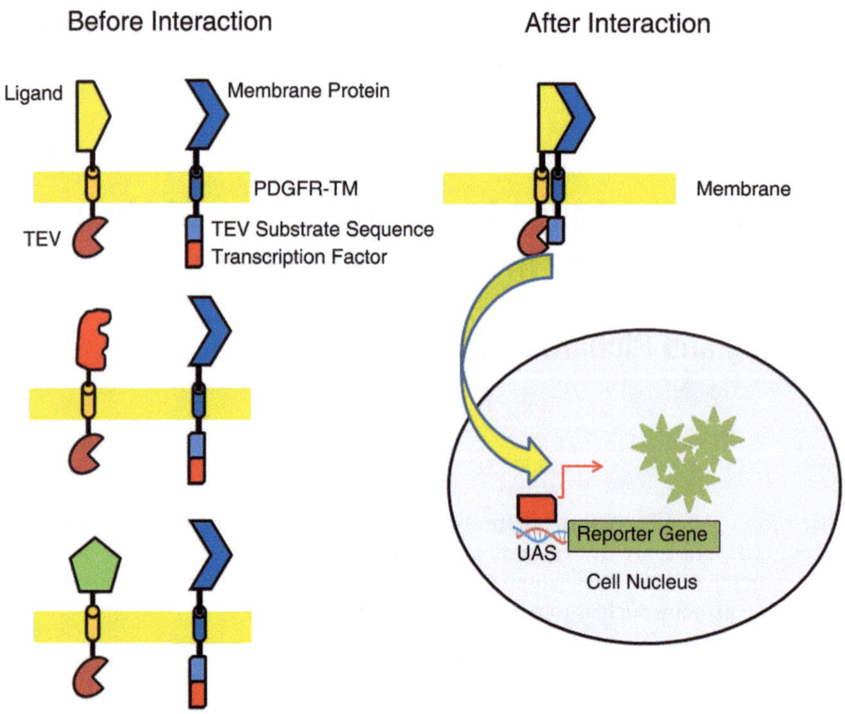

Fig. 1 Schematic of proximity-based method. Membrane tethered peptides are coupled to the TEV protease on their cytoplasmic sides. Membrane protein is fused to a TEV substrate sequence and an artificial transcription factor GAL4-VP16 on its cytoplasmic side. Interaction of co-located membrane protein and ligand approximates the TEV and its substrate sequence. Cleavage of the substrate sequence by TEV protease releases the transcription factor for expression of the reporter gene. (Reproduced from [8] with permission from Elsevier)

approximation of the TEV protease and its substrate sequence, resulting in increased catalytic reaction. After cleavage of the TEV substrate sequence, the transcription factor is released to activate the expression of the reporter gene (Fig. 1) [8].

2 Materials

The proximity-based assay can be set up with basic molecular laboratory instruments and reagents.

2.1 Cell Culture

1. HEK-293 (ATCC).
2. HEK-293/HEK-293FT Complete Growth Medium: 500 mL DMEM (high-glucose) (ThermoFisher Scientific) supplemented with 50 mL fetal bovine serum (ThermoFisher Scientific), 5 mL 10 mM nonessential amino acids (ThermoFisher Scientific), and 12.5 mL 1 M HEPES (ThermoFisher Scientific).
3. Puromycin (Invivogen).
4. Hygromycin B (Invivogen).

5. Opti-MEM I Reduced Serum Medium (ThermoFisher Scientific).
6. Lipofectamine 2000 Transfection Reagent (ThermoFisher Scientific).
7. Trypsin (ThermoFisher Scientific).
8. Lenti-X p24 Rapid Titer Kit (Clontech).
9. Polybrene (Sigma-Aldrich).
10. Luciferase Assay System (Promega).
11. QIAprep Spin Miniprep Kit (QIAGEN).
12. Infinite 200 PRO multimode plate reader (Tecan).

2.2 Assay Components

1. The pReceptor vector is used for expression of the membrane protein construct (Fig. 2a). The membrane protein of interest can be cloned into Multiple Cloning Site (MCS) of pReceptor to express a fusion protein with the TEV substrate sequence (*see* **Note 1**) and the artificial transcription factor at its C-terminus. The artificial transcription factor GAL4-VP16 is composed of *Saccharomyces cerevisiae* GAL4 DNA binding domain of Upstream Activated Sequence (UAS) (*see* **Note 2**) and herpesvirus protein VP16 transcriptional activation domain. The entire cassette is flanked by genomic insulator elements for stabilized expression of protein of interest.

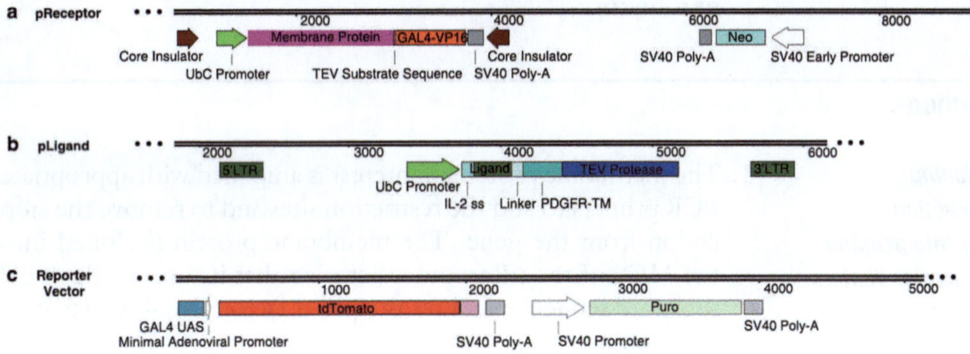

Fig. 2 Proximity-based assay vectors. (**a**) pReceptor vector is used for expression of the membrane protein in-frame fusion with an optimized TEV protease substrate sequence and GAL4-VP16 at its C-terminus under control of human Ubiquitin C (UbC) promoter. The entire cassette is flanked by genomic insulator elements for stabilized expression. Expression of neomycin resistance gene (Neo) enables selection of stable cell line with Geneticin. (**b**) pLigand lentivirus transfer vector is used for ligand expression and based on the third generation of lentiviral transfer vector. UbC promoter is used for low level expression of ligand in-frame fusion with PDGFR-TM and TEV protease at its C-terminus. IL-2 leader sequence targets the ligand construct to secretory pathway. PDGFR-TM anchors the ligand to the plasma membrane. (**c**) Reporter vector contains seven UAS for the GAL4 DNA binding domain upstream of a minimal adenoviral promoter that is designed for transcriptional activation of the reporter genes Luc2 by association of the artificial transcription factor GAL4-VP16. Expression of puromycin resistance gene (Puro) enables selection of stable cell line by puromycin

Expression of neomycin resistance gene under control of SV40 early promoter allows selection of stable cell line with Geneticin (*see* **Note 3**).

2. The pLigand lentivirus transfer plasmid is used for expression of the ligand construct and based on the third generation of lentiviral vector (Fig. 2b). The ligand expression cassette is flanked by long terminal repeats (LTRs) that facilitate integration into the genome. The ligand is cloned into MCS of pLigand to express a fusion protein with PDGFR-TM and TEV protease at its C-terminus (*see* **Note 4**). Murine Interleukin-2 (IL-2) leader sequence targets the ligand construct to the secretory pathway. PDGFR-TM anchors the ligand to the plasma membrane. Linker between the ligand and PDGFR-TM consists of Gly-Gly-Gly-Gly-Ser repeats (*see* **Note 5**). TEV protease releases the transcription factor GAL4-VP16 from the membrane protein.
3. The third generation lentivirus packaging vector contains all necessary viral structure proteins and the envelope vector expresses Vesicular Stomatitis Virus glycoprotein (VSVg).
4. Reporter vector contains seven UAS upstream of a minimal adenoviral promoter, which is designed for transcriptional activation of the reporter genes firefly luciferase gene, luc2 by association of the artificial transcription factor GAL4-VP16 (Fig. 2c). Expression of puromycin resistance gene under control of SV40 promoter allows selection of stable cell line by puromycin.

3 Methods

3.1 Cloning the Interaction Proteins into pLigand and pReceptor Vectors

1. The membrane protein of interest is amplified with appropriate PCR primers to add the restriction sites and to remove the stop codon from the gene. The membrane protein is cloned into the MCS of the pReceptor vector so that it fuses to the TEV substrate sequence and transcription factor GAL4-VP16 at its C-terminus.
2. The ligand is cloned into the MCS of the pLigand vector so that it fuses to PDGFR-TM and TEV protease at its C-terminus.
3. The plasmids are isolated for transfection using QIAprep Spin miniprep Kit according to the manufacturer's instructions (*see* **Note 6**).

3.2 Generation of the Stable Cell Line UAS-Luc2 HEK-293 Containing Stably Integrated Reporter Gene Firefly Luciferase Under Control of the UAS Response Element

1. Plate two wells of HEK-293 cells in complete growth medium in a 6-well plate so that the cell confluency reaches 90% by the second day.
2. Transfect cells with the reporter vector. Dilute 2 μg reporter vector with 250 μL Opti-MEM medium and mix gently. Gently mix 4 μL Lipofectamine 2000 to another 250 μL Opti-MEM medium and incubate at room temperature for 5 min. Gently mix diluted Lipofectamine and DNA and incubate the mixture at room temperature for 30 min. Add 500 μL Lipofectamine-DNA complexes drop-wise to each well of cells and gently rock the plate back and forth to ensure that the complexes are dispersed evenly. Place the plate in the 37 °C, 5% CO_2 incubator overnight.
3. Trypsinize cells the next day and transfer both wells of cells into one T125 flask. Place the flask at 37 °C, 5% CO_2 incubator overnight.
4. After cells attach to flask, aspirate the medium and replace with selective medium containing 4 μg/mL puromycin.
5. Exchange the medium and feed cells with fresh selective medium every 3–4 days for the next 2 weeks.

3.3 Generation of the Membrane Protein Expressing UAS-Luc2 HEK-293 Cell Line

1. Plate two wells of UAS-Luc2 HEK-293 cells in a 6-well plate so that the cell confluency reaches 90% the second day.
2. Transfect cells with pReceptor vector carrying the membrane protein of interest. Dilute 2 μg of DNA in 250 μL Opti-MEM medium and mix gently. Gently mix 4 μL of Lipofectamine with another 250 μL Opti-MEM Medium and incubate at room temperature for 5 min. Gently mix diluted Lipofectamine and DNA and incubate at room temperature for 30 min. Add the Lipofectamine-DNA complexes drop-wise to the well and gently rock the plate back and forth to ensure the complexes are dispersed evenly. Place the plate in the 37 °C, 5% CO_2 incubator overnight.
3. Trypsinize cells the next day and transfer 2 wells of cell into a T125 flask. Place the flask at 37 °C, 5% CO_2 incubator overnight.
4. After the cells attach to the flask, aspirate the medium and replace with selective medium containing 600 μg/mL Geneticin (*see* **Note 7**).
5. Exchange the medium and feed cells with fresh selective medium every 3–4 days for the next 2–3 weeks (*see* **Note 8**).

3.4 Preparation of Lentivirus

1. Plate two wells of HEK-293FT cells in a 6-well plate so that cells reach 80–90% confluency the second day (*see* **Note 9**).
2. Co-transfect the cells with the lentivirus transfer plasmid, packaging plasmid and envelop plasmid mixture. Dilute 3 μg

of DNA (equal mixture of 1 μg pLigand lentivirus transfer plasmid, 1 μg packaging plasmid pCMV, and 1 μg envelop plasmid pVSVg) in 250 μL Opti-MEM medium. Gently mix 8 μL of Lipofectamine with another 250 μL Opti-MEM medium and incubate at room temperature for 5 min. Gently mix dilute Lipofectamine and DNA and incubate at room temperature for 30 min. Add the Lipofectamine-DNA complexes drop-wise to the well and gently rock the plate back and forth to ensure the complexes are dispersed evenly. Place the plate in the 37 °C, 5% CO_2 incubator overnight.

3. Change the fresh growth medium the next morning (*see* **Notes 10** and **11**).
4. Collect the virus containing medium in sterile capped tubes 72 h post-transfection (*see* **Note 12**).
5. Centrifuge the virus containing medium at 1000 × *g* at 4 °C for 10 min to remove cell debris and filter the supernatant through 0.45 μm filter (*see* **Note 13**).
6. Filtered supernatant can be stored at 4 °C for up to 1 week or aliquoted and stored at −20 °C for 1 month. Minimize freeze-thaw cycles.

3.5 Assessing Membrane Protein–Ligand Interaction with Proximity-Based Assay

1. Plate the membrane protein expressing UAS-Luc2 HEK-293 cells at 20,000 cells per well in a 96-well plate.
2. The next day replace the growth medium with medium containing 4 μg/mL polybrene. Cells are infected with lentivirus carrying the to-be-tested ligand at multiplicity of infection (MOI) equal to 1 (*see* **Notes 14** and **15**).
3. Infected cells are cultured for 24–72 h before measuring the luciferase activity. Remove the growth medium from cultured cells.
4. Dispense 100 μL 1× lysis buffer per well. Pipette up and down a few times and incubate for 5 min to ensure complete lysis.
5. Mix 20 μL of cell lysate with 100 μL of luciferase substrate and measure the luminescence produced.

4 Notes

1. A systematic study demonstrated Glu-Asn-Phe-Tyr-Phe-Gln-Ser was the optimum TEV substrate sequence in proximity-based assay resulting in high signal-to-noise ratio of the reporter cell line (*see* ref. 8).
2. The regular expression of upstream activation sequence for *Saccharomyces cerevisiae* GAL4 recognition is 5′-CGGRNNRCYNYNYNCNCCG-3′ (R=G/A, Y=C/T,

N=A/T/C/G). It is a cis-acting regulatory sequence distinct from promoter and is considered to be analogous to enhancer due to its essential role in activating transcription.

3. Geneticin, commonly known as G418, can be used for stable cell selection based on HEK-293 cell. HEK-293T cell is G418 resistant due to the presence of neomycin resistance gene cassette.
4. PDGFR-TM domain corresponds to amino acids 514–561 of PDGFR.
5. The length of linker between the ligand and PDGFR-TM has to be tailor-made for individual receptor and channel. The length varies from 10 to 20 amino acids.
6. Plasmid DNA for transfection into HEK-293/HEK-293FT must be sterile and free from chemical contaminants.
7. Performing a kill curve experiment is recommended. Kill curve is a dose-response experiment where the cells are subjected to increasing amounts of antibiotic to determine the minimum antibiotic concentration needed to kill all the cells over the course of 2 weeks.
8. The non-stably transfected cells should die off within 5 days while the stably transfected cells should grow and they become visible to the naked eye as a colony at the end of selection.
9. Studies have shown that maximal lentivirus production in HEK-293FT cell.
10. The HEK-293FT cells detach very easily after transfection for lentivirus preparation. Carefully add the fresh medium to avoid detachment of the cells.
11. From this point on, supernatants contain infectious lentiviral particles. Follow the recommended guidelines for working with Biosafety Level 2 (BL2) organisms.
12. Expression of the VSV glycoprotein causes disruption of plasma membrane of HEK-293FT, large, multinucleated adherent cells known as syncytia can be observed. Meanwhile, many individual cells are suspended in the culture medium. This morphological change is a sign of high titer lentivirus generation.
13. Use cellulose acetate or polyethersulfone (PES) filters (low protein binding). Do not use nitrocellulose filters as nitrocellulose is known to bind lentivirus and reduce titers.
14. We recommend validating the reporter cell line with both positive control (if a peptide ligand exists) and negative control before the cell line is used to screen for the identification of unknown ligand partners of the membrane protein.
15. The lentivirus titer can be rapidly determined with p24 ELISA. P24 values can be used to determine the relative virus titers.

To calibrate a relationship between p24 level and infectivity, determine the p24 level of lentivirus required for MOI equal to 1 in fluorescence-based titering. The MOI is equal to 1 when the transduction efficiency of ~60% is achieved.

Acknowledgment

This work was supported by JPB foundation and Zebra Biologics.

References

1. Fields S, Song O (1989) A novel genetic system to detect protein–protein interactions. Nature 340:245–246
2. Johnsson N, Varshavsky A (1994) Split ubiquitin as a sensor of protein interactions in vivo. Proc Natl Acad Sci U S A 91:10340–10344
3. Barnea G, Strapps W, Herrada G, Berman Y, Ong J, Kloss B, Axel R, Lee KJ (2008) The genetic design of signaling cascades to record receptor activation. Proc Natl Acad Sci U S A 105:64–69. doi:10.1073/pnas.0710487105
4. Petschnigg J, Wong V, Snider J, Stagljar I (2012) Investigation of membrane protein interactions using the split-ubiquitin membrane yeast two-hybrid system. Methods Mol Biol 812:225–244. doi:10.1007/978-1-61779-455-1_13
5. Stumpf CR, Opperman L, Wickens M (2008) Analysis of RNA-protein interactions using a yeast three-hybrid system. Methods Enzymol 449:295–315. doi:10.1016/S0076-6879(08)02414-2
6. Chidley C, Haruki H, Pedersen MG, Muller E, Johnsson K (2011) A yeast-based screen reveals that sulfasalazine inhibits tetrahydrobiopterin biosynthesis. Nat Chem Biol 7:375–383. doi:10.1038/nchembio.557
7. Kirby AJ (1981) Effective Molarities for Intramolecular Reactions. Adv Phys Org Chem 17:183–278
8. Zhang H, Xie J, Lerner RA (2014) A proximity based general method for identification of ligand and receptor interactions in living cells. Biochem Biophys Res Commun 454:251–255. doi:10.1016/j.bbrc.2014.10.085

Chapter 14

A Biotin Ligase-Based Assay for the Quantification of the Cytosolic Delivery of Therapeutic Proteins

Wouter P.R. Verdurmen, Marigona Mazlami, and Andreas Plückthun

Abstract

The efficient delivery of external proteins from the external milieu to the cytosol of mammalian cells has great potential for both scientific investigations and future therapies. However, when assessing the cellular uptake of proteins, it is often difficult to distinguish between proteins that are stuck in the endosomes and those that have escaped into the cytosol. Here, we describe a method employing the prokaryotic enzyme biotin ligase that overcomes this problem and which can be employed for a highly sensitive quantification of cytosolic protein delivery.

Key words Biotin, Biotin ligase, Cellular internalization, Cytosolic protein delivery, Streptavidin, Western blot, Protein engineering

1 Introduction

Over the years, many different techniques have been proposed for the quantification of protein delivery into cells. The most popular ones rely either on fluorescence or biological readouts, which, however, each have their own serious drawbacks [1–4]. Fluorescence-based methods typically cannot distinguish well between fluorescence signals from the endosomes and fluorescence signals from the cytosol. Furthermore, cellular fluorescence is difficult to reliably quantify [5, 6]. Additionally, great care must be taken to prevent that the fluorescence microscopy itself might be skewing the result, either through the handling of the cells, especially fixation, or through contributing effects of the fluorophore to endosomal escape, chemically or photochemically. Biological readouts, if carefully controlled, can give direct information about cytosolic delivery, but are generally not quantitative either and require the delivery of an intact "assay" protein such as cre-recombinase [4] or ubiquitin [1]. For the former, only one active molecule can be sufficient and carry out one turnover to be detectable.

Thomas Tiller (ed.), *Synthetic Antibodies: Methods and Protocols*, Methods in Molecular Biology, vol. 1575,
DOI 10.1007/978-1-4939-6857-2_14, © Springer Science+Business Media LLC 2017

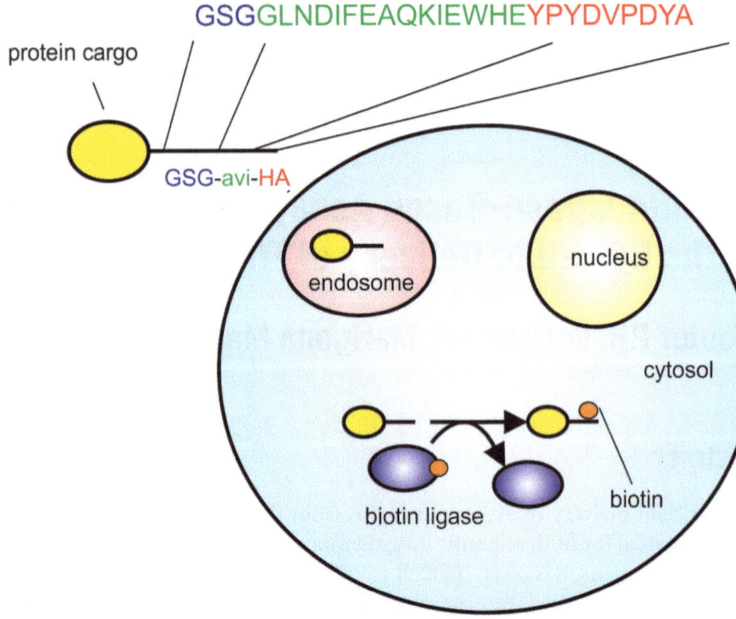

Fig. 1 Principle of the biotin ligase assay. The avi tag fused to a cargo protein gets biotinylated exclusively in the cytosol via an overexpressed bacterial biotin ligase. The biotinylated fraction of the cargo protein represents the fraction that has reached the cytosol. Additionally, the HA tag on the cargo protein remains unchanged during the internalization process and its presence in cell lysates thus reflects total cellular internalization (endosomes or cytosol)

To obtain reliable quantitative information on cytosolic protein delivery, we developed an assay that relies on the cytosolic overexpression of the prokaryotic enzyme biotin ligase and the fusion of a short peptide tag to a cargo protein (Fig. 1) [7]. While all cells express biotin ligases, only the bacterial enzyme BirA biotinylates a peptide of 15 amino acids, referred to as the avi tag, with a very high efficiency, and has no known mammalian targets [8, 9], and conversely, the avi tag is not a substrate of the mammalian enzymes. The assay is applicable to *any* protein and functions by incubating cargo proteins with cells that stably overexpress biotin ligase in the cytosol. The assay thus measures whether the substrate has reached the enzyme.

After the desired incubation time, cells are lysed under harsh conditions immediately, thus avoiding the potential occurrence of biotinylation events after lysis. Lysates are then analyzed by western blotting and probed for the presence of a cargo protein modified with biotin, using fluorescently labeled streptavidin. In parallel, the lysates are probed for the presence of an HA tag, also fused to the cargo protein, which does not get modified during or after transport and thus reflects the total amount of internalized protein. (We refer to "internalized" as the sum of protein in endosomes and the cytosol; in contradistinction to "cytosolic".)

Important controls that are performed in every single assay ensure that the biotinylation of cytosolically localized proteins is complete and that there is no detectable biotinylation after cell lysis. The biotin and HA-derived bands of the protein of interest on the western blots thus separately identify the cytosolic fraction and the total internalized fraction (Fig. 2). The method is highly sensitive, as low nanomolar cytosolic concentrations of delivered proteins can still be reliably quantitated.

Three key features enable the highly sensitive absolute quantitation that can be achieved: first, the exceptionally strong affinity between biotin and streptavidin, reported to be 10^{-15} M [10, 11]; second, the broad linear correlation between the amount of protein present and the signal intensity when using a fluorescence scanner [7], in contrast to detection systems that rely on an enzymatic signal amplification, which are generally less suitable for absolute quantification purposes; third, the SDS-PAGE-based separation of the biotinylated protein of interest from the very low

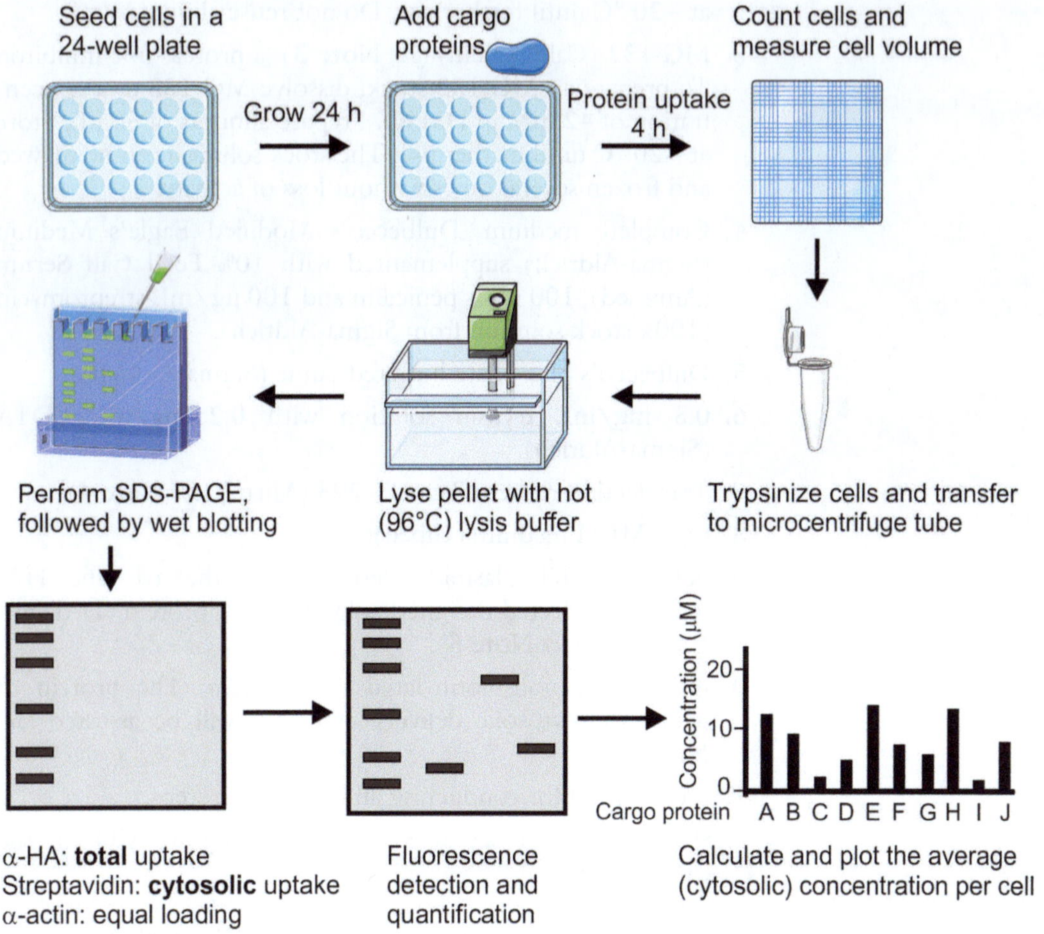

Fig. 2 Schematic representation of the most important steps of the biotin ligase assay

levels of endogenous biotinylated proteins. Altogether, the assay provides a robust way to gain a quantitative insight into the cytosolic delivery of any protein of interest.

2 Materials

All solutions that will be used for live cell culture should be sterile or sterilized before use. Solutions that are used after lysing the cells do not have to be sterile per se, but we highly recommend working with filtered buffers, as unseen particles may cause unwanted background problems in western blotting.

2.1 Cell Culture Reagents

1. Flp-In 293 cell line stably overexpressing biotin ligase (*see* **Note 1**).
2. D-Biotin. To prepare a D-Biotin stock, dissolve D-Biotin to a concentration of 4 mM in water (*see* **Note 2**). Sterile-filter the D-Biotin in a flow-hood. Prepare aliquots of 1 mL and store at −20 °C until further use. Do not reuse aliquots.
3. MG-132 (Calbiochem) (*see* **Note 3**), a proteasome inhibitor. To prepare an MG-132 stock, dissolve MG-132 to a concentration of 42 mM in DMSO. Prepare aliquots of 50 μL. Store at −20 °C until further use. The stock solution can be thawed and frozen several times without loss of activity.
4. Complete medium, Dulbecco's Modified Eagle's Medium (Sigma-Aldrich) supplemented with 10% Fetal Calf Serum (Amimed), 100 units penicillin and 100 μg/mL streptomycin (100× stock solution from Sigma-Aldrich).
5. Dulbecco's phosphate buffered saline (Sigma-Aldrich).
6. 0.5 mg/mL trypsin solution with 0.2 mg/mL EDTA (Sigma-Aldrich).
7. Transfection reagent TransIT 293 (Mirus) (*see* **Note 4**).
8. Opti-MEM medium (Gibco).
9. pcDNA5/FRT plasmid overexpressing the avi- and HA-tagged unselected designed ankyrin repeat protein (DARPin) E3_5 [12] (*see* **Note 5**).
10. Avi-tagged, non-biotinylated test protein: The protein of which the cytosolic delivery efficiency will be assessed (*see* **Note 6**).
11. Sterile hood for conducting all preparation steps.
12. Sterile incubator set at 37 °C containing 5% CO_2 in the atmosphere.

2.2 Lysis and Western Blotting Components

1. Stabilized avi tag peptide: (Ac-GGLNDIFEAQKIEWHED–NH_2). Purchased from EMC Microcollections. The peptide is N-terminally acetylated and C-terminally amidated to protect

against exoproteolytic degradation. It contains an additional amino acid at either terminus in addition to the 15 amino acids that make up the avi tag. The peptide is to be used as a competitive substrate. Prepare a 10 mM stock in DMSO.

2. Sodium dodecyl sulfate (SDS) lysis buffer: 100 mM Tris–HCl, pH 6.8, 4% (w/v) SDS, 10% (v/v) glycerol, 175 mM β-mercaptoethanol, and 0.02% (w/v) bromophenol blue.
3. Digitonin-based extraction buffer: 25, 50 or 100 μg/mL digitonin, 75 mM NaCl, 1 mM NaH_2PO_4, 8 mM Na_2HPO_4, 250 mM sucrose, 0.1 mM 4-(2-aminoethyl) benzenesulfonyl fluoride hydrochloride (AEBSF), 1 μM leupeptin, and 1.4 μM pepstatin-A.
4. A heating block that can reach temperatures of up to 96 °C.
5. 5× SDS-PAGE sample buffer: 175 mM Tris–HCl, pH 6.8, 10% (w/v) SDS, 50% (v/v) glycerol, 286 mM β-mercaptoethanol, 0.15% (w/v) bromophenol blue.
6. Non-denaturing lysis buffer: 0.5 M Tris–HCl, pH 6.8, 20% (v/v) glycerol, 0.02% (w/v) bromophenol blue.
7. Streptavidin (Promega). Dissolve to a concentration of 1 mg/mL in PBS. Shock-freeze in liquid nitrogen and store aliquots at −20 °C. Thaw aliquots only once and use on the same day.
8. Ammonium persulfate (APS) (Amresco). Prepare a 10% (w/v) solution in water. Store aliquots at −20 °C. A working aliquot may be kept at 4 °C for several months.
9. *N*,*N*,*N*′,*N*′-tetramethyl-ethylenediamine (TEMED) (Sigma-Aldrich).
10. Resolving gel 4× stock solution: 1.5 M Tris–HCl, pH 8.8. Store at 4 °C for up to 6 months.
11. Stacking gel 4× stock solution: 0.5 M Tris–HCl, pH 6.8. Store at 4 °C for up to 6 months.
12. Acrylamide solution: Rotiphorese® Gel 30 (37.5:1): 30% acrylamide/bisacrylamide, mixing ratio 37.5:1 (Roth).
13. SDS-PAGE running buffer: 25 mM Tris–HCl, 0.192 M glycine, 0.1% (w/v) SDS, pH 8.3.
14. PageRuler: Prestained protein ladder (Thermo Scientific).
15. Western blot transfer buffer: 25 mM Tris–HCl, 0.192 M glycine, 20% (v/v) methanol, pH 8.3 (*see* **Note 7**).
16. Blotting paper (Macherey-Nagel).
17. Immobilon-FL PDVF membrane (Millipore) (*see* **Note 8**).
18. 10× casein blocking buffer (Sigma-Aldrich). Prepare 15 mL aliquots and store them at −20 °C. Thaw only once and use within several weeks (*see* **Note 9**).
19. Phosphate-buffered saline (PBS): 137 mM NaCl, 12 mM phosphate, 2.7 mM KCl, pH 7.4. To make 1 L of 10× PBS, combine 80 g NaCl, 2 g KCl, 14.4 g Na_2HPO_4, and 2.4 g

KH_2PO_4. Dilute tenfold to make 1× PBS. The pH should be 6.8 before dilution, and close to 7.4 after dilution. Adjust if necessary.

20. PBS-T: PBS supplemented with 0.1% (v/v) Tween-20.
21. Antibody incubation buffer: PBS-T + 1× casein blocking buffer.
22. 10% (w/v) SDS. Dissolve 5 g SDS in 50 mL. Mix slowly to avoid foam formation. Store at room temperature up to 6 months before discarding.
23. Streptavidin incubation buffer: PBS-T + 1× casein blocking buffer + 0.1% (w/v) SDS.
24. Mini-PROTEAN® Tetra system.
25. Mini-PROTEAN® Tetra cell casting module.
26. Mini-PROTEAN® short plates and spacer plates.
27. 15-well 1 mm combs.

2.3 Antibodies and Detection Reagents

1. Streptavidin IRDye 680LT (LI-COR Biosciences).
2. Polyclonal anti-HA antibody (H6908, Sigma-Aldrich).
3. Mouse anti-actin (ab8224, Abcam).
4. Goat anti-rabbit Alexa Fluor 680 (Invitrogen).
5. Donkey anti-mouse IRDye 800 (Rockland Immunochemicals Inc.).
6. Goat anti-rabbit IRDye 800CW (LI-COR Biosciences).
7. Odyssey IR scanner (LI-COR Biosciences).

3 Methods

Until the point of cell lysis, all procedures should be performed with sterile materials in a sterile flow hood.

3.1 Incubation of Cells with Proteins

1. Seed 300,000 Flp-In 293 cells stably overexpressing biotin ligase in a 24-well plate (*see* **Note 1**). The cells will have a confluence of ~60–80% 24 h after seeding. Use 400 μL complete medium per well for seeding the cells.
2. 24 h after seeding, transiently transfect the cells from a single well on the 24-well plate with a plasmid containing the DARPin HA_E3_5_avi, which carries both an HA- and an avi-tag (*see* **Note 10**). For transfection, combine 1.5 μL TransIT 293 with 500 ng sterile-filtered plasmid DNA in 50 μL Opti-MEM medium. Incubate at room temperature for 30 min to allow complex formation. Add transfection mixture dropwise to the well. Do not exchange the medium in the well before or after the addition of the transfection mixture.

3. Prepare incubation medium: add 0.1 mM sterile-filtered D-Biotin from a 4 mM D-Biotin stock solution to the complete medium (*see* **Note 11**). Then, add the proteasome inhibitor MG-132 to a final concentration of 50 μM, 840× diluted from a 42 mM stock (*see* **Note 12**).
4. 48 h after seeding, incubate cells with cargo proteins to be tested for transport that contain both an avi tag and an HA tag (*see* **Note 13**). Dilute the protein before addition in incubation medium to the desired final concentration, reaching a total volume of 450 μL. Mix well. Fully exchange the medium in the wells with 400 μL of the incubation medium containing the cargo proteins of interest. As soon as the medium has been exchanged, place the 24-well plate back in the incubator at 37 °C for 4 h (*see* **Note 14**).

3.2 Cell Lysate Preparation

1. Add 20 μM of the stabilized avi tag peptide to the SDS lysis buffer (dilute 500-fold from a 10 mM stock). The stabilized avi tag peptide serves as a competitive biotin ligase substrate, such that if any ligase was not quenched completely after lysis with hot SDS-containing buffer, it would mostly work on the synthetic peptide substrate (*see* **Note 15** and Fig. 3). Preheat the peptide-containing SDS lysis buffer at 96 °C. Make sure the lid is tightly closed to avoid buffer evaporation. 50 μL SDS lysis buffer for each condition in a 24-well plate is needed. Prepare at least 20% more than what is strictly needed to cover the possibility of some minor evaporation.
2. After a 4 h incubation time, remove the 24-well plate from the incubator and wash the cells once with 1 mL PBS to remove the protein that is still in the medium.
3. Trypsinize the cells with 100 μL of a 0.5 mg/mL trypsin solution with 0.2 mg/mL EDTA for approximately 5 min, until the cells detach from the surface.
4. Add 800 μL fresh complete medium to dilute the trypsin, mix well, and transfer the contents of the well to a microcentrifuge tube.
5. Centrifuge the cells for 3 min at 300 × *g*.
6. Carefully remove the supernatant with a glass pipette and a suction system. Resuspend the cell pellet in 1 mL PBS. Centrifuge again.
7. Carefully remove the supernatant fully. Be careful not to disturb the pellet. You can now transfer the pellets to a non-sterile working area. Lyse the cell pellets in 50 μL preheated (96 °C) SDS lysis buffer containing the avi tag peptide. Heat the lysates *immediately* for 8 min at 96 °C to prevent biotin ligase from biotinylating avi-tagged proteins after lysis (*see* **Note 16**). Store at −20 °C until further use.

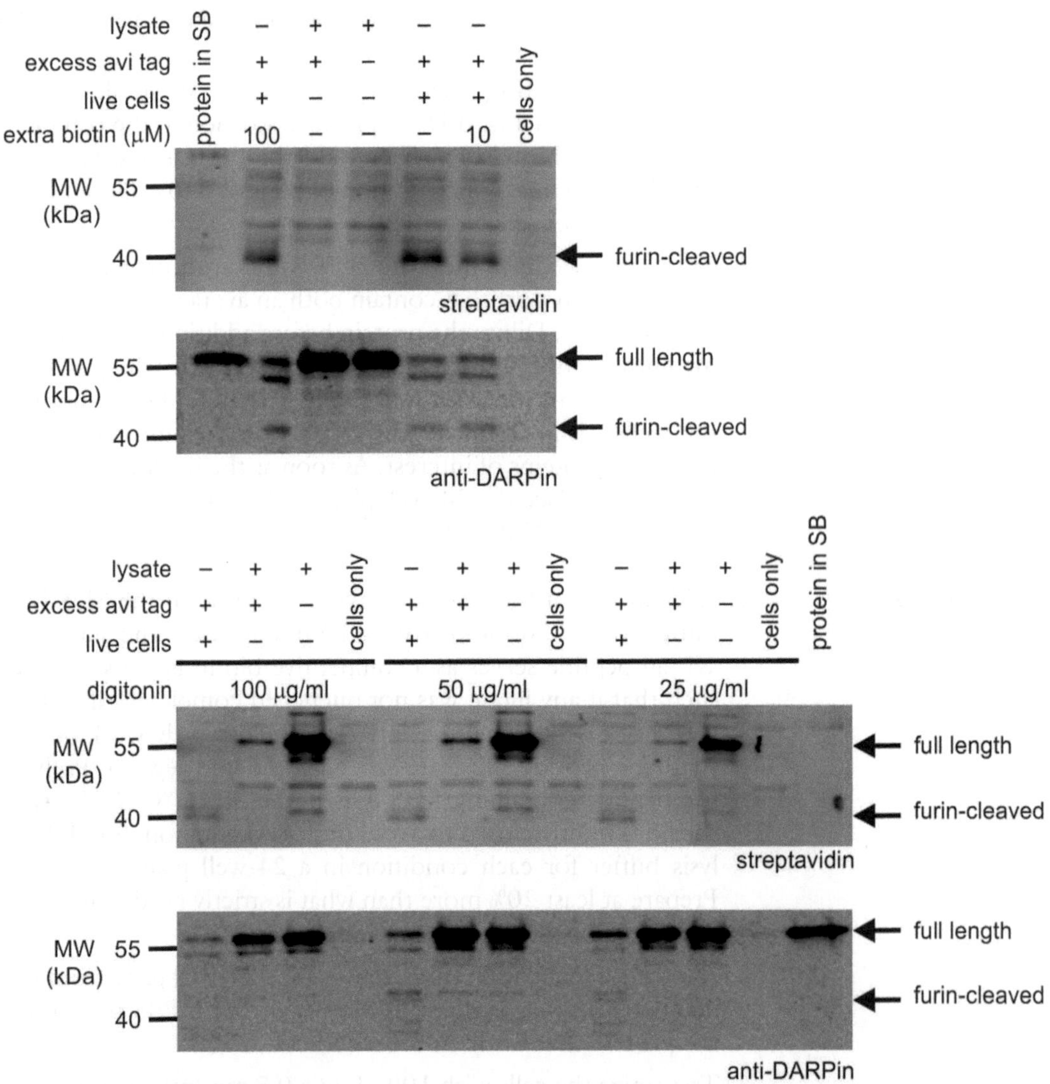

Fig. 3 The effect of adding D-Biotin and the avi tag peptide. The cargo protein Ec1-ETA(252-412)-NI$_2$C (full length; MW = 55 kDa), which reaches the cytosol only after being cleaved by furin (furin-cleaved; MW = 32 kDa) [7], was used as a model protein to study the role of the avi tag peptide and D-Biotin in the biotin ligase assay. The uncleaved fraction does not reach the cytosol and serves as a convenient internal control for any labeling by BirA after cell lysis. External unbiotinylated ETA(252-412)-NI$_2$C was either added to biotin ligase-expressing cells (Flp-In 293) at the moment of lysis (labeled "*lysate*") or incubated for 4 h with live cells before lysis (labeled "*live cells*"). In the upper panel, the SDS lysis buffer was used with or without added excess avi tag peptide (20 μM). Since there is no difference between the *lysate* lanes, biotin ligase seems to be completely deactivated in the lysis procedure, and the avi tag peptide is only an additional precaution. Live cells incubated with varying concentrations of extra D-Biotin in the DMEM medium did not produce more furin-cleaved streptavidin-stained band. In the lower panel, lysis was accomplished with a digitonin-based cytosol extraction buffer containing varying amounts of digitonin and also with or without excess avi tag peptide. The full-length protein is strongly labeled, indicating that this procedure does not quench the biotin ligase activity, but addition of avi tag peptide partially inhibits appearance of this band. Note that, even with the avi tag peptide present, the false-positive band of the uncleaved protein is still very intense compared to the *correct* live-cell band for the furin-cleaved protein, underscoring the necessity of using a lysis protocol that instantly abolishes biotin ligase activity (which the digitonin-based buffer evidently does not do). For clarity, the contrast of the blots has been individually adjusted. *Streptavidin*, detection via streptavidin LT680; *anti-DARPin*, detection via polyclonal anti-DARPin serum). *SB*, sample buffer

8. To check for successful immediate abolishing of biotin ligase activity, mix a purified avi-tagged protein with hot (96 °C) SDS lysis buffer containing the stabilized avi tag peptide. Mix briefly and add the hot mixture immediately to a pellet containing untreated biotin ligase-expressing cells from the same experiment. Resuspend well and heat directly for 8 min at 96 °C. Store at −20 °C until further use (*see* **Note 17**).
9. To check for complete biotinylation of cytosolically localized proteins, dilute the lysate from the cells expressing the avi-tagged protein HA_E3_5_avi fourfold with a non-denaturing, nonreducing lysis buffer. Incubate the diluted lysates for 2 h at 4 °C in the presence or absence of (unlabeled) streptavidin (final streptavidin concentration: 50 μg/mL, diluted from a 1 mg/mL stock in PBS; *see* **Note 18**).

3.3 SDS-PAGE and Western Blotting

1. Pour two 12% (w/v) resolving acrylamide gels with a 4% (w/v) stacking acrylamide gel. Use the Mini-PROTEAN® Tetra cell casting module and compatible short plates and spacer plates according to the manufacturer's instructions. Insert a 15-well comb in the stacking gels.
2. Load 12 μL of each cell lysate sample in the slots. Also include: (a) 1.5 μL of a prestained protein MW marker; (b) 6 μL of a 84 nM solution of a fully biotinylated test protein diluted in SDS lysis buffer that contains both the avi tag and an HA tag; (c) untreated biotin ligase-expressing cells; (d) the diluted lysates of cells expressing HA_E3_5_avi, pre-incubated in the absence or presence of streptavidin (see the previous section). Repeat the procedure for the second gel, thus generating two identically loaded gels.
3. Run the gels in a Mini-PROTEAN® Tetra system at 50 V until the samples enter the resolving gel. Increase the voltage to 150 V and run the gel for an additional 1–1.5 h until the dye front reaches the bottom of the resolving gel.
4. While the gels are running, cut PVDF-FL membranes to match the size of the gel (70 × 85 mm). Activate the membranes with methanol, wash them with ultrapure water, and keep them in transfer buffer until use.
5. When the dye front has reached the bottom of the gels, separate the gel plates, remove the stacking gels, and put the resolving gels in transfer buffer.
6. Prepare the blotting sandwich as follows. Place an absorbent membrane in the holder and add 3 Whatman no. 3 filter papers (80 × 90 mm). Add the blotting membranes. Place the gels carefully on the top of the membrane. Avoid moving the gels afterward, as this may cause smearing. Add three additional Whatman no. 3 filter papers, place a second absorbent membrane on top and close the blotting sandwich with a clamp.

Insert the blotting sandwich in the Mini-PROTEAN® Tetra system for wet blotting.

7. Run the protein transfer with a constant voltage of 100 V for 1 h.
8. Remove the blotting sandwich and inspect the blots. The protein marker should be visible on the PVDF-FL membranes, which is a first indication for successful transfer.
9. Place the membranes in 1× casein blocking buffer and block for 20 min.
10. Incubate one of the membranes for 1 h at ambient temperature with streptavidin IRDye 680LT (1:10,000; LI-COR Biosciences) in streptavidin incubation buffer (contains 0.1% (w/v) SDS). In parallel, incubate the other membrane for 1 h in antibody incubation buffer, which does not contain SDS, with a 1:1,000 dilution of the rabbit anti-HA antibody and a 1:5,000 dilution of the mouse anti-actin antibody (*see* **Note 19**).
11. After the incubation with the primary antibody, wash the membranes 4 × 5 min with PBS-T.
12. Incubation with a secondary antibody is only required for the membrane incubated with the anti-HA and anti-actin antibodies. Use a goat anti-rabbit Alexa Fluor 680 antibody for detecting the anti-HA antibody and a donkey anti-mouse IRDye 800 antibody for detecting the anti-actin antibody. A 1:5,000 dilution in antibody incubation buffer is suitable for both secondary antibodies. Incubate for 45 min at ambient temperature.
13. After the incubation with the secondary antibody, wash the membranes 4 × 5 min with PBS-T.
14. Fluorescent signals can be acquired with an Odyssey IR scanner (LI-COR Biosciences).
15. Use the Image Studio Lite (LI-COR) software to quantify band intensities.
16. On the basis of knowing the number of cells in the lysate, a measurement of the average cell volume, e.g., by the Coulter principle [13], an estimate of the volume fraction of the cytosol (related to the total volume of the cell, assumed to be approximately one), and the band intensities of the directly loaded biotinylated and HA-tagged control protein, it is straightforward to calculate back the amount of protein delivered into the cytosol for each of the test conditions.

4 Notes

1. We have the most experience with an Flp-In 293 cell line stably overexpressing BirA and EpCAM, but the method is in no way restricted to or adapted to this cell line. Experiments

in at least four other BirA-overexpressing cell lines have demonstrated that the presented method works well in all of them. Stable cell lines can be generated from single clones as well as from pools that are under the constant pressure of a selection agent [14].

2. D-Biotin is poorly soluble at low pH, but becomes soluble in dilute alkali (pH ~10). Mild heating (below 50 °C) also aids in the solubilization of D-Biotin. Make sure the pH of the D-Biotin solution is very carefully titrated, as D-Biotin is less stable at very basic pH values. Also the pH of the cell culture medium must be checked after the addition of the D-Biotin solution, since the base used for dissolving D-Biotin may shift the pH.
3. We chose MG-132 because it is a well-established, affordable proteasome inhibitor. Other proteasome inhibitors may work just as well. 50 μM MG-132 will be sufficient for most, if not all, cell lines. Lower concentrations can likely be used for certain cell lines, but the efficiency would have to be tested in each specific case.
4. For Flp-In 293 cells, we have good experiences with the TransIT 293 reagent from Mirus. Generally speaking, any transfection reagent suitable for the cell line being investigated will be appropriate for this step.
5. The cytosolic overexpression of an avi-tagged protein serves as a control showing that a cytosolically localized protein becomes fully biotinylated quickly after expression. Although we use HA-E3_5-avi for this purpose, any protein with both an HA and an avi tag will be suitable for this purpose (*see* **Note 10**).
6. There are many ways to produce a protein with an avi tag attached to it. Our standard approach is a genetic fusion that is overexpressed in *E. coli*. With this approach, care must be taken to remove the fraction that has already been biotinylated in *E. coli*. The biotin ligase gene is essential for survival [15], so this gene cannot be removed in a dedicated expression strain. The removal of the biotinylated fraction of the overexpressed protein can be accomplished with a commercially available streptavidin resin using the manufacturer's protocol. We have good experiences with the streptavidin resin from Genscript.
7. Western blot transfer buffer can be easily prepared from a tenfold stock solution (250 mM Tris–HCl; 1.92 M glycine) without methanol. Adjusting the pH should not be necessary, although checking the final pH is of course recommended, which should be 8.3.
8. Using a membrane that has been specifically fabricated for the use with fluorescent reagents is strongly recommended. In our experience, other membranes have given high background levels and thus resulted in a greatly reduced sensitivity.

9. Milk should be avoided when using detecting streptavidin–biotin interactions. Endogenous biotin present in milk will cause a high background signal [16].
10. To ensure that all the cells in the experiment express sufficient amounts of biotin ligase for the full biotinylation of a delivered cargo protein, we include in all experiments one well in which we overexpress a protein in the cytosol, which is both HA- and avi-tagged. The cytosolic localization of the entire fraction of the protein means that it should be 100% biotinylated. To check this, cell lysates of BirA-overexpressing cells in which this avi-tagged protein was overexpressed are divided into two aliquots after lysis. One aliquot is pre-incubated with streptavidin before SDS-PAGE, whereas the other is not. Since the streptavidin–biotin incubation survives SDS-PAGE, the full biotinylation of the overexpressed protein can be checked by a *complete* shift of the band representing the avi-tagged protein (i.e., from the size of the "unshifted" avi-tagged protein to the size of the "shifted" avi-tagged protein-streptavidin complex).
11. Although we add 0.1 mM D-Biotin in all our assays, we have not observed an evident reduction in biotinylation efficiency of delivered proteins when additional D-Biotin was omitted (Fig. 3). Nevertheless, it is still possible that in some instances the omission of extra D-Biotin might reduce the biotinylation efficiency, thus potentially underestimating the rate or amount of delivered protein. For this reason, we prefer to always include it, as we have not observed any negative effects of D-Biotin addition.
12. The amount of intact biotinylated protein that is present in the cytosol at any one time is a function of the rate of delivery and the rate of degradation. To minimize the rate of degradation in 4-h uptake experiments, we add the proteasome inhibitor MG-132. In a 4-h uptake experiment, proteasomal degradation is generally the main mechanism of protein breakdown in the cytosol. For longer incubations, the role of autophagy becomes more important. It is important to note that proteasome inhibition will induce apoptosis after extended times (e.g., 20 h) due to the aberrant accumulation of ubiquinated proteins.
13. The specific concentrations required for the incubation are not part of this protocol as they are a function of the protein that is to be delivered and of the receptor that is being targeted. It is to be advised to test a broad concentration range, if only few proteins are being investigated. If many proteins are being investigated, we would recommended that for proteins that bind a receptor with a high affinity ($K_D < 10$ nM), 200 nM is first examined, whereas for proteins that have a receptor-independent mechanism of uptake, a concentration of 1 μM can be used as a starting concentration.

14. The incubation time can of course be varied. The desirable time depends on many factors, including the nature of the protein transporter being investigated, the cell line in question and whether the rate of internalization or the total accumulation will be investigated.
15. The addition of an excess of a non-biotinylated stabilized avi tag peptide serves to quench any possible residual activity of biotin ligase. Although we have not seen a residual activity with direct lysis using a hot 4% (w/v) SDS-containing buffer, this activity can be seen with milder forms of lysis, such as using a digitonin-based cytosol extraction buffer (Fig. 3). Nevertheless, it is difficult to exclude a very low (i.e., undetectable) level of residual biotin ligase activity. For this reason, we include an excess of avi tag peptide as a competitive substrate in the SDS lysis buffer in all experiments as a precaution.
16. In order to prevent false-positive biotinylation events of avi-tagged proteins that occur after lysis, it is absolutely essential that biotin ligase activity is fully abolished at the moment of lysis (*see* **Note 15**).
17. The complete absence of biotinylation of the avi-tagged protein, as determined by western blotting, indicates that all biotin ligase activity had been stopped immediately at the moment of lysis.
18. To confirm biotinylation activity of biotin ligase overexpressing cells, the mixtures from this step are investigated for a complete band-shift via western blot of the transfected protein (avi tag and antibody-detectable tag should be at opposite termini to avoid interference). A complete shift confirms the activity of biotin ligase in the cells of choice (*see* **Notes 5** and **10**).
19. Streptavidin IRDye 680LT incubations should be performed in the presence of 0.1% (w/v) SDS to reduce the nonspecific signal intensity. In contrast, the antibody incubations should be performed in the absence of SDS, as this may decrease the signal intensities. Since three signals will be obtained from each cell lysate (e.g., from biotin, HA, and actin), it is better to combine the anti-HA and anti-actin antibodies on the same blot than combining one of the antibodies with streptavidin IRDye 680LT on the same blot, as the antibodies require similar incubation conditions.

Acknowledgments

The research leading to these methods has received support from the EU/EFPIA Innovative Medicines Initiative Joint Undertaking COMPACT grant no 115363. W.P.R.V. was supported by a

Rubicon grant (Project number: 825.12.002) from the Netherlands Organisation for Scientific Research and the Forschungskredit from the University of Zurich.

References

1. Cronican JJ, Thompson DB, Beier KT et al (2010) Potent delivery of functional proteins into Mammalian cells in vitro and in vivo using a supercharged protein. ACS Chem Biol 5:747–752
2. Mellert K, Lamla M, Scheffzek K et al (2012) Enhancing endosomal escape of transduced proteins by photochemical internalisation. PLoS One 7:e52473
3. Mohammed AF, Abdul-Wahid A, Huang EH et al (2012) The Pseudomonas aeruginosa exotoxin A translocation domain facilitates the routing of CPP-protein cargos to the cytosol of eukaryotic cells. J Control Release 164:58–64
4. Wadia JS, Stan RV, Dowdy SF (2004) Transducible TAT-HA fusogenic peptide enhances escape of TAT-fusion proteins after lipid raft macropinocytosis. Nat Med 10:310–315
5. Dunn KW, Mayor S, Myers JN et al (1994) Applications of ratio fluorescence microscopy in the study of cell physiology. FASEB J 8:573–582
6. Richard JP, Melikov K, Vives E et al (2003) Cell-penetrating peptides. A reevaluation of the mechanism of cellular uptake. J Biol Chem 278:585–590
7. Verdurmen WP, Luginbühl M, Honegger A et al (2015) Efficient cell-specific uptake of binding proteins into the cytoplasm through engineered modular transport systems. J Control Release 200:13–22
8. Barker DF, Campbell AM (1981) The birA gene of *Escherichia coli* encodes a biotin holoenzyme synthetase. J Mol Biol 146:451–467
9. Beckett D, Kovaleva E, Schatz PJ (1999) A minimal peptide substrate in biotin holoenzyme synthetase-catalyzed biotinylation. Protein Sci 8:921–929
10. Chaiet L, Wolf FJ (1964) The properties of Streptavidin, a biotin-binding protein produced by Streptomycetes. Arch Biochem Biophys 106:1–5
11. Weber PC, Ohlendorf DH, Wendoloski JJ et al (1989) Structural origins of high-affinity biotin binding to streptavidin. Science 243:85–88
12. Binz HK, Amstutz P, Kohl A et al (2004) High-affinity binders selected from designed ankyrin repeat protein libraries. Nat Biotechnol 22:575–582
13. Mattern CF, Brackett FS, Olson BJ (1957) Determination of number and size of particles by electrical gating: blood cells. J Appl Physiol 10:56–70
14. Gurtu V, Yan G, Zhang G (1996) IRES bicistronic expression vectors for efficient creation of stable mammalian cell lines. Biochem Biophys Res Commun 229:295–298
15. Chakravartty V, Cronan JE (2012) Altered regulation of *Escherichia coli* biotin biosynthesis in BirA superrepressor mutant strains. J Bacteriol 194:1113–1126
16. Hoffman WL, Jump AA (1989) Inhibition of the streptavidin-biotin interaction by milk. Anal Biochem 181:318–320

Chapter 15

Data-Driven Antibody Engineering Using Genedata Biologics™

Maria Wendt and Guido Cappuccilli

Abstract

Genedata Biologics is a novel informatics platform specifically designed for biologics R&D. Here, we discuss the main principles employed in designing such a platform, focusing on antibody engineering. To illustrate, we present a case study of how the platform effectively supports an antibody optimization workflow and ensures the successful integration and analysis of all relevant sequence, expression, assay, and analytics data.

Key words Antibody engineering, Antibody library design, Bioinformatics, Biologics database, Data visualization, Mutagenesis

1 Introduction

The discovery and development of biopharmaceuticals is a complex and time-consuming endeavor, owing to the size of the molecules and the complexity of biological production and subsequent evaluation. On an industrial scale, as the saying goes, "it takes a village" to produce a successful biopharmaceutical due to the different disciplines and laboratories (e.g., molecular biology, cell biology, assay development, among others) that generate and contribute various types of information (e.g., molecule, sequence, material, process, functional assay, and analytical characterization data). Many types of technologies that are growing in sophistication are employed at each stage, which these days also means generating more complex and larger volumes of data, e.g., the increased efficiency in generating protein libraries [1, 2]; the increased utilization of mass spectrometry for biologics characterization [3, 4]; to name a few. The capture and management of all data generated, as well as the subsequent interpretation and extraction of the encoded knowledge, are major challenges that need to be addressed in order to make biologics R&D more efficient. This requires a foundation of a highly organized data structure, as well

Thomas Tiller (ed.), *Synthetic Antibodies: Methods and Protocols*, Methods in Molecular Biology, vol. 1575,
DOI 10.1007/978-1-4939-6857-2_15,

as dedicated automated workflow support, data processing and data analysis tools.

In this chapter, we discuss the main principles of the design of our biologics data platform, Genedata Biologics, which serves as a central data capture and repository, workflow support and analysis system for biologics R&D. This was a result of multi-year collaborations with some of the leading biotech and biopharma laboratories in Europe and in the US. It is now an established database and software platform that is productively being used in the industry.

While the system has an end-to-end coverage of the biologics R&D workflow, applicable to both antibodies and non-antibodies, we will focus our discussion on the design and application of Genedata Biologics in guiding and enhancing antibody screening and engineering workflows.

2 The Genedata Biologics Platform for Antibody Screening and Engineering

2.1 Comprehensive Database and Workflow Support

Because the system has been designed from the very beginning specifically for biologics R&D, knowledge of underlying biological processes primarily guides the data model. While the workflow support component is commonly treated as a separate, sometimes optional objective when designing an informatics platform, in our view, in designing a database for biological processes, workflow support is a necessary integral component.

Complex workflows such as those employed in biologics R&D involve handling of various types of biological materials (e.g., antibody libraries, DNA constructs, cell lines, supernatant protein products, etc.) in different steps in the laboratory to which very heterogeneous types of data would need to be associated later. A necessary foundation for this is a formal molecule and material registry (database component) with the ability to automatically generate authoritative and unique molecule and material identifiers (workflow tool), store these including their physical genealogy (database component), using system business logic to infer and automatically set these genealogy whenever possible (workflow tool). Further on is the capture of data generated in the laboratory, including sequencing data, functional assays, and analytics, which requires knowledge of the sample to be analyzed (database component), ability to process complex and/or high volume data automatically (workflow component), and store this in a structured, queryable fashion (database component). This briefly illustrates the dependence as well as synergy between a database and the workflow support components of a successful biologics platform.

In Genedata Biologics, we have put in place a comprehensive data model and supporting workflow tools that reflect the actual processes in the laboratory. For example, in antibody screening, the system tracks the generation of the antigens, the generation of the antibody library, the clone selection, individualization of pooled clones onto individual wells of microtiter plates, the subsequent handling of the clones from plate to plate (as DNA constructs, plasmids, transfected cells, or supernatants), clone sequencing, and the performance of assays, among others.

Figure 1 is a simplified view of part of Genedata Biologics' database schema showing the most important entities participating in a typical antibody screening campaign, for both screening of initial antibody libraries or of optimized variants. The main entity, the Antibody Clone, represents the individualized isolate (i.e., within the molecule and material registry) that is being evaluated and to which all data can be subsequently traced or linked. Using this schema, it is possible to trace the origin of an individual Antibody Clone (e.g., specific output of a phage selection round of a specific Antibody Library and Antigen(s)) (a), the physical locations of its isolated materials at any point of the screening process (e.g., well positions and plate barcodes of supernatants for an ELISA assay) (b) and its assay (c) and sequence data (d). This powerful Antibody Clone-centric view is made possible by the ability of the system to automatically generate individual unique Antibody Clone IDs (or identities) during the clone individualization step in the laboratory such as colony-picking from agar plates, and to preserve this single identifier as material (e.g., DNA construct, supernatant) of the same clone is transferred from plate to plate. The latter is also known as "clone tracking" and is made possible by the system's ability to store plate information (e.g., plate barcodes, well layouts) and reflect plate operations that are happening in the laboratory, e.g., a plate stamping or a cherry-picking operation by a liquid handling instrument (LHS). This is accomplished by the system by automatically parsing so-called report files from any LHS instrument or (human) manual operation. As one can see, the success of the database is heavily encouraged by the presence of workflow support tools and the ability of the system to reflect that actual physical processes in the laboratory.

2.2 Integrated Sequence Management and Analysis Tools

DNA and protein sequences are primary descriptors of biological molecules and are naturally of highest relevance during antibody engineering. Sequence data not only provide information of predictive properties of a protein molecule that may contribute to the determination of an engineering strategy, but they are the necessary instruments in designing and creating new molecules. Sequence data can provide different pieces of information depending on the different stages of an antibody screening and engineering workflow. Thus, the platform needs to have not only the ability to parse sequence data files and store sequence data, but must also include diverse sequence processing engines that perform tasks that are relevant at particular stages of the workflow (e.g., DNA

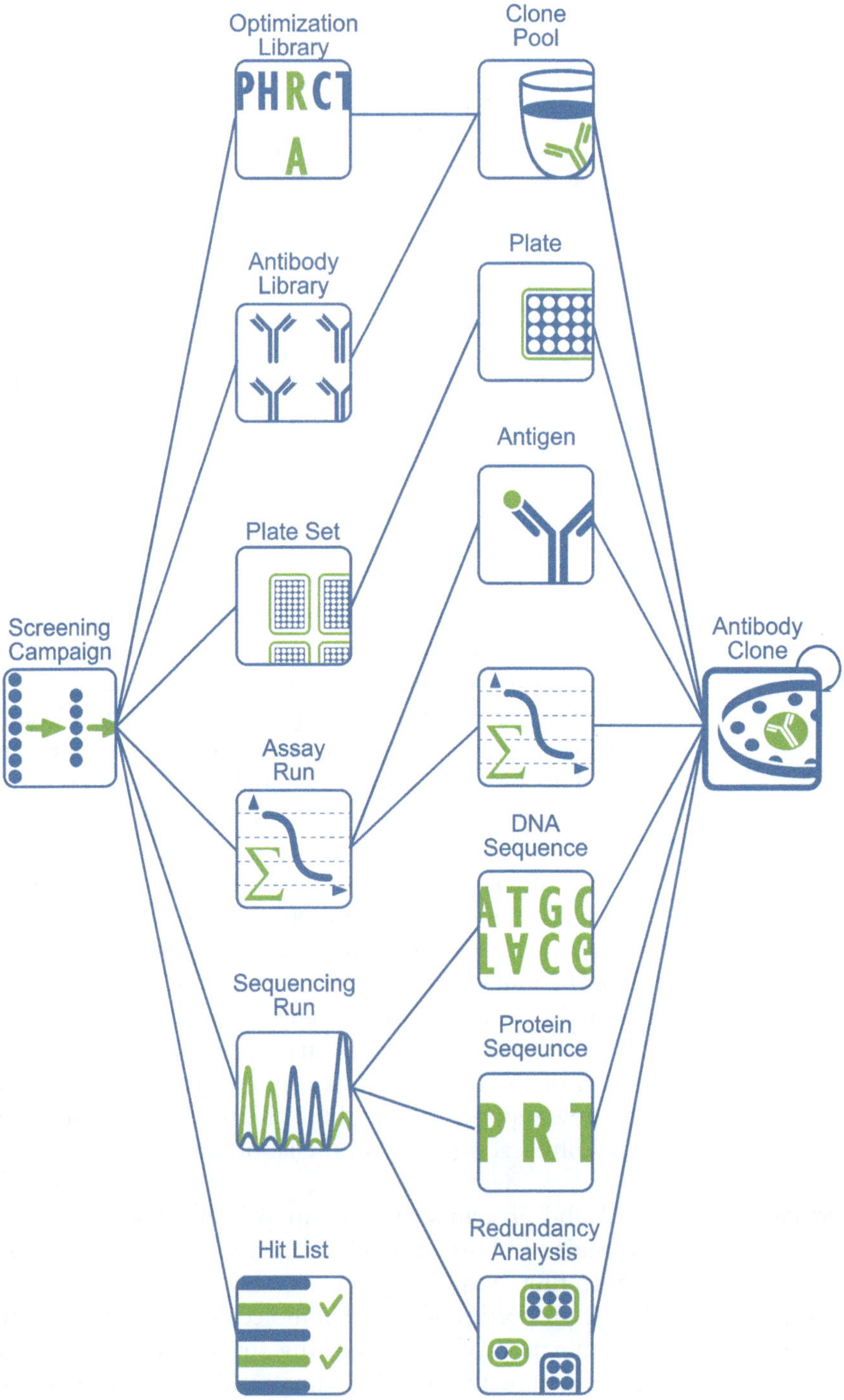

Fig. 1 High-level description of data model

synthesis verification, in silico developability assessment, etc.). This in turn requires an awareness within the system of the stage of the workflow at hand using the appropriate data model and business logic.

To support the registration of antibody molecules, Genedata Biologics includes a sequence database for storing and referencing all protein and DNA sequences as well as feature annotations. Figure 2a shows part of a hit selection table containing typical results of the automated sequence processing of a panel of Antibody Clones. The system includes a sequence annotation pipeline that automatically annotates all registered sequences according to configurable annotation rules (e.g., CDR/FW regions (a), potential posttranslational modification sites (c), mutation information). Similarly, DNA sequences are automatically scanned for relevant restriction sites, stop codons, etc. for vector design and automated expression construct generation. In addition, specific business logic is in place for engineering purposes, e.g., identifying closest germline reference sequences (b); automatically annotating sequence variations of an engineered child molecule against a parent; automatically calculating physicochemical properties of full-length molecules based only on partial sequence information and knowledge of the molecule format (e.g., calculation of the extinction coefficient of an IgG based only on the sequenced variable region using a human IgG1 molecule template) (d); and modular sequence redundancy analyses (e.g., grouping of clones with identical HCDR3s or sharing the same VH to identify over-representation). In addition, the system includes an ability to automatically incorporate annotations from external in silico evaluation tools (e.g., AGGRESCAN for identifying aggregation hotspots [5] (Fig. 2, (c)) to aid in early developability assessments. Protein and DNA sequence visualizers can be launched at many points in the software allowing further sequence analyses such as sequence alignments and clustering.

2.3 Integrated Assay Data Management and Hit Selection Tools

Many types of assays and analytics are routinely applied throughout the different stages of the biologics R&D process utilizing a wide range of technologies (i.e., various fluorescent, electrophoretic, chromatographic, and spectroscopic techniques). Thus, it is of utmost importance that the database is flexible and robust such that introduction of new assays and analytic technologies will not necessitate a system reprogramming and restriction of data submission.

The most powerful assay and analytics technologies usually generate the largest data sizes and require the most complex analysis, both of which can present real challenges in terms of analytical effort and system performance. It is important then that solutions provide as automated data capture as possible, with direct integration with laboratory equipment supported by high-performance, automated data processing. This requires great flexibility for the central system to be able to integrate with key equipment and laboratory IT infrastructures. This is possible by providing so-called application programming interfaces (or APIs). This also requires

Hit Selection for Screening Campaign SC-1 (Anti-PD-L1 phage display screening)

Assay Results for 17 Antibody Clones filtered by selected Hit List — Create Hit List

	Antibody Clone			Clone Pool		(b) Reference Sequence		(d) Antibody Clone Full Molecule Sequence		(a) Antibody Clone Sequence		(c) Feature Count	Quality Scores	
✓	ID ▲	Name	Format	ID	Name	VL Reference Sequence Name	VH Reference Sequence Name	Molecular Weight, MW [Da]	Molar Extinction Coefficient [1/(M*cm)]	LCDR3	HCDR3	PTM	VL Consensus Score Average	VH Consensus Score Average
	CL-7469	015IMGT-M014-G05	sFab	CP-3	Kappa R2 gradient 1:2.5	IGKV5-2*01_IGKJ2*01	IGHV1-46*01_IGHJ1*01	44,317.62	67,190	QQHYSYPW	IGAIERGRSM	10	285	289
	CL-7516	015IMGT-M015-C04	sFab	CP-3	Kappa R2 gradient 1:2.5	IGKV4-1*01_IGKJ2*01	IGHV1-46*01_IGHJ1*01	43,991.59	63,350	QQHYSLPW	KGLIIIFSV	9	289	314
	CL-7554	015IMGT-M015-F06	sFab	CP-3	Kappa R2 gradient 1:2.5	IGKV1-8*01_IGKJ2*01	IGHV1-46*01_IGHJ1*01	44,681.10	78,630	QQHYSTPW	LNKQYPWMDY	10	286	299
	CL-9007	015IMGT-M030-G07	sFab	CP-5	Lambda R2 gradient 1:2.5	IGLV6-57*01_IGLJ2*01	IGHV6-1*01_IGHJ1*01	44,385.43	72,940	QQHYSYPW	YPQNGNHPTAFDY	6	314	294
	CL-9028	015IMGT-M031-A04	sFab	CP-5	Lambda R2 gradient 1:2.5	IGLV1-44*01_IGLJ2*01	IGHV6-1*01_IGHJ1*01	43,307.39	70,380	QQHYSTPW	SQTQGVNE	3	288	305
	CL-9114	015IMGT-M031-H06	sFab	CP-5	Lambda R2 gradient 1:2.5	IGLV5-39*01_IGLJ2*01	IGHV2-5*01_IGHJ1*01	44,256.77	80,480	QQHYSTPW	GIPESIWRSMDY	6	291	289
	CL-9223	015IMGT-M033-A07	sFab	CP-5	Lambda R2 gradient 1:2.5	IGLV1-44*01_IGLJ2*01	IGHV3-72*01_IGHJ1*01	43,384.50	70,380	QQHYSTPW	AVRTHPFNE	6	294	281

to pane B ⇨

A

(a) ELISA AR-2 (PriS ELISA) AG-1 (hPD-L1 ECD) [TPP-1 (hPD-L1 ECD)]			ELISA AR-3 (SecS ELISA (hPD-L1)) AG-1 (hPD-L1 ECD) [TPP-1 (hPD-L1 ECD)]			ELISA AR-7 (SecS ELISA (competition)) AG-1 (hPD-L1 ECD) [TPP-1 (hPD-L1 ECD)]		(b) Biacore AR-8 (Fab Biacore assay) AG-1 (hPD-L1 ECD) [TPP-1 (hPD-L1 ECD)]				SDS-PAGE AR-57 (sFab SDS-PAGE)			FACS AR-58 (FACS) AG-1 (hPD-L1 ECD) [TPP-1 (hPD-L1 ECD)]				(c) AggreScan AR-66 (Aggregation Prediction)
S	B	S/NegC	S	B	S/NegC	S	Inhibition [%]	K_D [pM]	k_{on} [$^1/_{(M*s)}$]	k_{off} [$^1/_s$]	Sensorgram	Purity [%]	Mass [kDa]	Gel Image - Processed	G-Percent [%]	R-Percent [%]	FACS PLOT - FL1: FITC (G)	FACS PLOT - FL2: PE (R)	Number of Hot Spots
4.81	0.27	3.41	5.89	0.35	3.38	5.75	2.4	0.927	9.32E05	8.64E-04		99.6	45.32		55.5	21			5
1.40	0.25	7.75	2.02	0.27	1.54	2.01	0.8	0.939	1.11E05	1.04E-04		98.9	44.67		49.5	49.7			5
2.25	0.32	8.77	3.01	0.33	2.53	3.01	0.0	0.876	1.75E06	1.53E-03		97.8	46.05		45	41.4			3
4.05	0.31	3.91	4.35	0.25	1.36	4.27	2.0	0.913	9.32E05	8.51E-04		96.2	46.08		40.8	41			5
2.08	0.31	7.30	2.23	0.32	3.37	2.21	0.8	0.892	3.17E06	2.83E-03		98.0	43.93		20.2	28.1			5
3.33	0.25	5.62	4.31	0.26	5.62	4.29	0.5	0.927	7.43E05	6.89E-04		98.7	45.83		57.3	18.1			4
1.67	0.27	3.64	2.14	0.26	0.96	2.14	0.1	0.842	7.52E05	6.33E-04		99.7	44.06		50.4	42.5			5

⇦ to pane A

B

Fig. 2 (**a**) Hit selection table (pane A). (**b**) Hit selection table (pane B)

absolute knowledge of the analyte's identity and highlights the importance of the ready availability of authoritative molecule and material identifiers for autonomous and error-free submission of relevant data.

Genedata Biologics comes with a database for the storage of all relevant biologics screening assay and analytics data. Assay data can be imported in bulk, such as when derived from high-throughput, plate-based campaigns. The system's data processing architecture and data model has been designed such that it can be flexibly configured to process and store data from different assay types (e.g., ELISA, FACS, FMAT, Biacore), plate formats (e.g., 96-, 384-, 1536-well MTPs), data processing steps (e.g., Signal/Background, control handling), and data aggregation (e.g., replicate handling, IC50 curve-fitting). Aggregated assay data is stored in the database and automatically referenced against appropriate screening isolates. All assay data for a given isolate can be queried. Integrated tools for hit selection allow users to identify the best hits based on multiple criteria (e.g., cross-assay analyses combining ELISA and FACS readouts, HCDR3 sequence, PTMs). Figure 2b shows part of a hit selection table containing typical results of various assays performed during different stages of screening a panel of antibodies. In this example, ELISA data from primary and secondary screening have been reported (a) as well as Biacore confirmatory assays (b). Note the ability of the system to collect and align data collected along different stages of the screening workflow due to the central Antibody Clone registry and clone-tracking ability of the system. The system also supports management of hit lists.

2.4 Highly Structured, Integrated Database for Data-Mining

Due to the unique property to have in one database such varied data types as molecule DNA and protein sequences, assays, analytics and process information (such as antibody libraries, selection histories, and conditions), as well as sufficiently detailed and flexible sequence annotations (e.g., of subdomains, individual point mutations), in a highly structured schema, it is possible to interrogate the database on many levels. Figure 3 shows some of the visualization that can be invoked from direct queries to the database. From the top, it is possible to perform evaluations taking into account the full molecule (e.g., an investigation of expressibility of fully formatted antibodies) (a); to specific domains (e.g., the cross-reactivity of variable regions to human and mouse antigens) (b); to subdomains (e.g., the potential effect of HCDR3 length to binding; HCDR3 family analysis as they correlate with binding) (c); and individual amino acid (AA) positions and residues (e.g., for identifying critical residues and positions that correlate with improved binding) (d).

2.5 Architecture

Genedata Biologics is based on an industry standard, 3-tier multi-layered, client-server architecture. The system is a web-based application, based on a central relational Oracle RDBMS, that can be

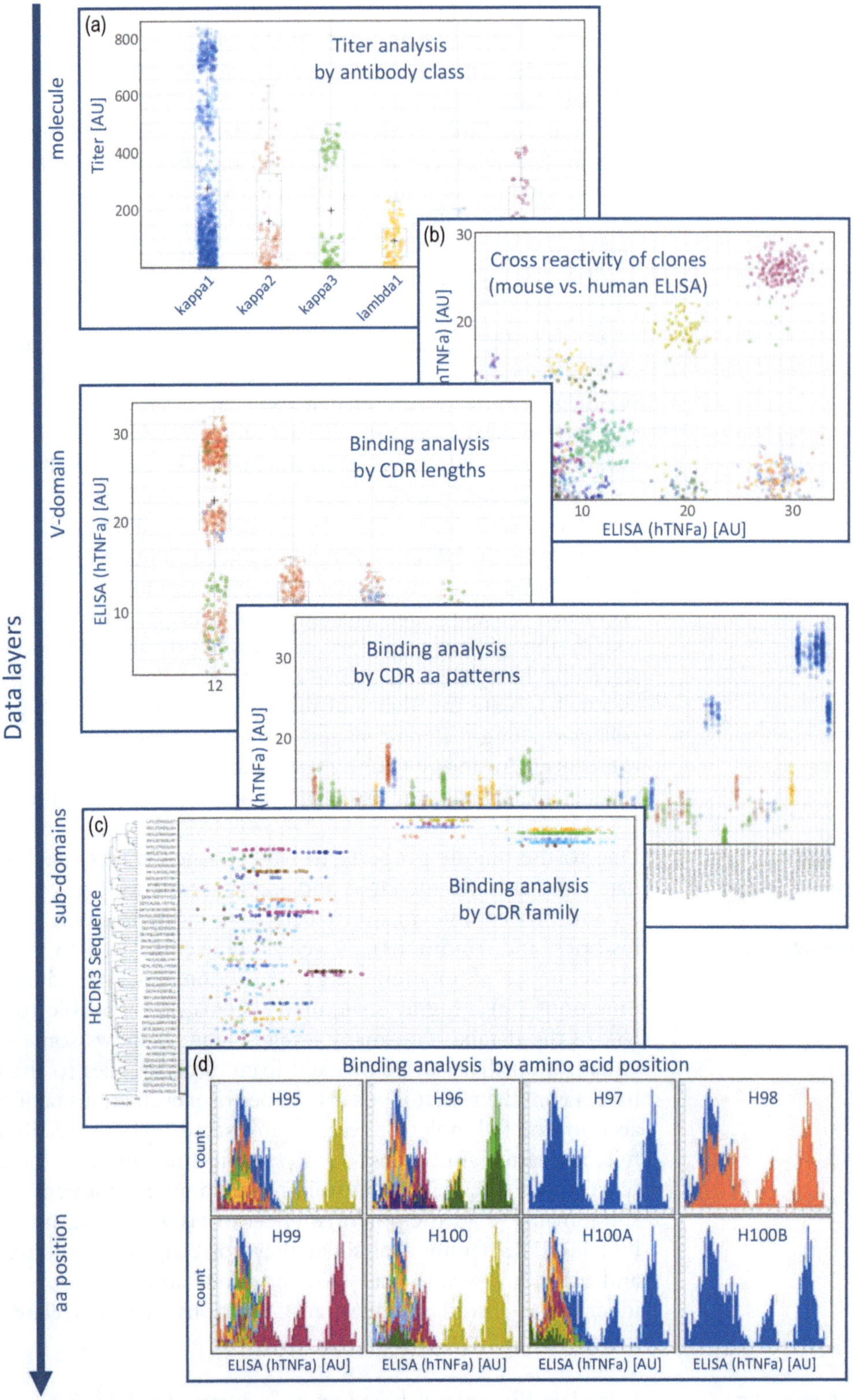

Fig. 3 Top-down approach to information extraction

easily accessed by any client computer connected to an institution's intranet from a standard web browser. As an open system, it enables seamless interfacing and integration with existing IT infrastructures. The system has been designed and optimized for processing, storing, and querying large and complex data volumes. As discussed, the system also comes with a significant built-in business logic layer and works out-of-the box in its default configuration and can be further configured to support laboratory-specific variants of the biologics R&D workflows.

3 Case Study: A Two-Step Optimization Approach for the Identification of High-Affinity Anti-TNFa Antibodies

3.1 The Antibody Engineering Strategy

In this section, we illustrate how the Genedata Biologics platform enables an antibody optimization workflow. Here, we shed light on common workflow challenges in antibody engineering aside from data management and analysis, and show how it is both beneficial and necessary to simultaneously and synergistically address these aspects to have an effective platform solution.

Many strategies have been identified for efficiently and successfully optimizing the binding ability of an antibody. Chaparro-Riggers et al. [6] provided a comprehensive overview of various data-driven protein engineering approaches, exhaustively cataloging important studies that successfully employed alternatives to rational design and combinatorial engineering. The authors discussed the particular challenges of designing and evaluating antibody optimization libraries and measures for controlling library size through the use of data.

Rajpal et al. [7] successfully further optimized an already highly optimized (Kd 1 nM) anti-TNF-alpha antibody to about a 500–870-fold improvement in affinity via a two-step approach that relied on the generation of large panels of antibodies and binding data. The first step via a Look-Through Mutagenesis (LTM) method simultaneously assessed and optimized combinations of mutations of selected amino acids, wherein clones were generated with a mutation in at most one position of the complementarity determining region (CDR) domains in at most 3 of the 6 CDRs of the VH-VL pair at a time to control library size. Furthermore, only a subset of all possible amino acid residues to vary toward was selected based on their representation of specific amino acid side chain chemistries. After positive binding selections and screening, 38 substitutions in 21 CDR positions were identified to confer higher binding affinities. In a second step, these beneficial mutations were combined to generate a new combinatorial mutagenesis library (Combinatorial Beneficial Mutations, CBM) that was further screened to reveal clones with vastly enhanced TNF-alpha binding affinities.

In the following sections, we illustrate how the Genedata Biologics platform enables this full LTM + CBM process and laboratory workflow using artificial data approximating the data flow used in Rajpal et al. [7].

3.2 An Antibody Optimization Workflow in Genedata Biologics

Figure 4 shows an outline of the various workflow support tools and data entry analysis steps in Genedata Biologics. The full workflow is divided into main arms, the first arm corresponding to the first LTM step that utilizes a clone-by-clone optimization workflow where each clone is assigned an individual design, and the second arm corresponding to the second CBM step where a full set of clones is assigned one combinatorial library design.

The workflow begins by the registration of the parental Antibody Clone to be optimized (a), in this case, the anti-TNF-alpha antibody D2E7 [8]. The protein sequence of this antibody is provided to the system (e.g., via cut-and-paste or any other standard sequence file format such as FASTA or Genpept). For the first round (LTM step), each variant would encode a single position of any one to three CDRs that need to be varied. The individual mutational variants to be synthesized in the laboratory are then pre-registered in the system with individual design information (b), described in comparison to the parental antibody's

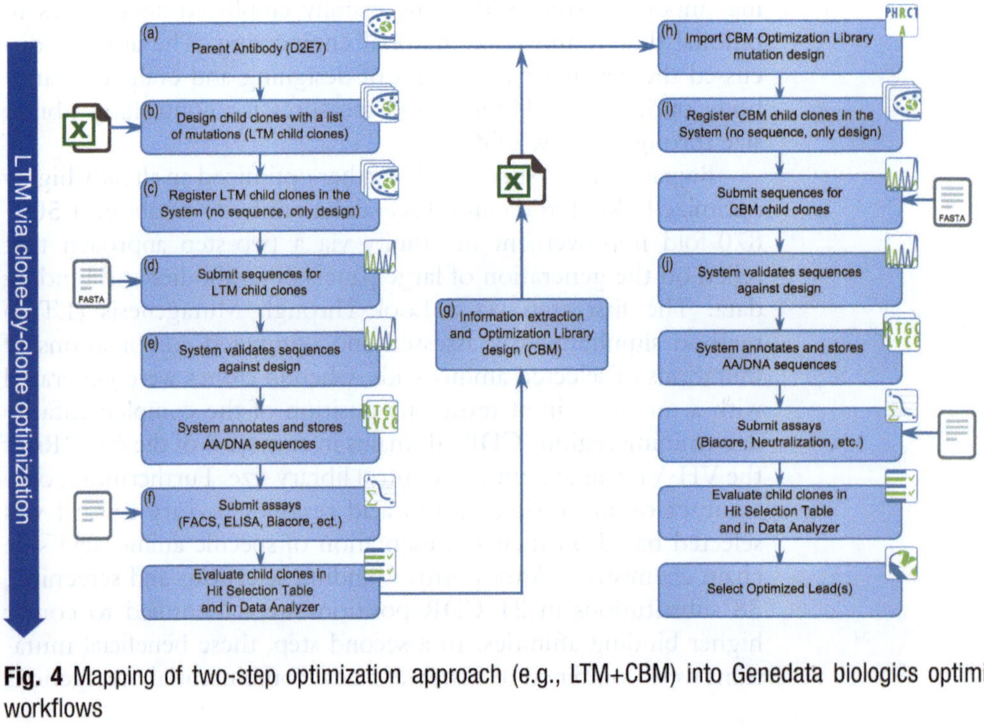

Fig. 4 Mapping of two-step optimization approach (e.g., LTM+CBM) into Genedata biologics optimization workflows

protein sequence (e.g., "M34L" indicating the Methionine (M) at position 34 will be mutated to Leucine (L)). At this point the children mutational variant clones are created by the system and assigned an individual identifier (i.e., the Antibody Clone ID) and a relational link to the parental clone to be optimized, preserving genealogy information (c). Also at this point, each mutational variant has its own design information that can be used to verify the sequences of the synthesized DNA. Once the DNA has been synthesized and sequenced, sequences are submitted to the respective antibody clones using the Antibody Clone ID (d). The system automatically invokes a processor that compares the translated protein sequence to the design information and either confirms a match (i.e., that a successful synthesis against the design, with all expected mutations, with the specific AA residues at the correct positions, occurred) or identifies a mismatch and annotates the position at which an unexpected mutation occurred (e). Thus, the system helps ensure that the overall library design strategy is accomplished via comparison of individual clones to its individual design (i.e., clone-by-clone). Once this is done, all data generated during plate-based screening is captured in the system as described earlier, including all plate locations and assay data (f). This results in a table shown in Fig. 5 where the panel of individual variants is displayed along with sequence library design information, very importantly, the position of the mutations (a) and the collected assays (b), a type of a "Structure-Activity Relationship" or SAR view. This is easily accomplished due to the level of integration and systematic structure of the database. It is then possible to investigate which residues at which positions correlate with higher affinities, and thus identify a set of potentially beneficial mutations. One example illustrated here is a box plot for position 27 (VL), where substitutions of Tyrosine (Y), Serine (S), and Lysine (K) appear to show higher average ELISA binding scores than the native amino acid Glutamine (Q) (c). Provided the data structure, statistical calculations (such as the Fischer's Exact test) can also be applied to identify mutations that are relatively more abundant in clones with improved affinities.

This is a critical step during which knowledge has to be extracted from multidimensional data and then used to establish a set of rules that guide the design of the combinatorial strategy to be implemented in the next step. The complexity and the importance of this step requires the use of ad-hoc methods and tools for both visualization and rigorous analysis of the available data, without which one can only rely on the skills and experience of the scientists. Without such specialized "tools" the success of the whole process would be exposed to the risk of subjective interpretation of the data, risk that rapidly increases with size, complexity, and heterogeneity of the data.

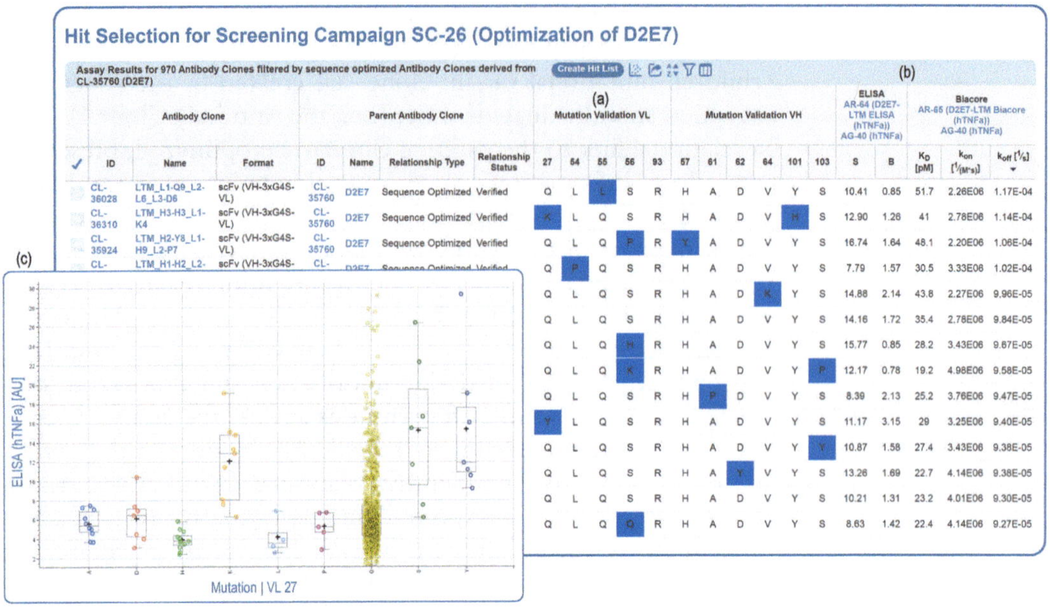

Fig. 5 Multi-parametric selection of beneficial mutations via hit selection table with SAR view and visualization in Data Analyzer

Figure 4 again shows an outline of the workflow support tools for the second round of the optimization using a combinatorial (CBM) library. Following the process described in Rajpal, et al., after identifying the set of beneficial mutations, a combinatorial library is designed (g). In this second step, there is no limit on the number of mutated positions as opposed to the first round of the LTM where a maximum of three mutations for the entire antibody is expected. This design is registered in Genedata Biologics as an entity called an Optimization Library where permitted residues at each position are described as well as the corresponding parental Antibody Clone (i.e., clone D2E7) (h). Once the designed clones are synthesized these clones can be registered in the system (i), for example via a plating operation mimicking the individualization of the pool of engineered clones in the laboratory. Once sequenced, the individual sequences are submitted and automatically annotated against the overall design of the library (j). This is in contrast to the Clone-by-Clone Optimization workflow where each clone is compared against its individual design information. In Fig. 6, shown is the library design based on the result of the LTM round as described in the paper (a) and the corresponding optimization library design stored in the system (b). The hit selection table displays the result of this automated engineering-focused analysis performed by sequence processing engine of the platform. An overall assessment of each clone against the design is reported by the system (c). As this is for a combinatorial library, more specific information of the evaluation of the synthesis is shown in (d): gray means it is a design position with the residue conserved, blue means it is a design position with a confirmed expected mutation (allowed residue in correct position), red means an unexpected mutation in a position not included in the mutation design (no mutation expected), yellow means an unexpected mutation (a non-allowed residue in a design position). This allows qualifying the overall quality of the library synthesis (in terms of meeting design goals and diversity). Decisions can then be made whether to proceed with the synthesized library, to resynthesize, or to reevaluate the library synthesis method. Data from plate-based screening is again captured as described before with all assay data aligned with the previously identified mutation information (e). Use of the data visualizers and filtering allow the identification of the top engineered candidates.

4 Conclusion

The case study provides a concrete example of how Genedata Biologics successfully enables data-driven antibody engineering, by allowing all relevant data—in this case antibody genealogy, sequence, and binding data—to be utilized at every step of the antibody engineering process; from the initial library design, generation and synthesis, until the identification of the best engineered antibody. This also illustrates the benefits and synergies of taking into account the antibody engineering workflow that is to be

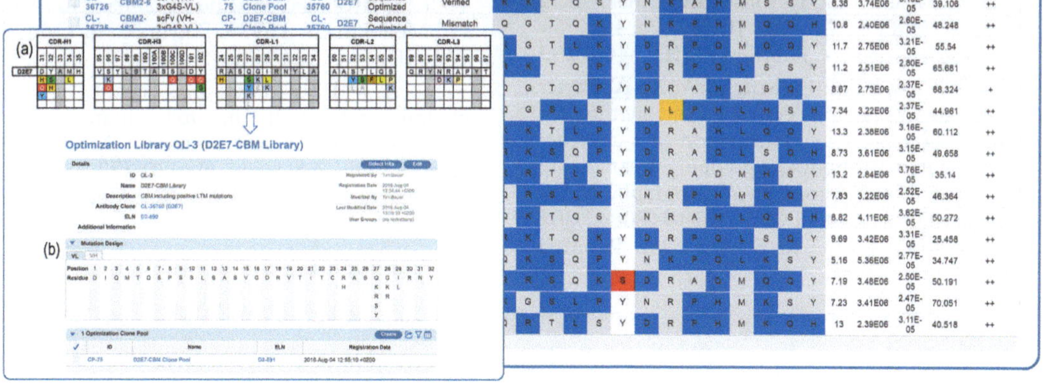

Fig. 6 Optimization Library (CBM) and optimized hits selection

enabled by the data management system. The workflow tools facilitate the more efficient handling of clones in the laboratory and data processing, and directly contribute to the quality, integrity, and completeness of the resulting data. The database in turn provides the relevant antibody genealogy and the highly structured framework needed for the data-driven analysis.

As the database also provides a persistent data repository where data can accumulate over time across projects, targets, antibody libraries, and antibody formats (e.g., bispecific, multi-specific), among others, such a platform provides a very good foundation not only for making immediate decisions such as identifying engineering leads or defining engineering strategies, but also for identifying potential process improvements (e.g., library synthesis) and development of prediction models—various measures for harnessing knowledge from data that will further drive the increased efficiency and success of antibody engineering efforts.

References

1. Schlinkmann KM, Hillenbrand M, Rittner A, Künz M, Strohner R, Plückthun A (2012) Maximizing detergent stability and functional expression of a GPCR by exhaustive recombination and evolution. J Mol Biol 422(3):414–428. doi:10.1016/j.jmb.2012.05.039 Epub 2012 Jun 6. PubMed PMID: 22683350
2. Dokarry M, Laurendon C, O'Maille PE (2012) Automating gene library synthesis by structure-based combinatorial protein engineering: examples from plant sesquiterpene synthases. Methods Enzymol 515:21–42. doi:10.1016/B978-0-12-394290-6.00002-1 PubMed PMID: 22999168
3. Debaene F, Wagner-Rousset E, Colas O, Ayoub D, Corvaïa N, Van Dorsselaer A, Beck A, Cianférani S (2013) Time resolved native ion-mobility mass spectrometry to monitor dynamics of IgG4 Fab arm exchange and "bispecific" monoclonal antibody formation. Anal Chem 85(20):9785–9792. doi:10.1021/ac402237v Epub 2013 Sep 26 . PubMed PMID: 24007193
4. Birdsall RE, Shion H, Kotch FW, Xu A, Porter TJ, Chen W (2015) A rapid on-line method for mass spectrometric confirmation of a cysteine-conjugated antibody-drug-conjugate structure using multidimensional chromatography. MAbs 7(6):1036–1044. doi:10.1080/19420862.2015.1083665 Epub 2015 Aug 25. PubMed PMID: 26305867; PubMed Central PMCID: PMC4966495
5. Zambrano R, Jamroz M, Szczasiuk A, Pujols J, Kmiecik S, Ventura S (2015) AGGRESCAN3D (A3D): server for prediction of aggregation properties of protein structures. Nucleic Acids Res 43(W1):W306–W313. doi:10.1093/nar/gkv359 Epub 2015 Apr 16. PubMed PMID: 25883144; PubMed Central PMCID: PMC4489226.
6. Chaparro-Riggers JF, Polizzi KM, Bommarius AS (2007) Better library design: data-driven protein engineering. Biotechnol J 2(2):180–191 Review PubMed PMID: 17183506
7. Rajpal A, Beyaz N, Haber L, Cappuccilli G, Yee H, Bhatt RR, Takeuchi T, Lerner RA, Crea R (2005) A general method for greatly improving the affinity of antibodies by using combinatorial libraries. Proc Natl Acad Sci U S A 102(24):8466–8471 Epub 2005 Jun 6. PubMed PMID: 15939870; PubMed Central PMCID: PMC1143585
8. Kempeni J (1999) Preliminary results of early clinical trials with the fully human anti-TNFalpha monoclonal antibody D2E7. Ann Rheum Dis 58(Suppl 1):I70–I72 Review. PubMed PMID: 10577977; PubMed Central PMCID: PMC1766582

Part II

Nucleotide-Based Synthetic Antibodies: Aptamers

Chapter 16

Selection of Aptamers Against Whole Living Cells: From Cell-SELEX to Identification of Biomarkers

Nam Nguyen Quang, Anna Miodek, Agnes Cibiel, and Frédéric Ducongé

Abstract

Aptamer selection protocols, named cell-SELEX, have been developed to target trans-membrane proteins using whole living cells as target. This technique presents several advantages. (1) It does not necessitate the use of purified proteins. (2) Aptamers are selected against membrane proteins in their native conformation. (3) Cell-SELEX can be performed to identify aptamers against biomarkers differentially expressed between different cell lines without prior knowledge of the targets. (4) Aptamers identified by cell-SELEX can be further used to purify their targets and to identify new biomarkers. Here, we provide a protocol of cell-SELEX including the preparation of an oligonucleotide library, next-generation sequencing and radioactive binding assays. Furthermore, we also provide a protocol to purify and identify the target of these aptamers. These protocols could be useful for the discovery of lead therapeutic compounds and diagnostic cell-surface biomarkers.

Key words Aptamer, Cell-SELEX, Oligonucleotides, Biomarkers

1 Introduction

Nucleic acids can simultaneously store genetic information, perform catalytic activity and interact specifically with many partners such as metabolites or proteins. Since 1990, several strategies of directed molecular evolution have been developed to use these properties and to develop nonnatural ligands or nucleic acids-based enzymes. These methods were popularized by Tuerk and Gold using the term "SELEX" for Systematic Evolution of Ligands by EXponential enrichment [1] and nucleic acid-based ligands identified by SELEX were named "aptamers" by Ellington and Szostak, from the Latin *aptus* meaning to fit [2]. Aptamers are sometimes compared as chemical antibodies, since, like them, they can be used to neutralize or activate proteins upon binding [3, 4] and they can be used to detect them using ELISA, microarray, microscopy or flow cytometry [5–9].

Thomas Tiller (ed.), *Synthetic Antibodies: Methods and Protocols*, Methods in Molecular Biology, vol. 1575,
DOI 10.1007/978-1-4939-6857-2_16, © Springer Science+Business Media LLC 2017

The SELEX consists in a succession of several steps. It always starts with a library of oligonucleotide sequences named candidates. Two fixed regions allowing PCR amplification flank these sequences. Usually, the library is composed of 10^{14}–10^{15} different sequences that can fold into a vast array of different 3D structures. These candidates are incubated with a target and those that can bind to it are recovered while the others are removed. These "winning" sequences are then amplified by PCR (or RT-PCR and in vitro transcription for RNA libraries) before another round of in vitro selection. At each round, there is a competition between selected candidates and mutations, which could appear during the amplification steps, may be selected if they are advantageous. After several rounds (4 to 20) the library is sequenced and the sequences predominantly amplified are evaluated for binding to the target in order to identify aptamers.

Thousands of aptamers have been identified for the past 25 years. They have been used for several applications including sensors, biochips, chromatography, microscopy, flow cytometry and many more. Furthermore, several of them have already been enrolled in clinical trials, and one is a commercially available treatment for age-related macular degeneration [4]. Currently, an increasing number of aptamers are being identified against cell surface biomarkers [3, 10–12]. Cell surface biomarkers represent an important class of targets since overexpression or mutations of membrane proteins are associated with several pathologies. However, identification of aptamers against such targets could be challenging because the structure of membrane proteins is dependent of both their inclusion in lipid bi-layers and their interaction with the intracellular and extracellular matrix. Thereby, our group and others have been developing strategies of SELEX during the past 15 years to solve these drawbacks and to select aptamers directly against whole living cells. Such strategies have been named cell-SELEX [3, 13–20].

During cell-SELEX, aptamer selection is conducted directly on whole living cells meaning towards a mixture of targets present in the plasma membrane (Fig. 1). Therefore, thousands of aptamers could theoretically coevolve against thousands of different membrane proteins. To circumvent this drawback, negative selection steps are often used during cell-SELEX using mock cells to favor the selection of aptamers against targets that are specifically expressed on a cell of interest and not on mock cells. Such strategy could be used without prior knowledge of the targets leading to identify aptamers against biomarkers differentially expressed between different cell lines [21–24]. Such aptamers can further be used to purify their targets and identify these biomarkers using mass spectrometry [20, 25–28]. Here, we provide a protocol of cell-SELEX that was optimized during more than 15 years by our

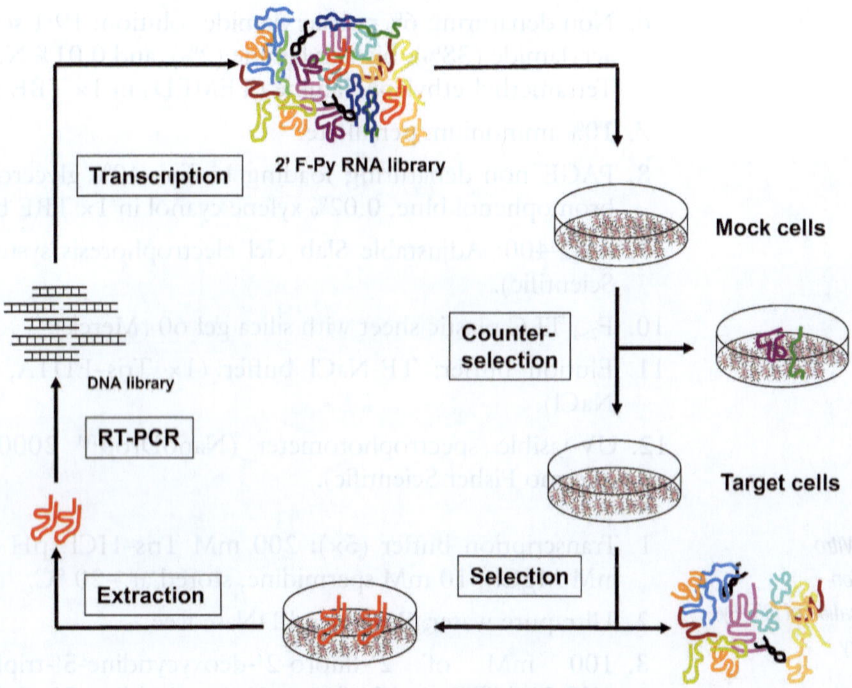

Fig. 1 Schema of cell-SELEX

group [13, 16, 18, 19] in order to identify simultaneously aptamers and biomarkers. This protocol was recently used to identify an aptamer against Annexin A2 [19].

2 Materials

2.1 Library Preparation

2.1.1 Synthesis and Purification of Double Strand DNA Template

1. The DNA library used for selection consisted of a 50-nucleotide random region (N50) flanked by two constant regions: 5′-GGGAGAUGAUCCGUUGAUGCGAG-(N50)-AAGUCGUCGUUCGUAGGCAGAAUC-3′ (*see* **Note 1**). The primer used for the elongation of the library is P70 (5′-CTCGAG<u>TAATACGACTCACTATA</u>GGGAGATGATCCGTTGATGCGAG-3′), the sequence underlined corresponds to the promoter of the T7 RNA polymerase. All oligonucleotides are purchased at Eurogentec.
2. Amicon YM 30 ultra filter (Millipore).
3. Absolute ethanol.
4. Sodium acetate solution (3 M, pH 5.2).
5. Migration buffer: 1× TBE (89 mM Tris-borate and 2 mM EDTA, pH 8.3).

6. Non denaturing 6% polyacrylamide solution: 19:1 solution of acrylamide (38%), bis-acrylamide (2%), and 0.01% N,N,N,N′-Tetramethyl-ethylenediamine (TEMED) in 1× TBE buffer.
7. 10% ammonium persulfate.
8. PAGE non-denaturing loading buffer: 40% glycerol, 0.02% bromophenol blue, 0.02% xylene cyanol in 1× TBE buffer.
9. ASU-400: Adjustable Slab Gel electrophoresis system (CBS Scientific).
10. F_{254} TLC plastic sheet with silica gel 60 (Merck).
11. Elution buffer: TE-NaCl buffer (1× Tris–EDTA, 25 mM NaCl).
12. UV-visible spectrophotometer (NanoDrop™ 2000/2000c, Thermo Fisher Scientific).

2.1.2 *In Vitro* Transcription and Purification of (2′F-Py) RNA Library

1. Transcription buffer (5×): 200 mM Tris–HCl (pH 8.0), 30 mM $MgCl_2$, 10 mM spermidine, stored at −20 °C.
2. Ultrapure water, RNase and DNase free.
3. 100 mM of 2′-fluoro-2′-deoxycytidine-5′-triphosphate (2’F-2′-dCTP) and 2′-fluoro-2′-deoxyuridine-5′-triphosphate (2’F-2′-dUTP) (TriLink Biotechnologies) (*see* **Note 2**).
4. 100 mM of Adenosine 5′-triphosphate (ATP) and guanosine 5′-triphosphate (GTP) (GE Healthcare).
5. 1 M 1,4-Dithiothreitol (DTT).
6. Inorganic pyrophosphatase (1 mg/mL) (Roche Diagnostics) (*see* **Note 3**).
7. $T7^{Y639F}$ RNA polymerase (20×) purified from a bacterial strain kindly provided by R. Souza (*see* **Note 4**).
8. DNase I, RNase-free solution (Sigma).
9. Dry Block Heaters.
10. Denaturing 6% polyacrylamide solution: 19:1 solution of acrylamide (38%), bis-acrylamide (2%), 6 M urea, and 0.01% N,N,N,N′-Tetramethyl-ethylenediamine (TEMED) in 1× TBE buffer mixed.
11. PAGE denaturation loading buffer: 7 M urea, 0.02% bromophenol blue, 0.02% xylene cyanol in 1× TBE buffer.

2.2 Cell-SELEX

1. CHO-K1 (ECACC, Wiltshire, UK) and CHO-ETBR cells (CHO-K1 cells stably transfected with vector expressing human ETBR) were donated by the laboratory of Frédéric Ducancel (CEA, Saclay, France). These cells are an epithelial cell line derived from the ovary of the Chinese hamster and are grown in Nutrient Mixture F-12 HAM supplemented with 10% heat-inactivated fetal bovine serum (FBS).

2. MCF-7 cells derived from human breast adenocarcinoma are purchased from ATCC (Manassas, VA). They are grown in RPMI 1640 GlutaMAX medium supplemented with 10% heat-inactivated FBS.
3. All cells are cultured at 37 °C with 5% CO_2 in the presence of 1% antibiotic/antifungal solution (Sigma).
4. Trireagent (Sigma).
5. Chloroform.
6. DNA primer P60 (5′-GATTCTGCCTACGAACGAC GACTT-3′) from Eurogentec.
7. Ambion, DNA free kit.
8. SuperScriptTM II Reverse Transcriptase kit (Invitrogen).
9. dNTP mix (Sigma).
10. DyNAzyme EXT DNA Polymerase (1 U/μL) kit (Thermo Fisher Scientific).
11. Ultrapure water, RNase and DNase free.
12. PCR instrument.
13. 3% agarose gel: 3 g of ultrapure agarose 1000 (Invitrogen) with 100 mL 1× TBE buffer.
14. 1× GelRed (Nucleic Acid Gel Stain, Interchim).

2.3 Restriction Fragment Length Polymorphism (RFLP)

1. EasyTides Adenosine 5′-triphosphate, [gamma-32P] (222 TBq/mmol, PerkinElmer, reference BLU502Z001MC).
2. T4 kinase (Life Technologies).
3. Bio-Spin Columns with Bio-Gel P-6, in SSC buffer (Bio-Rad).
4. Restriction endonucleases *Rsa*I, *Alu*I, *Hae*III, *Hin*PI (New England Biolabs).
5. NEBuffer2 10× (New England Biolabs).
6. Storm® gel and blot imaging system from Molecular Dynamics.

2.4 Next-Generation Sequencing

1. 10× buffer with $MgCl_2$ (Finnzymes, F-514).
2. dNTPs (Sigma, D7295).
3. Ultrapure water.
4. $MgCl_2$ 50 mM.
5. Taq Polymerase (Finnzymes, Dynazyme EXT DNA polymerase, F-505 L).
6. PCR machine (Applied Biosystems).
7. Specific primers (20 μM) for deep sequencing on Illumina system: p1: 5′-AAT-GAT-ACG-GCG-ACC-ACC-GAG-ATC-TAC-ACT-CTT-TCC-CTA-CAC-GAC-GCT-CTT-CCG-ATC-Tga-ttc-tgc-cta-cga-acg-acg-act-t-3′ and p2:

5′-CAA-GCA-GAA-GAC-GGC-ATA-CGA-GA-index-GTG-ACT-GGA-GTT-CAG-ACG-TGT-GCT-CTT-CCG-ATC-Tgg-gag-atg-atc-cgt-tga-tgc-gag-3′. Uppercase corresponds to the sequences used to amplification and sequencing on the flow-cell. Lowercase corresponds to the primers from our library. Several p2 primer are used with different index sequence for every rounds of SELEX (T-CGT-GAT, T-ACA-TCG, T-GCC-TAA, T-TGG-TCA, T-ATT-GGC, T-CAC-TGT, T-GAT-CTG, T-TCA-AGT, T-CTG-ATC, T-AAG-CTA, T-GTA-GCC, T-TAC-AAG, G-TTG-ACT, C-GGA-ACT, C-TGA-CAT, G-GGA-CGG, respectively for round 0–15).

8. 1× TBE (Tris–Borate–EDTA buffer).
9. GelRed Nucleic Acid Gel Stain 10,000× (Interchim).
10. Agarose gel 3% (3 g of ultrapure agarose 1000—Invitrogen—with 100 mL 1× TBE and 1× GelRed).
11. Imaging system Fusion FX7 (Thermo Fisher Scientific).
12. RunOne electrophoresis system (Embi Tec).
13. Centrifuge system 5417R (Eppendorf).
14. Centrifuge system 4–16 K (Sigma).
15. Surgical blades (Swann-morton).
16. Qubit 2.0 fluorometer (Thermo Fisher Scientific).

2.5 Radioactive Binding Assay

1. Apex heat-labile alkaline phosphatase (Epicentre).
2. Heating block.
3. T4 kinase (Thermo Fisher Scientific).
4. [gamma-^{32}P] ATP (PerkinElmer, reference BLU502Z001MC, 222 TBq/mmol).
5. Bio-Spin columns with Bio-Gel P-6, in SSC buffer (Biorad).
6. 24-well plate (Greiner).
7. Microlab Starlet automate (Hamilton).
8. RPMI 1640 with GlutaMAX.
9. tRNA 10 mg/mL (Sigma).
10. Polyinosinic acid (polyI) that is a homopolymer of inosine (Sigma).
11. Lysis buffer: 50 mM Hepes PH 7.5, 150 mM NaCl, 10% glycerol, 1% Triton X-100, 2 mM $MgCl_2$, and 5 mM EGTA.
12. BCA protein assay (Pierce).
13. Spectrofluorimeter Xenius XC (Safas).
14. MicroBeta TriLux counter (Wallac).
15. Accutase solution (Sigma).
16. BD Accuri C6 (BD Biosciences).

2.6 Target Identification

1. Biotinylated aptamers and scramble sequence (*see* **Note 5**).
2. PBS containing 5 mM EDTA.
3. D-PBS (Dulbecco's phosphate buffered saline).
4. D-PBS (Dulbecco's phosphate buffered saline) with calcium and magnesium.
5. D-PBS supplemented with BSA (0.5 mg/mL).
6. DNA single strand (DNAss) from salmon sperm (Sigma); transfer RNA (tRNA) isolated from baker's yeast (Sigma) and polyinosinic acid (polyI) that is a homopolymer of inosine (Sigma).
7. Dynabeads M-280 streptavidin (Invitrogen).
8. Magnetic stand (Invitrogen).
9. Lysis buffer (D-PBS supplemented with 0.1% Triton X-100).
10. 8 M urea solution.
11. 10% SDS-PAGE gels.
12. Proteosilver stain kit (Sigma).
13. Cell scraper (Nunc).
14. Scepter™ 2.0 Cell Counter (Millipore).

3 Methods

3.1 Library Preparation

A chemically synthesized single-strand DNA pool is first elongated in double-strand DNA template before being in vitro-transcribed in 2'F-Py RNA pool (*see* **Note 2**). During these steps several gel purifications are used to remove products from failed synthesis.

3.1.1 Synthesis and Purification of Double Strand DNA Template

1. Mix 500 μL of PCR buffer containing $MgCl_2$ (10×), 150 μL of $MgCl_2$ (50 mM), 250 μM of DMSO (100%), 100 μL of dNTP mix (10 mM), 125 μL of Taq polymerase (1 U/μL), 10 nmol of chemically synthesized single strand DNA library, 10 nmol of P70 primer and complete to a final volume of 5 mL with nuclease-free water.
2. Chemically synthesized single-strand DNA pool is elongated in double-strand DNA template in a thermo-cycler with the following settings: 5 min 95 °C, 5 min 53 °C, 30 min 72 °C.
3. Obtained double stranded DNA template is concentrated using YM30 Amicon Ultra filter-tube by centrifugation at 4000 × *g* (20 min, 4 °C), which gives final volume of around 200 μL.
4. The concentrated solution is precipitated by addition of 20 μL of sodium acetate solution (3 M, pH 5.2) and 1 mL of pure ethanol (HPLC grade). After vortexing and centrifugation at 20,000 × *g* (15 min, 4 °C), supernatant is removed and the

pellet is dissolved in 50 μL of nuclease-free ultrapure water containing 1 mM of $MgCl_2$. The tube should be left on the air for at least 15 min, in order to evaporate the residues of ethanol.

5. Double stranded DNA library is then purified by non-denaturing polyacrylamide gel electrophoresis. For this purpose DNA pool is mixed with 20 μL of non-denaturing loading buffer and directly placed on a long slice (around 30 cm) of polyacrylamide gel, without prior denaturation step.
6. Purification starts after application of a current and migration should be performed with a low voltage value of 500 V. This process can be terminated when bromophenol blue from loading buffer begins to leave the gel.
7. The gel is unmolded on a Saran film and F_{254} TLC plastic sheet with silica gel 60 is placed below the gel. After illumination of the gel with a UV lamp, a shadow is observed that corresponds to the presence of nucleic acids. This part of gel is cut with a sterilized scalpel and DNA is passively eluted in 10 mL of TENaCl elution buffer, by placing the tube overnight in rotator at 4 °C.
8. The 10 mL of TENaCl containing the purified DNA library is concentrated using YM30 Amicon Ultra filter-tube and precipitated, as described in **steps 3** and **4**.
9. The DNA concentration is determined by measuring the absorbance at 260 nm using a UV-visible spectrophotometer and calculated based on relation: 1 DO unit is equal to 50 μg/mL of double stranded DNA.

3.1.2 *In Vitro Transcription and Purification of (2′ F-Py) RNA Library*

1. Around 140 μg (approximately 2 nmoles) of double-strand DNA template is in vitro-transcribed in 2′F-Py RNA pool at 37 °C overnight in 2 mL of reaction mix containing: 10 μL of 2′F-2′-dCTP (100 mM), 10 μL of 2′F-2′-dUTP (100 mM), 10 μL ATP (100 mM), 10 μL GTP (100 mM), 200 μL of transcription buffer (5×), 10 μL of DTT (1 M), 20 μL of inorganic pyrophosphatase (1 mg/mL), 50 μL of $T7^{Y639F}$ RNA polymerase (20×), and nuclease-free water (*see* **Note 6**).
2. Add 10 μL of DNase I solution per mL of transcription mix and incubate at 37 °C for 30 min (*see* **Note 7**).
3. 2′F-Py RNA library is concentrated using YM30 Amicon Ultra filter-tube and precipitated, as described in **steps 3** and **4** of Subheading 3.1.1.
4. The pellet is dissolved in 40 μL of denaturing loading buffer. The tube should be left on the air for at least 15 min, in order to evaporate the residues of ethanol. 2′F-Py RNA library is then denatured on a heating block for 5 min at 85 °C before being purified on a denaturing polyacrylamide gel electrophoresis at 35 W as described in **steps 6–8** of section 3.1.1.

5. The 2′F-Py RNA concentration is determined by measuring the absorbance at 260 nm using a UV-visible spectrophotometer and calculated based on the Lambert-Beer law, which is: Concentration (mol/L) = Absorbance (260 nm)/($\varepsilon_{\lambda=260}$ × path length (cm)).
6. The 2′F-Py RNA library should be stored at −20 °C.

3.2 Cell-SELEX

Cell-SELEX consists of several rounds of in vitro selection (Fig. 1). In each round, the 2′F-Py RNA library is incubated with mock cells (counterselection step) to remove unwanted aptamers. Sequences that do not bind to mock cells are incubated on cells of interest and, after several washings, sequences bound to these cells are recovered and amplified by RT-PCR and in vitro transcription before another round of in vitro selection. After several rounds of in vitro selection, sequences with the best-inherited traits (affinity for cells of interest) will gradually accumulate in the population (*see* **Note 8**). In the following procedure, we provide a protocol of cell-SELEX that was used to select aptamers against Endothelin type B receptor (ETBR). The cell-SELEX was performed against the hamster CHO-K1 cell line stably transformed to overexpress ETBR (CHO-ETBR). The wild type CHO-K1 cell line was used in negative-selection steps [19].

1. 2′F-Py RNA library (5 nmol) is denatured at 85 °C for 5 min following immediate cooling at 4 °C (ice block) for 5 min and warming up to 37 °C (*see* **Note 9**).
2. Two steps of counterselection are then performed. 2′F-Py RNA library is incubated with 10^7 adherent CHO-K1 cells in 3 mL RPMI 1640 for 30 min at 37 °C, with shaking each 5 min. The supernatant containing all unbound sequences is placed a second time on 10^7 adherent CHO-K1 cells for 30 min at 37 °C, with shaking each 5 min.
3. The supernatant containing all unbound sequences is then placed on 10^7 adherent CHO-ETBR cells for 30 min at 37 °C, with shaking each 5 min. The supernatant containing unbound sequences is removed and the cells are washed several times with 5 mL of RPMI 1640 to remove sequences that are non-specifically attached to the surface of cells or that are bound with weak affinities.
4. Sequences that are bound to cells or internalized are recovered by trizol-extraction. The cells are resuspended in 4 mL (4 × 1 mL) of trireagant (Sigma T9424) and centrifuged at 12,000 × *g* (10 min, 4 °C). Supernatant is recovered, mixed with chloroform (200 μL of chloroform per 1 mL of trireagent) and after few minutes the tubes are centrifuged at 12,000 × *g* (15 min, 4 °C). Subsequently, isopropanol is added to the supernatant (500 μL of isopropanol per 1 mL of trireagent) and

mixture is centrifuged again at 12,000 × *g* (10 min, 4 °C). After solvents elimination, the pellet is washed with 1 mL of ethanol (70%) and centrifuged at 7500 × *g* (5 min, 4 °C). Ethanol is then removed, 2′F-Py RNA is dissolved in RNase and DNase ultrapure water (20 μL per 1 mL of trireagent) and left on air for around 20 min. The tubes with 2′F-Py RNA are then pooled.

5. In order to eliminate any DNA contamination, the sample is treated with DNase (Ambion, DNA free kit). The extracted 2′F-Py RNA solution is incubated in DNA free buffer (1×) with 1 μL of DNase, at 37 °C for 30 min. Afterwards 10 μL of inactivating beads is added and homogenized for 2 min, following centrifugation at 10,000 × *g* (2 min, room temperature).
6. The supernatant with 2’F-Py RNA is collected and precipitated as described in **step 4** of Subheading 3.1.1. The pellet is resuspended in 40 μL of DNase, RNase ultrapure water and left on air for around 20 min. The quantity of extracted RNA is determined by measuring the absorbance at 260 nm using UV-visible spectrophotometer and calculated based on relation: 1 DO unit is equal to 40 μg/mL of RNA (*see* **Note 10**).
7. 2′F-Py RNA is then reverse transcribed in DNA using Reverse Transcription (RT) in a final volume of 200 μL (*see* **Note 11**). The volume of extracted RNA is first completed to a final volume of 100 μL with nuclease-free water and then mixed with 24 μL of p60 primer (20 μM). The mixture is incubated in PCR instrument at 70 °C for 10 min and 55 °C for 2 min.
8. Subsequently the following solutions, previously mixed, are added: 40 μL of RT buffer (5×), 20 μL of DTT (0.1 M), 10 μL of dNTP (10 mM) and 10 μL of reverse transcriptase (200 U/μL).
9. The reverse transcription is then performed at 55 °C for 50 min.
10. cDNA is then amplified by PCR in a final volume of 1 mL. RT mixture (200 μL) is mixed with PCR mix containing 100 μL of PCR buffer containing $MgCl_2$ (10×), additional 30 μL of $MgCl_2$ (50 mM), 50 μM of DMSO (100%), 100 μL of p60/p70 primers mixture (10 μM each), 20 μL of dNTP mix (10 mM), 25 μL of Taq polymerase (1 U/μL). As a control, 50 μL of PCR mix without DNA matrix is also prepared.
11. PCR is performed by subsequent incubations of the sample: 30 s at 94 °C, 1 min at 53 °C and 1.5 min at 72 °C for 10 cycles (*see* **Note 12**). The efficiency of PCR amplification and the absence of DNA contamination are verified on agarose gel (3%).
12. Aliquot of PCR (50 μL) is kept apart and stored at −20 °C for sequencing procedure.

13. The remaining 950 μL of PCR is concentrated to 200 μL using YM30 Amicon Ultra as described in **step 3** of Subheading 3.1.1 before being used as template to produce a new pool of 2'F-Py RNA by 1 mL of transcription as described in Subheading 3.1.2.
14. The **steps 1–12** represent one round of cell-SELEX that are repeated several times (*see* **Note 8**). During successive rounds, the selective pressure is gradually increased by decreasing the time of incubation on cells, decreasing the concentration of the 2′F-Py RNA pool, increasing the number of washings, increasing the time of washing (incubated cells on medium for several minutes before aspirating the medium), decreasing the number of cells, adding nonspecific competitors such as yeast total RNA (100 μg/mL) or inversing counterselection and selection steps (*see* Table 1).

3.3 Restriction Fragment Length Polymorphism (RFLP)

If some sequences are predominantly amplified it may lead to the appearance of specific restriction sites in the population that can be detected by restriction fragment length polymorphism (RFLP) analysis [16, 29]. In such experiment, a mixture of restriction enzymes that specifically cut a sequence of four bases degrades the pool. Such short restriction sites have a high probability to be present at least one time in most of the sequences.

1. 200 pmol of primer P60 is 5′-end-radiolabeled at 37 °C during 30 min in the presence of 1.8×10^6 Bq of [gamma-^{32}P] ATP, 20 U of T4 kinase and forward buffer (1×) in final volume of 50 μL.
2. The radioactive oligonucleotide is purified with Bio-Spin columns with Bio-Gel P-6 according to the provider's instruction.
3. 1 μL of PCR from each round of selection is amplified by 8 cycles of PCR in a final volume of 50 μL completed with 1 μL of radioactive primer as previously described in **steps 9** and **10** of Subheading 3.2.
4. The radiolabeled PCR products are digested at 37 °C during 1 h by a combination of *Rsa*I, *Alu*I, *Hae*III, and *Hin*PI endonuclease (10 units each) in NEB2 buffer.
5. After an ethanol precipitation, as described in **step 4** of Subheading 3.1.1., the digested samples are subsequently analyzed by electrophoresis on a denaturing polyacrylamide gel (6% acrylamide/bis acrylamide 19:1, 7 M urea, 1× TBE).
6. After the electrophoresis, the dried gel was exposed on imaging plate and analyzed using Storm® gel and blot imaging system from Molecular Dynamics (Fig. 2).

Table 1

Evolution of selection pressure during cell-SELEX directed against CHO-ETBR

Round	Millions of cells during CS	Millions of cells during S	Concentration of the pool (μM)	Incubation time (min)	Number of washing	Increased time of washing	Addition of nonspecific competitor tRNA	Inversion of CS and S
1	10	10	2	30	2			
2	10	10	1	30	3			
3	10	10	1	25	3			
4	5	5	1	25	3			
6	5	5	1	20	4			
7	5	5	0.5	20	5			
8	5	5	0.5	15	5			
9	5	5	0.5	15	5	5 min for W5		
10	5	5	0.5	10	5	5 min for W5		
11	5	5	0.5	10	5	5 min for W5	100 μg/mL	
12	5	5	0.5	10	5	5 min for W5	100 μg/mL	
13	5	5	0.5	10	5	5 min for W4&W5	100 μg/mL	
14	5	5	0.5	10	5	5 min for W4&W5	100 μg/mL	+
15	5	5	0.5	10	5	5 min for W4&W5	100 μg/mL	+

The selective pressure is gradually increased during cell-SELEX. *CS* counterselection step, *S* selection step, *W5* the fifth washing, *W4* the fourth washing

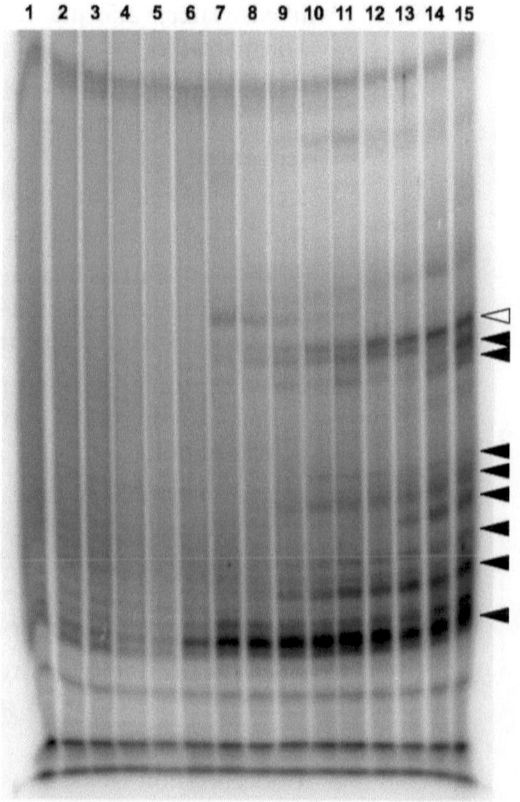

Fig. 2 Estimation of the pool complexity during each round of SELEX by RFLP. [32P] 5′-end-labeled PCR from each of the indicated rounds are digested with a combination of endonucleases and analyzed by electrophoresis on a denaturing polyacrylamide gel. During the six first rounds, the pool is highly heterogeneous and lead to a smearing RFLP profile. After seven rounds of selection, some sequences are predominantly amplified, resulting in discrete restriction bands. One restriction band (*white arrow*) disappears between round seven and nine, while others appear and are amplified during the cell-SELEX (*black arrows*). During the three last rounds, RFLP profiles seem stabilized suggested that the population does not change

3.4 Next-Generation Sequencing

After cell-SELEX, it is necessary to sequence libraries in order to know which sequences have been enriched. It is now common to use next-generation sequencing instead of cloning and Sanger sequencing. Next-generation sequencing allows analyzing a larger part of the population (millions of sequences) and to analyze every round of the SELEX in parallel. Here, we describe the preparation of libraries for deep sequencing on Illumina system using PCR for the addition of adapter sequences and indexing every round.

1. 10 μL of DNA libraries (1.5 μM) previously aliquoted from each round of selection are mixed with PCR reagents (10 μL of 10× buffer, 3 μL of $MgCl_2$, 2 μL of dNTP, 62.5 μL of dis-

tilled water; 2.5 μL of p1 primer and 2.5 μL of Taq polymerase).

2. Each sample is then mixed with 2.5 μL of different p2 primers (20 μM) in order to amplify each library with a different index sequence.
3. 7 cycles of PCR (30 s 94 °C, 1 min 53 °C, 1.5 min 72 °C) are performed for all samples.
4. At the end, 5 μL is analyzed by electrophoresis on 3% agarose gel. Intensity of expected band (size around 200 bp) in each sample is measured with ImageJ using ROI Manager.
5. Based on intensity, appropriate volume is calculated for each sample to make a mix of every library at equimolar concentration.
6. After concentration using YM30 Amicon Ultra filter-tube and precipitation, as described in **steps 3** and **4** of Subheading 3.1.1., the mix of amplified pools is purified by non-denaturing 3% agarose gel electrophoresis using RunOne electrophoresis system. Migration is performed at 50 V during 1 h.
7. The band of interest is revealed by UV before being cut in order to recover DNA as described in **steps** 7 and **8** of Subheading 3.1.1.
8. Concentration of DNA is measured with the Qubit system, following supplier's instructions. The minimum required to continue with the Illumina's protocol for sequencing is 20 fmol in 10 μL.
9. The pool is sequenced on a single lane of the microfluidic cell using 80 sequencing cycles in single read and 7 cycles for the index sequence.
10. Using *bcl2fastq Conversion Software v1.8.4*, base calling files are converted in fastq files for each round of selection, recognized by the index sequence.
11. To analyze data from deep sequencing, several toolkits are developed by different group such as *Galaxy project*, a web interface (https://usegalaxy.org) that can manipulate many file formats and count the frequency of each sequence in a file. In the lab, we use *Patternity-seq*, a homemade software that allows to cluster sequences based on Levenstein distance and to monitor the enrichment of sequences and predicted secondary structures during the cell-SELEX (access to this software can be asked at MIRCEN Plateforms http://i2bm.cea.fr/drf/i2bm/english/Pages/MIRCen/Services-Collaborations.aspx). This analysis helps to select some sequences (predominantly amplified or with the highest enrichment) to be evaluated by binding experiments (Fig. 3).

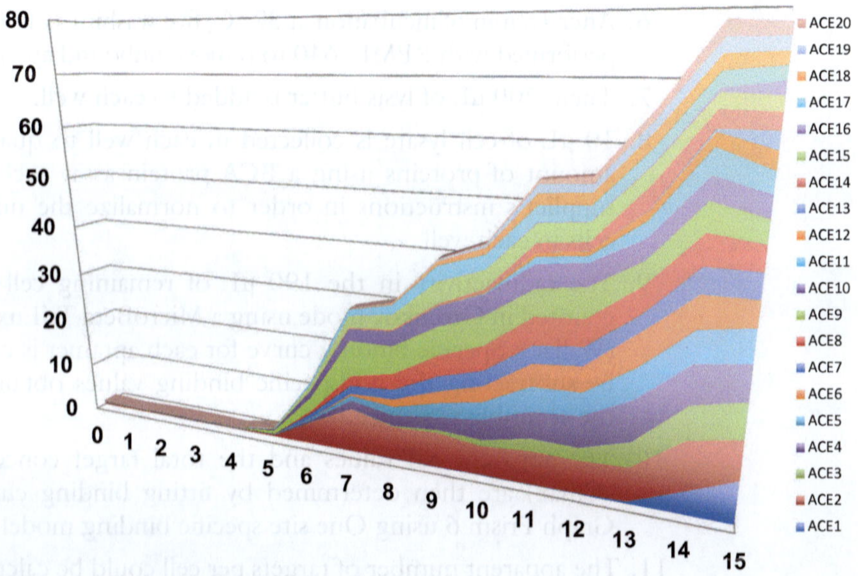

Fig. 3 Evolution of sequences during cell-SELEX measured by next-generation sequencing. The evolution of the 20 most abundant sequences from rounds 15 is presented. The graph displays the percentage of each sequence in the pool at each round of cell-SELEX. These sequences start to be amplified mostly after five rounds and represent much around 80% of the pool after 14 rounds

3.5 Radioactive Binding Assay

Radioactive binding assay is performed in order to identify the sequences with the highest affinity for cells. We previously described in detail this method focusing on different parameters that have to be optimized [30].

1. Every sequence to be tested and a negative control scramble sequence are first ordered as DNA templates before being transcribed and purified in 2′F-Py-RNA as described in **steps 9–12** of Subheading 3.2.
2. 5′ end of 2′F-Py RNA is dephosphorylated using Apex Heat-Labile Alkaline phosphatase. 60 pmol of sequence is mixed with 1 μL of phosphatase, 2 μL of 10× phosphatase buffer. The reaction volume is completed with distilled water to have 20 μL.
3. Sample is incubated 30 min at 37 °C before inactivation of the phosphatase by denaturation at 90 °C during 5 min.
4. Aptamer is then radiolabeled as described in **steps 1** and **2** of Subheading 3.3 and stored at −20 °C before use.
5. MCF-7 cells (approximately 100,000 cells per well) are incubated in 200 μL of RPMI 1640 with six different concentrations of radioactive oligonucleotides (1, 2, 5, 10, 20, and 50 nM) in the presence of tRNA and PolyI (100 μg/mL each, to reduce nonspecific binding).

6. After 15 min of incubation at 37 °C, five washings of 500 μL are performed with RPMI 1640 to remove unbound aptamers.
7. Then, 200 μL of lysis buffer is added to each well.
8. 10 μL of cell lysate is collected in each well to quantify the amount of proteins using a BCA protein assay according to supplier's instructions in order to normalize the number of cells in each well.
9. The radioactivity in the 190 μL of remaining cell lysate is counted in Cerenkov mode using a MicroBeta TriLux counter (Wallac). Specific binding curve for each aptamer is calculated by subtracting the nonspecific binding values obtained with the scramble sequence.
10. The apparent Kd values and the total target concentration (Cmax) are then determined by fitting binding curve with Graph Prism 6 using One site specific binding model.
11. The apparent number of targets per cell could be calculated by the formula: Avogadro number × (Cmax × volume of incubation)/number of cells per well.

3.6 Target Identification

Cell-SELEX can identify aptamers without prior knowledge of the targets. In this case, these aptamers can be a posteriori used to purify their targets and as a consequence to identify the latter. Here, we presented a slightly modified protocol previously described by Berezovski et al. [28].

1. 20 millions of adherent MCF-7 cells are washed twice with PBS and harvested on ice with PBS containing 5 mM EDTA using a cell scraper (*see* **Note 13**).
2. Cells are individualized by serial aspiration in a Pasteur pipette and counted using a Scepter Cell Counter.
3. Cells are pelleted by centrifugation at 300 × *g*, 5 min. From that step, all the procedures are realized at 4 °C or on ice.
4. 15 millions of cells are suspended in 3 mL D-PBS with calcium and magnesium and then mixed with 30 μL tRNA (10 μg/μL), 30 μL ssDNA (10 μg/μL), and 30 μL polyI (10 μg/μL).
5. Cells are then separated in three tubes. The tube 1 will be used as negative control for extraction without biotinylated oligonucleotide. 20 pmol of biotinylated scramble sequence are added in tube 2 that will also be used as negative control. 20 pmol of biotinylated ACE4 aptamer is added to tube 3 (*see* **Note 14**).
6. All the tubes are incubated for 30 min on ice. Then cells are pelleted and washed twice by centrifugation (300 × *g*, 5 min) with 1 mL of D-PBS supplemented with BSA (0.5 mg/mL).
7. Cells are resuspended in 1 mL of D-PBS supplemented with BSA (0.5 mg/mL) and mixed with 1 mg of Dynabeads M-280

streptavidin. After 15 min of incubation on ice, bead-cell-aptamer complexes are recovered using a magnetic stand and washed twice with 1 mL of D-PBS (*see* **Note 15**).

8. Bead-cell-aptamer complexes are then resuspended in 1 mL of lysis buffer supplemented with tRNA, ssDNA, and polyI at 100 mg/mL each. After 30 min of cell lysis on ice, the cell lysate is eliminated using a magnetic stand and the beads are washed twice with lysis buffer without nucleic acid competitors.
9. Targets bound to aptamers are eluted from the beads in 30 μL of 8 M urea. After 15 min of incubation at 60 °C. The supernatants for all the tubes are recovered from the beads using a magnetic stand and loaded on 10% SDS-PAGE gels.
10. After electrophoresis, the gel is stained using Proteosilver stain kit. Proteins that have a substantial higher intensity in the condition with aptamer (tube 3) compared to the two other conditions (tubes 1 and 2) are detected (Fig. 4). These bands are cut and analyzed by nano-LC-MS/MS mass spectrometry (*see* **Note 16**).

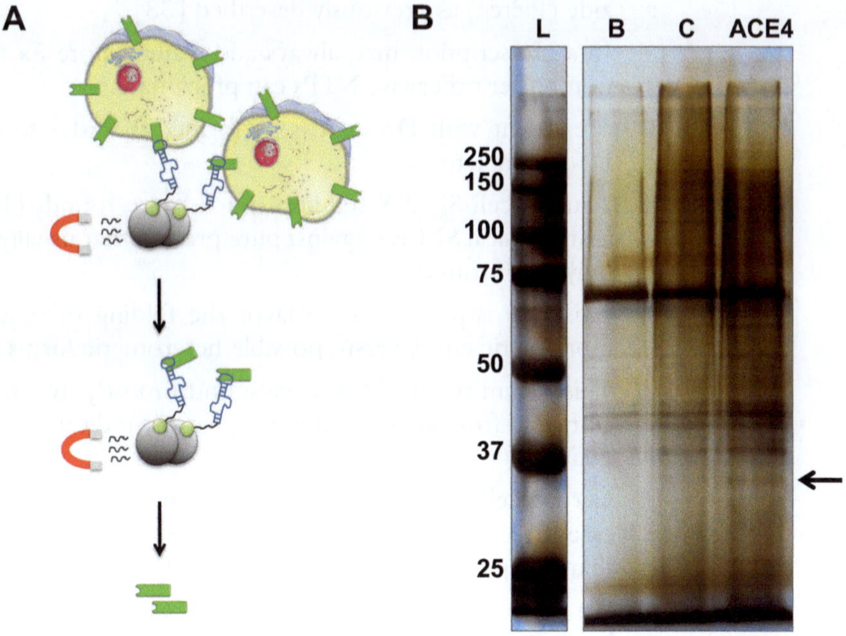

Fig. 4 Aptamer-mediated target purification. (**a**) Schema of the purification. Magnetic streptavidin beads are used to recover cells bound to biotinylated aptamers (*blue*). Then, cells are lysed to recover the target (*green*) bound to the aptamers. (**b**) After elution, the extracted proteins are separated by SDS-PAGE electrophoresis before being silver-stained in a polyacrylamide gel. *Lane L.*, molecular marker. *Lane B*, extracted proteins with empty beads. *Lane C*, extracted proteins with a biotinylated scramble sequence. *Lane ACE4*, extracted proteins with the biotinylated anti-annexin A2 aptamer (ACE4). The *arrow* indicates a band with a substantial higher intensity in lane ACE4 compared to *lane B* and *C*. This band has been cut and analyzed by nano-LC-MS/MS mass spectrometry. The target identified using this method for ACE4 aptamer is Annexin2. *Figure from the original research paper* [19]

4 Notes

1. Chemically synthesized single strand DNA library can contain few remaining protecting groups that may inhibit polymerases. Therefore, it is important to purify it by PAGE purification before used such library can be ordered purified by denaturing PAGE purification or such purification can be performed as described in **steps 4** and **5** of Subheading 3.1.2.
2. We used 2′Fluro-modified pyrimidines in order to increase the resistance to nucleases.
3. Inorganic pyrophosphatase improves the yield of in vitro transcription reactions by degrading pyrophosphates that are generated during transcription and inhibit the T7 polymerase [31].
4. $T7^{Y639F}$ RNA polymerase is a mutated form of the T7 polymerase that has a better yield of incorporation of 2′F-Py [32]. If the polymerase cannot be produced in the lab it can be ordered from Epicentre.
5. Biotinylated oligonucleotides can be chemically synthesized or the biotin can be introduced using biotinamidocaproyl hydrazide (Pierce) as previously described [33].
6. In a transcription mix, always add water before 5× transcription buffer otherwise NTPs can precipitate.
7. Treatment with DNase I is performed in order to eliminate DNA template
8. Usually, cell-SELEX requires much more rounds (10 to 20) than classical SELEX against pure protein that usually requires only 4–10 rounds.
9. This step is performed to favor the folding of sequences as monomeric entity versus possible heteromeric forms.
10. This quantity of RNA corresponds mostly to total RNA extracted from cells and doesn't give any indication about the quantity of 2′F-RNA.
11. Certain cells provide a very high amount of total RNA, in that case the volume of RT has to be adjusted in order to use less than 1.5 μg of total RNA per μL of RT.
12. We reduce the number of PCR cycles as much as possible because a high number of cycles may favor the amplification of sequences that are preferentially amplified by PCR instead of aptamers.
13. We do not use trypsin to detach cells in order to ensure that the target of aptamer is not removed from the cell surface during the detachment from the plate.
14. We suggest to use a final concentration of aptamer corresponding to ten times the Kd in order to saturate the target.

15. A clear pellet of cells on magnetic beads can be observed with aptamer in contrast to the other conditions. It represents a positive sign that the aptamer interacts tightly with its target.
16. This protocol is not really stringent to ensure that aptamer-target complexes are not dissociated during the purification. Accordingly, a high number of proteins are visualized in the gel, which can be a problem to detect the interesting bands. In that case, beads can be washed using increasing concentration of NaCl before lysis to reduce nonspecific interaction.

Acknowledgments

This protocol of cell-SELEX was first developed in the lab of Domenico Libri before being further optimized in the lab of Bertrand Tavitian and now in the Neurodegenerative Diseases Laboratory, we thank all our collaborators from these labs for help and fruitful discussions, especially Carine Pestourie, Karine Gombert, Benoit Jego, Isabelle Janssens, Jocelyne Boulay, Mohamed Aissouni, Bertrand Tavitian, and Domenico Libri. We are also grateful to Dr. Rui Sousa (University of Texas, San Antonio) for his generous gift of a T7Y639F RNA polymerase-expressing plasmid. Studies relating to selection of aptamers in our laboratories were supported by grants from the "Agence Nationale pour la Recherche" [projects ANR-RNTS TomoFluo3D, ANR-PNANO nanorings and under the frame of EuroNanoMed (project META)]; the FMT-XCT European program [Grant agreement no. 201792] and the European Molecular Imaging Laboratory (EMIL) network [EU contract LSH-2004-503569].

References

1. Tuerk C, Gold L (1990) Systematic evolution of ligands by exponential enrichment: RNA ligands to bacteriophage T4 DNA polymerase. Science 249:505–510
2. Ellington AD, Szostak JW (1990) In vitro selection of RNA molecules that bind specific ligands. Nature 346:818–822
3. Cibiel A, Dupont DM, Duconge F (2011) Methods to identify aptamers against cell surface biomarkers. Pharmaceuticals 4:1216–1235
4. Keefe AD, Pai S, Ellington A (2010) Aptamers as therapeutics. Nat Rev Drug Discov 9:537–550
5. Opazo F, Levy M, Byrom M, Schafer C, Geisler C, Groemer TW, Ellington AD, Rizzoli SO (2012) Aptamers as potential tools for super-resolution microscopy. Nat Methods 9:938–939
6. Zhang P, Zhao N, Zeng Z, Chang CC, Zu Y (2010) Combination of an aptamer probe to CD4 and antibodies for multicolored cell phenotyping. Am J Clin Pathol 134:586–593
7. Meyer M, Scheper T, Walter JG (2013) Aptamers: versatile probes for flow cytometry. Appl Microbiol Biotechnol 97(16):7097–7109
8. Mayer G (2009) The chemical biology of aptamers. Angew Chem Int Ed Engl 48:2672–2689
9. Keeney TR, Bock C, Gold L, Kraemer S, Lollo B, Nikrad M, Stanton M, Stewart A, Vaught JD, Walker JJ (2009) Automation of the somalogic proteomics assay: a platform for biomarker discovery. J Assoc Lab Automa 14:360–366
10. Pestourie C, Tavitian B, Duconge F (2005) Aptamers against extracellular targets for in vivo applications. Biochimie 87:921–930

11. Cibiel A, Pestourie C, Duconge F (2012) In vivo uses of aptamers selected against cell surface biomarkers for therapy and molecular imaging. Biochimie 94:1595–1606
12. Dickinson H, Lukasser M, Mayer G, Huttenhofer A (2015) Cell-SELEX: in vitro selection of synthetic small specific ligands. Methods Mol Biol 1296:213–224
13. Cerchia L, Duconge F, Pestourie C, Boulay J, Aissouni Y, Gombert K, Tavitian B, de Franciscis V, Libri D (2005) Neutralizing aptamers from whole-cell SELEX inhibit the RET receptor tyrosine kinase. PLoS Biol 3:e123
14. Ohuchi, S.P., Ohtsu, T. and Nakamura, Y. (2005) A novel method to generate aptamers against recombinant targets displayed on the cell surface. Nucleic Acids Symp Ser (Oxf), 49:351–352
15. Ohuchi SP, Ohtsu T, Nakamura Y (2006) Selection of RNA aptamers against recombinant transforming growth factor-beta type III receptor displayed on cell surface. Biochimie 88(7):897–904
16. Pestourie C, Cerchia L, Gombert K, Aissouni Y, Boulay J, De Franciscis V, Libri D, Tavitian B, Duconge F (2006) Comparison of different strategies to select aptamers against a transmembrane protein target. Oligonucleotides 16:323–335
17. Meyer S, Maufort JP, Nie J, Stewart R, McIntosh BE, Conti LR, Ahmad KM, Soh HT, Thomson JA (2013) Development of an efficient targeted cell-SELEX procedure for DNA aptamer reagents. PLoS One 8:e71798
18. Zueva E, Rubio LI, Duconge F, Tavitian B (2011) Metastasis-focused cell-based SELEX generates aptamers inhibiting cell migration and invasion. Int J Cancer 128:797–804
19. Cibiel A, Quang NN, Gombert K, Theze B, Garofalakis A, Duconge F (2014) From ugly duckling to swan: unexpected identification from cell-SELEX of an anti-Annexin A2 aptamer targeting tumors. PLoS One 9:e87002
20. Blank M, Weinschenk T, Priemer M, Schluesener H (2001) Systematic evolution of a DNA aptamer binding to rat brain tumor microvessels. selective targeting of endothelial regulatory protein pigpen. J Biol Chem 276:16464–16468
21. Wang C, Zhang M, Yang G, Zhang D, Ding H, Wang H, Fan M, Shen B, Shao N (2003) Single-stranded DNA aptamers that bind differentiated but not parental cells: subtractive systematic evolution of ligands by exponential enrichment. J Biotechnol 102:15–22
22. Shangguan D, Meng L, Cao ZC, Xiao Z, Fang X, Li Y, Cardona D, Witek RP, Liu C, Tan W (2008) Identification of liver cancer-specific aptamers using whole live cells. Anal Chem 80:721–728
23. Parekh P, Tang Z, Turner PC, Moyer RW, Tan W (2010) Aptamers recognizing glycosylated hemagglutinin expressed on the surface of vaccinia virus-infected cells. Anal Chem 82:8642–8649
24. Zueva E, Rubio LI, Duconge F, Tavitian B (2010) Metastasis-focused cell-based SELEX generates aptamers inhibiting cell migration and invasion. Int J Cancer 128:797–804
25. Daniels DA, Chen H, Hicke BJ, Swiderek KM, Gold L (2003) A tenascin-C aptamer identified by tumor cell SELEX: systematic evolution of ligands by exponential enrichment. Proc Natl Acad Sci U S A 100:15416–15421
26. Mallikaratchy P, Tang Z, Kwame S, Meng L, Shangguan D, Tan W (2007) Aptamer directly evolved from live cells recognizes membrane bound immunoglobin heavy mu chain in Burkitt's lymphoma cells. Mol Cell Proteomics 6:2230–2238
27. Ulrich H, Wrenger C (2009) Disease-specific biomarker discovery by aptamers. Cytometry A 75(9):727–733
28. Berezovski MV, Lechmann M, Musheev MU, Mak TW, Krylov SN (2008) Aptamer-Facilitated Biomarker Discovery (AptaBiD). J Am Chem Soc 130(28):9137–9143
29. Bartel DP, Szostak JW (1993) Isolation of new ribozymes from a large pool of random sequences [see comment]. Science 261:1411–1418
30. Quang NN, Pestourie C, Cibiel A, Ducongé F (2016) How to measure the affinity of aptamers for membrane proteins expressed on the surface of living adherent cells. Methods 97:35–43
31. Cunningham PR, Ofengand J (1990) Use of inorganic pyrophosphatase to improve the yield of in vitro transcription reactions catalyzed by T7 RNA polymerase. Biotechniques 9:713–714
32. Padilla R, Sousa R (1999) Efficient synthesis of nucleic acids heavily modified with non-canonical ribose 2′-groups using a mutantT7 RNA polymerase (RNAP). Nucleic Acids Res 27:1561–1563
33. Hansske F, Cramer F (1979) Modification of the 3′ terminus of tRNA by periodate oxidation and subsequent reaction with hydrazides. Methods Enzymol 59:172–181

Chapter 17

Rapid Selection of RNA Aptamers that Activate Fluorescence of Small Molecules

Grigory S. Filonov*

Abstract

RNA aptamers can serve as valuable tools for studying and manipulating live cells. Fluorescent aptamers are the ones that bind to and turn on fluorescence of small-molecule dyes (fluorogens). Similarly to fluorescent proteins, fluorescent RNA aptamers can be used to image spatial and temporal RNA dynamics in live cells. Additionally, these aptamers can serve as a basis for engineering genetically encoded fluorescent biosensors. This chapter presents a protocol for rapid and efficient screening of RNA aptamer libraries to isolate fluorescent aptamers. The protocol describes how to design, clone, and express RNA aptamer library in bacterial cells and how to screen the bacteria to find aptamers with the desired fluorescent properties.

Key words RNA, Aptamer, Fluorescence, High-throughput screening, Directed evolution, FACS

1 Introduction

RNA aptamers are a diverse class of synthetic molecules that have found numerous applications in modern biotechnology and biology [1]. Introduced in early 1990s, aptamers are now widely used as protein inhibitors, drug delivery vehicles or biosensors, among other applications [2–5]. The success of aptamers is a result of their advantageous properties such as small size, high affinity, high specificity of interaction with a cognate ligand, and relative easiness of aptamer production [3].

Fluorescent RNA aptamers are a relatively recent expansion of the aptamer application range [6, 7]. These aptamers bind to and turn on fluorescence of a small-molecule dye (fluorogen) that is normally nonfluorescent in solution. The first fluorescent aptamer was introduced more than 10 years ago [6]. However, that aptamer

The original version of this chapter was revised. The erratum to this chapter is available at: DOI 10.1007/978-1-4939-6857-2_27

*** Current affiliation:** Essen Bioscience, Ann Arbor, MI, USA
Previous affiliation: Department of Pharmacology, Weill Medical College, Cornell University, New York, NY, USA

Thomas Tiller (ed.), *Synthetic Antibodies: Methods and Protocols*, Methods in Molecular Biology, vol. 1575, DOI 10.1007/978-1-4939-6857-2_17, © Springer Science+Business Media LLC 2017

activated fluorescence of malachite green, the fluorogen which is not compatible with live-cell imaging due to high cytotoxicity and non-specific binding to cellular components. A fluorescent aptamer that was termed Spinach was the first one that was suitable for live-cell RNA imaging [7]. Spinach is a ~100 nucleotide long RNA that binds to and activates fluorescence of a fluorogen called DFHBI (3,5-difluoro-4-hydroxybenzylidene imidazolinone). DFHBI is a small-molecule mimic of the chromophore of GFP (green fluorescent protein). DFHBI is dim in solution but becomes highly fluorescent upon Spinach binding. More importantly, DFHBI has virtually no fluorescence background in cells and is not toxic. This makes Spinach-DFHBI complex fully compatible with live-cell imaging. The Spinach family of aptamers later expanded and now includes such aptamers as Spinach2 and Broccoli [8, 9]. The latter is currently the brightest green fluorescent RNA for intracellular imaging.

One of the major applications of fluorescent aptamers is to serve as fluorescent tags for RNA visualization in live cells. Currently, Spinach and aptamers that belong to the Spinach family have been used to track different noncoding RNAs and mRNAs in bacteria, yeast, and mammalian cells [7–13]. Another broad application of these aptamers is to serve as a platform for engineering of fluorescent biosensors that can be used both in basic biology and industrial biotechnology. Numerous Spinach-based biosensors have been reported that signal the presence of small-molecule metabolites and proteins [14–18]. Thus, fluorescent RNAs are an important class of RNA aptamers.

Conventional aptamer selection techniques are not optimal for fluorescent aptamer isolation. Until recently, isolation of fluorescent aptamers posed a challenge. Indeed, SELEX (systematic evolution of ligands by exponential enrichment) is an in vitro approach that is ideally fit to produce aptamers with the highest affinity towards the desired target [19]. The result of multiple rounds of SELEX is normally an aptamer that has outcompeted the rest of the library and thus has the lowest dissociation constant. However, binding to a fluorogen does not necessarily mean rendering it fluorescent. Affinity towards the fluorogen is necessary but not sufficient for fluorescence activation. The latter is achieved by precise positioning and movement suppression of the fluorogen within the RNA and may not necessarily be associated with the highest possible affinity. Thus, to efficiently isolate fluorescent aptamer modified selection protocols are required.

The solution to this problem is selecting aptamers based on their fluorescence, rather than on their affinity. This has been already achieved for fluorescent proteins by expressing a fluorescent protein library in bacterial cells and sorting cells using FACS (fluorescence-activated cell sorting) [20]. Isolated brightest cells provided the genes of the brightest fluorescent proteins.

In principle, a RNA aptamer library can also be expressed inside bacteria and the resulting bacterial library can be screened for fluorescence in presence of a respective fluorogen. However, there is a major difference between fluorescent protein and RNA screening. Fluorescent protein libraries are normally created by mutagenesis of an existing fluorescent protein and so each library member comes from a single ancestor. The size of this library may be large, but the substantial part of the library is expected to be fluorescent. Selection of fluorescent aptamers for a new fluorogen normally starts with a completely random library. This library's complexity is usually enormous ($>10^{14}$ different species) and the vast majority of the sequences have no or very low affinity for the target. Thus, in order to pre-enrich the library for the binders it is crucial to have an initial selection step that is done using a conventional SELEX protocol. Only after the library complexity is reduced to a manageable level ($<10^7$ different species) the library can be transferred to cells for fluorescence-based screening.

This chapter describes a protocol that has been successfully used to isolate Broccoli which is currently the brightest green fluorescent aptamer [9]. The chapter will cover only the steps that involve aptamer library expression, screening and analysis in bacteria. As noted above, the RNA library should be selected in vitro before bacterial expression. To select for DFHBI-binding aptamers, we routinely utilize conventional SELEX on DFHBI-agarose beads (resin preparation and selection procedure are extensively described in ref. 7). If a fluorogen of interest cannot be immobilized on the resin, it is possible to use other selection approaches, such as "capture SELEX" [21] or "ribozyme SELEX" [22]. In general, any selection protocol can be combined with the subsequent screening in bacteria. Application of the protocol outlined below should result in isolation of novel or improved fluorescent aptamers for a fluorogen of interest.

2 Materials

Prepare all solutions using ultrapure water and analytical grade reagents. Prepare and store all reagents at room temperature (unless indicated otherwise). Follow all waste disposal regulations when disposing materials.

2.1 Plasmids

1. pETDuet-5-tSpinach-eqFP670 (ref. 9). This plasmid allows to have a fluorescent protein expressed off the same plasmid. Fluorescent signal of the fluorescent protein is used to normalize aptamer fluorescence to plasmid copy number and thus aptamer expression level. This allows more precise discrimination between fluorescent aptamers with different levels of brightness. If fluorescent aptamers are expected to have spectral properties similar to those of eqFP670, then the fluorescent protein should be changed to another one with nonoverlapping emission and excitation spectra.

Alternatively, consider inserting the library into the pET28c-tRNA-Spinach plasmid [23] that does not bear the gene of a fluorescent protein.

2. pETDuet-1 (Novagen) or pET28c (EMD Biosciences) are used for transformation into bacteria to generate negative control cells.

2.2 Molecular Biology Reagents and Equipment

1. Restriction enzymes: *Eag*I and *Sac*II (e.g., New England Biolabs (NEB)).
2. Antarctic phosphatase or alkaline phosphatase (e.g., NEB).
3. T4 DNA ligation kit (NEB) or Quick Ligation Kit (NEB).
5. 1.5 mL microcentrifuge tubes (e.g., USA Scientific).
6. 37 °C air incubator or heat block.
7. Molecular biology grade agarose.
8. Microcentrifuge (Eppendorf).
9. Electrophoresis chamber and power supply.

2.3 Bacterial Expression Reagents and Equipment

1. Electrocompetent *E. coli* cells: Acella (EdgeBio) or BL21 Star (DE3) (Invitrogen) made electrocompetent in-house (see numerous protocols, such as available at Journal of Visualized Experiments website [24]).
2. LB medium with carbenicillin or kanamycin: 10 g tryptone, 5 g yeast extract, 10 g NaCl, add deionized water up to 1 L, sterilize by autoclaving. Supplement with 100 μg/mL carbenicillin or 50 μg/mL of kanamycin sulfate when culturing cells. Store up to several months at 4 °C.
3. LB/agar dishes carbenicillin or kanamycin: 10 g tryptone, 5 g yeast extract, 10 g NaCl, 20 g agar, add deionized water up to 1 L. Sterilize by autoclaving. Supplement with 100 μg/mL carbenicillin or 50 μg/mL of kanamycin sulfate. Pour into 10 cm petri dishes, 25 mL of medium per dish. Store up to several months at 4 °C.
4. Super Optimum Culture Media (e.g., Life Technologies).
5. 1 M IPTG solution.
6. Electroporation cuvettes (e.g., Gene Pulser Cuvette, Bio-Rad).
7. Electroporator (e.g., Nucleofector II, Amaxa).
8. Petri dishes (e.g., VWR International).
9. Culture tubes (e.g., VWR International).
10. 25 mL Erlenmeyer flasks with baffles (e.g., Corning).
11. Water bath or heat block.
12. 37 °C air incubator.
13. Orbital shaker.

2.4 Reagents and Instruments for Bacterial Library Sorting and Imaging on the Plates

1. 40 mM stock of DFHBI (Lucerna) or the stock of a fluorogen of interest (*see* **Notes 1** and **2**).
2. 1× phosphate buffered saline: 100 mL of 10× PBS (e.g., Gibco), 900 mL of water. Filter-sterilize. Store up to several months at 4 °C.
3. Round-bottom FACS sample tubes with cell strainer cap (e.g., VWR).
4. Round bottom FACS collection tubes (e.g., VWR).
5. Cell sorter (e.g., FACSaria II, BD biosciences).
6. Orbital shaker.
7. Gel imager (e.g., ChemiDoc MP imager Bio-Rad) (*see* **Note 3**).

3 Methods

3.1 In Vitro Pre-selection of RNAs that Bind to the Fluorogen of Interest.

During this step, the initial random RNA pool is enriched for the specific binders regardless of their ability to activate fluorescence. This is achieved using the classical SELEX approach and typically runs for four to six rounds. The major goal of this step is to eliminate aptamers with no or low affinity towards the fluorogen and to determine the right moment to switch to bacterial screening.

1. Run 4–6 rounds of SELEX using the protocol of your choice. We had success in isolation of Spinach and Broccoli when using agarose bead-immobilized DFHBI [7, 9]. However, other selection approaches, such as capture SELEX [21] or ribozyme SELEX [22], can be utilized as well. These alternative protocols can be especially useful when fluorogen derivatization needed for subsequent conjugation to the resin is problematic or laborious.
2. Start fluorescence measuring of RNA-fluorogen pool after every round of selection, beginning round three. Measuring the fluorescence of the RNA pool–fluorogen mixture is the best way to determine if the selection process is successful and when to switch to bacteria screening. To measure RNA pool fluorescence during selection of DFHBI binders we mixed 20 μM RNA and 20 μM DFHBI in the buffer containing 40 mM HEPES, pH 7.4, 100 mM KCl, and 1–5 mM $MgCl_2$ and measured fluorescence exciting at 480 nm and collecting emission at 500 nm. For selections based on other fluorogen it is necessary to determine excitation and emission maxima of the free fluorogen. Measure fluorescence of RNA pool–fluorogen mixture with excitation light corresponding to the free fluorogen excitation maximum and collecting emission at the wavelength corresponding to the free fluorogen emission maximum.

3.2 Cloning of the Library

Once the RNA pool–fluorogen mixture exhibits weak fluorescence, in vitro selection is stopped and the library can be transferred to bacteria. For our DFHBI-based selection this happened after round six [9].

1. Design and order the primers that allow cloning into the pETDuet-5-tSpinach-eqFP670 or pET28c-tRNA-Spinach plasmids using *Eag*I and *Sac*II restriction sites. Both plasmids encode a gene of the tRNA scaffold and cloning of the library using these restriction sites places the library members inside the scaffold. The tRNA scaffold is a natural RNA structure that confers resistance to intracellular degradation and provides increased folding to RNAs that are inserted into it. Utilization of a scaffold is highly recommended when screening for functional aptamers in cells (*see* **Note 4**). The majority of random libraries are designed so they have constant and largely complementary 5′ and 3′ flanking sequences. When designing the PCR primers, ensure that *Eag*I and *Sac*II sites are introduced into the flanking sequences in a way so the stem connecting the library member and the scaffold is devoid of bulges. Figure 1 shows the 29-1 aptamer [9] inserted into the tRNA scaffold so the connector stem remains fully helical.
2. Reverse transcribe and PCR amplify the RNA library following your standard protocol. As in conventional SELEX, use low number of PCR cycles (5–8) to prevent diversity loss. Alternatively, consider using emulsion PCR that allow standard number of PCR cycles (25–30) while preserving library diversity [25].

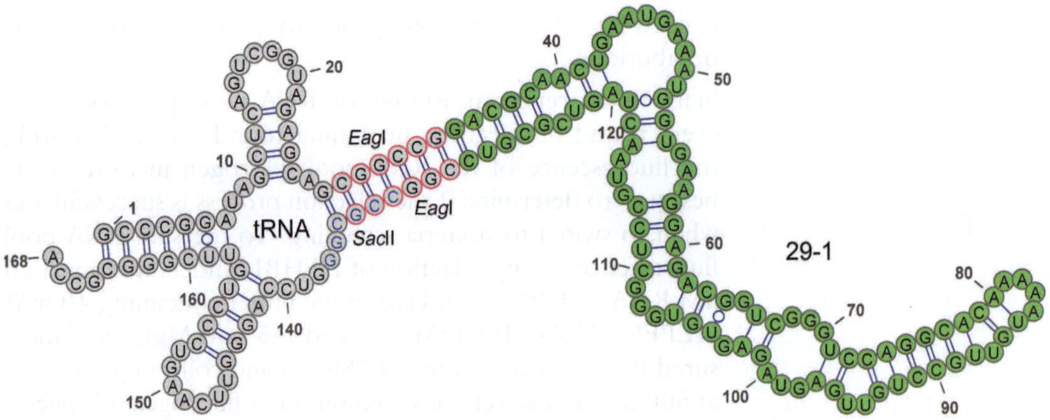

Fig. 1. mFold-predicted secondary structure of the aptamer 29–1 (*green*) in the tRNA scaffold (*grey*). 29-1 was the brightest aptamer isolated from the random library following the protocol described in this chapter [9]. 29-1 then served as a basis to engineer Broccoli. The stem that connects tRNA and 29-1 has two *Eag*I sites (red outline) and one *Sac*II site (*blue letters*). Two *Eag*I sites are complementary and form a stem that should then be continued by the aptamer sequence for optimal folding of the whole construct

3. Digest the plasmid and the PCR product with *Eag*I and *Sac*II for 2–3 h. Prepare the reaction mixture according to the manufacturer's recommendations. Please note that in order to maintain full complementarity of the flanking sequences and thus to have a stable helix, there should be two *Eag*I sites within the PCR product and two *Eag*I sites within the plasmid (Fig. 1). Thus, restriction enzyme digest should be done sequentially, with the *Sac*II enzyme applied first. Aim to have at least 1 μg of vector and 100 ng of insert upon purification.
4. Treat the digested vector with the phosphatase, such as Antarctic phosphatase (NEB) for 1 h at 37 °C. Prepare reaction mixture according to the manufacturer's recommendations. This will prevent vector ligation in case of incomplete digestion.
5. Gel-purify the vector by electrophoresis using a 0.8% agarose gel. Gel-purify the digested PCR products using 1.5–2% agarose gel, depending on the insert size. Purify the DNA using an appropriate gel purification kit (Qiagen). Consider using spin columns that allow concentrating DNA samples in a small volume, e.g., the MiniElute family of kits (Qiagen).
6. Ligate 1 μg of vector and threefold molar excess of the insert utilizing the T4 DNA ligation kit (NEB). Prepare the reaction mixture according to the manufacturer's recommendations and incubate the mixture at 16 °C overnight. If in hurry, consider using a quick ligation protocol, such as reaction using Quick Ligation Kit (NEB). Follow the manufacturer's recommendations when preparing the reaction mix, incubate for 15 min at RT.
7. Purify the ligation mixture using a spin column-based kit or ethanol precipitation. To concentrate the ligation mix, we routinely utilize MiniElute Reaction Cleanup Kit (Qiagen). Elute the ligation mix with 10 μL of elution buffer. After spinning down, repeat the elution with 10 μL more of elution buffer to maximize the yield.

3.3 Bacteria Transformation and Aptamer Library Expression

Since the library is designed to be expressed in bacteria off the T7 promoter (*see* **Note 5**), it is critical to ensure that the bacteria strain used produces T7 polymerase. We have successfully used two strains: commercially available electrocompetent Acella *E. coli* cells (EdgeBio) or Bl21 Star (DE3) *E. coli* strain that was made electrocompetent in-house. In principle, other *E. coli* strains can be used, however, it is necessary to verify that they produce the T7 polymerase by checking the genetic background of the strain: conventionally, these strains are designated "(DE3)".

1. Thaw ~110 μL of the competent cells on ice.
2. Combine each 10 μL of the prechilled ligation mix with 50 μL of the competent *E. coli* cells and mix by gently flicking the tubes.
3. Mix 10 μL of the competent *E. coli* cells with 1–10 ng of the negative control plasmid (e.g., pET28c). Use also a positive control plasmid when screening for an aptamer that activates fluorescence of a fluorogen that already has a cognate fluorescent aptamer. For example, if the fluorogen in study is DFHBI, use Spinach or Broccoli-expressing plasmid, such as pET28c-F30-Broccoli (Addgene).
4. Transfer bacteria–DNA mix into prechilled electroporation cuvettes, use two cuvettes for the library and one cuvette for the control.
5. Electroporate using an available electroporator, such as Nucleofector II (Amaxa), for the latter use preset bacteria protocol 5 (*see* **Note 6**).
6. Immediately after electroporation add 450 μL of SOC medium and transfer to a 16 mL tube. Combine library cells at this point, so there will be two tubes: one with ~1 mL of library cells and another with ~450 μL of control cells.
7. Incubate the tubes at 37 °C with 220 rpm shaking for 1 h. Pre-warm an LB-agar plate with the corresponding antibiotic to 37 °C during this time.
8. Pour the library cells into 100 mL of LB with the corresponding antibiotic in 500 mL–1 L flask. Inoculate 10 μL of the control cells into ~1 mL of LB medium with the corresponding antibiotic.
9. Take 10 μL of the cell suspension from the flask with the library cells, resuspend in 300 μL of SOC or LB medium. Spread all 310 μL onto the pre-warmed LB-agar plate and incubate the plate at 37 °C overnight. This plate will allow to determine the library size.
10. Incubate the flask with the library and the tube with the negative control cells overnight at 37 °C with 220 rpm shaking.
11. Next morning, 2 h before the sorting, dilute the library cells and the negative control cells 1/10 in 3 mL of fresh LB medium in a 16 mL tube. Add the appropriate antibiotic and IPTG to the 1 mM final concentration. Incubate at 37 °C with shaking for two more hours.
12. Count the number of bacterial colonies on the plate and multiply the number by 10,000. This is your library size, i.e., the number of different clones in your library. Typical library sizes range between five to fifty million individual members.

13. Prepare library glycerol stock. Spin down 2 mL of the night culture at 2000–3000 × *g* for 5 min and remove the medium. Add 100 μL of SOC medium with 15% glycerol and resuspend thoroughly. Immediately transfer to −70 °C and store there until further use. When reusing the library, inoculate 10 μL of the frozen stock into 50 mL of LB medium with antibiotic and proceed as described in **step 10** above (*see* **Note** 7).

3.4 Bacteria Cell Sorting

FACS sorting is the most efficient way to identify fluorescent aptamers. While in principle it is possible to screen the bacterial library on the agar plates, the throughput of this approach is very low since one plate can resolve only 1000–5000 colonies. Thus, a large number of plates should be screened to cover the whole library diversity. FACS, on contrary, allows screening tens of thousands cells per seconds and thus the whole library can be analyzed within an hour or two.

1. Dilute 50 μL of the induced library cells or control cells with 1 mL of PBS containing the fluorogen in study (*see* **Note 8**).
2. Filter the bacterial suspension through the cell strainer into the FACS sample tube, bring the samples to the sorter on ice. Also bring one to three collection tubes with 1 mL of SOC medium.
3. Decide on the FACS excitation laser wavelength and emission filter. If the RNA pool–fluorogen mixture has already shown fluorescent signal, then the excitation laser wavelength should be chosen to be close to the RNA-fluorogen fluorescence excitation maximum, and FACS emission filter should collect as much as possible of the RNA-fluorogen emission. If, however, RNA pool does not show noticeable fluorescence, then choose FACS parameters based on the fluorescent properties of the free fluorogen. For DFHBI-based screening we used 488 nm laser and 525 ± 25 emission filter. If the library is made based on the pETDuet vector, then monitor the eqFP670 fluorescent protein in the PE-Cy5 channel (561 nm excitation and 660 ± 10 nm emission).
4. Decide on the temperature of the sample compartment (if controllable). There may be two options for the temperature regime of the library tube: 4 °C and 37 °C. The former may be required when the positive fluorescent signal is weak. We have noticed that lower temperature increases the signal, possibly by increasing the percentage of the folded aptamer. However, if the signal is bright already, higher temperature may be desirable since it will allow discriminating between more and less thermostable aptamers.
5. Run the negative control to establish fluorescence background signal. If you have a positive control cells, run them next to

establish the expected positive fluorescent signal. Make sure to flush the FACS fluidic line with bleach and/or water to remove any fluorescent bacteria before running the library.

6. Run the library. If the library contains high percentage of fluorescent aptamers, then you should be able to see a separate population of positive cells (e.g., as shown in Fig. 2). In this case set the sorting gate to collect the brightest ~0.1% of the positive population. If the library consists of mostly background fluorescence cells, then collect the brightest cells of the main, dim population. In general, the gate should be positioned in a way that takes into account two important numbers. The first number is the library size. To screen every possible library member one should run a number of cells that is roughly equal to ten library sizes (e.g., if the library size is 5 million then run 50 million cells). The second number is how many colonies you plan to analyze. We usually collect 1000–2000 brightest cells and spread them onto 3–6 plates, so each plate has ~330 colonies. The sorting gate thus should be positioned so 1000–2000 cells will be collected after running ten library sizes cells.
7. Start sorting and collect the cells into the tubes with SOC medium. Aim to collect ~1000 cells into each 1 mL of SOC medium.

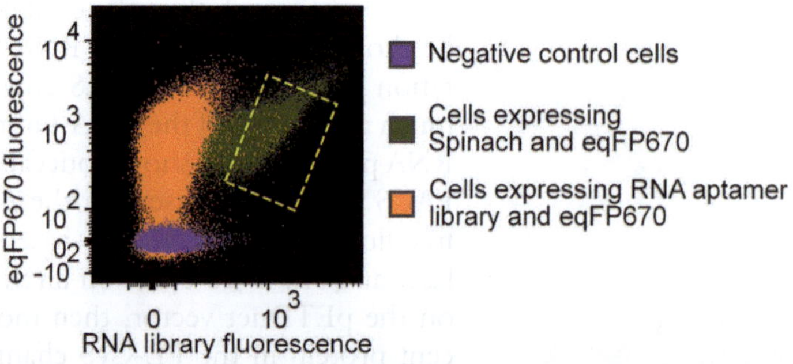

Fig. 2. *Dot plot* of the FACS data for the aptamer library that was expressed in bacteria. eqFP670 is synthesized in bacteria as well to serve for normalization of the aptamer expression level. In presence of DFHBI the library clearly exhibited fluorescent cells that are comparable in brightness (or brighter) than the control cells expressing Spinach and eqFP670. The brightest library cells were sorted using the gate indicated with the *yellow box*. The box is positioned along the main dim population to avoid the cells that also express eqFP670 at high level. These bright double positive cells likely has increased plasmid number which causes higher than average RNA aptamer expression level. This way seemingly bright fluorescence of these cells is a result of higher RNA aptamer copy number and not brighter RNA aptamers. Aptamer library fluorescence was collected using 488 nm laser and 525 ± 25 emission filter. eqFP670 fluorescence was collected using 561 nm excitation and 660 ± 10 nm emission

8. After cell sorting is complete, gently vortex the collection tubes to rinse down the cells that landed on the tube walls. Quickly spin down the tubes and rescue the collected cells by shaking at 37 °C for 1 h.

9. Plate the collected cells onto pre-warmed LB/agar plates supplemented with the respective antibiotic. If the fluorogen in use has no noticeable cytotoxicity (as in case of DFHBI) then the fluorogen can be added to the plates before spreading bacteria (or even mixed with the molten LB-agar before pouring onto the plates). The final concentration of the fluorogen should again be determined empirically, ideally using a positive control (e.g., Spinach or Broccoli for DFHBI). When a positive control is not available, the starting concentration can also be 40 μM unless it generates too strong fluorescence background.

10. Incubate the plates overnight at 37 °C.

3.5 Bacteria Plate Imaging

1. Induce the expression of aptamers in sorted bacteria the next morning. Prepare an induction solution of IPTG in water. Applying of 300 μL of the solution to the plate should result in 1 mM final concentration of IPTG. Since a typical agar slab volume in a plate is 25 mL then one plate will require 300 μL of 83 mM IPTG. If the fluorogen in use is cytotoxic then it can be added to the plate at this stage till the final concentration of 40 μM as well (see discussion above). To apply the induction solution, use a thin metal spatula to gently lift the agar slab. The solution is then poured underneath the agar and then the spatula is moved along the edge of the agar slab to complete a full circle and spread the solution evenly (Fig. 3). Incubate the plates at 37 °C.

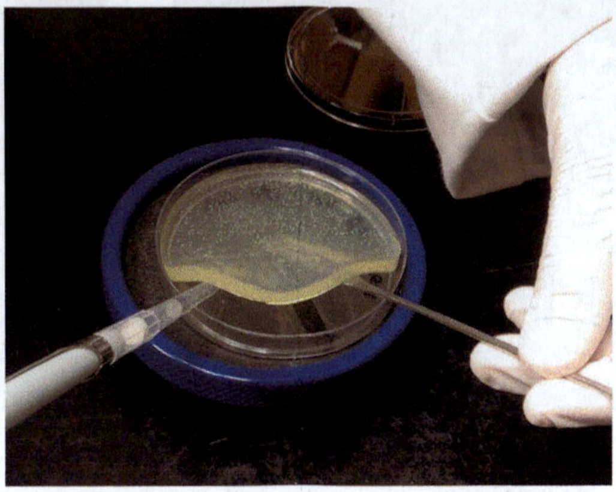

Fig. 3. Induction of the grown bacteria colonies on the plate. The agar slab is gently lifted using a spatula and the induction solution is squirted underneath. Then the spatula is moved around to spread the liquid evenly. If needed, the fluorogen can also be added at this step

2. Image the plates 2–4 h after induction. Use any suitable gel imager for that. We routinely use a ChemiDoc MP imager (Bio-Rad). For DFHBI binders, aptamer library fluorescence is collected in a channel with 470 ± 15 nm excitation and 532 ± 14 nm emission. To normalize the specific fluorescent signal, eqFP670 fluorescent protein signal should be collected with 630±15 nm excitation and 697 ± 22.5 nm emission. Figure 4 presents a typical image of a bacterial plate expressing DFHBI-binding fluorescent aptamer library and eqFP670. If the expressing plasmid does not contain a fluorescent protein gene, then image the plates using a fluorescent channel that is the farthest from the one where the actual signal is. On a ChemiDoc MP the same channel with 630 ± 15 nm excitation and 697 ± 22.5 nm emission can be used for that purpose. All two-channel images should be taken with the same zoom and plate position.

3.6 Processing of the Bacteria Plate Images

1. Export images from the plate imager and save them as tiff files.
2. Open these files in ImageJ (NIH) or Fiji (http://fiji.sc/).
3. Subtract background—go to Process/Subtract background. We routinely use rolling ball radius of 100 pixels.

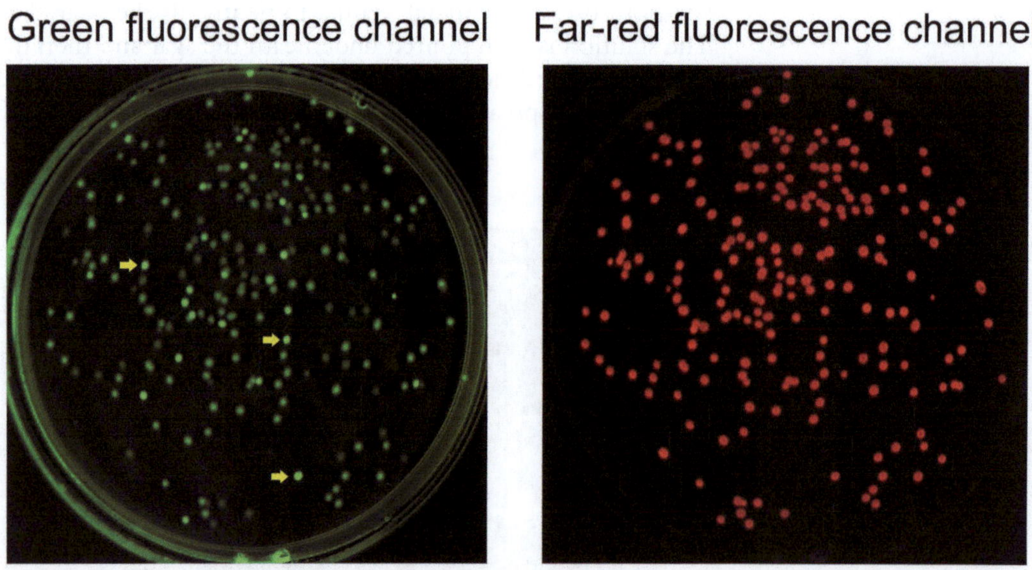

Fig. 4. Images of the bacterial colonies on the plate. Upon sorting the colonies were grown overnight and DFHBI and 1 mM IPTG were added underneath the agar layer in the morning. The plates were imaged ~3 h later using ChemiDoc MP. Two fluorescence channels were used: 470±15 nm excitation and 532±14 nm emission for DFHBI binders (green fluorescence channel) and 630±15 nm excitation and 697±22.5 nm emission for eqFP670 (far-red fluorescence channel). The library clearly shows a diversity of colony fluorescence intensity in the green fluorescence channel with some of the brightest colonies indicated with *yellow arrows*

4. Create a mask image based on one of the images, use the one that has higher contrast between signal and background. For that go to Image/Adjust/Threshold. Mark "dark background" and move the bars until your actual colonies are all red while the rest part of the image is black. Once you apply changes you will create a binary image with the colony signal intensity value of 1 and the rest of the field having intensity value 0. This will allow to quantify the signal for the colonies only. Save the mask image separately.
5. Normalize your actual signal to colony size. For that, divide the image in the aptamer channel by image in the colony size channel. To do that, go to Process/Image calculator. Choose appropriate images (they should be already open in ImageJ) and "divide" operation. Mark both "create new window" and "32-bit (float) results".
6. To clear the resulting image, multiply it by the mask image that was made before. This will generate the image with the normalized colony signal only. Go to Process/Image calculator, choose appropriate images and "multiply" operation.
7. Use the resulting image to compare colony brightness. For that, create an ROI within the colony of interest and use Analyze/Measure operation to get an average intensity for this colony. Repeat that for 20–30 colonies that appear to be the brightest. Make the final decision based on the resulting values.

3.7 Further Steps: Fluorescent Aptamer Characterization and Optimization

Once the brightest colonies are chosen, inoculate 10–20 colonies and isolate plasmids. The plasmids should be first sequenced since a number of them may have the same sequence. Once the individual sequences are identified, the plasmids can be used as templates for in vitro transcription. This individual RNAs are then mixed with the fluorogen to confirm, compare and characterize their fluorescent properties. Depending on the envisioned application, the best aptamers can be chosen based on the total brightness (combination of high extinction coefficient and high quantum yield), affinity towards the fluorogen, fluorescence spectrum or photostability. It also should be noted, that the brightest fluorescent aptamers in vitro are not necessarily the brightest ones in the cells (bacteria). Indeed, in-cell performance of fluorescent aptamers largely depends on aptamer in vivo folding efficiency and/or resistance to RNases and these properties may not correlate to aptamer in vitro performance.

After the best aptamer (aptamers) is chosen, it may be necessary to perform a round of directed evolution to further improve the desired properties. Normally, a "doped" library is created on the basis of the best aptamer. This library is designed in a way that it contains all possible single-mutation sequences, two-mutation sequences, three-mutation sequences and so on, up to seven(or more)-mutation sequences (depending on the aptamer length). For more details on the doped library engineering please refer to Filonov et al. [9]. The doped library is subjected to 2–3 rounds of SELEX to eliminate nonfunctional sequences and then transferred to and screened in bacteria, essentially as described in this protocol. Normally, one round of directed evolution is enough to isolate a desired fluorescent aptamer.

4 Notes

1. If the screening is intended to find brighter DFHBI binders, note that there is an advanced version of this fluorogen termed DFHBI-1 T (3,5-difluoro-4-hydroxybenzylidene 1-trifluoroet hyl-imidazolinone). It is a modified version of DFHBI that has higher extinction coefficient, lower fluorescent background in cells, and excitation and emission maxima (472 nm and 507 nm, respectively) that are optimized for standard microscope FITC filter cubes [26]. While this fluorogen cannot be immobilized on the agarose beads, it still can be used to elute the binders off the resin and later for screening steps in bacteria.
2. When using a different fluorogen, make sure to characterize its properties before starting any selection. If the desired RNA–fluorophore complex is intended for in vitro assays, it is crucial that the fluorogen has minimal fluorescence background while also has a potential to increase its fluorescent signal substantially. Low fluorescence background can be verified by measuring spectra of the free fluorogen at the expected working concentration. Fluorescence activation can be assessed by comparing the fluorescent signal before and after freezing the fluorogen solution in ethanol in liquid nitrogen (add 0.1% (v/v) of trimethylamine if the fluorogen requires deprotonation, as in case of DFHBI). Low fluorescence activation indicates low fluorogenic potential of the small molecule chosen. If the RNA–fluorogen complex is envisioned to work in live cells, few other important properties should be additionally verified. The fluorogen should be able to penetrate cells. If comprehensive cell-permeability assays are not available, the small molecule should at least be examined if it complies with Lipinski's five rules of cell permeability [27]. Another important consideration is low intracellular background fluorescence of the fluorogen chosen. Indeed, some small-molecules have an intrinsic affinity towards intracellular components, such as nucleic acids or lipids, and binding to these biomolecules can

lead to undesirable fluorescence activation of the fluorogen. Thus, if the fluorogen is cell-permeable, it is critical that it also has a very low nonspecific fluorescence activation in cells.

3. There may be other imager options for bacterial plate imaging. These include instruments as advanced as small animal imaging station (e.g., IVIS Spectrum (PerkinElmer)) or as simple as UV-transilluminator. The latter can be used if the RNA–fluorogen complex has excitation maximum in UV-violet part of the spectrum.

4. F30 is another scaffold option for library expression in vivo. F30 scaffold was engineered based on a stable natural three-way junction [28] and is more preferable for aptamer expression for some applications. The major problem with the tRNA scaffold is that it is recognized and processed in cells as a genuine tRNA [28]. This results in cleavage of the 5′ and 3′ flanking sequences which essentially prohibits tRNA-aptamer usage as an RNA fluorescent tag. The F30 does not induce cleavage of RNAs that are linked to it and so F30-aptamer fusions can be used as imaging tags.

   More complicated cloning scheme is the major problem with having F30-aptamer libraries. The F30 scaffold does not have convenient cloning sites to insert the library members. Thus, F30 should be made the part of the library from the first round of SELEX, or F30 should be added using long primers during the final PCR before cloning into a bacterial expression vector. We have inserted F30-aptamer library members into pET28c-Broccoli plasmid [28] using *Bgl*II and *Nhe*I restriction sites. However, it should be noted that aptamers are connected to either tRNA or F30 in a similar way, i.e., using a stem. Thus, the aptamers can be isolated using the tRNA scaffold and then transferred to F30.

5. T7 polymerase is very efficient and produces amounts of RNA that reach micromolar concentration in bacteria. While this is clearly beneficial for efficient screening, this high RNA concentration is apparently toxic. In our experience, if the cells in suspension are induced with IPTG for longer than 2 h, the resulting cellular viability after the FACS may drop significantly. Two-hour expression should ensure good cellular viability after FACS. If, however, the sorted cells do not survive, another bacterial strain can be tested. Another way to increase cell viability is to use a different promoter. We have successfully utilized the araBAD promoter that is less efficient, compared to T7 [9]. If the viability issues persist, one should refer to Filonov et al. [9] for more details on the araBAD plasmid utilization.

6. The most common problem with bacteria electroporation is generating an arc discharge in a cuvette. If this happens, the cells are almost all dead and transformation efficiency is minimal. The major reason for the arc discharge is presence of salts in the ligation mix. To minimize salt presence ensure that DNA is properly washed on a spin column or in the pellet. In general, it is advisable to wash spin columns or DNA pellet twice with the respective wash buffer (e.g., Qiagen PE buffer and 70% ethanol, respectively). Another way to avoid arc discharge is to dilute DNA–cell mixture with 30–50 μL of prechilled ultrapure water.
7. While the glycerol stock can be used to rescreen the library, it is not advisable to generate a stock few days before sorting and then use the stock to inoculate the library just before the screening. In our experience, freshly made library provides the strongest expression level and thus fluorescent signal.
8. The optimal fluorogen concentration may need to be determined empirically; however, it is recommended to keep it at least tenfold higher than the expected aptamer–fluorogen dissociation constant. We routinely use 40 μM DFHBI in our cell sorting experiment and this concentration can be used as a starting point for experiments with another fluorogen. Too high fluorogen concentration should be avoided since it will produce strong fluorescence background and thus will complicate screening. Another important point to consider is the time for the fluorogen to enter the cells. DFHBI is very cell-permeable and binds all RNA inside bacteria within 2 min. Other fluorogens may take longer. Thus, if not sure, incubate the cells with the fluorogen of interest for 10–30 min at room temperature before proceeding to sorting.

Acknowledgment

The author thanks Prof. Samie R. Jaffrey for his generous support and scientific guidance. Flow cytometry experiments were performed with the help of J. McCormick and S.Z. Merlin (Department of Pathology and Laboratory Medicine cell sorter core). The author is also grateful to the members of the Jaffrey lab for their feedback as they used this protocol. This work was supported by NIH grants to Prof. Samie R. Jaffrey (R01 NS064516 and R01 EB010249).

References

1. Famulok M, Hartig JS, Mayer G (2007) Functional aptamers and aptazymes in biotechnology, diagnostics, and therapy. Chem Rev 107:3715–3743
2. Thiel KW, Giangrande PH (2009) Therapeutic applications of DNA and RNA aptamers. Oligonucleotides 19:209–222
3. Bruno JG (2015) Predicting the uncertain future of aptamer-based diagnostics and therapeutics. Molecules 20:6866–6887
4. Kaur G, Roy I (2008) Therapeutic applications of aptamers. Expert Opin Investig Drugs 17:43–60
5. Keefe AD, Pai S, Ellington A (2010) Aptamers as therapeutics. Nat Rev Drug Discov 9:537–550
6. Babendure JR, Adams SR, Tsien RY (2003) Aptamers switch on fluorescence of triphenylmethane dyes. J Am Chem Soc 125:14716–14717
7. Paige JS, Wu KY, Jaffrey SR (2011) RNA mimics of green fluorescent protein. Science 333:642–646
8. Strack RL, Disney MD, Jaffrey SR (2013) A superfolding Spinach2 reveals the dynamic nature of trinucleotide repeat RNA. Nat Methods 10:1219–1224
9. Filonov GS, Moon JD, Svensen N, Jaffrey SR (2014) Broccoli: rapid selection of an RNA mimic of green fluorescent protein by fluorescence-based selection and directed evolution. J Am Chem Soc 136(46):16299–16308
10. Guet D, Burns LT, Maji S, Boulanger J, Hersen P, Wente SR et al (2015) Combining Spinach-tagged RNA and gene localization to image gene expression in live yeast. Nat Commun 6:8882
11. Zhang J, Fei J, Leslie BJ, Han KY, Kuhlman TE, Ha T (2015) Tandem spinach array for mRNA imaging in living bacterial cells. Sci Rep 5:17295
12. Lu Z, Filonov GS, Noto JJ, Schmidt CA, Hatkevich TL, Wen Y et al (2015) Metazoan tRNA introns generate stable circular RNAs in vivo. RNA 21:1554–1565
13. Pothoulakis G, Ceroni F, Reeve B, Ellis T (2014) The spinach RNA aptamer as a characterization tool for synthetic biology. ACS Synth Biol 3:182–187
14. Paige JS, Nguyen-Duc T, Song W, Jaffrey SR (2012) Fluorescence imaging of cellular metabolites with RNA. Science 335:1194
15. Song W, Strack RL, Jaffrey SR (2013) Imaging bacterial protein expression using genetically encoded RNA sensors. Nat Methods 10:873–875
16. You M, Litke JL, Jaffrey SR (2015) Imaging metabolite dynamics in living cells using a spinach-based riboswitch. Proc Natl Acad Sci U S A 112:E2756–E2765
17. Kellenberger CA, Chen C, Whiteley AT, Portnoy DA, Hammond MC (2015) RNA-based fluorescent biosensors for live cell imaging of second messenger cyclic di-AMP. J Am Chem Soc 137:6432–6435
18. Kellenberger CA, Wilson SC, Sales-Lee J, Hammond MC (2013) RNA-based fluorescent biosensors for live cell imaging of second messengers cyclic di-GMP and cyclic AMP-GMP. J Am Chem Soc 135:4906–4909
19. Stoltenburg R, Reinemann C, Strehlitz B (2007) SELEX--a (r)evolutionary method to generate high-affinity nucleic acid ligands. Biomol Eng 24:381–403
20. Piatkevich KD, Verkhusha VV (2010) Advances in engineering of fluorescent proteins and photoactivatable proteins with red emission. Curr Opin Chem Biol 14:23–29
21. Stoltenburg R, Nikolaus N, Strehlitz B (2012) Capture-SELEX: selection of DNA aptamers for aminoglycoside antibiotics. J Anal Methods Chem 2012:415697
22. Koizumi M, Soukup GA, Kerr JN, Breaker RR (1999) Allosteric selection of ribozymes that respond to the second messengers cGMP and cAMP. Nat Struct Biol 6:1062–1071
23. Strack RL, Song W, Jaffrey SR (2014) Using spinach-based sensors for fluorescence imaging of intracellular metabolites and proteins in living bacteria. Nat Protoc 9:146–155
24. Gonzales, M.F., Brooks, T., Pukatzki, S.U., and Provenzano, D. (2013) Rapid protocol for preparation of electrocompetent Escherichia coli and Vibrio cholerae. J Vis Exp 80:e50684.
25. Shao K, Ding W, Wang F, Li H, Ma D, Wang H (2011) Emulsion PCR: a high efficient way of PCR amplification of random DNA libraries in aptamer selection. PLoS One 6:e24910
26. Song W, Strack RL, Svensen N, Jaffrey SR (2014) Plug-and-play fluorophores extend the spectral properties of Spinach. J Am Chem Soc 136:1198–1201
27. Lipinski CA, Lombardo F, Dominy BW, Feeney PJ (2001) Experimental and computational approaches to estimate solubility and permeability in drug discovery and development settings. Adv Drug Deliv Rev 46:3–26
28. Filonov GS, Kam CW, Song W, Jaffrey SR (2015) In-gel imaging of RNA processing using broccoli reveals optimal aptamer expression strategies. Chem Biol 22:649–660

Chapter 18

An Enzyme-Linked Aptamer Sorbent Assay to Evaluate Aptamer Binding

Matthew D. Moore, Blanca I. Escudero-Abarca, and Lee-Ann Jaykus

Abstract

Nucleic acid aptamers are a class of alternative ligands increasingly growing in importance in the face of contemporary detection challenges. Aptamers offer multiple advantages over traditional ligands like antibodies; however, their ability to specifically bind target molecules must first be confirmed after their generation. Use of a plate-based enzyme-linked aptamer sorbent assay (ELASA) is a generally rapid way to screen and characterize aptamer binding to protein targets. ELASA involves directly plating a protein target onto a nonspecific (polystyrene) surface and assessing binding of functionalized (biotinylated) aptamers to those plated proteins using an enzyme conjugate that recognizes the aptamers. Here, we describe an ELASA that was designed and used to evaluate and compare binding of ssDNA aptamers against the capsids of different strains of human norovirus.

Key words Aptamer, Plate assay, Binding evaluation, Enzyme-linked aptamer sorbent assay, DNA-protein binding, Aptamer binding

1 Introduction

Nucleic acid aptamers are short strands of ssDNA or RNA (~20–80 nucleotides) that fold into unique structures as a consequence of their sequence and bind specific targets. Unlike antibodies or other alternative ligands, aptamers have the advantages of being less costly, easily chemically synthesized and modified, more shelf stable, easily purified, and do not evoke an immune response. Aptamers are generated in a highly customizable process known as SELEX (Systematic Evolution of Ligands by EXponential enrichment). In this process, a library of many unique different sequences is generated, and this library is subjected to iterative binding steps to enrich the sequences that bind the target. Additionally, a counterselection against unwanted matrices or targets can also be performed in the process to remove nonspecifically binding sequences. The result is a pool of candidate sequences with affinity for the target of interest [1].

Thomas Tiller (ed.), *Synthetic Antibodies: Methods and Protocols*, Methods in Molecular Biology, vol. 1575,
DOI 10.1007/978-1-4939-6857-2_18, © Springer Science+Business Media LLC 2017

Once identified, screening and characterization of the most promising sequences from the SELEX pool must be performed. For this purpose, the enzyme-linked aptamer sorbent assay (ELASA) was designed. In fact, enzyme-linked immunosorbent assays (ELISA) have long been a favorable, widely used method to study protein-protein interactions [2, 3], and the ELASA described here is based on such assays [4–7]. Specifically, the target protein (viral capsid in this case) is directly applied to a plate and biotinylated aptamer is then incubated with the target. This is followed by incubation with a streptavidin-horseradish peroxidase complex that then catalyzes a colorimetric signal in the presence of substrate.

The assay presented here has utility to serve as a standard design that is easily open to modification, largely due to the ease with which aptamers can be functionalized. For instance, the assay could be designed to bypass the enzyme-catalyzed development step by substituting a fluorophore in place of the biotin label on the aptamer. The general structure of the ELASA could also be modified to evaluate the binding properties of other ligands. For example, synthetic carbohydrate receptors specific for human norovirus binding were used in place of aptamers in the same assay design as described here [8]. Our ELASA is highly versatile, allowing for use of many different aptamers and protein targets. Specifically, this protocol was utilized for screening aptamers generated against human norovirus purified capsids (known as virus-like particles, or VLPs) and characterizing the cross-reactivity of various aptamers with different human norovirus strains and other enteric viruses. The assay was also used to determine relative binding affinity of aptamers to human norovirus in partially purified clinical (stool) samples [6, 7]. In short, this rapid, reproducible, and inexpensive method facilitates the screening of target binding to a large number of aptamer candidates, or alternatively, a large number of different potential targets, all in a single assay.

2 Materials

Be certain to use nuclease-free reagents and minimize the risk of exposing the reagents to potential nuclease contamination. The use of commercially prepared buffers and water is recommended for assay consistency. Human norovirus aptamers that are biotinylated on the 5′ end (Table 1) and HPLC-purified were obtained from Integrated DNA Technologies (IDT, Coralville, IA) but could be ordered from a variety of reputable vendors. Aptamers should be reconstituted in nuclease-free water to make 100 μM stocks and stored at −20 °C for near term use (within 3 months),

Table 1

Relevant aptamer and oligonulceotide sequences used

Name	Sequence (5′→3′)	Reference
Aptamer M6–2	Biotin-AGTATACGTATTACCTGCAGCTGGGAAGAGGTCCGGTAAATGCAGGGTCAGCCCGGAGAGCGATATCTCGG AGATCTTGC	[7]
hlyQF	CATGGCACCACCAGCATCT	[11]
hlyQR	ATCCGCGTGTTTCTTTTCGA	[11]
L23SQF	AGGATAGGGAATCGCACGAA	[11]
L23SQR	TTCGCGAGAAGCGGATTT	[11]

and −80 °C for long-term storage. Follow all standard recommended chemical and biohazard waste disposal methods when using any potentially hazardous reagents in this or any related assays.

2.1 Buffers and Solutions

1. Nuclease-free water: Sterile filtered water that is treated with 0.1% diethylpyrocarbonate (DEPC) and autoclaved can be prepared or purchased (e.g., nuclease-free water, Thermo Fisher, Waltham, MA).
2. 1× Phosphate-buffered saline (PBS): Dissolve 0.210 g/L KH_2PO_4 (1.54 mM), 9.000 g/L NaCl (155.17 mM), and 0.726 g/L. $Na_2HPO_4{\bullet}7H_2O$ (2.71 mM) in nuclease-free water. Adjust pH to 7.2 and autoclave. Store at room temperature.
3. Target solution (VLPs): Assembled, purified virus-like particles (VLPs) consisting of the human norovirus major capsid protein were obtained in concentrated form (>1.0 mg/mL) in PBS courtesy of R. Atmar (Baylor College of Medicine, Houston, TX) and stored at 4 °C until use (*see* **Note 1**). Concentrated stock solutions of VLPs were diluted to 3 μg/mL in PBS. Diluted VLP solutions should be prepared immediately before use at room temperature.
4. PBS-Tween 20 (PBST): Addition of 0.05% (v/v) Tween-20 (Sigma-Aldrich, St. Louis, MO) to PBS. This solution is stored at room temperature out of direct light.
5. Blocking buffer: Dissolve 5% (w/v) skim milk solids (Thermo Fisher) in PBST. Add an equimolar mixture of unrelated ssDNA sequences (hlyQF, hlyQR, L23SQF, L23SQR; Table 1; *see* **Note 2**) to reach a 10 nM final concentration. New blocking buffer should be prepared at room temperature every time the assay is performed.
6. Biotinylated aptamer solution: Dilute thawed stock of 100 μM biotinylated aptamer(s) to 1 μM (*see* **Note 3**) in nuclease-free water (*see* **Note 4**). This solution should also be prepared at room temperature immediately before each assay is performed.
7. Enzyme conjugate solution: Dilute a streptavidin-horseradish peroxidase conjugate (1.0 mg/mL; Invitrogen, Carlsbad, CA) 1:5000 in PBS. Prepare right before use and keep at room temperature until use.
8. Peroxidase substrate solution: Obtain the two components of 3,3′,5,5′-Tetramethylbenzidine (TMB) microwell peroxidase substrate from KPL (Kirkegaard & Perry Laboratories, Inc., Gaithersburg, MD). Mix both components 1:1 per manufacturer's instructions and store at room temperature until use.

Prepare a fresh solution at least 3 h prior to each assay to ensure that the substrate is at room temperature before use (*see* **Note 5**).

9. TMB stop solution: Commercially prepared TMB stop solution was used in this assay (KPL, Inc.) and stored at room temperature. Alternatively, a 1 M H_3PO_4 solution can be used. Be cautious when preparing and handling the stop solution as it is highly acidic.

2.2 ELASA Assay Components and Required Equipment

1. Flat bottom, clear medium-binding polystyrene 96 well enzyme immunoassay (EIA) plates (Fisher Scientific, Pittsburgh, PA) (*see* **Note 6**).
2. Universal microplate lids (Fisher Scientific) to cover plate.
3. Parafilm M (Sigma-Aldrich, St. Louis, MO) to wrap the side of the plate.
4. An orbital shaker for gentle agitation of plates.
5. A microplate reader [in our study, we used a Tecan Infinite M200pro (Tecan Group Ltd., Männedorf, Switzerland) but other brands are acceptable] to record well absorbances.
6. A multichannel pipette [e.g., Eppendorf Research Plus 8-channel pipette (Eppendorf, Hamburg, Germany)].

3 Methods

The following methods are performed at room temperature unless otherwise stated.

3.1 Depositing Target Protein on Plate

1. Place 100 μL per well of the prepared VLP (target) solutions into each plate well. Use at least two wells per specific aptamer–VLP pair to be tested (*see* **Note 7**). Apply 100 μL of PBS to two wells to serve as "no VLP" negative controls (*see* **Note 8**). Positive control wells should be included using a VLP known to bind to one or more ligand(s) being tested.
2. Place lid on plate and wrap sides of plate with Parafilm to prevent evaporation (Fig. 1).
3. Incubate plate on orbital shaker with gentle rotation (*see* **Note 9**) overnight at 4 °C.

3.2 Blocking and Aptamer Binding

1. Remove VLP solutions from the wells using a multichannel pipette set to 200 μL, being sure to remove all residual liquid from each well (*see* **Note 10**).
2. Wash by placing 200 μL of PBST into each well using the multichannel pipette and immediately removing the solution.

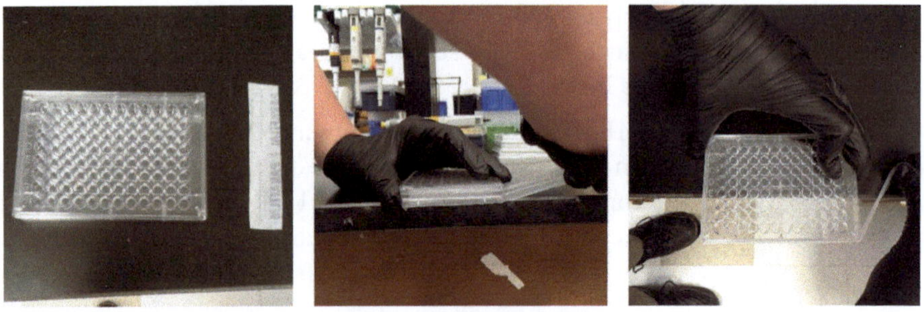

Fig. 1 Application of Parafilm to seal plate. Cut a 2–2.5 cm piece of Parafilm and stretch along side of plate to reduce the possibility of evaporation

3. Repeat wash step two more times, being sure to remove all residual buffer after the last wash (*see* **Note 11**).
4. Apply 200 μL of blocking solution to each well and incubate for 2 h (*see* **Note 12**) with gentle shaking on an orbital shaker.
5. Discard blocking solution, removing as much residual buffer as possible, and wash plates three times with 200 μL of PBST as described above.
6. Apply 100 μL of aptamer solution to each well and incubate for 1 h with gentle shaking.
7. Discard residual aptamer solution by multichannel pipette and wash each well four times with 200 μL PBST (*see* **Note 13**).

3.3 Substrate Binding and Production of Signal

1. Apply 100 μL of the enzyme conjugate solution per well, being careful to expel liquid directly into bottom of each well and not along the walls of the wells (*see* **Note 14**). Incubate for 15 min with gentle shaking (*see* **Note 15**).
2. Remove enzyme conjugate solution by multichannel pipette and wash each well three times with 200 μL of PBST as described above (*see* **Note 16**).
3. Apply 100 μL of the room temperature TMB substrate solution by multichannel pipette to each well and incubate for 3–20 min with gentle shaking (*see* **Note 17**); a blue color should develop in the positive wells (Fig. 2).
4. Apply 100 μL of TMB stop solution per well (*see* **Note 18**) and gently tap the microtiter plate to ensure solution is mixed and development reaction quenched. Positive wells should change from blue to yellow (Fig. 3).
5. Immediately read plate on the plate reader set at a 450 nm wavelength (*see* **Note 19**).

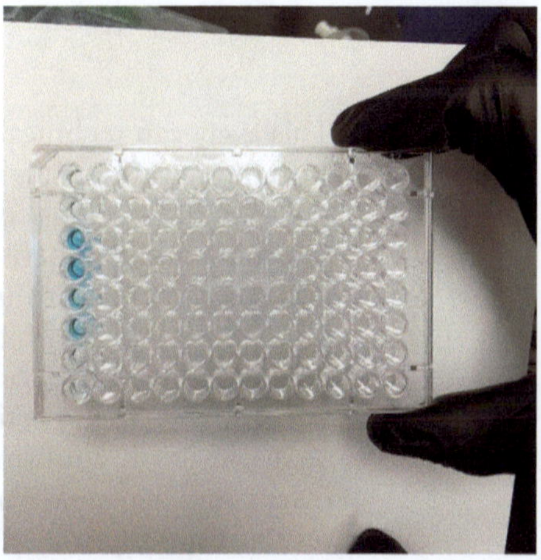

Fig. 2 Developed plate. Biotinylated aptamer M6-2 was used in the ELASA assay against (in order from top to bottom of column 1, 2 wells/VLP) GI.8, GII.2 Snow Mountain, and GII.4 Grimsby VLPs along with no VLP control wells. TMB substrate was added and plate developed at room temperature for 6.5 min before the photo was taken

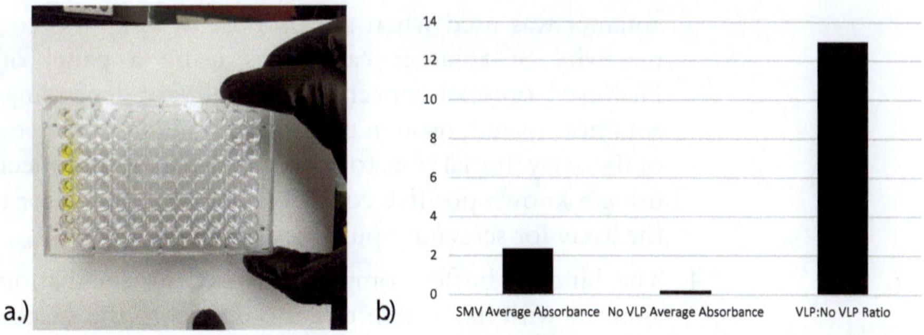

Fig. 3 Appearance of a quenched ELASA plate. (**a**) Biotinylated aptamer M6-2 was used against GI.8, GII.2 Snow Mountain, and GII.4 Grimsby VLPs using the assay above. Plate was quenched with TMB stop solution after 7.5 min of development at room temperature. (**b**) Raw average absorbance values and consequent VLP:no VLP ratio of the pictured ELASA plate

3.4 Data Analysis and Interpretation

1. Once absorbances are recorded, average the two wells for each aptamer–VLP pair. Also average the two wells corresponding to each aptamer–no VLP pair (negative control).
2. Divide the average absorbance for each aptamer–VLP combination and divide by the average absorbance of the negative control to generate a positive–negative (VLP/no VLP) ratio for each aptamer–VLP pair (*see* **Note 20**).
3. Traditionally, positive–negative ratios greater than 2.0 are considered as indicative of binding between the VLP and the aptamer [6, 7, 9, 10] (*see* **Note 21**).

4 Notes

1. This assay can technically be adapted to other protein targets, with modification. It is important to consider characteristics of the protein target based on the purpose of the assay. For initial screening and comparison of aptamer candidates from a pool, it is recommended to use highly purified protein targets such as were used here with the VLPs. However, this assay may also be done using more complex sample matrices. For instance, in our study we used partially purified human stool samples that were previously confirmed positive or negative for human norovirus and having virus concentrations approximated using reverse transcription-qPCR [6, 7].
2. Our lab has found that the addition of unrelated oligonucleotides to the blocking buffer reduces nonspecific background signal. This is probably due to the similarities between chemical structure of unrelated oligonucleotides and aptamers, the former of which quench nonspecific binding without interfering with the binding specificity of the aptamers.
3. For our studies, a 1 μM concentration of the biotinylated aptamer was used when the purpose of assay was comparing reactivity of aptamer candidates using a panel of VLPs. However, optimal concentration will vary depending on the aptamer created, protein target, and binding buffer conditions of the assay. Initial tests to optimize the aptamer concentration using a known positive control should be done prior to using the assay for screening purposes.
4. The binding buffer composition used for incubation of the aptamer with its protein target is crucial to the success of the assay and may be the most important factor impacting the assay performance. Care should be taken to select the best buffer or solution for this step, as aptamers are sensitive to alterations in salt concentration and pH. The buffer used in the SELEX process is often suitable for use in this assay. If modifying the assay for other uses, it is advisable to first optimize the binding buffer.
5. Ensuring that the TMB substrate solution is freshly prepared on the day of the assay is a good quality control practice. More importantly, ensuring that the substrate solution has been brought to room temperature is important to ensure comparable development times across experimental replicates.
6. Different types of 96 well plates may be chosen depending on the properties of the protein reagent used. We have found that medium-binding polystyrene plates worked the best with our protein target, as enough hydrophobic residues were exposed

to allow stable deposition of the VLPs onto the plate. Treated, higher-binding plates resulted in increased background signals because the aptamer bound naturally to the plates. However, such plates may be required for analysis of some protein targets.

7. For this example, the purpose of the assay was to examine binding specificity of a selected aptamer to a panel of VLPs corresponding to different human norovirus strains. One of the major advantages of the ELASA is the ability to screen a large number of aptamer candidates and protein targets at the same time. However, care should be taken to use at least two wells for each aptamer–protein interaction so as to identify assay inconsistency that would indicate a need for reanalysis. Also, we have found that use of all plate wells at the same time tends to increase the chance of inconsistency, especially for novices to the assay. Care should be taken to not over-exceed the precision capabilities of any one user.
8. The purpose of the ELASA should be carefully considered when selecting negative controls. For the purposes of this study, a no VLP (negative) control well was used. However, experiments may (and should) be performed using unrelated proteins or matrices to serve as controls to demonstrate that aptamer binding is specific. For example, partially purified (chloroform-extracted) norovirus-negative stool samples were used in the ELASA that was applied to naturally contaminated clinical samples [6, 7].
9. Ideally, it is best to have access to two orbital shakers; one always kept at room temperature and one always kept at 4 °C. If only one shaker is available, one may skip orbital rotation for the overnight incubation at 4 °C. Rotation speed should be slow enough so that liquid does not get expelled from the wells (for example, 200 rpm for the model shaker used in this study).
10. Place the multichannel pipette with tips at 45° angle relative to the bottom of plate. Place the ends of the tips up against the side of the left wall and slowly slide tips down into the bottom left edge of the wells. Proceed to remove all liquid from the wells while slowly raising tips to nearly perpendicular against the bottom left edge of the wells to remove as much residual liquid as possible. This pipetting procedure should be used throughout the assay. Never touch the tips of the pipette to any other part of the bottom of the well other than the bottom left edge. For left-handed users, do everything against the bottom right edge of the wells. Replace tips after removal of each column (do not reuse tips on more than one set of wells).

11. Remove as much buffer as possible when washing. To assure complete removal of wash buffer during the last wash associated with any one assay step, expel most of the liquid removed from a set of wells, then place the pipette tips back into the same wells and lift to complete volume, sucking up any residual liquid at the bottom of the wells. After the wash steps, tap the plate upside down onto a fresh paper towel to knock out any liquid that may have remained on the walls of the wells. If working with potentially hazardous biological or chemical targets, be certain that all precautions (i.e., use of a biosafety or chemical containment hood, respectively) are taken as this step could potentially result in aerosolization. Do not perform this technique without advising environmental health officials if working with a microbe or chemical for which aerosolization is a known route of transmission.
12. This blocking step can actually be completed overnight at 4 °C following the same steps as done for application of the VLPs. In some cases this reduces background signal slightly, but not substantially.
13. It is important to be sure that all liquid is removed and that washing is done carefully and thoroughly, as residual aptamer at this step increases the background signal of the assay.
14. As with the aptamer step, avoid residual liquids, as residual enzyme that is not removed will lead to increased background signal.
15. Only 15 min is needed for this step due to the strong streptavidin-biotin interaction. A longer incubation can be performed, but it may result in slightly higher background signal.
16. Assure that the wash solution covers the walls of the wells and that all residual wash solution has been removed after the last wash step. Pat the plate fairly vigorously upside down on a paper towel as done for the other steps to remove residual liquid. This ensures the lowest background possible.
17. Depending on the aptamer and target used, the time to optimal signal development will vary. Ideally, reactions should be stopped after a reasonable amount of time ($\leq$20 min) is allowed for color development. Be sure to stop the reaction before the color is too dark because the resultant absorbance may exceed the range that the plate reader can measure (for example, the plate reader used in our study does not read above 4 absorbance units). If signal develops extremely rapidly, dilution of aptamer or target is recommended. Do not let the development of solution go much longer than 20 min using the substrate system discussed here, as nonspecific color development may occur.

18. Application of stop solution amplifies the degree of signal. Consider this factor when choosing a time to allow for development prior to stopping. Once a time is chosen, keep that development time consistent between experimental runs to facilitate comparison to other plate replicates.
19. Although the stop solution drastically slows the reaction, it does not completely stop it. Because this is an immediate endpoint reading, and reading absorbances takes less than a minute, this last step should be done immediately after the stop solution is added.
20. For example, for aptamer M6-2 binding to VLP Snow Mountain Virus (SMV); take the average absorbance units of the aptamer M6-2-SMV wells (average = 2.33; Fig. 3). Then, divide that average by the average of the no VLP-aptamer M6-2 wells (negative control average = 0.18; Fig. 3). The VLP/no VLP ratio is then obtained (in this case it is 12.94) (Fig. 3).
21. Sample positivity is determined based on an arbitrary positive–negative ratio, and is of course dependent on the strength of the negative control signal. Although ≥2.0 has been used relatively consistently in the literature [6, 9, 10], this cut-off ratio can be altered depending on the ligands and targets being screened. For more conservative estimations specifically accounting for nonspecific binding of aptamer (negatively charged) to positively charged amino acids, a negative control that contains an unrelated protein (such as bovine serum albumin) or the sample matrix itself may be used.

Acknowledgments

This work was supported by the Agriculture and Food Research Initiative Competitive Grant no. 2011-68003-30395 from the US Department of Agriculture, National Institute of Food and Agriculture through the NoroCORE project. The authors would like to thank R. Atmar (Baylor College of Medicine, Houston, TX) for providing the VLPs and S. Rupp for his input on the figures.

References

1. Tombelli S, Minunni M, Mascini M (2007) Aptamers-based assays for diagnostics, environmental and food analysis. Biomol Eng 24:191–200 http://www.ncbi.nlm.nih.gov/pubmed/17434340. Accessed 7 Nov 2013
2. Lequin RM (2005) Enzyme immunoassay (EIA)/enzyme-linked immunosorbent assay (ELISA). Clin Chem 51:2415–2418. doi:10.1373/clinchem.2005.051532
3. Schuurs AHWM, Van Weemen BK (1980) Enzyme-immunoassay: a powerful analytical tool. J Immunoassay 1:229–249 http://www.tandfonline.com/doi/abs/10.1080/01971528008055786
4. Moe CL, Sair A, Lindesmith L, Estes MK, Jaykus L (2004) Diagnosis of Norwalk virus infection by indirect enzyme immunoassay detection of salivary antibodies to recombinant Norwalk virus antigen. Clin Vaccine Immunol 11:1028–1034. doi:10.1128/CDLI.11.6.1028
5. Rogers JD, Ajami NJ, Fryszczyn BG, Estes MK, Atmar RL et al (2013) Identification and

characterization of a peptide affinity reagent for detection of noroviruses in clinical samples. J Clin Microbiol 51:1803–1808 http://www.pubmedcentral.nih.gov/articlerender.fcgi?artid=3716098&tool=pmcentrez&rendertype=abstract. Accessed 20 Mar 2014

6. Escudero-Abarca BI, Suh SH, Moore MD, Dwivedi HP, Jaykus L-A (2014) Selection, characterization and application of nucleic acid aptamers for the capture and detection of human norovirus strains. PLoS ONE 9:e106805 http://www.ncbi.nlm.nih.gov/pubmed/25192421. Accessed 8 Sept 2014
7. Moore MD, Escudero-Abarca BI, Suh SH, Jaykus L-A (2015) Generation and characterization of nucleic acid aptamers targeting the capsid P domain of a human norovirus GII.4 strain. J Biotechnol 209:41–49 http://linkinghub.elsevier.com/retrieve/pii/S0168165615300183
8. Manuel CS, Moore MD, Jaykus LA (2015) Destruction of the capsid and genome of GII.4 human norovirus occurs during exposure to metal alloys containing copper. Appl Environ Microbiol 81:4940–4946 http://aem.asm.org/lookup/doi/10.1128/AEM.00388-15
9. Ebel GD, Dupuis AP, Nicholas D, Young D, Maffei J et al (2002) Detection by enzyme-linked immunosorbent assay of antibodies to West Nile virus in birds. Emerg Infect Dis 8:979–982 http://www.pubmedcentral.nih.gov/articlerender.fcgi?artid=2732549&tool=pmcentrez&rendertype=abstract
10. Hirneisen KA, Kniel KE (2012) Comparison of ELISA attachment and infectivity assays for murine norovirus. J Virol Methods 186:14–20. http://www.ncbi.nlm.nih.gov/pubmed/22820074. Accessed 8 Sept 2014.
11. Rodríguez-Lázaro D, Hernández M, Scortti M, Esteve T, Vázquez-boland JA et al (2004) Quantitative detection of *Listeria monocytogenes* and *Listeria innocua* by real-time PCR: assessment of hly, iap, and lin02483 targets and AmpliFluor technology. Appl Environ Microbiol. doi:10.1128/AEM.70.3.1366

Chapter 19

Incorporating Aptamers in the Multiple Analyte Profiling Assays (xMAP): Detection of C-Reactive Protein

Elyse D. Bernard, Kathy C. Nguyen, Maria C. DeRosa, Azam F. Tayabali, and Rocio Aranda-Rodriguez

Abstract

Aptamers are short oligonucleotide sequences used in detection systems because of their high affinity binding to a variety of macromolecules. With the introduction of aptamers over 25 years ago came the exploration of their use in many different applications as a substitute for antibodies. Aptamers have several advantages; they are easy to synthesize, can bind to analytes for which it is difficult to obtain antibodies, and in some cases bind better than antibodies. As such, aptamer applications have significantly expanded as an adjunct to a variety of different immunoassay designs. The Multiple-Analyte Profiling (xMAP) technology developed by Luminex Corporation commonly uses antibodies for the detection of analytes in small sample volumes through the use of fluorescently coded microbeads. This technology permits the simultaneous detection of multiple analytes in each sample tested and hence could be applied in many research fields. Although little work has been performed adapting this technology for use with apatmers, optimizing aptamer-based xMAP assays would dramatically increase the versatility of analyte detection. We report herein on the development of an xMAP bead-based aptamer/antibody sandwich assay for a biomarker of inflammation (C-reactive protein or CRP). Protocols for the coupling of aptamers to xMAP beads, validation of coupling, and for an aptamer/antibody sandwich-type assay for CRP are detailed. The optimized conditions, protocols and findings described in this research could serve as a starting point for the development of new aptamer-based xMAP assays.

Key words Aptamer assay, Multiple-analyte profiling, Luminex, xMAP, CRP, Sandwich assay, Bead-based assay

1 Introduction

Immunoassays have been used for several decades to measure analytes of interest [1]. Due to the variety of antibodies available as well as many assay formats to work with, immunoassays have become one of the most common methodologies used in clinical diagnostics and life-science research [1]. The options available for immunoassay development have now expanded further with the development of other suitable detection systems, such as aptamers. Aptamers are short, synthetic oligonucleotides that are generated

Thomas Tiller (ed.), *Synthetic Antibodies: Methods and Protocols*, Methods in Molecular Biology, vol. 1575,
DOI 10.1007/978-1-4939-6857-2_19, © Springer Science+Business Media LLC 2017

through an iterative in vitro process to bind to target molecules with high specificity and affinity [2, 3]. Much research has been done in the last two decades to incorporate aptamers into a variety of immunoassay formats both alone and in conjunction with antibodies [4].

There are many advantages to the use of aptamers in immunoassays when compared to antibodies. For example, since aptamers are produced by chemical synthesis and efficiently purified following production, there is no batch-to-batch variation in their quality. The use of chemical synthesis for their preparation also permits modifications into aptamer sequences at precise locations [5]. Aptamers have a longer shelf-life than antibodies and can be denatured and regenerated many times without significant degradation [5]. Finally, the in vitro process used in the development of aptamers, called SELEX (systematic evolution of ligands by exponential enrichment), allows for aptamers to be selected to bind to targets for which the development of antibodies is difficult or impossible. This means that the incorporation of aptamers into applications that typically use antibodies could lead to the development of assays for targets that would otherwise not be feasible: compounds with a high toxicity, poor immunogenicity, or that are too small for antibodies to be readily obtained [6].

Technologies that use the fundamental aspects of immunoassays have been developed for a variety of different applications [1]. The work described herein uses Multi-Analyte Profiling (xMAP) technology developed by Luminex Corporation [7]. The xMAP platform is bead-based, giving it significant advantages over traditional two-dimensional platforms. For example, xMAP technology permits the simultaneous detection of multiple analytes (i.e., multiplexed analysis), the use of lower sample volumes, and shorter incubation times due to near-liquid-phase binding kinetics [7, 8]. Little work has thus far been performed on the incorporation of aptamers into the xMAP platform even though xMAP technology is an open platform that permits researchers to develop and design their own assays using readily available components (i.e., beads, buffers) and suggested protocols (i.e., the coupling of macromolecules to beads, assay protocols) [8, 9]. Since the incorporation of aptamers into this platform could lead to the development of xMAP assays for novel targets, this is an area worth exploring.

xMAP technology uses beads that have been doped with a mixture of fluorescent dyes that give each bead region (i.e., bead type) a unique spectral signature and macromolecules (i.e., antibodies, antigens, oligonucleotides) can be covalently or noncovalently bound to these beads [8]. Unique bead regions are used for each analyte to be measured in an xMAP assay and the fluorescent signal of the bead is measured by the xMAP analyzer in order to identify the target; this permits multiplexed analysis. The specific xMAP analyzer used determines how many different bead regions can theoretically be measured in a single assay and this value ranges from 100 to 500 [8]. As in traditional immunoassays there are many factors, such as buffer conditions and antibody concentrations, that must be considered when determining the optimal conditions for an xMAP assay, and these will impact the detection sensitivity and how many analytes can ultimately be measured in a multiplexed assay. So far, existing kits for multiplexed analysis of samples using xMAP technology can measure up to about 30 analytes [10].

xMAP assay design can vary depending on the outcome desired and the reagents selected. For example, a sandwich-type format is often used in which a capture antibody and a detection antibody sandwich the analyte to be measured. The capture antibody is bound to the bead surface and selectively binds the analyte present in a sample solution. The detection antibody that binds to a different epitope than that of the capture antibody is then used to measure the amount of target in the sample. In multiplex assays, specifically labeling the capture antibody identifies the analyte being detected and this is achieved in xMAP assays through the use of the fluorescently doped beads described above. The detection antibody might be directly labeled with a fluorescent dye, such as phycoerythrin (PE), or it could be labeled with biotin which would then be detected using streptavidin–phycoerythrin (SAPE) [8]. The signal generated by the detection antibody is compared to a calibration curve to quantify the amount of analyte detected [8].

The first use of aptamers in xMAP technology was demonstrated a decade ago by Porschewski and coworkers when they used the thrombin aptamer in both sandwich-type and competitive assay formats [11]. The thrombin aptamer [12], however, is a very well-studied aptamer that is often used in proof-of-principle studies even though aptamers have been developed for many more clinically relevant targets [4, 13]. In order to increase the usefulness and prevalence of aptamers in clinical assays and biosensors, a greater variety of aptamers must be incorporated into technologies such as the xMAP platform for use in conjunction with or alongside antibody-based assays [13].

For our work on the development of an aptamer/antibody xMAP assay [14], we chose to use an RNA aptamer for C-reactive protein (CRP) that had previously been incorporated into sandwich-type assays using electrochemical [15] and absorbance detection [16]. CRP is a biomarker of inflammation and its serum levels can increase 1000-fold following an injury or inflammation [17]. Serum levels of CRP have also been associated with the likelihood of cardiovascular events [17, 18]. The successful development of a bead-based aptamer/antibody assay for CRP as shown in this work demonstrated that an aptamer for a target of interest in human health could be incorporated into the xMAP platform and that the use of aptamers in xMAP assays should be further explored.

2 Materials

Micropipettes (single and/or multichannel) and pipette tips are needed for most of the following protocols. Pipette tips should be suitable for use with oligonucleotides and proteins (sterile, RNase- and DNase-free).

2.1 Preparation of Sterile, RNase-Free Deionized Water

Deionized water [diH_2O, 18 MΩ] treated using diethyl pyrocarbonate (DEPC) to denature RNases should be used for preparation of all buffers when an RNA aptamer is used in the detection system (*see* **Note 1**).

1. Deionized water [diH_2O, 18 MΩ].
2. Diethyl pyrocarbonate (DEPC).
3. Detergent, such as Liqui-nox (Sigma-Aldrich).
4. Equipment and materials: autoclavable glass bottles, sterile glass pipette, incubator (set to 37 °C), autoclave.

2.2 Coupling of Oligonucleotide to Bio-Plex Pro Magnetic COOH Beads

Buffers used during the coupling of oligonucleotides to beads and for hybridization assays should be prepared and stored as suggested by Luminex Corporation [19]. All buffer components must be appropriate for use with oligonucleotides (i.e., molecular biology grade).

1. Oligonucleotide Coupling Buffer: 0.1 M MES (2[N-Morpholino] ethanesulfonic acid), pH 4.5. Add 4.88 grams of MES (Bioshop Canada) to about 240 mL of DEPC-treated diH_2O in a sterile bottle (*see* **Note 2**). Add about 5 drops of 5 N NaOH then add DEPC-treated water to a total volume of 250 mL. Seal and shake the bottle until the MES has been dissolved. Measure the pH and adjust if necessary (*see* **Note 3**). Filter-sterilize the buffer (*see* **Note 4**) and store at 4 °C.
2. Wash Buffer I: 0.02% Tween 20. Add 50 μL of Tween 20 to 250 mL of DEPC-treated diH_2O in a sterile bottle (*see* **Note 2**). Shake to mix then filter-sterilize the solution (*see* **Note 4**) and store it at room temperature.
3. Wash Buffer II: 0.1% SDS (sodium lauryl sulfate). Add 2.5 mL of a 10% SDS solution to about 240 mL of DEPC-treated diH_2O in a sterile bottle (*see* **Note 2**). Make up the total volume to 250 mL using DEPC-treated diH_2O then shake to mix. Filter-sterilize the solution (*see* **Note 4**) and store at room temperature.
4. Sample Diluent: TE (Tris–ethylenediaminetetraacetic acid [EDTA]) buffer (pH 8.0) with 0.1% (*v/v*) Tween 20. Add 2.5 mL of a Tris–EDTA buffer, pH 8.0 (*see* **Note 5**) and 250 μL of Tween 20 to about 240 mL of DEPC-treated diH_2O in a

sterile bottle (*see* **Note 2**). Make up the buffer to a total volume of 250 mL using DEPC-treated diH_2O then shake to mix. Filter-sterilize the buffer (*see* **Note 4**) and store at room temperature.

5. Bio-Plex Pro Magnetic COOH beads or similar product (i.e., Luminex Magnetic COOH beads). Use one bead region per target (*see* **Note 6**).
6. *N*-(3-dimethylaminopropyl)-*N'*-ethylcarbodiimide hydrochloride (EDC) (Pierce). Store at −20 °C (*see* **Note 7**). Prepare solutions of 10 mg/mL EDC in DEPC-treated diH_2O immediately prior to use.
7. 5′ Amino-modified aptamer and control oligonucleotide sequences composed of the same number of nucleotides (*see* **Note 8**). Prepare a 0.1 nmol/μL solution of each oligonucleotide sequence in DEPC-treated diH_2O.
8. Equipment and materials: balance, sterile RNase-free 1.5 mL microcentrifuge tubes, magnetic separator for 1.5 mL tubes, aluminum foil, vortex mixer, shaker for 1.5 mL tubes, sonicator bath.

2.3 Determination of Oligonucleotide-Coupled Bead Concentration

1. Oligonucleotide-coupled beads.
2. DEPC-treated diH_2O.
3. Equipment: sterile RNase-free 1.5 mL tubes, hemocytometer, light microscope.

2.4 Detection of Beads and Data Analysis

1. xMAP analyzer (*see* **Note 9**).
2. Deionized H_2O.
3. 70% (*v/v*) ethanol or 70% (*v/v*) isopropanol.
4. xMAP Sheath fluid (*see* **Note 9**).
5. xMAP Calibration Kit (*see* **Note 9**).
6. 10% (*v/v*) bleach solution.
7. Bio-Plex Manager software (*see* **Note 9**).

2.5 Validation of RNA Coupling: Hybridization Assays

1. Hybridization Bead Diluent (1.5× TMAC (tetramethylammonium chloride) solution): 225 mL 5 M TMAC solution; 1.88 mL 20% Sarkosyl solution; 18.75 mL 1 M Tris–HCl, pH 8.0 solution; 3.0 mL 0.5 M EDTA, pH 8.0 solution; and 1.37 mL of DEPC-treated diH_2O. Shake solution to mix and store it at room temperature. Use a sterilized glass bottle for preparation and storage of this solution. Concentrated TMAC and the resulting buffer should be handled in a fumehood.
2. Sample Diluent: TE (Tris–ethylenediaminetetraacetic acid [EDTA]) buffer (pH 8.0) with 0.1% (*v/v*) Tween 20. Prepare as described in Subheading 2.2.

3. Working Microsphere Mixture: Resuspend the oligonucleotide-coupled beads by vortexing and sonicating (~20 s each) immediately prior to diluting to a final concentration of 150 beads per microliter using the Hybridization Bead Diluent (1.5× TMAC solution). Thirty-three microliters of the Working Microsphere Mixture is required per well (*see* **Note 10**). Remember to work in a fumehood and to properly dispose of waste.
4. Complementary Oligonucleotide Stock Solutions (*see* **Note 11**): Dilute the appropriate oligo sequence (labeled with biotin at the 5′ terminus) in DEPC-treated diH_2O or Sample Diluent [TE buffer with 0.1% (*v/v*) Tween 20] for the oligo-coupled bead set under study.
5. Reporter Solution: Prepare a 4 μg/mL solution of streptavidin–R-phycoerythrin (SAPE) in TE with 0.1% (*v/v*) Tween 20. Seventy-five microliters of this solution is required for each well (*see* **Note 12**).
6. Equipment and materials: vortex mixer, sonicator bath, polypropylene Greiner 96-well plates or similar product (must be able to withstand being heated to 95 °C), multi well plate sealing film or cover, microtiter plate shaker, heating block, sterile RNase-free 1.5 mL microcentrifuge tubes, magnetic separator for 96-well plates, Millipore Multiscreen HTS 96-well filter plates or similar product (*see* **Note 13**), vacuum manifold for 96-well filter plates.

2.6 Bead-Based Aptamer/Antibody Assay and Spiked Recovery Experiments

1. Assay Bead Diluent: 100 mM Tris (pH 7.5), 100 mM NaCl, 10 mM $CaCl_2$, 0.1% (*w/v*) bovine serum albumin (BSA), 0.1% (*v/v*) Tween 20 (*see* **Notes 14** and **15**). Add 40 mL of a 1 M Tris–HCl solution (pH 7.5) (*see* **Note 16**) to about 350 mL of DEPC-treated diH_2O in a sterile bottle (*see* **Note 2**). Add 2.34 g of NaCl, 0.59 g of $CaCl_2$•$2H_2O$, 0.4 g of BSA, and 400 μL of Tween 20 to the solution. Make up the volume to 400 mL using DEPC-treated diH_2O then shake to thoroughly mix all components. Filter-sterilize the solution (*see* **Note 4**) and store at 4 °C.
2. Blocking Buffer: 100 mM Tris (pH 7.5), 100 mM NaCl, 10 mM $CaCl_2$, 0.1% (*w/v*) bovine serum albumin (BSA) (*see* **Notes 14** and **15**). Add 40 mL of a 1 M Tris-HCl solution (pH 7.5) (*see* **Note 16**) to about 350 mL of DEPC-treated diH_2O in a sterile bottle (*see* **Note 2**). Add 2.34 g of NaCl, 0.59 g of $CaCl_2$•$2H_2O$, and 0.4 g of BSA to the solution. Make up the volume to 400 mL using DEPC-treated diH_2O then shake to thoroughly mix all components. Filter-sterilize the solution (*see* **Note 4**) and store at 4 °C.
3. Assay Wash Buffer: 100 mM Tris (pH 7.5), 100 mM NaCl, 10 mM $CaCl_2$, 0.1% (*v/v*) Tween 20 (*see* **Notes 14** and **15**). Add 40 mL of a 1 M Tris–HCl solution (pH 7.5) (*see* **Note 16**) to

about 350 mL of DEPC-treated diH_2O in a sterile bottle (*see* **Note 2**). Add 2.34 g of NaCl, 0.59 g of $CaCl_2$•$2H_2O$, and 400 μL of Tween 20 to the solution. Make up the volume to 400 mL using DEPC-treated diH_2O then shake to thoroughly mix all components. Filter-sterilize the solution (*see* **Note 4**) and store at room temperature.

4. Serum or other applicable sample matrix for recovery experiments (*see* **Note 17**).
5. Working Microsphere Mixture: Resuspend the oligonucleotide-coupled beads by vortexing and sonicating (~20 s each) immediately prior to diluting to a final concentration of 100 beads per microliter using Assay Bead Diluent. 50 μL of this suspension will be needed per well (*see* **Note 10**).
6. Calibration Solutions (*see* **Note 18**): Dilute a stock solution of the analyte to be measured in Blocking Buffer and/or the appropriate matrix (*see* **Note 17**). 50 μL of a calibration solution is required per well (*see* **Note 10**).
7. Sample Solutions or Spiked Samples: Prepare samples in Blocking Buffer and/or the appropriate matrix (*see* **Note 17**). Fifty microliters of a calibration solution is required per well (*see* **Note 10**).
8. Equipment and materials: vortex mixer, sonicator bath, sterile RNase-free 1.5 mL microcentrifuge tubes, microtiter plate shaker, Millipore Multiscreen HTS 96-well filter plates or similar product (*see* **Note 13**), vacuum manifold for 96-well filter plates.
9. Detection Antibody Solution (*see* **Note 19**): Prepare a 4 μg/mL solution of the detection antibody using Blocking Buffer as diluent (*see* **Notes 20** and **21**). 50 μL of this solution is required per well.
10. Reporter Solution (if needed; *see* **Note 19**): Prepare a 4 μg/mL solution of streptavidin–R-phycoerythrin (SAPE) using Blocking Buffer as diluent (*see* **Notes 20** and **21**). Fifty microliters of this solution is required per well.

3 Methods

In this assay for CRP, the aptamer is coupled to the xMAP beads and used to capture CRP from solution while a labeled antibody is used for detection (*see* Fig. 1). The following protocols reflect this assay design, but recommended protocols for protein coupling to beads and validation of this coupling are readily available from Luminex Corporation [20] if a different configuration is preferable for other aptamer/antibody xMAP assays. Figure 1 summarizes the stages of preparation of the aptamer/antibody assay: coupling of the aptamer to xMAP beads, validation of the coupling reaction, and the final assay for CRP. These methods are adapted from suggested protocols published by Luminex Corporation in their xMAP Cookbook [20]. It is important that the beads be shielded from prolonged exposure to light throughout all protocols.

3.1 Preparation of Sterile, RNase-Free Deionized Water

1. Clean several autoclavable glass bottles using a detergent and diH_2O then rinse thoroughly with diH_2O to remove all traces of the detergent. Fill the bottles with diH_2O once they are clean.
2. Add DEPC to the diH_2O in the autoclavable glass bottles to a final DEPC concentration of 0.1% (*v/v*). Use a sterile pipette for addition of DEPC and handle this chemical carefully in a fumehood. Thoroughly mix the solution.
3. Heat this solution in an incubator at 37 °C for 12 h.
4. Autoclave the DEPC-treated diH_2O to deactivate the DEPC.

3.2 Coupling of RNA to Bio-Plex Pro Magnetic COOH Beads

1. Let EDC warm to room temperature in a desiccator (*see* **Note 7**).
2. Aliquot beads for coupling: Select the bead region to be used for coupling (*see* **Note 6**) and resuspend the beads in the stock suspension by vortexing and sonicating (~20 s each). Pipette

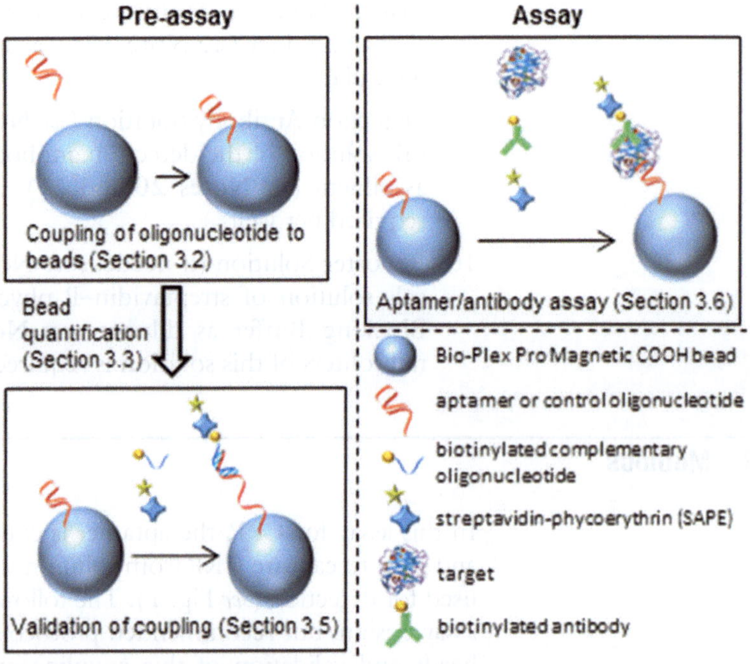

Fig. 1 Schematic representations of the major pre-assay experiments and the final assay format. For clarity, only one oligonucleotide strand is illustrated coupled to a bead

200 μL of the bead suspension into a sterile, RNase-free 1.5 mL tube to obtain an aliquot of 2.5×10^6 beads (*see* **Notes 22** and **23**).

3. Place the tube in a magnetic separator for ~60 s (*see* **Note 24**) and then carefully remove the supernatant using a pipette without disturbing the beads.
4. Remove the tube from the magnetic separator and add 25 μL of Oligonucleotide Coupling Buffer. Resuspend the beads by vortexing and sonicating the tube for ~20 s each.
5. Add 0.16 nmol (1.6 μL) of the appropriate amino-modified oligonucleotide to the aliquot of 2.5×10^6 beads (*see* **Notes 22** and **23**).
6. Add 2 μL of 10 mg/mL EDC (final concentration of EDC of ~0.8 mg/mL) and vortex to mix. Incubate the tube for 30 min at room temperature in the dark.
7. Prepare a new 10 mg/mL EDC solution and add 2 μL of this freshly prepared solution to the bead suspension. Vortex to mix and then incubate for 30 min at room temperature in the dark.
8. Add 500 μL of Wash Buffer I (0.02% Tween 20) to the coupled beads. Put the tube in the magnetic separator for ~60 s (*see* **Note 18**) and then carefully remove the supernatant via pipette.
9. Remove the tube from the magnetic separator and add 500 μL of Wash Buffer II (0.1% SDS). Vortex the tube to resuspend the beads.
10. Put the tube in the magnetic separator for ~60 s (*see* **Note 24**) and then carefully remove the supernatant via pipette.
11. Remove the tube from the magnetic separator and resuspend the coupled beads in 50 μL of Sample Diluent.
12. Store coupled beads at 4 °C in the dark.

3.3 Determination of Oligonucleotide-Coupled Bead Concentration

A hemocytometer was used to enumerate the beads following the oligonucleotide-coupling protocol.

1. Dilution of beads for counting: Resuspend the oligonucleotide-coupled beads by vortexing and sonicating (~20 s each). Dilute the resuspended beads 100-fold in DEPC-treated diH_2O and mix dilute bead suspension thoroughly by vortex.
2. Transfer 10 μL of the dilute bead suspension to a clean hemocytometer slide. Count the beads within the four large corners of the hemocytometer grid.
3. Use the sum of the beads in these four large corners to calculate the number of beads per microliter as per the following equation, where the dilution factor is 100:

$$\frac{\text{Beads}}{\mu L} = (\textit{Sum} \text{ of beads in 4 large corners}) \times 2.5 \times \text{dilution factor}$$

3.4 Detection of Beads and Data Analysis

About 45 min before samples are ready for analysis, the xMAP analyzer should be powered on and daily start-up protocols should be followed as indicated in the instrument or software user guide (for example, *see* Subheading 11, Bio-Plex Protein Array System Operation in ref. 21). Prior to using the instrument, ensure that the sheath fluid is adequately full and the waste reservoir is empty. A rinse with deionized water and a 70% (*v/v*) alcohol solution and a 30 min warm-up period are required prior to calibration of the system. Once the system has been successfully calibrated, samples may be analyzed. After the run has completed, perform the shutdown protocol to clean the system using deionized water and a 10% (*v/v*) bleach solution.

3.5 Validation of RNA Coupling: Hybridization Assays

1. Prepare solutions of the appropriate complementary oligo sequence at fmol/µL concentrations through dilution of their Complementary Oligonucleotide Stock Solutions using Sample Diluent. A volume of 17 µL of complement solution is needed per well (*see* **Notes 10** and **25**).
2. Obtain a heat-resistant 96-well plate (*see* **Note 26**).
3. Resuspend the Working Microsphere Mixture by vortexing and sonicating (~20 s each) immediately prior to adding 33 µL of the suspension to each well.
4. To each background well (i.e., 0 fmol of complementary oligo), add 17 µL of Sample Diluent.
5. To each sample well, add 17 µL of the appropriate biotinylated complementary oligo solution.
6. Cover the plate and mix samples on a shaker for 5 min at room temperature.
7. Incubate the plate at 95 °C for 5 min while shielding the samples from light. Let the plate cool to room temperature then place it on a shaker for 15 min (*see* **Note 27**).
8. Place the plate into a magnetic separator for 96-well plates for ~60 s (*see* **Note 24**). Hold the plate in the separator, invert it, and gently tap out the supernatant on a paper towel.
9. Remove the plate from the magnetic separator and add 75 µL of the Reporter Solution to each reaction well. Place the plate on a shaker for 15 min.
10. Pre-wet a 96-well filter plate by adding 100 µL of Sample Diluent to each well and then removing the liquid using a vacuum manifold. Use a multichannel pipette to transfer the samples to the prewet filter plate.
11. Analyze 50 µL of each sample at room temperature on the xMAP analyzer. Export and plot the data (*see* Fig. 2 for sample data from the CRP assay) (*see* **Note 28**).

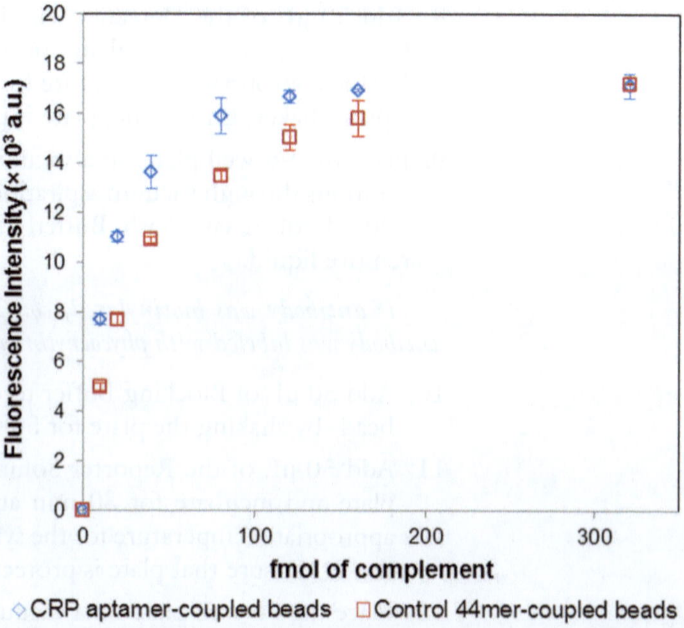

Fig. 2 Sample hybridization assay data. Data points represent the means of three replicates ± standard deviation. These results indicate that both the aptamer and the control oligonucleotide were successfully coupled to separate aliquots of COOH beads with similar coupling efficiencies. Reprinted from [14] with permission from Elsevier

3.6 Bead-Based Aptamer/Antibody Assay and Spiked Recovery Experiments

1. Pre-wet a 96-well filter plate by adding 100 μL of Assay Bead Diluent to each well and then using vacuum aspiration to remove the liquid (*see* **Note 21**).
2. Resuspend the beads in the Working Microsphere Mixture by vortexing and sonicating (~20 s each) immediately prior to adding 50 μL of the suspension to each well.
3. To each background well, add 50 μL of the diluent used for preparation of the calibration solutions.
4. Add 50 μL of the appropriate calibration or sample solution to each well.
5. Cover the plate and incubate for 30 min at room temperature (or the appropriate temperature for the system under study) on a plate shaker. Ensure that plate is protected from light.
6. Place the 96-well plate on a vacuum manifold and remove the solutions through vacuum aspiration. Wash each well twice with 100 μL portions of Assay Wash Buffer, using vacuum aspiration to remove liquid.
7. Add 50 μL of Blocking Buffer to each well and resuspend the beads by shaking the plate for 5 min.

8. Add 50 μL of the Detection Antibody Solution to each well. Cover the plate and incubate for 30 min at room temperature (or the appropriate temperature for the system under study) on a plate shaker. Ensure that plate is protected from light.
9. Place the 96-well plate on a vacuum manifold and remove the solutions through vacuum aspiration. Wash each well twice with 100 μL of Assay Wash Buffer, using vacuum aspiration to remove liquid.

If antibody was biotinylated, continue through the protocol. If antibody was labeled with phycoerythrin, skip to step 13.

10. Add 50 μL of Blocking Buffer to each well and resuspend the beads by shaking the plate for 5 min.
11. Add 50 μL of the Reporter Solution to each well. Cover the plate and incubate for 30 min at room temperature (or the appropriate temperature for the system under study) on a plate shaker. Ensure that plate is protected from light.
12. Place the 96-well plate on a vacuum manifold and remove the solutions through vacuum aspiration. Wash each well twice with 100 μL portions of Assay Wash Buffer, using vacuum aspiration to remove liquid.
13. Add 100 μL of Assay Bead Diluent to each well and resuspend the beads by shaking vigorously for 15 min at room temperature (or the appropriate temperature for the system under study). Ensure that plate is protected from light.
14. Analyze 50 μL of each sample at the appropriate temperature on the xMAP analyzer. The instrument's software can be used to generate the calibration curve and analyze the sample data. *See* Fig. 3 for sample data from a CRP detection assay.

4 Notes

1. Autoclaved diH_2O without DEPC treatment may be suitable for buffer preparation when a DNA aptamer is used instead of an RNA aptamer, but this has not been tested by the authors.
2. The maintenance of buffer sterility was a higher priority than the accuracy of buffer concentration; therefore buffers were prepared directly in the bottles used for DEPC-treatment of diH_2O. We used autoclavable, graduated bottles with a capacity of 500 mL and the graduations were used to approximate the total volumes of the buffers as they were prepared.
3. In order to maintain the sterility of the prepared buffers during pH adjustment, a small portion (a few millilitres) of the solu-

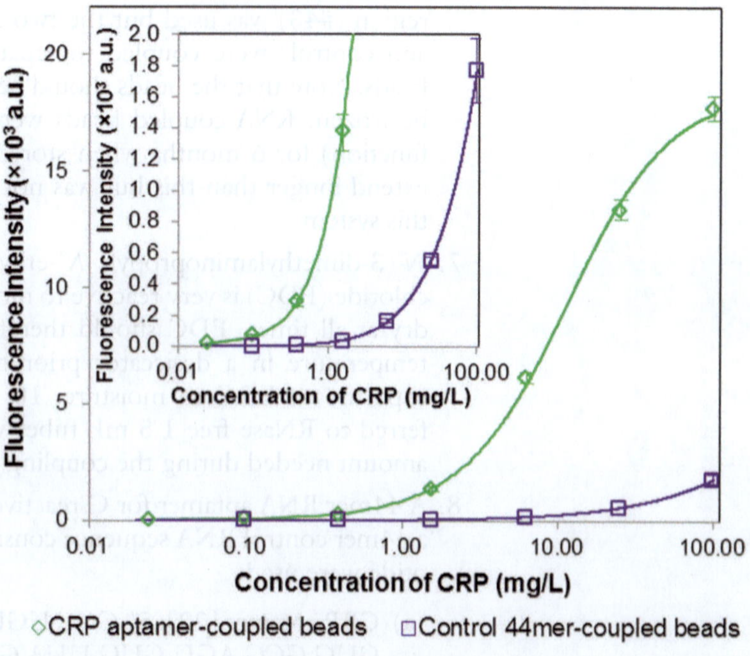

Fig. 3 Sample data from the xMAP aptamer/antibody assay for CRP detection performed using optimized conditions in aqueous solution. Data points represent the means of three replicates ± standard deviation and a 5PL regression was performed using Bio-Plex Manager software. The inset focuses on data points at the lowest concentrations of CRP tested. The specificity of the aptamer's interaction with CRP is illustrated here through the comparison of assays in which the aptamer- and control oligonucleotide-coupled beads were tested separately under identical conditions. Reprinted from [14] with permission from Elsevier

tion was removed from the bulk, tested using a pH meter, and then discarded. Additional acid or base was added to the bottle if needed, the solution was mixed, and this process was repeated until the pH was adjusted to the desired value.

4. In order to filter-sterilize several of the buffers, Corning Disposable Filter Systems with a pore size of 0.22 μm and cellulose acetate membranes were used.

5. A Tris–EDTA buffer solution from Sigma-Aldrich (catalog number T9285) was used. It contains 1 M Tris–HCl at approximately pH 8.0 and 0.1 M EDTA.

6. The beads used in for xMAP-based assays are doped with fluorescent dyes so that each numbered bead region has a unique spectral signature. A different bead region should be used for each target to be measured so that the detection systems can be used simultaneously and they should be compatible with the specific analyzer to be used (e.g., Bio-Plex 100 versus Bio-Plex MAGPIX Multiplex Reader). In this work, one bead

region (#43) was used but the two RNA sequences (aptamer and control) were coupled to separate aliquots of the same beads. Note that the beads should be stored at 4 °C and never be frozen. RNA-coupled beads were stable (i.e., maintained function) for 6 months when stored at 4 °C; stability might extend longer than this but was not tested past 6 months for this system.

7. *N*-(3-dimethylaminopropyl)-*N′*-ethylcarbodiimide hydrochloride (EDC) is very reactive to moisture and should be kept dry at all times. EDC should therefore be brought to room temperature in a desiccator prior to use. To minimize the exposure of EDC to moisture, 10 mg portions were transferred to RNase-free 1.5 mL tubes which corresponds to the amount needed during the coupling protocol.
8. A 44mer RNA aptamer for C-reactive protein (CRP) [22] and a 44mer control RNA sequence consisting of 11 of each nucleotide were used:
   (a) CRP aptamer [22]: 5′-GCC UGU AAG GUG GUC GGU GUG GCG AGU GUG UUA GGA GAG AUU GC-3′.
   (b) Control 44mer: 5′-GAU CCA UGA GUC AGG CUA GUA CCG UAG UCA UUC GCA GUA CCA GU-3′.

The control sequence should demonstrate little to no binding with the target under study. Both sequences are modified at the 5′ terminus with a triethylene glycol spacer followed by a C6 amino modifier. Fluoropyrimidines are used to enhance the stability of these RNAs; however, this might interfere with target binding for some aptamers. Some of these modifications are based on those used in a reported biosensor that used the CRP aptamer [15]. It might be necessary to test different spacers for other aptamer systems to determine the optimal spacing between the microsphere surface and the oligonucleotide.

9. We used a Bio-Plex 100 analyzer during the development of our aptamer/antibody assay for CRP along with associated products, such as Sheath Fluid, the Bio-Plex Calibration Kit, and Bio-Plex Manager software, from Bio-Rad. Other analyzer models from Bio-Rad or Luminex Corporation are available for the detection and analysis of xMAP beads.
10. We recommend running triplicate samples to increase the precision of quantification.
11. Hybridization experiments should be used to confirm that the coupling reaction worked. The RNA aptamer for CRP and the RNA control sequence that were coupled to the beads were 44 nucleotides long and were coupled through their 5′ termini. We used 21mer DNA strands that were complementary to the 21 nucleotides on the 3′ ends of the aptamer and control oligo sequences in the validation experiments. The 5′ termini

of these complementary strands were labeled with biotin to permit detection of successfully hybridized strands using streptavidin–R-phycoerythrin (SAPE). The complements we used were:

(a) CRP aptamer complement: 5′-GCA ATC TCT CCT AAC ACA CTC-3′.

(b) Control 44mer complement: 5′-ACT GGT ACT GCG AAT GAC TAC-3′.

12. Luminex Corporation recommends diluting SAPE in a TMAC solution for this experiment, but there were problems with bead aggregation when this buffer was used for the analysis stage of the assay. Using TE+0.1% (*v/v*) Tween 20 as diluent helped decrease the amount of bead aggregation reported by the Bio-Plex instrument during analysis.

13. Samples were transferred to a pre-wet Millipore Multiscreen HTS 96-well filter plate from the polypropylene Greiner 96-well plate for analysis on the Bio-Plex 100 since the needle height on the Bio-Plex 100 was adjusted for use with the filter plates. Given that other researchers using the shared Bio-Plex 100 all used filter plates, it was determined that this was preferable to readjusting the needle height for this one experiment.

14. The specific buffer composition and pH should be chosen based on the aptamer being used since some aptamers bind their targets under very specific buffer conditions. For example, the RNA aptamer for CRP that was used in our study [22] binds more efficiently to CRP in the presence of calcium(II) ions [16, 22]. The conditions used during aptamer selection or during a previously developed assay using the aptamer as well as possible incompatibilities with the beads (i.e., organic solvents [19]) should be considered.

15. The use of additives such as proteins or detergents in the buffer used during the assay was found to be critical for the reduction of nonspecific binding and bead aggregation. Bovine serum albumin (BSA) was used at 0.1% (*w/v*) in the Assay Bead Diluent and in the Blocking Buffer used to dilute samples and resuspend the beads following filtration. Tween 20 was used at 0.1% (*v/v*) in the Assay Bead Diluent and in the Assay Wash Buffer that was used to rinse the beads. It might be necessary to optimize additive levels included in the buffer for the detection system being developed.

16. A 1 M solution of Tris–HCl at pH 8.0 (appropriate for use in molecular biology) was purchased. In a sterile bottle, mix 45 mL of the 1 M Tris–HCl pH 8.0 solution with 0.6 mL of concentration HCl to bring the pH to 7.5. The pH of this solution should be tested as described in **Note 3** and adjusted further if required.

17. Human serum is the matrix in which our target, CRP, is typically measured. Several challenges were encountered when working on spiked recovery samples using serum. Matrix effects were observed for the detection system when undiluted serum and 10- and 100-fold diluted serum were spiked with CRP when compared to the same amount of CRP spiked into Blocking Buffer (serum was diluted with Blocking Buffer where appropriate), though 1000-fold diluted serum showed no appreciable matrix effects. In order to measure biologically relevant serum levels of CRP (2–1000 mg/L), samples diluted 100-fold would overlap with our system's analytical range of 0.4–100 mg/L. in 100-fold diluted serum. Calibration solutions were therefore prepared in serum diluted 100-fold with Blocking Buffer for our spiked recovery samples and the samples themselves were also prepared in serum diluted 100-fold with Blocking Buffer. It is recommended that if a bead-based aptamer/antibody detection system is being developed to measure a target in a complicated matrix like serum, that the system is optimized to work in a concentration range that permits dilution of the samples to help mitigate or eliminate matrix effects.
18. Preliminary experiments need to be performed in order to determine the concentration range over which the assay will work. It was determined that a Fourfold dilution series with a top concentration in the range or 100–200 mg/L. of CRP was appropriate for the detection system in buffer and in the 100-fold diluted serum used in recovery experiments.
19. When selecting an antibody and aptamer pair for a sandwich-type assay, it should be considered that the antibody and aptamer being used must bind to different epitopes on the target and must be able to bind at the same time. It might be necessary to test multiple antibodies for each target in order to determine which one will work best in a specific detection system. The antibody used has been proven to work in sandwich-type assays with the CRP aptamer in studies by Pultar et al. [16] and Centi et al. [15]. It is a biotinylated anti-CRP antibody, and therefore it has been used in conjunction with streptavidin–R-phycoerythrin (SAPE) to detect the CRP captured by the aptamer strands on the beads. It might be possible to use a phycoerythrin-labeled antibody for some systems instead of a biotinylated antibody paired with SAPE. However, using the two-step detection might help to amplify the signal if the antibody is labeled with more than one biotin since more than one SAPE would be captured for every target molecule captured [23]. It would also be possible to use a biotinylated or fluorescently labeled aptamer for an aptamer-based sandwich assay if a second aptamer that binds to a different epitope of the target is available.

20. The biotinylated anti-CRP antibody was diluted to 4 μg/mL; this is a concentration recommended by Luminex Corporation in their antibody assay protocols [24]. It might be necessary to optimize the concentration of the antibody used depending on the system and how the antibody is labeled (i.e., biotinylated or fluorescently labeled). The concentration of SAPE, if used, may also need to be optimized.

21. Antibody and SAPE concentrations are only two of many conditions that could possibly need to be optimized for a particular bead-based aptamer/antibody detection system. Different aptamer/antibody or aptamer/aptamer pairs might be tested (*see* **Note 19**), the particular buffer and additives used need to be determined (*see* **Notes 14** and **15**), and the oligo to beads ratio during coupling must be carefully considered (*see* **Notes 22** and **23**). Once oligo-coupled beads are prepared, it is recommended that several different conditions be tested simultaneously in an optimization experiment. For example, many more incidents of bead aggregation were detected by the Bio-Plex Manager software for the CRP aptamer/antibody assay than in antibody-only assays. Bead aggregation can lead to inaccurate fluorescence readings during analysis so changes to the protocol were investigated in an effort to diminish the occurrence of aggregation. It was hypothesized that hybridization interactions of the RNA coupled to the beads could be causing bead aggregation and that holding the 96-well plate on a magnetic separator might be contributing to this problem. The use of vacuum aspiration to remove the supernatant from 96-well filter plates was compared with the use of a magnetic separator and tapping the supernatant out of the plate and vacuum aspiration was found to help reduce bead aggregation. Furthermore, a significant increase in fluorescence intensity between blank and 50 mg/L. CRP samples was obtained when using vaccum aspiration. The selection of optimal buffers was also conducted through the comparison of different buffer conditions. By considering many different factors all at once, an optimization experiment can aid greatly in the development of a successful bead-based aptamer/antibody detection system. Optimal conditions will result in the target being specifically captured for detection by aptamer-coupled beads while low nonspecific binding is observed for control oligonucleotide-coupled beads (*see* Fig. 3 for sample data).

22. Bead stocks are sold at a concentration of 12.5×10^6 beads per millilitre. The protocol for coupling RNA to the beads is given for a portion of 2.5×10^6 beads and was based on information published by Luminex Corporation [25]. Luminex Corporation provides suggested methods and protocols for both protein- and oligonucleotide-based xMAP assays in

"The xMAP Cookbook" on their website [20]. This includes information on how to scale coupling reactions based on the number of beads to be used.

23. A separate aliquot of beads should be used for each RNA sequence being used. In the present work separate aliquots were used for the CRP aptamer and a 44mer control sequence, respectively. Separate aliquots can also be used if different coupling ratios are being tested (i.e., ratio of the oligonucleotide to the total number of beads). Two oligonucleotide masses (0.08 and 0.16 nmol) were coupled with 2.5×10^6 beads each. The use of 0.16 nmol of amino-modified oligonucleotide with 2.5×10^6 beads was optimal for the CRP detection system since these beads gave the most reproducible hybridization assay results when comparing CRP aptamer-coupled beads with beads coupled to the control 44mer sequence. Aside from hybridization assay results, a full aptamer/antibody assay for the desired target could also be performed to compare beads prepared with varying coupling ratios.
24. Particular attention should be paid to the instructions in the use for the particular magnetic separator utilized and the waiting time should be adjusted as necessary.
25. Solutions were prepared so that 0, 10, 20, 40, 80, 120, 160, and 320 fmol of the appropriate complementary oligo were added per well (17 μL). Luminex Corporation recommends femtomolar values of 200 and lower [26]. In the present work saturation was achieved at 320 fmol. *See* Fig. 2.
26. A polypropylene Greiner 96-well plate or similar product should be used for at least the initial portion of a hybridization experiment. Once the oligo-coupled beads and the complementary oligos are pipetted into the wells, the plate will be covered and heated to an appropriately high temperature to denature the oligo strands (i.e., disrupt hydrogen bonds formed if the strands folded into intramolecular structures, thus linearizing the sequences). The plate will then be cooled back to room temperature to promote hybridization of the strands coupled to the beads and the complementary oligos added in solution.
27. The 96-well plate was heated on a heating block at 95 °C for 5 min then set the temperature of the block to 24 °C and left the plate to cool at the rate of the block. A PCR thermal cycler could be used to control the drop in temperature more precisely, if desired.
28. Successful coupling will be indicated by hybridization assay results that demonstrate an increase in fluorescence with an increase in the concentration of the oligonucleotide complement. If there is little or no increase in fluorescence over a

range of complement masses (*see* **Note 25**), the coupling reaction might have failed and possible reasons for this should be explored (e.g., incorrect or insufficient labeling of the coupled oligonucleotide, inactive EDC). If hybridization results obtained for aptamer-coupled beads and for control oligonucleotide-coupled beads show the same trend and comparable fluorescence intensities, as in Fig. 2, it is implied that there was little difference in coupling efficiency between the two sequences.

References

1. Wild D (ed) (2013) The immunoassay handbook, 4th edn. Oxford, Elsevier Ltd.
2. Ellington AD, Szostak JW (1990) *In vitro* selection of RNA molecules that bind specific ligands. Nature 346:818–822
3. Tuerk C, Gold L (1990) Systematic evolution of ligands by exponential enrichment: RNA ligands to bacteriophage T4 DNA polymerase. Science 249:505–510. doi:10.1126/science.2200121
4. Mascini M, Palchetti I, Tombelli S (2012) Nucleic acid and peptide aptamers: fundamentals and bioanalytical aspects. Angew Chem Int Ed 51:1316–1332. doi:10.1002/anie.201006630
5. Tombelli S, Minunni M, Mascini M (2005) Analytical applications of aptamers. Biosens Bioelectron 20:2424–2434. doi:10.1016/j.bios.2004.11.006
6. Liu J, Cao Z, Lu Y (2009) Functional nucleic acid sensors. Chem Rev 109:1948–1998. doi:10.1021/cr030183i
7. Luminex Corporation (2015) xMAP® technology. Luminex Corporation, Austin, TX https://www.luminexcorp.com/research/our-technology/xmap-technology/. Accessed 29 Dec 2015
8. Dunbar S, Hoffmeyer MR (2013) Microsphere-based multiplex immunoassays: development and applications using. Luminex xMAP Technology. In: Wild D (ed) The immunoassay handbook, 4th edn. Elsevier Ltd, Oxford, pp 157–174
9. Luminex Corporation (2015) https://www.luminexcorp.com/. Accessed 29 Dec 2015
10. Tayabali AF, Nguyen KC, Seligy VL (2007) Antibody arrays and their application to screening level risk assessment of biotechnology-related bacteria. In: Kumarathasan P, Vincent R (eds) Recent research developments in analytical biochemistry: applications in environmental toxicology. Research Signpost, Trivandrum, Kerala, pp 157–176
11. Porschewski P, Grättinger MA, Klenzke K, Erpenbach A, Blind MR, Schäfer F (2006) Using aptamer as capture reagents in bead-based assay systems for diagnostics and hit identification. J Biomol Screen 11:773–781
12. Bock LC, Griffin LC, Latham JA, Vermaas EH, Toole JJ (1992) Selection of single-stranded DNA molecules that bind and inhibit human thrombin. Nature 355:564–566
13. Baird GS (2010) Where are all the aptamers? Am J Clin Pathol 134:529–531
14. Bernard ED, Nguyen KC, DeRosa MC, Tayabali AF, Aranda-Rodriguez R (2015) Development of a bead-based aptamer/antibody detection system for C-reactive protein. Anal Biochem 472:67–74. doi:10.1016/j.ab.2014.11.017
15. Centi S, Sanmartin LB, Tombelli S, Palchetti I, Mascini M (2009) Detection of C reactive protein (CRP) in serum by an electrochemical aptamer-based sandwich assay. Electroanalysis 21:1309–1315
16. Pultar J, Sauer U, Domnanich P, Preininger C (2009) Aptamer-antibody on-chip sandwich immunoassay for detection of CRP in spiked serum. Biosens Bioelectron 24:1456–1461
17. Ansar W, Ghosh S (2013) C-Reactive protein and the biology of disease. Immunol Res 56:131–142. doi:10.1007/s12026-013-8384-0
18. Myers GL, Rifai N, Tracy RP, Roberts WL, Alexander RW, Biasucci LM, Catravas JD, Cole TG, Cooper GR, Khan BV, Kimberly MM, Stein EA, Taubert KA, Warnick GR, Waymack PP (2004) CDC/AHA workshop on markers of inflammation and cardiovascular disease: application to clinical and public health practice:

report from the Laboratory Science Discussion Group. Circulation 110:e545–e549. doi:10.1161/01.CIR.0000148980.87579.5E

19. Angeloni S, Cordes R, Dunbar S, Garcia C, Gibson G, Martin C, Stone V (2014) Appendix A: common buffers used in xMAP protocols. In: Luminex (eds) The xMAP cookbook, 2nd edn. Luminex Corporation, Austin, TX. http://info.luminexcorp.com/download-the-xmap-cookbook. Accessed 10 Dec 2015
20. Angeloni S, Cordes R, Dunbar S, Garcia C, Gibson G, Martin C, Stone V (2014) The xMAP cookbook, 2nd edn. Luminex Corporation, Austin, TX. http://info.luminexcorp.com/download-the-xmap-cookbook. Accessed 10 Dec 2015
21. Bio-Rad Laboratories Inc. 2013 Bio-Plex Pro™ Magnetic COOH Beads, Bio-Plex Beads® amine coupling kit, instruction manual, Rev C. http://www.bio-rad.com/webroot/web/pdf/lsr/literature/4110012C.pdf. Accessed 27 Nov 2013
22. Bini A, Centi S, Tombelli S, Minunni M, Mascini M (2008) Development of an optical RNA-based aptasensor for C-reactive protein. Anal Bioanal Chem 390:1077–1086
23. Weeks I, Kricka LJ, Wild D (2013) Signal generation and detection systems (excluding homogeneous assays). In: Wild D (ed) The immunoassay handbook, 4th edn. Elsevier Ltd., Oxford, pp 267–285
24. Angeloni S, Cordes R, Dunbar S, Garcia C, Gibson G, Martin C, Stone V (2014) Capture sandwich immunoassay. In: The xMAP cookbook, 2nd edn. Luminex Corporation, Austin, TX. http://info.luminexcorp.com/download-the-xmap-cookbook. Accessed 10 Dec 2015
25. Angeloni S, Cordes R, Dunbar S, Garcia C, Gibson G, Martin C, Stone V (2014) Standard nucleic acid coupling to xMAP microspheres. In: The xMAP cookbook, 2nd edn. Luminex Corporation, Austin, TX. http://info.luminexcorp.com/download-the-xmap-cookbook. Accessed 10 Dec 2015
26. Angeloni S, Cordes R, Dunbar S, Garcia C, Gibson G, Martin C, Stone V (2014) Oligonucleotide coupling confirmation. In: The xMAP cookbook, 2nd edn. Luminex Corporation, Austin, TX. http://info.luminexcorp.com/download-the-xmap-cookbook. Accessed 10 Dec 2015

Part III

Molecular Imprinted Polymers

Chapter 20

Transferring the Selectivity of a Natural Antibody into a Molecularly Imprinted Polymer

Romana Schirhagl

Abstract

Natural antibodies are widely used for their unprecedented reproducibility and the remarkable selectivity for a wide range of analytes. However, biodegradability and the need to work in biocompatible environments limit their applications. Molecularly imprinted polymers are a robust alternative. While molecularly imprinted polymers have shown remarkable selectivities for small molecules, large structures as proteins, viruses or entire cells are still problematic and flexible structures are virtually impossible to imprint. We have developed a method to form a polymeric copy of the antibodies instead. This book chapter aims to summarize the progress with this technique. To make it easier for other scientists to use this methods I critically discuss advantages and drawbacks of the method compared to alternative techniques. The discussion should help to identify for which applications this technique would be valuable. Finally, I provide a practical guide to use this new method. I highlight potential problems and give hints for possible improvements or adaptations for different applications.

Key words Imprinting, Molecularly imprinted polymers, Synthetic antibodies, Artificial binding sites

1 Introduction

Natural antibodies are widely used due to a set of remarkable properties. They are available for a wide variety of different targets, the antigens, and show remarkable selectivity. Furthermore, they can be produced with unprecedented reproducibility.

However, they are time consuming to make and thus expensive. Since they are glycoproteins biodegradation by microorganisms or air oxidation limits their lifetime [1]. Also they are degraded in harsh environments as acids, bases, or organic solvents, which further limits their applications [2]. These reasons were the driving force to develop artificial antibodies during the last decades. One way to achieve this goal is molecular imprinting. To generate a molecularly imprinted polymer (MIP) one has to polymerize a prepolymer with high cross linker content in presence of a template molecule (= the desired analyte). When the polymer cures, two

Thomas Tiller (ed.), *Synthetic Antibodies: Methods and Protocols*, Methods in Molecular Biology, vol. 1575,
DOI 10.1007/978-1-4939-6857-2_20, © Springer Science+Business Media LLC 2017

things happen simultaneously. First, functional molecules or groups in the polymer move close to their counterparts on the analyte. This complex formation has been described in several articles [3–5]. This can for instance be a positive charge in a polymer chain which is pulled close to an electronegative area on the analyte and form a binding site. However, not only charges are an option. In principle any type of non-covalent interaction (as ππ interactions [6], dipoles [7], van der Waals interactions [8]) can form binding sites. Second, the prepolymer is cross-linked. As a result the binding sites (negative copies of the template) remain even if the template itself is removed. The remaining holes in the polymer will preferentially bind the template molecule over similar molecules due to the geometric fit and the spatial orientation of surface groups. This strategy successfully produced artificial binding sites for small molecules [9–11], proteins [12, 13], viruses [14–17], or even entire cells [18–20]. Molecularly imprinted polymers are way more robust than their natural counterparts and thus have longer shelf life and can usually be reused for almost indefinite amounts of times (at least compared to natural antibodies which are usually discarded after one time use). They can be used in harsh condition and show decent selectivities. They are especially useful to detect small molecules which often pose a problem for antibody detection since triggering an immune response towards small molecules is much less effective than for instance for a protein. For more information about molecularly imprinted polymers in general I would like to refer the reader to the other chapters in the section on molecular imprinting in this book as well as recent reviews [21–24].

However, MIPs could not solve all the problems either. Especially large analytes remained a challenge for MIPs. While very high selectivities where achieved for small molecules the results for large molecules like proteins or viruses were less promising. Especially, mobile analyte surfaces which change their orientation or location on the analyte pose a serious problem. This is for instance the case in cells with a flexible cell membrane where molecules can float around quite rapidly. Large templates also are more difficult to remove from a cured MIP [25]. Another disadvantage of using MIPS is that the polymer composition has to be adapted for each new template molecule.

In this protocol I am describing the fabrication of a MIP based receptor, which is better suited for large analytes but has the robustness of a MIP. These receptors are generated by a double imprinting strategy. A similar principle of using two imprinting steps to generate a copy of the initial template has before only been used in the Mosbach group [26, 27]. The authors coined the term anti-idiotypic imprinting to describe their method. The purpose of their work was to find new enzyme inhibitors.

2 Materials

1. Prepolymer 1: *N,N'*-(1,2-dihydroxyethylene) bisacrylamide (DHEBA), methacrylic acid, 20 mg vinylpyrrolidone, distilled water.
2. 1 M KOH.
3. Antibody against the desired analyte.
4. Potassium peroxydisulfate.
5. Acetonitrile (as receiver solvent to precipitate particles).
6. Prepolymer 2: *N,N'*-(1,2-dihydroxyethylene) bisacrylamide (DHEBA), methacrylic acid, vinylpyrrolidone, distilled water.
7. Equipment used for characterization (optional): Fluorescence microscope, atomic force microscope, quartz crystal microbalance, IR spectrometer (ATR).
8. Brilliant gold paste (Heraeus, Germany).
9. Quartz discs (10 MHz, AT-cut, 15.5 mm diameter).
10. Equipment for sensor measurement: custom-built oscillator circuits [28], HP53131A frequency counter, HP-IB bus, computer.

3 Methods

3.1 Generating Polymeric Antibody Replicas via Double Imprinting

A schematic for the method described in this protocol is shown in Fig. 1. As starting material natural antibodies for the desired analyte are needed. In a first imprinting step negative copies of these are formed on the surface of polymer particles. When the antibodies are removed, cavities remain which are negative copies of the respective antibodies. The resulting polymer particles can be used in a second imprinting step. This leads to antibody copies which resemble their natural counterparts in size and shape.

3.2 First Imprinting Step: Imprinted Nanoparticles

The first fabrication step is the formation of antibody imprinted nanoparticles. As a polymer material we choose an acrylate based polymer since this composition has been successfully used to make antibody imprinted polymers in preliminary studies. To prepare the prepolymer a procedure described in [29] has been used.

1. Dilute 60 mg of *N,N'*-(1,2-dihydroxyethylene) bisacrylamide (DHEBA) as crosslinker, 50 mg methacrylic acid, and 20 mg vinylpyrrolidone in 800 mL distilled water.
2. Add 1.5 mg potassium peroxydisulfate, as a radical initiator.
3. Prepolymerize at 70 °C until the gel point is reached (*see* **Note 1**).

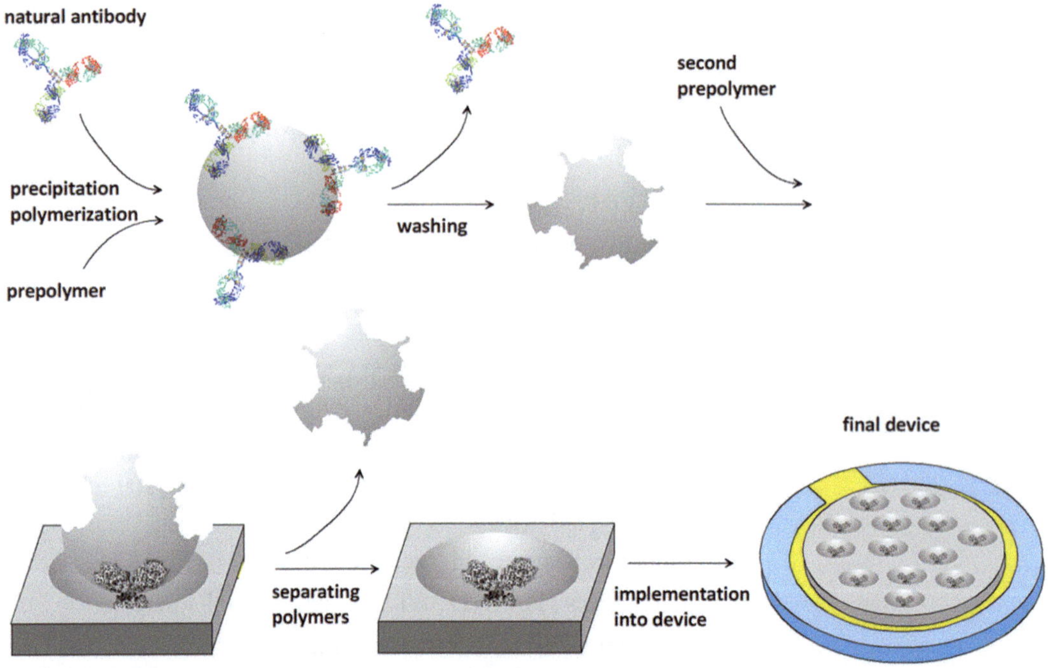

Fig. 1 Schematic procedure of the fabrication process: The procedure consists of two imprinting steps. First polymer particles are formed which are imprinted with natural antibodies. Then these particles are used in a second imprinting step. This results in copies of the original antibodies. Reprinted with permission from [44]

4. Adjust the pH to 7.0 with 1 M KOH (*see* **Note 2**).
5. Dilute the solution 1:2 (v/v) with water and add 5 mg antibody. (For the non-imprinted control the antibody is left out.)
6. Perform precipitation polymerization to form particles. Therefore inject 200 μL of the resulting solution drop wise into 20 mL acetonitrile while stirring rapidly (*see* **Note 3**). During this process the solution turns from transparent to milkish white (*see* Fig. 2).
7. Continue stirring overnight to complete the polymerization. As a result of this procedure, particles precipitated which was confirmed by atomic force microscopy (AFM). Figure 3 shows the typical results of such an AFM measurement. The size distribution which was obtained based on height measurements using AFM (*see* **Note 4**).
8. Centrifuge at 4300 × *g* for 5 min to remove the antibodies.
9. Then discard the supernatant liquid and re-suspend the pellet in water using an ultrasonic bath for 1 min.
10. Repeat the **steps 8** and **9** several times to obtain optimal cleaning. The resulting particles have cavities which are negative copies of the antibody starting material (*see* **Note 5**).

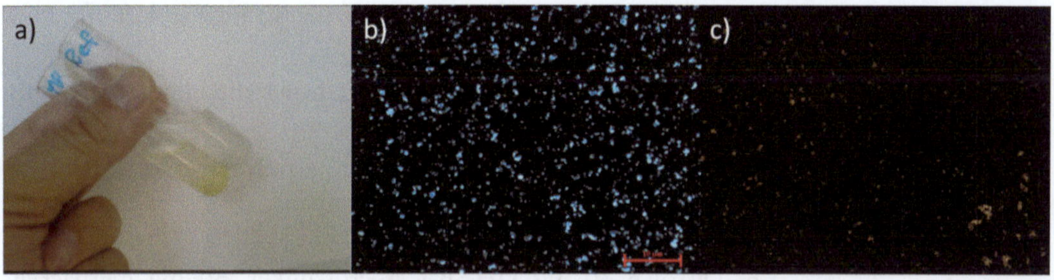

Fig. 2 Confirming absence or presence of antibodies on the particles: (**a**) shows the xanthoprotein reaction (a reaction with HNO3 which colors proteins *yellow*) on particles with (MIP) and without (Ref) proteins. In (**b**) and (**c**) we used a nonspecific protein label called dansyl chloride to show the presence of antibodies before washing (**b**) and the absence after (**c**) (the *red* scale bar is 10 μm)

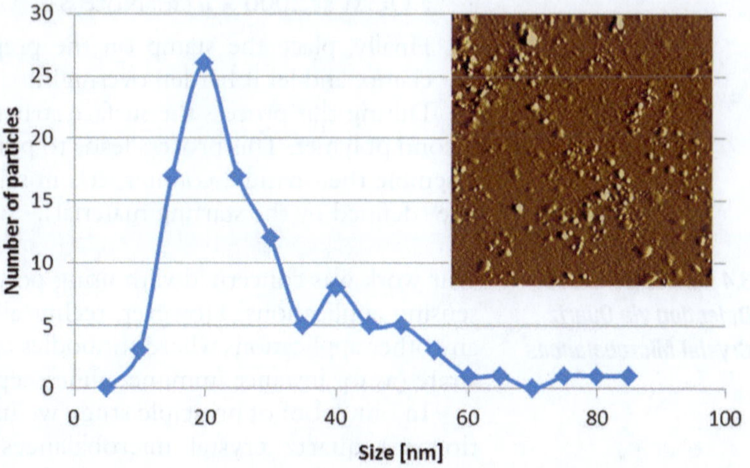

Fig. 3 Atomic force microscopy analysis of antibody imprinted nanoparticles. The graph shows the particle size distribution determined by AFM (contact mode). The histogram was generated using the *z*-values of over 100 particles. The inset shows a typical AFM scan covering a 10 × 10 μm area

The presence and absence of antibodies on the particles before and after washing was confirmed by different analytical techniques. These include mass spectrometry on the particles, the so-called xanthoprotein reaction or fluorescently labeling proteins with dansyl chloride [30].

3.3 Second Imprinting Step: Antibody Copies

In a second imprinting step these particles, with a negative copy of the antibody imprinted in their surface are pressed into another prepolymer. This is done via so-called stamp coating.

1. Apply the washed particle suspension (5 μL) onto microscope slides (5 x 5 mm).

2. Leave the slides in a refrigerator for 30 min at 4 °C to allow particle sedimentation.
3. Spin off the surplus liquid and leave the slides to dry for 30 min.
4. Press these stamps into a second prepolymer on a surface (*see* **Note 6**). *See* Fig. 4 for a schematic representation of the process.
5. Dissolve 30 mg DHEBA as crosslinker, 50 mg methacrylic acid, and 20 mg vinylpyrrolidone in 800 μL distilled water (*see* **Note 7**).
6. Add the initiator 1,1′-azobisisobutyronitrile (AIBN) and prepolymerize the mixture at 70 °C until the gel point is reached.
7. Dilute the prepolymer 1:2 (v/v) in water and spin-coate onto a QCM at 2000 × *g* (*see* **Note 8**).
8. Finally, place the stamp on the prepolymer film, fix it by a clamp, and let it harden overnight.

During this process the surface structure is imprinted into the second polymer. This process leads to polymeric antibodies, which resemble their natural counterparts in selectivity for a certain analyte (defined by the starting material).

3.4 Mass Sensitive Detection via Quartz Crystal Microbalances

Our work was concerned with using polymeric antibody copies in sensing applications. However, technically one could use also for any other applications where antibodies are adhered to a solid substrate (as for instance immune-affinity separations).

In our proof of principle study we used mass sensitive detection via quartz crystal microbalances (QCM). The method makes use of the piezoelectrical effect in quartz. The quartz, sandwiched between two electrodes can be triggered to oscillate

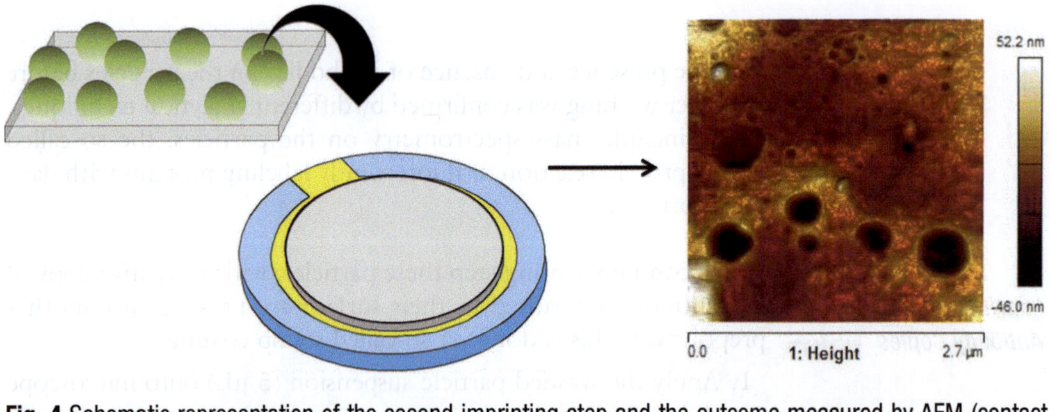

Fig. 4 Schematic representation of the second imprinting step and the outcome measured by AFM (contact mode): Particles are first adhered to a glass plate. Then they are stamped into a second prepolymer to perform the second imprinting step

at a special frequency (the quartzes resonance frequency which is given by its thickness). When anything with a mass adheres to the gold surface (or in our case the biomimetic antibody on the gold surface) this resonance frequency changes (*see* **Note 9**). This frequency shift is proportional to the analyte mass on the device. Thus, virtually any analyte can be detected in principle. Additionally, the method is relatively sensitive and quantities down to ng range can be detected theoretically. On a 10 MHz quartz 1 Hz shift in frequency roughly corresponds to 1 ng. Nonrigid or nonuniform polymers or presence of salts in the solution increase the noise level and lead to the detection limits presented in the different articles (around the μg/ml range for stopped flow measurements). In our studies we used custom-made dual electrode geometries. Over commercially available QCMs, this geometry has the advantage that a non-imprinted reference can be implemented on the same quartz. Q uartz crystal microbalances with one measuring electrodeare also available commercially.

1. Screen-print the electrodes (brilliant gold paste from Heraeus, Germany) onto quartz discs (10 MHz, AT-cut, 15.5 mm diameter) (*see* **Note 10**).
2. Burn the discs for 2 h at 400 °C [28]. The left side of Fig. 5 shows the simple (homemade) device which was used to screen print the electrodes.
3. Print the second side by repeating **steps 1** and **2** on the other side).

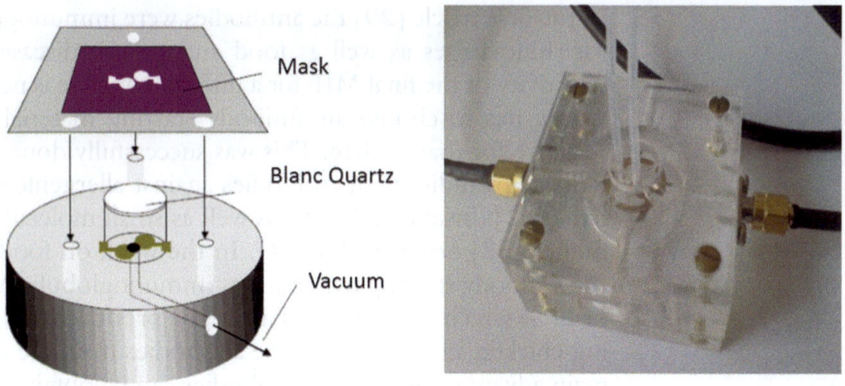

Fig. 5 *Left image*: Homemade device to screen print electrodes on the surface of blank quartz plates. The quartz is held in place by vacuum. Then a mask is placed above and gold paste is applied through the mask. *Right image*: Measuring cell used for QCM measurements in liquid media. The electrodes facing the aqueous phase were electrically grounded. Their diameters were 5 mm whereas the electrodes oriented towards the gas phase were 4 mm in diameter

4. Mount the QCMs horizontally in flow cells with a volume of 150 μL (*see* **Note 11**). The measuring cell is shown in the left side of Fig. 5.
5. Use a custom-built oscillator circuits [31] (*see* **Note 12**). The oscillator frequency was measured with a HP53131A frequency counter, and the data was transferred via a HP-IB bus to a computer.

4 Notes

1. This step can be tricky since the prepolymer hardens very quickly after that point is reached. The gel point is indicated by a change in viscosity (from liquid to viscous) but also accompanied by a slight color change (from transparent into blueish white).
2. Neutral pH is required to prevent the antibody being denatured by the acidic monomers.
3. During a precipitation polymerization a prepolymer is added in a solution where monomers are soluble but oligomers are not. The solubility determines the size of particles that precipitate.
4. This recipe worked best for our applications; however, the particle size can be tuned to some extent. Increasing the amount of crosslinker leads to increased particle sizes. Additionally, the relative amount of polymer used in the precipitation polymerization influences the particle size. If more polymer starting material is used particles tend to be larger (up to a few hundred nanometers).
5. In the first article [29] the antibodies were immunoglobulin G for rhinoviruses as well as food and mouth disease virus. If selectivity of the final MIP for a different analyte is needed one simply has to choose an antibody starting material which is selective for that analyte. This was successfully done in several follow-up studies with antibodies against allergenic food proteins [32], insulin [33, 39], as well as small molecules as hormones [33] or pesticides [34]. In the work on food proteins the antibody starting material was immunoglobulin Y (instead of IgG) which is a structural analogue produced by immunizing chicken and harvesting the antibodies from the eggs. The main advantage is that the antibodies can be obtained without sacrificing the animal. For generating polymeric antibodies, no significant differences between the two antibody starting materials were observed. Here an advantage over conventional MIPs unfolds. Although we use very different analytes we do not have to change the procedure significantly. While conventional

MIPs have to be adapted for each new analyte, here the imprinting steps only involve the antibody for the analyte. As a result it is easier to adapt the procedure for a new target.

6. To separate the stamp from the final artificial antibodies the devices are submerged in water for a few minutes. Due to different swelling in the two polymers and penetration of water in the space between the two polymers the stamp is released. It has to be noted that the size of the stamp (5 × 5 mm) is crucial. While smaller stamps can be made without any problems significantly increasing the stamp size can be problematic. (This problem is not specific for this method but arises generally in stamp coating.) The problem arises when the stamp should be separated from the polymer substrate. Large stamps (around 10 × 10 mm) can be very difficult to remove since water cannot penetrate the space in between the polymers sufficiently enough.
7. As alternative polymer for the second imprinting step we also experimented on polyurethane (composition: bisphenol A: 46.8 mg, phloroglucin: 56 mg, methyl diphenyl diisocyanate: 97.2 mg in 200 μL THF. When the gel point is reached 800 μL THF and finally 10–15 μL pyridine are added to start the reaction. A film is spin-coated and then the procedure is continued as for the abovementioned recipe). This recipe also was successful for the detection of sesame proteins (unpublished) and thus might be a valuable alternative if more rigid or smooth surfaces are desired. However, the polyurethane based biomimetic antibodies incorporated significantly less analyte and thus some optimization might be needed.
8. Forming the final polymer on the surface of a quartz crystal microbalance is optional. For a different application the polymer might be formed on a different surface.
9. It has to be noted that using a QCM puts some restriction on the (second) polymer. It has to be in the form of a thin film which is rigid enough and stick to the gold surface. In a different application where these restrictions are not present a different polymer might perform better.
10. These QCMs are usually slightly more reproducible. The homemade option offers more flexibility with respect to the electrode design. The standard group design for instance was a dual electrode geometry which allows to implement two measuring electrodes on one quartz. As a result one can compensate for non-specific adsorption or frequency changes due to changes in electric field strength. This way also three [35, 36] or even up to four electrodes [37, 38] have been implemented on one quartz measuring different components. As a result the device becomes smaller and easier to handle then if each

electrode would be on a separate quartz. Furthermore, to prevent non-specific sensor responses from humidity heating coils can be added on the quartz [39]. On the downside: the entire quartz has to be exchanged if one electrode fails. Also cross talk between channels can occur and thus the devices are also less reliable than commercially available crystals which usually only have one measuring electrode.

11. This was a convenient sample volume for us (we did not have microfabrication for smaller volumes available). We experimented with different volumes between 40 μL and 100 ml. As expected we did observe faster responses with smaller sample volumes due to smaller diffusion ways.
12. There are also commercially available systems available for QCM. These are usually more reliable. However, the home-built version is more flexible and a bit cheaper.

5 Conclusion and Discussion

Figure 6 shows typical measuring curves obtained with our artificial antibodies mounted on a QCM device. When the correct analyte is present (here sesame protein, a common food allergen as example) it binds to the artificial antibody and is thus weighted by the QCM. The respective frequency shifts are plotted in Fig. 6. When the solution is replaced by water again the frequency drops back to the original level, indicating that the analyte is detached from the surface. This behavior is different from natural antibodies. They bind significantly stronger to their target and thus are not detached by simply adding water. Depending on the application this can either be an advantage or a disadvantage. The advantage

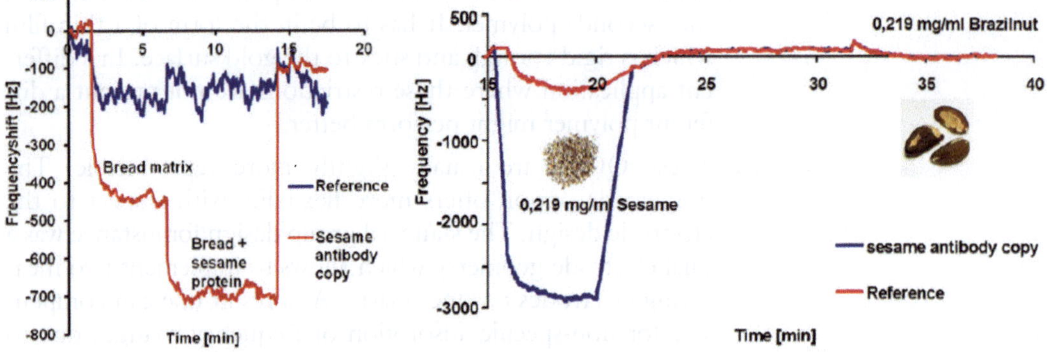

Fig. 6 *Left*: A typical sensor measurement curve obtained with artificial antibodies mounted on the gold surface of a QCM. In this case the analyte was sesame protein which was detected in a bread. The unknown concentration of sesame in the bread can be determined by the standard addition method. *Right*: The frequency shifts on an artificial antibody subjected to sesame or Brazil nuts are shown. Reprinted with permission from [33]

of reversible binding is that the sensor is easy to recover. As a result the polymeric antibodies can be used many times while antibodies are usually just used once. The disadvantage unfolds when very high selectivities are needed. After natural antibodies have been provided with their antigen they can be washed with water. While the desired target still binds nonspecifically adsorbed molecules can be removed. Although we only tested mass sensitive detection, the artificial antibodies could also be implemented in different sensing techniques as optical detection, electrochemical sensing, or others.

5.1 Comparison Between Conventional MIPS, Antibody Copies and Natural Antibodies

This section aims to explain the differences and similarities between natural antibodies, their polymeric counterparts and the conventional MIPs. It should aid the reader in deciding whether or not artificial antibodies would be valuable for their applications. I shortly compare the fundamental differences in binding and then discuss the consequences. The first difference which I would like to highlight is depicted in Fig. 7.

Figure 7 shows an example of a rhino virus as analyte. The upper row in the figure shows the respective receptors and the area which is responsible for the recognition in red. Natural antibodies (a) recognize their target with two small regions at the end of their arms. This region is relatively small and thus also recognizes only a relatively small part of the analyte (the so-called epitope, probably a peptide (a small part of the full protein) which is a few nm in

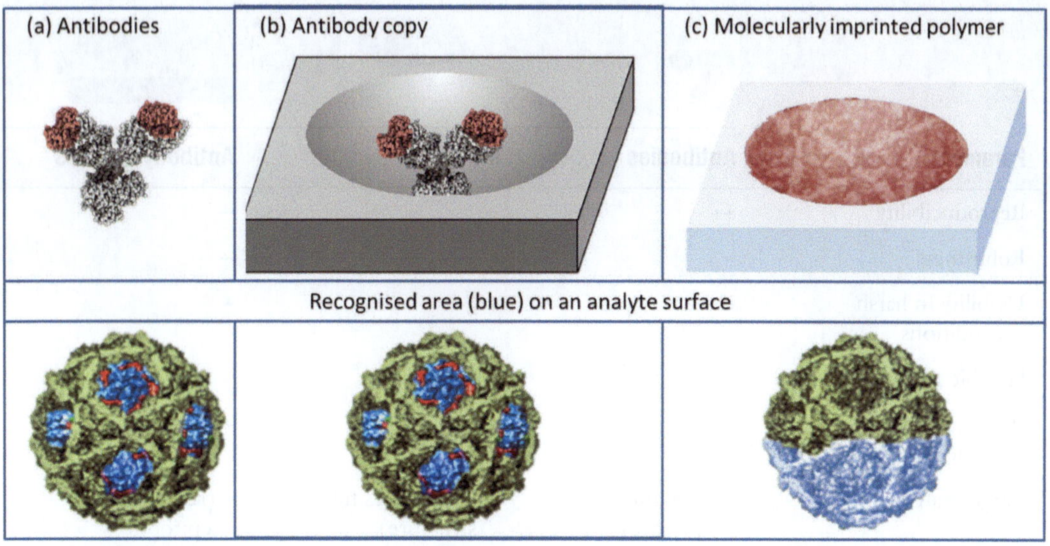

Fig. 7 Comparison between different binding characteristics in (**a**) natural antibodies, (**b**) artificial antibody copies and (**c**) traditional MIPs the area which is responsible for recognizing the target is market in *red*. The images below indicate which areas on the surface of an analyte (the example shown here is the rhino virus used in [33] (**a**), (**b**) and [45] (**c**). (The *blue areas* do not indicate a specific area but should highlight only the principle of targeting a small specific area instead of the whole structure)

size). An example for an area which is recognized is shown in blue on the virus below. The MIP (c) on the contrary creates a binding site using the full area which is in contact with the prepolymer. For surface imprinting this is up to half the surface. To make a polymeric antibody copy (b) the structure of an antibody is transferred into a polymer. Thus it resembles the binding characteristics of the natural antibody more closely than the ones of the MIP. Similar to a natural antibodies the receptors generated from them recognize an epitope not the full structure as the MIP.

This fundamental difference between antibodies/antibody copies and MIPs has several consequences which are summarized in Table 1. Each of the methods covers a niche and depending on the analyte and binding conditions different methods excel.

Small analyte molecules usually are inefficient to trigger immune reactions and thus not so many antibodies for them are available. For these, MIPs which use their entire surface for recognition are suited perfectly. If the analytes are large the fact that MIPs recognize the entire surface becomes problematic. This is even more so if the surface is flexible. In a flexible analyte, as for instance a cell, the surface molecules are not necessaryly at the same position as they were during the imprinting process. Here is

Table 1

Summarized comparison between natural antibodies, MIPs, and antibody replicae

			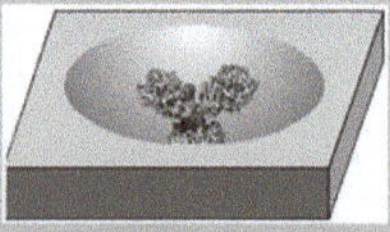
Parameter	**Antibodies**	**MIPs**	**Antibody replicas**
Reproducibility	++	–	–
Robustness	–	+	–
Usability in harsh conditions	–	+	+
Suitable for small analytes	–	+	–
Analytes with flexible surfaces	+	–	+
Large analytes	+ (recognizes epitopes)	– (recognizes full structure)	+ (recognizes epitopes)
Production cost/time	–	++	–
Adapting to a new target	+	–	+
Easy to fabricate	+	++	–

More detailed comparison description of these factors can be found throughout this chapter

an extreme example: If a negative charge was at location A during the imprinting process the binding site will have a positive charge there. If the analyte is too flexible, however, the negative charge can be in a different location when the analyte should rebind. Worst case a positive charge might have moved to location A. This results in repulsion and not attraction. Thus the larger and the more flexible the analyte the worse MIPs perform. Recognizing an epitope, which antibodies as well as their artificial counterparts do is the key to success for such analytes. Both antibodies recognize only a certain area on the analyte. These areas are mostly parts of proteins. While the surface structure, the cell membrane, of the whole cell allows diffusion of different components, proteins usually stay intact. As a result recognizing a smaller structure is favorable for these analytes.

Reproducibility is probably the largest advantage of natural antibodies. Monoclonal antibodies against a certain target are all exactly the same. Binding sites in both MIP based methods can vary substantially due to local differences. The result is that there is a distribution of binding affinities associated with a MIP rather than one binding affinity that is exactly the same for each cavity. If a large number of binding sites are available the distributions of binding affinities for different experiments are expected to be similar.

Both a conventional MIP as well as the antibody copy are made out of a polymer and thus far more robust than natural antibodies. Under conditions where robustness is crucial, as under harsh conditions, the MIP based methods are the only options. The antibody replicas have been shown to be useful already in organic solvents [18]. This is very useful if the analyte has very low solubility in water or many water soluble contaminants. In these cases a simple extraction step can be done to pre-concentrate the sample.

Cost and production time are also important factors. Here the conventional MIPs outperform both other techniques. Polymerizing a MIP is a relatively simple synthesis. Unless very special monomers are used the components which are needed are commercially available and almost negligibly cheap. On the contrary, making an antibody is so complicated that up to now only nature has been able to do it exactly. It is still necessary to use animals and/or cell cultures to generate antibodies. Fabricating an antibody copy requires not one but two imprinting steps and is thus more elaborate and time consuming. However, the robustness of MIP particles allows to produce them in larger quantities and store them until use. The most important drawback of using antibody copies is that the natural antibody is required as starting material. Fortunately, the particle stamps are reusable (for at least 3 times without significant changes in the final product) which reduces this drawback. In some cases (for instance when the template is a very dangerous pathogen, or is otherwise difficult to handle or to get) it might be an advantage that one does not need the analyte. If incompletely removed analyte molecules can compromise the analysis, the antibody copies also offer an elegant way to avoid the problem.

A direct comparison of sensor performance between the three materials has been shown in [33, 40] for insulin detection. An alternative strategy is the so-called epitope imprinting [41–43]. To achieve epitope imprinting one simply imprints only with a certain substructure of the analyte (typically a part of a protein). The main drawback of this technique, however, is that a prior knowledge on the surface structure of the analyte is needed to identify which part is exposed on the surface.

References

1. Ionescu R, Vlasak J (2010) Kinetics of chemical degradation in monoclonal antibodies: relationship between rates at the molecular and peptide levels. Anal Chem 82:3198–3206
2. Ye L, Mosbach K (2008) Molecular imprinting: synthetic materials as substitutes for biological antibodies and receptors. Chem Mater 20:859–868
3. O'Mahony J, Karlsson BCG, Mizaikoff B, Nicholls IA (2007) Correlated theoretical, spectroscopic and X-ray crystallographic studies of a non-covalent molecularly imprinted polymerisation system. Analyst 132:1161–1168
4. Wei S, Jakusch M, Mizaikoff B (2007) Investigating the mechanisms of 17β-estradiol imprinting by computational prediction and spectroscopic analysis. Anal Bioanal Chem 389:423–431
5. Singh K, Balasubramanian S, Amitha Rani BE (2012) Computational and experimental studies of molecularly imprinted polymers for organochlorine pesticides heptachlor and DDT. Curr Anal Chem 8(4):562–568
6. Ho W-L, Liu Y-Y, Lin T-C (2011) Development of molecular imprinted polymer for selective adsorption of benz[a]pyrene among airborne polycyclic aromatic hydrocarbon compounds. Environ Eng Sci 28(6):421–434
7. Andersson L, Mosbach K (1990) Enantiomeric resolution on molecularly imprinted polymers prepared with only non-covalent and non-ionic interactions. J Chromatogr 516:313–322
8. Whitcombe, M. J.; Rodriguez, M. E.; Villar, P.; Vulfson, E. N., A new method for the introduction of recognition site functionality into polymers prepared by molecular imprinting: synthesis and characterization of polymeric receptors for cholesterol, J Am Chem Soc 1995, 117, 7105–7111
9. Vlatakis G, Andersson LI, Müller RI, Mosbach K (1993) Drug assay using antibody mimics made by molecular imprinting. Nature 361:645–647
10. Kellens E, Bove H, Conradi M, D'Olieslaeger L, Wagner P, Landfester K, Junkers T, Ethirajan A (2016) Improved molecular imprinting based on colloidal particles made from miniemulsion: a case study on testosterone and its structural analogues. Macromolecules 49:2559–2567
11. Peeters M, Kobben S, Jimenez-Monroy KL, Modesto L, Kraus M, Vandenryt T, Gaulke A, van Grinsven B, Ingebrandt S, Junkers T, Wagner P (2014) Thermal detection of histamine with a graphene oxide based molecularly imprinted polymer platform prepared by reversible addition–fragmentation chain transfer polymerization. Sens Actuators B Chem 203:527–535
12. Liu J, Deng Q, Tao D, Yang K, Zhang L, Liang Z, Zhang Y (2014) Preparation of protein imprinted materials by hierarchical imprinting techniques and application in selective depletion of albumin from human serum. Sci Rep 4:5487
13. Wangchareansak T, Sangma C, Choowongkomon K, Dickert FL, Lieberzeit P (2011) Surface molecular imprints of WGA lectin as artificial receptors for mass-sensitive binding studies. Anal Bioanal Chem 400:2499–2506
14. Bolisay LD, Culver JN, Kofinas P (2007) Optimization of virus imprinting methods to improve selectivity and reduce nonspecific binding. Biomacromolecules 8(12):3893–3899

15. Birnbaumer GM, Lieberzeit PA, Richter L, Schirhagl R, Milnera M, Dickert FL, Bailey A, Ertl P (2009) Detection of viruses with molecularly imprinted polymers integrated on a microfluidic biochip using contact-less dielectric microsensors. Lab Chip 6:3549–3556
16. Cumbo A, Lorber B, Corvini PF-X, Meier W, Shahgaldian P (2012) A synthetic nanomaterial for virus recognition produced by surface imprinting. Nat Commun 4:1503
17. Bai W, Spivak DA (2014) A double imprinted diffraction grating sensor based on a virus responsive super aptamer hydrogel derived from an impure extract. Angew Chem Int Ed 53:2095–2098
18. Schirhagl R, Hall EW, Fuereder I, Zare RN (2012) Separation of bacteria with imprinted polymeric films. Analyst 137(6):1495–1499
19. Darder M, Aranda P, Burgos-Asperilla L, Llobera A, Cadarso VJ, Sanchez CF, Ruiz-Hitzky E (2010) Algae–silica systems as functional hybrid materials. J Mater Chem 20:9362–9369
20. Eersels K, van Grinsven B, Khorshid M, Somers V, Püttmann C, Stein C, Barth S, Dilien H, Bos GMJ, Germeraad WTV, Cleij TJ, Thoelen R, De Ceuninck W, Wagner P (2015) Heat-transfer-method-based cell culture quality assay through cell detection by surface imprinted polymers. Langmuir 31:2043–2050
21. Schirhagl R, Ren K, Zare RN (2012) Surface-imprinted polymers in microfluidic devices. Sci China Chem 55(4):469–483
22. Schirhagl R (2014) Bioapplications for molecularly imprinted polymers. Anal Chem 86(1):250–261
23. Sindhu Sharma P, Iskierko Z, Pietrzyk-Le A, D'Souza F, Kutner W (2015) Bioinspired intelligent molecularly imprinted polymerss for chemosensing: a mini review. Electrochem Commun 50:81–87
24. Haupt K, Mosbach K (2000) Molecularly imprinted polymers and their use in biomimetic sensors. Chem Rev 100(7):2495–2504
25. Lorenzo RA, Carro AM, Alvarez-Lorenzo C, Concheiro A (2011) To remove or not to remove? The challenge of extracting the template to make the cavities available in molecularly imprinted polymers (MIPs). Int J Mol Sci 12(7):4327–4347
26. Mosbach K, Yu Y, Andersch J, Ye L (2001) Generation of new enzyme inhibitors using imprinted binding sites: the anti-idiotypic approach, a step toward the next generation of molecular imprinting. J Am Chem Soc 123:12420–12421
27. Patent application, WO 1995021673 A1, preparation and application of artificial anti-idiotypic imprints
28. Eichelbaum F, Borngraber R, Schroder J, Lucklum R, Hauptmann P (1999) Interface circuits for quartz-crystal-microbalance sensors. Rev Sci Instrum 70:2537
29. Schirhagl R, Lieberzeit PA, Dickert FL (2010) Chemosensors for viruses based on artificial immunoglobulin copies. Adv Mater 22:2078–2081
30. Sikorski AF, Daczyńska RB (1982) Fluorescent labelling of proteins on thin layers of solid dansyl chloride. Biochim Biophys Acta Biomembr 690(2):302–305
31. Dickert FL, Hayden O, Bindeus R, Mann K-J, Blaas D, Waigmann E (2004) Bioimprinted QCM sensors for virus detection-screening of plant sap. Anal Bioanal Chem 378:1929
32. Schirhagl R, Seifner A, Husain FT, Cichna-Markl M, Lieberzeit PA, Dickert FL (2010) Antibodies and their replicae in microfluidic sensor systems—labelfree quality assessment in food chemistry and medicine. Sensor Lett 8:399–404
33. Schirhagl R, Qian J, Dickert FL (2012) Immunosensing with artificial antibodies in organic solvents or complex matrices. Sens Actuators B 173:585–590
34. Schirhagl R, Latif U, Dickert FL (2011) Atrazine detection based on antibody replicas. J Mater Chem 21(38):14594–14598
35. Dickert FL, Lieberzeit PA, Achatz P, Palfinger C, Fassnauer M, Schmid E, Werther W, Horner G (2004) QCM array for on-line-monitoring of composting procedures. Analyst 129:432–437
36. Iqbal N, Mustafa G, Rehman A, Biedermann A, Najafi B, Lieberzeit PA, Dickert FL (2010) QCM-arrays for sensing terpenes in fresh and dried herbs via bio-mimetic MIP layers. Sensors 10(7):6361–6376
37. Latif U, Mujahid A, Afzal A, Sikorski R, Lieberzeit PA, Dickert FL (2011) Dual and tetraelectrode QCMs using imprinted polymers as receptors for ions and neutral analytes. Anal Bioanal Chem 400:2507–2515
38. Seidler K, Polreichova M, Lieberzeit PA, Dickert FL (2009) Biomimetic yeast cell typing—application of QCMs. Sensors 9(10):8146–8157
39. Lieberzeit PA, Rehman A, Najafi B, Dickert FL (2008) Real-life application of a QCM-based

e-nose: quantitative characterization of different plant-degradation processes. Anal Bioanal Chem 391:2897–2903

40. Schirhagl R, Podlipna D, Lieberzeit PA, Dickert FL (2010) Comparing biomimetic and biological receptors for insulin sensing. Chem Commun 46:3128–3130
41. Nishino H, Huang CS, Shea KJ (2006) Selective protein capture by epitope imprinting. Angew Chem Int Ed Engl 45(15):2392–2396
42. Zhang Y, Deng C, Liu S, Wu J, Chen Z, Li C, Lu W (2015) Active targeting of tumors through conformational epitope imprinting. Angew Chem Int Ed Engl 54(17):5157–5160
43. Zhao X-L, Li D-Y, He X-W, Li W-Y, Zhang Y-K (2014) An epitope imprinting method on the surface of magnetic nanoparticles for specific recognition of bovine serum album. J Mater Chem B 2:7575–7582
44. Wackerlig J, Schirhagl R (2016) Applications of molecularly imprinted polymer nanoparticles and their advances toward industrial use: a review. Anal Chem 88(1):250–261
45. Jenik M, Schirhagl R, Schirk C, Hayden O, Lieberzeit P, Blaas D, Paul G, Dickert FL (2009) Sensing picornaviruses using molecular imprinting techniques on a quartz crystal microbalance. Anal Chem 81(13):5320–5532

Chapter 21

Preparation of Molecularly Imprinted Microspheres by Precipitation Polymerization

Tibor Renkecz and Viola Horvath

Abstract

Molecularly imprinted polymers (MIPs) gained an expansively growing interest in the past few decades. After an initial, explorative period of preparing MIPs exclusively with bulk polymerization, new polymer synthesis routes have been adapted to overcome the drawbacks of the traditional method. Among these the most appealing is precipitation polymerization that results in nano- and microspheres with narrow size distribution and makes the production of MIPs more straightforward. Here, we describe a precipitation polymerization protocol for a common small molecule template, propranolol that is carried out in the conventional way, in dilute monomer solution. Moreover, a modified precipitation polymerization protocol from concentrated monomer solution is presented for a diclofenac imprinted polymer which makes the synthesis even more versatile and circumvents the disadvantages of the dilute solution conditions.

Key words Molecularly imprinted polymer, Polymer microsphere, Precipitation polymerization, Monodisperse, Modified precipitation polymerization, Propranolol, Diclofenac

1 Introduction

1.1 Overview

Molecularly imprinted polymers can be prepared by various polymerization techniques (e.g., sol-gel synthesis, electropolymerization), but the vast majority is synthesized by free radical polymerization. The traditional and still widely applied method is "bulk polymerization" whereby a solution of the functional and crosslinking monomers is polymerized in the presence of the template, so strictly speaking this method is solution polymerization. Since the monomer solution is quite concentrated (a typical solvent/monomer ratio is around 1.5) the product of this process is a solid polymer monolith, which needs further processing, i.e., crushing, grinding, and sieving to an appropriate particle size range. During polymer workup only 20–50% of the polymer can be recovered and the procedure is rather time consuming and laborious. The resultant particles are irregular in size and shape, which is a disadvantage in many applications. The straightforward

Thomas Tiller (ed.), *Synthetic Antibodies: Methods and Protocols*, Methods in Molecular Biology, vol. 1575,
DOI 10.1007/978-1-4939-6857-2_21, © Springer Science+Business Media LLC 2017

preparation of beaded polymers especially with uniform size distribution overcomes the above-mentioned difficulties of the bulk synthesis route. The formed polymer particles can be separated by a simple filtration or centrifugation after polymerization. Uniform, spherical beads impart improved hydrodynamic properties and efficiency to chromatographic and capillary electrochromatographic stationary phases, solid phase extraction sorbents, and can be readily embedded into composite membranes, chemical sensors and used in ligand-binding assays.

As of the turn of the century various heterogeneous radical polymerization techniques [1] such as suspension polymerization [2–4], emulsion polymerization [5, 6], seeded polymerization [7], dispersion [8], and precipitation polymerization [9, 10] have been adapted for molecular imprinting to directly produce polymeric beads. Special considerations had to be taken into account when applying these well-established techniques to the preparation of MIPs. The aqueous dispersing media and various polymerization components (e.g., surfactants, emulsifiers, additives) may all interfere with the relatively weak template-functional monomer interactions prevalent in the overwhelmingly used non-covalent imprinting approach.

Among the above-mentioned heterogeneous techniques precipitation polymerization has proven to be the most compatible with molecular imprinting.

Precipitation polymerization is started from a homogeneous mixture of monomers, initiator, and solvent. The solvent selected is a good solvent for the monomers but a bad solvent for the forming polymer. Therefore, during polymerization phase separation occurs resulting in a stable colloidal solution of the polymer particles. The reaction can either proceed enthalpically when the growing polymer chains reach their solubility limit and precipitate from the solution, or entropically when increasing crosslinking density expels the solvent from the polymer chains and the particle collapses. A prerequisite for precipitation polymerization is the use of dilute monomer solutions (typical monomer concentrations are less than 5 w/v %) to maintain colloidal stability otherwise particles can coagulate/aggregate, or macrogelation occurs. Uniformly sized microspheres can be prepared when the crosslinking density is high [11]. In this case, particle initiation is restricted to the early phase of the polymerization. The primary particles grow by capturing soluble oligomeric species from the solution via their pendant vinyl groups. As the surface is less crosslinked than the core of the particle, it is swollen with solvent and the swollen gel layer functions as a steric stabilizer hindering particle coagulation. This mechanism accounts for the observed monodispersity. However, the autosteric stabilization mechanism is functioning only in dilute solutions.

The lack of interfering additives and dispersion medium makes precipitation polymerization an excellent choice for the preparation of molecularly imprinted beads. The requirement of high crosslinking density in precipitation polymerization meets the demands of molecular imprinting, whereby this feature is necessary to conserve the 3-D structure of the binding sites.

However, the highly dilute medium counteracts the complexation of the template and functional monomer in the prepolymerization mixture, thereby suppressing the formation of the selective binding sites.

We have shown that this problem can be tackled by using carefully chosen mixtures of a nonsolvent and a co-solvent as the polymerization diluent [12–14] and introduced a *modified precipitation polymerization* method. Monomer concentrations up to 25–30 w/v % still resulted in uniformly sized microspheres and full conversion of monomers could be obtained. We could explain this behavior by the fact that the co-solvent was enriched in the growing particles and swelled them providing a steric stabilization against coagulation [14].

1.2 General Considerations in Precipitation Polymerization Protocols

There are several key factors to be considered when setting up a precipitation polymerization protocol for the synthesis of MIP microspheres. From one side there are the typical considerations for efficient imprinting, i.e., the proper choice of the functional monomer and crosslinker, their molar ratio, the ratio of the template and the functional monomer, and the choice of an appropriate solvent. On the other side, there are parameters that influence the morphology of the particles, i.e., their size, polydispersity, and porosity.

In this context, the right choice of the polymerization solvent or solvent mixture is of utmost importance as the solvency of the reaction medium determines whether particles will form or not [14, 15]. The use of acetonitrile or acetonitrile/toluene mixtures in divinylbenzene- and methacrylate-based systems is almost exclusive because in other solvents spherical beads could not be produced [11, 16–18]. Nevertheless, since acetonitrile is rather polar it is not an optimal solvent for the synthesis of MIPs based on H-bonding interactions [19]. The *modified precipitation polymerization* method allows for the use of a wide variety of solvents for example chlorinated hydrocarbons that were not amenable in the *high dilution method* [14].

Particle size: In general, polymer beads in the 0.1–10 μm range can be produced by precipitation polymerization. The particle size is heavily dependent on the type of crosslinker. By using mixtures of crosslinkers the size can be tuned according to the ratio of the crosslinkers [16, 20]. The particle size slightly increases with the monomer concentration. In *dilute medium* increased initiator concentration will lead to a bigger bead size [11, 18], while in the *modified precipitation polymerization* the effect is the opposite [14]. Interestingly, polymers prepared in the presence or absence of the template might show a pronounced difference in the particle size. The template, depending on its chemical nature, can increase or decrease the particle diameter compared to the non-imprinted polymer. In the *modified precipitation polymerization* increasing the volume of the reaction medium also increases the particle size probably due to the wall effect (unpublished results), while in dilute polymerization media the opposite effect was observed [21].

Polydispersity: In *dilute precipitation polymerization* gentle agitation is usually necessary to obtain monodisperse beads. No agitation, or too high stirring speeds will lead to particle aggregation or coagulation increasing the polydispersity [21]. In the *modified precipitation method* no agitation is performed since the viscosity of the reaction medium becomes very high toward the end of the polymerization. The presence of the template also increases particle uniformity.

Porosity: The porosity of the beads is mainly determined by the solvency of the medium. In *dilute precipitation polymerization* it can be enhanced by adding toluene as co-solvent to the acetonitrile [22]. The porosity might also be affected by the presence of the template. Substantially higher specific surface area can be obtained for the imprinted than for the non-imprinted beads.

Due to the above-mentioned considerations precipitation polymerization protocols should be optimized for each targeted template. This might require the assessment of multiple polymerization conditions that can be most effectively carried out by experimental design [16].

In the followings, we give examples of the above-mentioned two precipitation polymerization methods. The first method illustrates how to prepare uniform spherical imprinted polymer particles for a beta blocker drug, propranolol under *high dilution* conditions. The second one gives an example on the synthesis of imprinted beads with *modified precipitation polymerization* for the selective binding of diclofenac, a non-steroidal anti-inflammatory drug.

2 Materials

2.1 Monomers

The monomers are usually shipped with an inhibitor, which prevents inadvertent polymerization processes. Store the monomers in the presence of air since most inhibitors require oxygen to function. Store the monomers separately from chemicals that might induce polymerization. Inhibitors are usually removed prior to polymerization (*see* **Note 1**). The most frequently used inhibitors are hydroquinone (HQ), and monomethyl ether hydroquinone (MEHQ) for acrylate/methacrylate monomers and 4-*tert*-butylcatechol (TBC) for styrene monomers. Ready to use, disposable prepacked columns and replacement packings for refilling are commercially available for HQ/MEHQ and for TBC (*see* **Notes 2** and **3**). Monomers after inhibitor removal can be stored in the freezer below −20 °C for several weeks or months until polymerization occurs.

1. Methacrylic acid (MAA) 99% contains 250 ppm monomethyl ether hydroquinone (MEHQ) as an inhibitor. Pass MAA through a HQ/MEHQ inhibitor remover column (SigmaAldrich, St. Louis, MO, USA). Store in the freezer below −20 °C.
2. Divinylbenzene (DVB-80) technical grade (80% DVB isomers and 20% ethylvinylbenzene) contains 1000 ppm TBC as an inhibitor. Pass DVB-80 through a TBC inhibitor remover column (SigmaAldrich, St. Louis, MO, USA). Store in the freezer below −20 °C.
3. Trimethylolpropane trimethacrylate (TRIM) technical grade contains 250 ppm MEHQ as an inhibitor. Pass TRIM through a HQ/MEHQ inhibitor remover column. Store in the freezer below −20 °C.
4. 4-vinylpyridine (4-VPy) 95% contains 100 ppm hydroquinone as an inhibitor. Distil 4-VPy freshly prior to use in vacuo (*see* **Note 4**). Store in the freezer below −20 °C under inert atmosphere until discoloration occurs.

2.2 Initiator

1. 2,2′-azobis(2-methylpropionitrile) (2,2′-azobisisobutyronitrile, AIBN) 98%. Re-crystallize AIBN from methanol. Dry under vacuum and store below 10 °C in the refrigerator, protected from light.

2.3 Templates

1. (±)-Propranolol hydrochloride, 99%. Prepare the free base form of propranolol by dissolving the salt in water that contains excess amounts of sodium hydroxide and extract it with *tert*-butyl methyl ether. Separate the organic layer and recover the free base by evaporating the solvent.
2. Diclofenac sodium, analytical standard. Prepare the free acid form of diclofenac by dissolving the salt in water that contains excess amounts of HCl and extract it with *tert*-butyl methyl ether. Separate the organic layer and recover the free acid by evaporating the solvent.

2.4 Solvents

1. Acetic acid glacial, ≥99.7%.
2. Acetonitrile, HPLC gradient grade.
3. Chloroform, HPLC grade.

4. Methanol, HPLC gradient grade.
5. Paraffin oil (PO), Ph. Eur.
6. *tert*-butyl methyl ether, HPLC grade.
7. Toluene, HPLC grade.

2.5 Gases

1. Nitrogen or argon, 99.995%.

3 Methods

3.1 Propranolol MIP and Control Beads by Precipitation Polymerization Under High Dilution Conditions

To prepare the non-imprinted control polymer (*see* **Note 5**) follow the recipe below but omit the template.

1. Weigh 38.9 mg (0.15 mmol) template propranolol (free base form) and 31.2 mg AIBN initiator (3 mol% of the total double bonds) into a screw-cap borosilicate glass vial and add 50.9 μL (0.60 mmol) MAA functional monomer and 513 μL (2.88 mmol) DVB-80 crosslinker (*see* **Notes 6** and **7**).
2. Solubilize the template, monomers, and initiator in the mixture of 3.2 mL toluene and 9.6 mL acetonitrile (25/75 v/v %). The total monomer concentration is 4 w/v %.
3. Homogenize the pre-polymerization solution by placing the vial into an ultrasonic bath for 5 min.
4. Remove the oxygen from the solution by purging with high purity (oxygen-free) argon or nitrogen for 13 min and stirring the vial on a magnetic stirrer meantime (*see* **Note 8**).
5. Place the polymerization vessel into an oil bath thermostated at 60 °C for 24 h. Gently agitate the mixture with a magnetic stir bar at a stirring rate of 120 rpm (*see* **Note 9**).
6. Centrifuge the reaction mixture at 1370 × *g* for 25 min to collect the precipitated particles (*see* **Note 10**).
7. Remove the template (*see* **Note 11**) by incubating the particles with 20 mL methanol/acetic acid = 90/10 v/v % washing solvent for 15–20 min, centrifuge the particles for 10 min at 1370 × *g*, discard the supernatant, and repeat the procedure with a fresh aliquot 10–12 times (*see* **Note 12**). Check the efficiency of the template removal by quantifying the amount of propranolol in the washing solvent with HPLC-UV at 215 nm (*see* **Note 13**).
8. Finally, wash the particles three times with 20 mL methanol to remove traces of the acid.
9. Dry the particles in vacuo overnight (*see* **Note 14** and Fig. 1).

3.2 Diclofenac MIP and Control Beads by the Modified Precipitation Polymerization Technique

To prepare the non-imprinted control polymer (*see* **Note 5**) follow the recipe below without addition of the template.

1. Weigh 47.7 mg (0.15 mmol) template diclofenac (free acid form) and 10.8 mg AIBN initiator (1 w/w% of monomers) into a screw-cap borosilicate glass vial (*see* **Notes 6** and 7).
2. Add 64.7 μL (0.6 mmol) 4-VPy and 958 μL trimethylolpropane trimethacrylate (3 mmol).
3. Deoxygenate ~2 mL chloroform and ~2 mL paraffin oil separately by bubbling argon or nitrogen gas through them for 2 min each (*see* **Note 8**).
4. Dissolve the monomer mixture (~25 w/v %) by adding 1.6 mL chloroform and then 1.6 mL paraffin oil (chloroform/ paraffin oil = 50/50 v/v %) (*see* **Note 15**).
5. Place the vial into an ultrasonic bath for 5 min to homogenize it.
6. Deoxygenate the pre-polymerization solution by bubbling argon or nitrogen through it for 0.5 min (*see* **Note 16**).
7. Place the polymerization vessel into an oil bath thermostated at 60 °C for 24 h (*see* **Note 17**).
8. Transfer the reaction mixture into a centrifuge tube with toluene and centrifuge the reaction mixture at 1370 × *g* for 15 min to collect the precipitated particles (*see* **Note 18**).
9. Wash the collected particles with 30 mL toluene three times to remove the paraffin oil.
10. Wash the particles 12–15 times with 30–30 mL methanol/ acetic acid = 90/10 v/v %, to remove the template (*see* **Notes 11** and **12**). Check the efficiency of template removal by measuring the UV absorbance of diclofenac in the washing solvent at 276 nm.

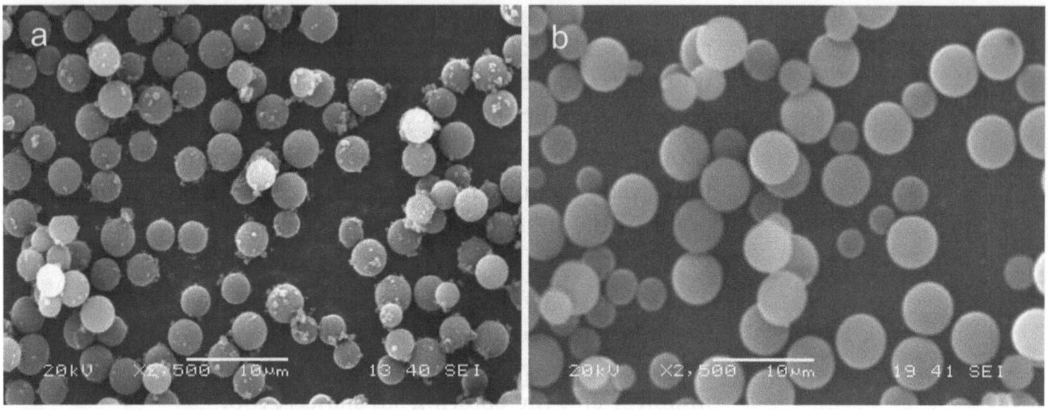

Fig. 1 Propranolol imprinted (**a**) and non-imprinted (**b**) polymer particles prepared by precipitation polymerization under *high dilution conditions*

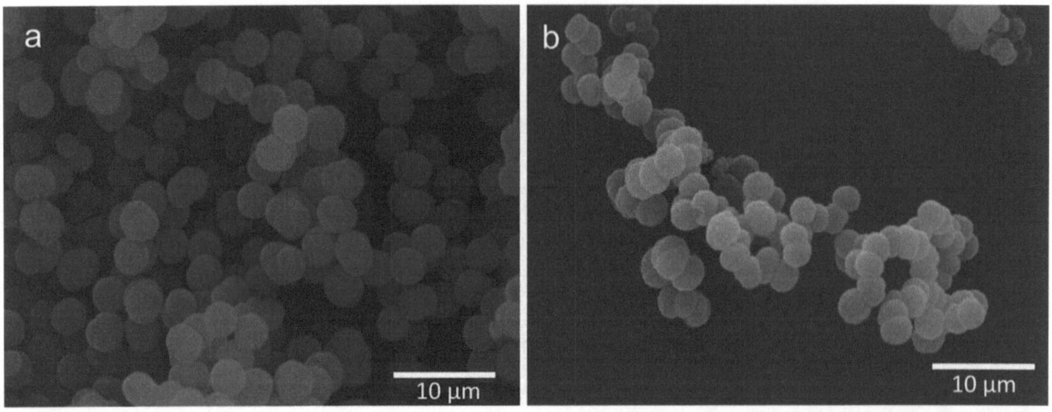

Fig. 2 Diclofenac imprinted (**a**) and non-imprinted (**b**) polymer particles prepared by the *modified precipitation polymerization*

11. Finally, wash the particles three times with 30 mL methanol to remove traces of the acid.
12. Dry the particles in vacuo overnight (*see* **Note 19** and Fig. 2).

4 Notes

1. As the inhibitor is present in low amounts in the monomers it is not always necessary to remove it. The inhibitor will cause some delay in the onset of the polymerization since it will scavenge the first radicals and finally it will be consumed. In the worst case an extra amount of initiator can be added to sustain an acceptable polymerization rate.
2. The commercial inhibitor remover column contains 20–30 g of packing material which is enough to purify 3–4 L of monomer. In our laboratory, we usually need much smaller volumes of monomer. Therefore, we fill portions of the packing material into a smaller column according to the volume of monomers to be purified at a time and dispose them after use according to the manufacturer's instructions.
3. Monomers with low enough boiling point can be alternatively purified by vacuum distillation. Make sure to check the boiling point of the inhibitor whether it can be removed by this way or not.
4. The fresh 4-VPy distillate is a colorless liquid that turns to yellow and later red and brown in the presence of traces of oxygen after a few weeks of storage. Distillation removes the inhibitor and the coloring substances, too.
5. To prove the effectiveness of molecular imprinting a so-called non-imprinted polymer is prepared and the template-binding

property of the MIP is compared to this reference material. As an alternative to the non-imprinted polymer, a control polymer using a non-related template can be prepared. This is a better reference material than the non-imprinted polymer since the template can have a pronounced effect on the polymerization process [23]. It might alter the size of the particles, increase their porosity and uniformity, and suppress the dimerization of the functional monomer [24]. All these will influence the adsorption capacity of the imprinted polymer in a way that is not related to the specific binding sites. Therefore, comparison to the non-imprinted polymer might be misleading.

6. To prepare larger quantities increase the amount of all components proportionally. Bear in mind that this might affect the particle size of the final product.
7. Mixing the template with the functional monomer and the solvent first helps to solubilize templates (or functional monomers) that are otherwise difficult to dissolve in the pre-polymerization mixture. This is due to the complex formation between the template and the functional monomer.
8. Free-radical polymerization is inhibited or retarded by the presence of oxygen that reacts with initiator, primary, and growing polymer radicals and essentially terminates or consumes them. This is manifested in an induction period in the polymerization reaction.

   Deoxygenation can be carried out by purging the pre-polymerization mixture with an inert gas for sufficient time. Agitation will increase the efficiency of this process. As a rule of thumb, we use 1 min purging for each mL of the polymerization solution. One has to keep in mind that evaporation of the solvents is also taking place and this can be problematic in case of volatile solvents or mixtures of nonvolatile and volatile solvents. Cooling the monomer mixture on ice while purging reduces the evaporation loss. When working with volatile solvents the best way is to purge them separately before adding them to the monomer mixture and apply only a short purge afterward.

   An alternative way to remove the oxygen is using repeated freeze-pump thaw cycles in a Schlenk line. In this procedure the pre-polymerization solution is put into a Schlenk tube and frozen in a liquid nitrogen bath while the gas-inlet tab is closed. A vacuum ($<10^{-3}$ mbar) is applied to the frozen system for a few minutes, after which the liquid nitrogen bath is removed. The system is placed under inert gas via opening of the gas inlet tab and subsequently is allowed to defrost. This is repeated two times.

9. Alternatively, the polymerization can be carried out either in a round-bottomed flask connected to a rotavapor (rotavapor method) or in a sealed high-density polyethylene bottle, or glass tube, rolled about its long axis on a low-profile roller in an incubator (low-profile roller method) [19].
10. The result of the polymerization is a white, milky suspension. Alternatively, the particles can be separated by vacuum filtration through a 0.2 μm nylon membrane.
11. The non-imprinted polymer does not need to be subjected to an extensive washing procedure since only the residual initiator, unreacted monomers, oligomers, and polymerization solvent have to be removed. We wash the non-imprinted polymer three to five times with methanol.
12. Alternative methods for template removal are Soxhlet extraction, microwave-assisted extraction, supercritical fluid extraction, pressurized liquid extraction, ultrasound-assisted extraction [25], or washing after vacuum filtration on a membrane filter. Other washing solvents including pure acids (e.g., acetic acid, trifluoroacetic acid, formic acid) or bases (0.01 M sodium hydroxide) can also be used and might be more efficient depending on the nature of the template-monomer system. Do not use more concentrated bases in case of acrylate monomers as they will hydrolyze the ester bonds.
13. As the washing solvent contains acetic acid that strongly absorbs at 215 nm, it is indispensable to separate propranolol from it by chromatography.
14. The yield is around 50–60%. As can be seen from Fig. 1, the MIP particles contain a few secondary particles with much smaller diameter. This was observed by others also using propranolol as a template [20].
15. It is important to add the chloroform first; otherwise, it is difficult to solubilize the monomer solution.
16. Be careful not to purge the pre-polymerization mixture for an extended time as evaporation of the highly volatile chloroform will markedly change the ratio of chloroform to paraffin oil.
17. Agitation is not carried out in the *modified precipitation polymerization* since the polymerization solution becomes very viscous toward the end of the polymerization.
18. The reaction mixture after polymerization is a lard-like, off-white material with a very little liquid phase above the polymer. The product has to be resuspended in toluene to remove the paraffin oil from the particles.
19. The yield is above 95%.

Acknowledgment

The authors gratefully acknowledge the financial support of the Hungarian Scientific Fund (Grant K104724).

References

1. van Herk AM, Monteiro M (2002) Heterogeneous systems. In: Matyjaszewski K, Davis TP (eds) Handbook of radical polymerization. John Wiley and Sons, Inc., Hoboken
2. Flores A, Cunliff D, Whitcombe MJ, Vulfson EN (2000) Imprinted polymers prepared by aqueous suspension polymerization. J Appl Polym Sci 77(8):1841–1850
3. Kempe H, Kempe M (2004) Novel method for the synthesis of molecularly imprinted polymer bead libraries. Macromol Rapid Commun 25(1):315–320
4. Mayes AG, Mosbach K (1996) Molecularly imprinted polymer beads: Suspension polymerization using a liquid perfluorocarbon as the dispersing phase. Anal Chem 68(21):3769–3774
5. Perez N, Whitcombe MJ, Vulfson EN (2000) Molecularly imprinted nanoparticles prepared by core-shell emulsion polymerization. J Appl PolymSci77(8):1851–1859.doi:10.1002/1097-4628(20000822)77:81851::aid-app23>3.0.co;2-j
6. Vaihinger D, Landfester K, Krauter I, Brunner H, Tovar GEM (2002) Molecularly imprinted polymer nanospheres as synthetic affinity receptors obtained by miniemulsion polymerisation. Macromol Chem Phys 203(13):1965–1973. doi:10.1002/1521-3935(200209)203:13<1965::aid-macp1965>3.0.co;2-c
7. Hosoya K, Yoshizako K, Tanaka N, Kimata K, Araki T, Haginaka J (1994) Uniform-size macroporous polymer-based stationary-phase for HPLC prepared through molecular imprinting technique. Chem Lett 8:1437–1438
8. Sellergren B (1994) Imprinted dispersion polymers—a new class of easily accessible affinity stationary phases. J Chromatogr A 673(1): 133–141
9. Ye L, Cormack PAG, Mosbach K (1999) Molecularly imprinted monodisperse microspheres for competitive radioassay. Anal Commun 36(2):35–38
10. Ye L, Weiss R, Mosbach K (2000) Synthesis and characterization of molecularly imprinted microspheres. Macromolecules 33(22): 8239–8245
11. Li K, Stover HDH (1993) Synthesis of monodisperse poly(divibylbenzene) microspheres. J Polym Sci Part A: Polym Chem 31(13):3257–3263. doi:10.1002/pola.1993.080311313
12. Horvath V, Lorantfy B, Toth B, Bognar J, Laszlo K, Horvai G (2009) Preparation of terbutylazine imprinted polymer microspheres using viscous polymerization solvents. J Sep Sci 32(19):3347–3358. doi:10.1002/jssc.2009 00230
13. Renkecz T, Laszlo K, Horvath V (2012) In situ synthesis of molecularly imprinted nanoparticles in porous support membranes using high-viscosity polymerization solvents. J Mol Recognit 25(6):320–329. doi:10.1002/jmr.2153
14. Renkecz T, László K, Horvath V (2014) Molecularly imprinted microspheres prepared by precipitation polymerization at high monomer concentrations. Mol Imprintig 2:1–17. doi:10.2478/molim-2014-0001
15. Goh ECC, Stover HDH (2002) Cross-linked poly(methacrylic acid-co-poly(ethylene oxide) methyl ether methacrylate) microspheres and microgels prepared by precipitation polymerization: a morphology study. Macromolecules 35(27):9983–9989. doi:10.1021/ma0211028
16. Benito-Pena E, Navarro-Villoslada F, Carrasco S, Jockusch S, Ottaviani MF, Moreno-Bondi MC (2015) Experimental mixture design as a tool for the synthesis of antimicrobial selective molecularly imprinted monodisperse microbeads. ACS Appl Mater Interfaces 7(20):10966–10976. doi:10.1021/acsami.5b02238
17. Chen LX, Xu SF, Li JH (2011) Recent advances in molecular imprinting technology: current status, challenges and highlighted applications. Chem Soc Rev 40(5):2922–2942. doi:10.1039/c0cs00084a
18. Shim SE, Yang S, Jin MJ, Chang YH, Choe S (2004) Effect of the polymerization parameters on the morphology and spherical particle size of poly(styrene-co-divinylbenzene) prepared

by precipitation polymerization. Colloid Polym Sci 283(1):41–48. doi:10.1007/s00396-004-1086-3

19. Wang JF, Cormack PAG, Sherrington DC, Khoshdel E (2007) Synthesis and characterization of micrometer-sized molecularly imprinted spherical polymer particulates prepared via precipitation polymerization. Pure Appl Chem 79(9):1505–1519. doi:10.1351/pac200779091505
20. Yoshimatsu K, Reimhult K, Krozer A, Mosbach K, Sode K, Ye L (2007) Uniform molecularly imprinted microspheres and nanoparticles prepared by precipitation polymerization: The control of particle size suitable for different analytical applications. Anal Chim Acta 584(1):112–121
21. Jiang JS, Zhang Y, Guo XZ, Zhang HQ (2011) Narrow or monodisperse, highly cross-linked, and "living" polymer microspheres by atom transfer radical precipitation polymerization. Macromolecules 44(15):5893–5904. doi:10.1021/ma201038e
22. Li WH, Stover HDH (1998) Porous monodisperse poly(divinylbenzene) microspheres by precipitation polymerization. J Polym Sci Part A Polym Chem 36(10):1543–1551. doi:10.1002/(sici)1099-0518(19980730)36:10<1543::aid-pola7>3.0.co;2-r
23. Kempe H, Kempe M (2010) Influence of salt ions on binding to molecularly imprinted polymers. Anal Bioanal Chem 396(4):1599–1606. doi:10.1007/s00216-009-3329-0
24. Zhang YG, Song D, Lanni LM, Shimizu KD (2010) Importance of functional monomer dimerization in the molecular imprinting process. Macromolecules 43(15):6284–6294. doi:10.1021/ma101013c
25. Moreno-Bondi MC, Urraca JL, Carrasco S, Navarro-Villoslada F (2013) Preparation of molecularly imprinted polymers. In: Alvarez-Lorenzo C, Concheiro A (eds) Handbook of molecularly imprinted polymers. Smithers Rapra, Shawbury, Shrewsbury, pp 23–86

Chapter 22

Generation of Janus Molecularly Imprinted Polymer Particles

Xiantao Shen, Chuixiu Huang, and Lei Ye

Abstract

Janus particles, which have two or more distinct physical surfaces, provide a number of potential applications in many fields. However, it is difficult to selectively load chemical or biological components onto the Janus particles. Here, the approach utilizing molecular imprinting and Pickering emulsion technologies for the synthesis of molecularly imprinted polymer (MIP)-based Janus particles (which showed high affinity to the targets) is described stepwise. As self-propelled microengines possessing specific molecular recognition ability, Janus molecularly imprinted polymer particles show high potential for specific and well-controlled drug delivery.

Key words Janus particles, Molecular imprinting, Pickering emulsion, Self-propelled microengine, Drug delivery, Colloidosomes

1 Introduction

Janus particles are a special class of colloidal particles possessing two different surface chemical compositions, which endow us with insights into nature [1]. Janus particles were first addressed by P.G. de Gennes in his Nobel Prize lecture in 1992 [2], and now Janus particles have been widely applied in various areas as solid stabilizers, targeting sensors, interfacial catalysts, imaging probes, and photonic materials because of the anisotropic dual functions [3].

Several methods have been developed to synthesize particles with proper compositions and structures, including surface modification, core–shell synthesis, layer-by-layer coating, and specific polymerization [4–6]. Recently, a molten wax in water Pickering emulsion system has been used to fabricate large quantities of Janus particles with controlled HLB (hydrophilic–lipophilic balance) [7]. Nevertheless, using the above synthesis strategies, Janus particles can be produced in large quantities and with well-controlled dual function.

Thomas Tiller (ed.), *Synthetic Antibodies: Methods and Protocols*, Methods in Molecular Biology, vol. 1575, DOI 10.1007/978-1-4939-6857-2_22, © Springer Science+Business Media LLC 2017

Molecular imprinting technology can endow specific binding cavities on homogenous polymeric particles [8, 9]. The specific binding cavities were created by introducing template molecules during the polymerization and by removing the template molecules afterwards [10–12]. Therefore, molecularly imprinted polymers (MIPs) have been used in many fields such as plastic antibody, separation and purification, enzyme-like catalysis, chemical sensors, drug delivery, and biological receptor systems. Pickering emulsion has attracted increasing attention in molecular imprinting [13], and molecularly imprinted beads have also been prepared using Pickering emulsion [14, 15].

Very recently, Janus MIP particles have been synthesized by a wax-in-water Pickering emulsion, which is stabilized by molecularly imprinted core–shell particles, and the schematic illustration of the step-by-step fabrication of Janus particles is reprinted in Fig. 1 [16]. In this work, monodisperse molecularly imprinted core (MIP-core) particles are synthesized first as presented in step 1 in Fig. 1a using precipitation polymerization described elsewhere

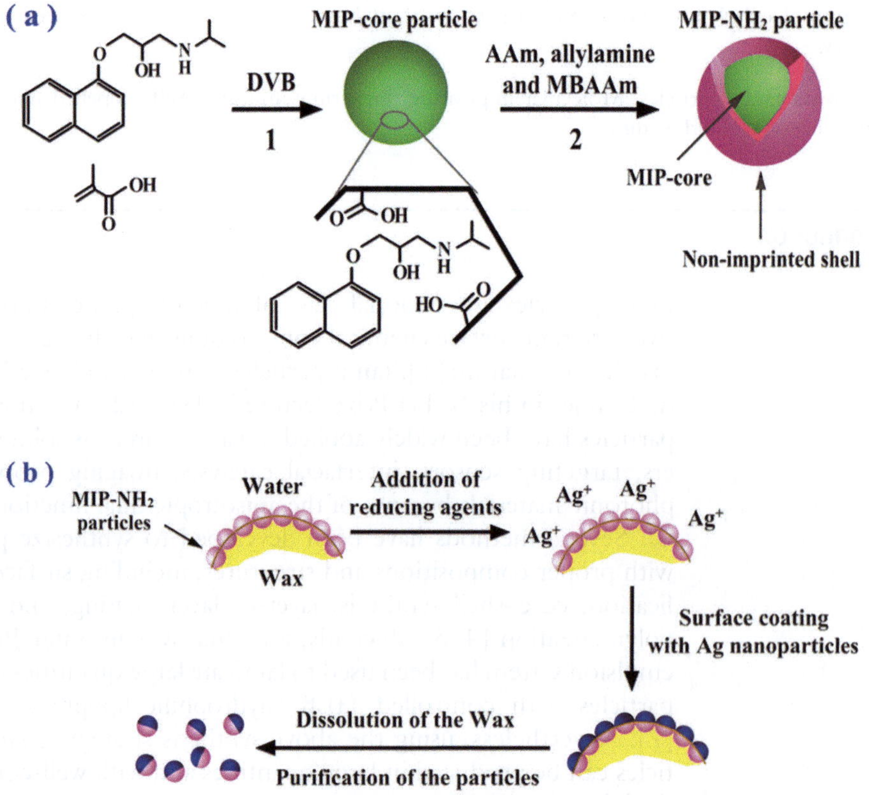

Fig. 1 (**a**) Schematic illustration of the synthesis of monodispersed MIP-NH_2 microspheres. Step 1: generation of propranolol-imprinted sites by precipitation polymerization. Step 2: formation of hydrophilic shell by subsequent copolymerization. (**b**) Step-by-step fabrication of Janus colloidal particles through Pickering emulsion process. Reprinted with permission from (*see* ref. 16) ©The Royal Society of Chemistry 2014

[17]. The amine groups on the MIP particles could facilitate the coating of silver nanoparticles (NPs) onto MIP particles [18]. Thus, after synthesis of MIP-core particles, hydrophilic core–shell molecularly imprinted particles (MIP-NH_2) are produced as shown in step 2 in Fig. 1a. Finally, Ag NPs-coated Janus particles are obtained by oil-in-water Pickering emulsion using MIP-NH_2 as the stabilizer (Fig. 1b).

2 Materials

Milli-Q water is produced by a Milli-Q purification system from EMD *Millipore* Corporation (*Billerica*, MA, *USA*). The reagents and chemicals are stored at room temperature, and are used without further purification (unless specified otherwise). All waste materials are disposed of following the related waste disposal regulations.

2.1 Chemicals

1. Methacrylic acid (MAA, 98.5%) (ACROS, Geel, Belgium). Store at 4 °C.
2. Aluminum oxide (Al_2O_3, 98%) (Sigma-Aldrich, Steinheim, Germany).
3. Divinylbenzene (DVB, technical grade, 55%, mixture of isomers) (Sigma Aldrich, Dorset, UK). Prior to use, DVB is purified by passing through an aluminum oxide column to remove the polymerization inhibitor (*see* **Note 1**). Store at 4 °C.
4. Azobisisobutyronitrile (AIBN, 98%) (Merck, Darmstadt, Germany). AIBN is recrystallized from methanol before use (*see* **Note 2**). Store at 4 °C.
5. Allylamine (98%) (Sigma-Aldrich, Steinheim, Germany). Store at 4 °C.
6. N-isopropylacrylamide (NIPA, 95%) (Monomer-Polymer Laboratories, Windham, USA). Store at 4 °C.
7. N, N'-methylene-bis-acrylamide (MBAAm) (Monomer-Polymer Laboratories, Windham, USA). Store at 4 °C.
8. Acrylamide (98%) (Fluka, Dorset, UK). Store at 4 °C.
9. (R, S)-propranolol hydrochloride (99%) (Fluka, Dorset, UK). It is converted into free base form before use (*see* **Note 3**).
10. (*S*)-propranolol hydrochloride (99%) (Fluka, Dorset, UK).
11. (*R*)-propranolol hydrochloride (99%) (Fluka, Dorset, UK).
12. Atenolol (98%) (Sigma-Aldrich, Gillingham, UK).
13. [3H]-(S)-Propranolol (specific activity 555 GBq $mmol^{-1}$, 66.7 μM in ethanol solution) (NEN Life Science Products, Inc., Boston, MA). Store at 4 °C.

14. Scintillation liquid, Ecoscint A (National Diagnostics, Atlanta, GA). Store at 4 °C.
15. Acetonitrile (99.7%) (Merck, Darmstadt, Germany).
16. Acetic acid (glacial, 100%) (Merck, Darmstadt, Germany).
17. Methanol (HPLC grade) (Merck, Darmstadt, Germany).
18. Paraffin wax with a melting point of 53–57 °C (Sigma-Aldrich, Gillingham, UK).
19. Hydrogen peroxide (H_2O_2, 30%) (Merck, Darmstadt, Germany). Store at 4 °C.

2.2 Solutions

1. 25 mM citric acid: weigh 2.63 g citric acid monohydrate and transfer into 500 mL flask, and make up to 500 mL with Milli-Q water.
2. 25 mM trisodium citrate: weigh 7.35 g trisodium citrate dihydrate and transfer into 1000 mL flask, and make up to 1000 mL with Milli-Q water.
3. 25 mM citrate buffer pH 6: use 25 mM citric acid to adjust the pH of 25 mM trisodium citrate solution to pH 6.
4. Binding buffer: mix 25 mM citrate buffer pH 6 with equal volume of acetonitrile.
5. Washing solution: mix 100 mL acetic acid with 900 mL methanol.
6. 3 M NaOH: weigh 120 g NaOH and dissolve into 1 L Milli-Q water (*see* **Note 4**).
7. $AgNO_3$ (99%, Sigma-Aldrich, Gillingham, UK) solution: weigh 5 mg $AgNO_3$ and dissolve into 5 mL Milli-Q water. Store at 4 °C and protect by light (*see* **Note 5**).
8. $NaBH_4$ (98%, Sigma-Aldrich, Gillingham, UK) solution: weigh 24 mg $NaBH_4$ and dissolve into 12 mL Milli-Q water. Store at 4 °C.
9. Propranolol stock solution (67 μM): weigh 173 mg free-base propranolol and dissolve it into 10 mL Milli-Q water. Store at 4 °C.
10. Atenolol stock solution (67 μM): weigh 178 mg atenolol and dissolve it into 10 mL Milli-Q water. Store at 4 °C.

2.3 Equipment

1. Borosilicate glass tubes (size 25 × 150 mm) with closure caps (Sigma-Aldrich, Dorset, UK).
2. Stovall HO-10 Hybridization Oven (Greensboro, NC, USA).
3. Vacuum chamber (a glass desiccator) connected to a vacuum system.
4. SCILOGEX SK-0330-Pro rotoshaker (SCILOGEX, LLC, Berlin, CT, USA).

3 Methods

3.1 Generation of Molecularly Imprinted Microspheres

1. Weigh 137 mg (0.53 mmol) of the template free-based (R, S)-propranolol and dissolve it in 40 mL of acetonitrile in a borosilicate glass tube equipped with a screw cap.
2. Pipet 110 μL (1.31 mmol) of the functional monomer MAA and 750 μL (5.25 mmol) (*see* **Note 6**) of the purified cross-linking monomer DVB (*see* **Note 1**), and weigh 28 mg of the initiator AIBN. Add all of ingredients into the borosilicate glass tube containing templates solution. Shake the mixture by hand until all ingredients have been dissolved.
3. Non-imprinted microspheres is prepared in 40 mL neat acetonitrile with the same amount of monomers, cross-linking, initiators in another a borosilicate glass tube but without adding the templates.
4. The clear mixture solutions are purged with N_2 for 15 min at room temperature under a gentle flow to remove O_2 and other unwanted elements from the air (*see* **Note 7**).
5. Meanwhile, turn on a hybridization oven (e.g., Stovall HO-10) and set the temperature to 60 °C.
6. After purging with N_2, the two borosilicate glass tubes are capped immediately and fixed horizontally in the hybridization oven, which has been preheated to 60 °C.
7. Start the rotation with a speed of 20 rpm (the effect of centrifugal force can be neglected), and the polymerization is carried out for 24 h at 60 °C.
8. After the desired polymerization time, the two tubes are taken out of the oven, and the polymeric particles are separated by centrifugation with a speed of 9270 × *g* for 10 min.
9. To extract the templates or other elements, the particles (with and without templates) are washed using washing solution (methanol containing 10% acetic acid, v/v) for 8 h. 40 mL of washing solution is used in each step, and the washing steps repeated 6 times.
10. Monitor the templates in the washing solvent from the particles by spectrometric measurement at a wavelength of 278 nm.
11. When no templates could be detected, wash the MIPs and NIPs with acetonitrile for three times (*see* **Note 8**).
12. Dry the MIPs and NIPs in a vacuum chamber (*see* **Note 9**).

3.2 Generation of Hydrophilic Core–Shell Microspheres

1. Pipet 188 μL (2.51 mmol) of allylamine, weigh 566 mg (5.00 mmol) of NIPA, 77.2 mg (0.50 mmol) of MBAAm (*see* **Note 10**), 24 mg of AIBN, and 1000 mg of the MIP (or NIP) particles.

2. Add all into the borosilicate glass tube containing 40 mL acetonitrile.
3. Sonicate the solutions for 3 min.
4. The suspensions are purged with N_2 for 15 min at room temperature under a gentle flow to remove O_2 and other unwanted elements from the air (*see* **Note 7**).
5. Meanwhile, turn on a hybridization oven (e.g., Stovall HO-10) and set the temperature to 60 °C.
6. After purging with N_2, the two borosilicate glass tubes are capped immediately and fixed horizontally in the hybridization oven, which has been preheated to 60 °C.
7. Start the rotation with a speed of 20 rpm (the effect of centrifugal force can be neglected), and the polymerization is carried out for 48 h at 60 °C.
8. After the desired polymerization time, the two tubes are taken out of the oven, and the polymeric core–shell particles are separated by centrifugation with a speed of 9270 × *g* for 10 min.
9. Finally, the core–shell particles (MIP-NH2 and NIP-NH2 particles) are washed with methanol and acetone (*see* **Note 11**).
10. Dry the particles (MIP-NH2 and NIP-NH2 particles) in a vacuum chamber (*see* **Note 9**).

3.3 Generation of Colloidosomes Using Pickering Emulsion

1. Mix 20 mg MIP-NH_2 or NIP-NH_2 particles and 1 mL of acetonitrile into a borosilicate glass tube equipped with a screw cap.
2. Sonicate for 2 min to disperse the MIP-NH_2 or NIP-NH_2 particles in acetonitrile.
3. Add 2 mL of Milli-Q water and 200 μL of 3 M NaOH into the borosilicate glass tube.
4. Sonicate for 2 min to disperse the MIP-NH_2 or NIP-NH_2 particles in the water phase for the Pickering emulsion.
5. Mix the water phase (containing MIP-NH_2 or NIP-NH_2 particles) with 1.5 mL of melt wax (oil phase).
6. Insert the borosilicate glass tube into a water bath preset to 80 °C (*see* **Note 12**).
7. After the wax has completely melted, the hot suspension is shaken vigorously for ~1 min by hand to make a stable Pickering emulsion.
8. Cool down the Pickering emulsion at room temperature (*see* **Note 13**).
9. Harvest the wax colloidosomes containing MIP-NH_2 or NIP-NH_2 particles.
10. Wash with water to neutral pH (*see* **Note 14**).
11. Dry in a vacuum chamber (*see* **Note 9**).

3.4 Generation of Janus Molecularly Imprinted POLYMERS Particles

1. Mix 1 mL of $AgNO_3$ solution (2 mg/mL), 200 mg of wax colloidosomes, and 9 mL of water (*see* **Note 15**).
2. Let the mixture stand still for 2 h at room temperature (*see* **Note 5**).
3. Add 5 mL of NaBH4 solution (2 mg/mL) to the above mixture dropwise (*see* **Note 16**). The suspended mixture immediately turns yellow-green indicating the formation of the Ag nanoparticles.
4. The reaction system is shaken for 5 h at room temperature on a horizontal shaker.
5. Wash the silver-capped colloidosomes with 30 mL water for three times following by washing with 30 mL methanol.
6. Dry the silver-capped colloidosomes in a vacuum chamber (*see* **Note 9**).
7. Add 30 mL of chloroform to the silver-capped colloidosomes to dissolve the wax.
8. Harvest the Janus particles by centrifugation with a speed of $9270 \times g$ for 10 min.
9. Wash the Janus particles with methanol.
10. Dry the Janus particles in a vacuum chamber (*see* **Note 9**).
11. Confirm the coating of Ag NPs on MIP-NH_2 particles by SEM (Fig. 2), and other techniques such as UV–vis analysis and thermogravimetric analysis (TGA) (*see* **Note 17**) (*see* ref. 18).
12. Confirm molecular recognition ability by equilibrium and displacement radioligand binding analysis (Fig. 3) (*see* **Note 18**).

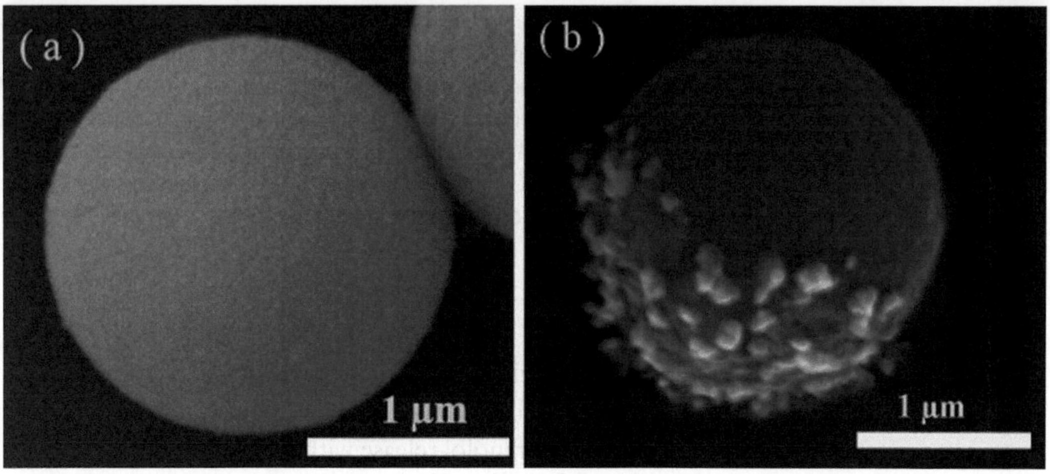

Fig. 2 SEM images of (**a**) MIP-NH_2 and (**b**) Janus MIP particles. Reprinted with permission from (*see* ref. 16) ©The Royal Society of Chemistry 2014

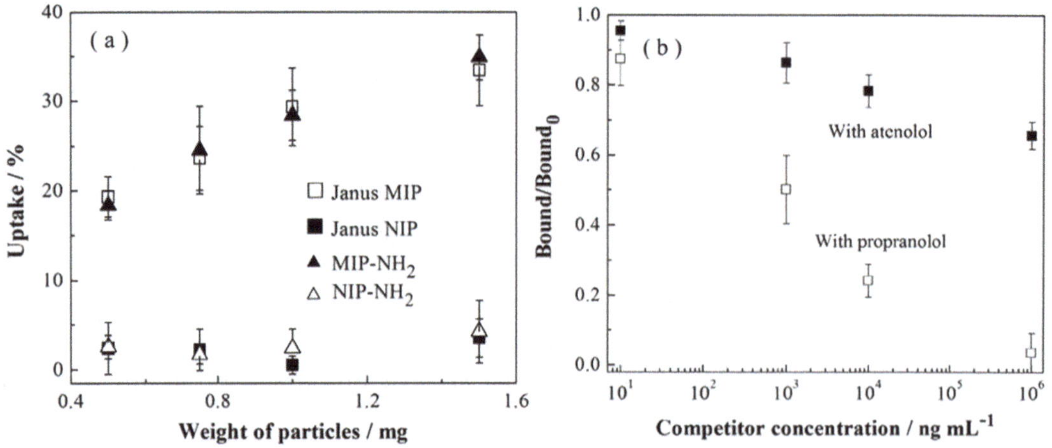

Fig. 3 (**a**) Equilibrium and (**b**) displacement radioligand binding analysis. Reprinted with permission from (*see* ref. 16) ©The Royal Society of Chemistry 2014

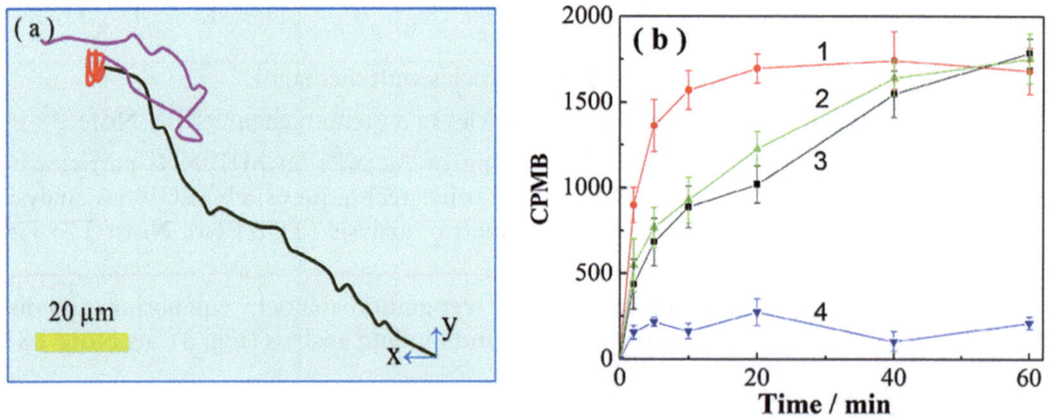

Fig. 4 (**a**) Recorded path of Janus MIP particles in the presence of 2.5% v/v H_2O_2 using real optical microscopy (Green with UV illumination: 0–30 s; Red without UV illumination: 31–40 s; Pink with UV illumination: 41–70 s.) (**b**) Propranolol release curve for the Janus MIP particles in the presence of H_2O_2 and with UV illumination (1), from the Janus MIP particles in the presence of H_2O_2 and without UV illumination (2), from the MIP–NH_2 particles in the presence of H_2O_2 and with UV illumination (3), and from the Janus NIP particles in the presence of H_2O_2 and with UV illumination (4). Counts per minute B (CPMB) is a measurement of the detection rate of ionization events per minute in region B. Reprinted with permission from (ref. 16) ©The Royal Society of Chemistry 2014

13. Study self-propelled movement by real-time optical microscopy (*see* ref. 16) (Fig. 4).
14. Investigate propranolol release with and without UV illumination (*see* ref. 16) (Fig. 4).

4 Notes

1. A syringe (20 mL) coupled with two 0.45 μm Teflon filters. An Al_2O_3 column is prepared by adding 5 g of Al_2O_3 in this syringe. To remove the inhibitors in DVB, 5 mL of DVB is passed through the column.
2. 50 mL of methanol is heated to 55 °C in an Erlenmeyer flask with cover. In this solvent, 5 g of AIBN is transferred. The mixture is stirred for 1 h at room temperature. Then a filtration is performed using filter paper. The undissolved parts are discarded while the filtrate is placed in the refrigerator for 24 h. The crystals of AIBN are separated by a new filtration and dried in the desiccator.
3. Add 1 g of (R, S)-propranolol hydrochloride (99%) into 50 mL saturated Na_2CO_3. Extract free-based propranolol using 30 mL of chloroform (three times). To separate the free-based propranolol, the oil phase is dried with rotary evaporator.
4. The dissociation reaction is highly exothermic and can cause burns if you are not careful. Adding much NaOH rapidly will result in a very hot solution that easily can cause burns. Thus, you should add the NaOH in portions to the water. In addition, wear gloves, goggles, and a lab coat.
5. Generally, silver salts can be decomposed to silver by light and become brown.
6. The amount of cross-linkers (DVB) is 80% of the total monomers to create specific and rigid cavities.
7. To avoid unwanted oxidation of some of the ingredients.
8. No clear peak of propranolol at ~295 nm is found in the UV-vis spectra.
9. The chamber is placed at room temperature. The degree of the vacuum is <40 KPa.
10. The amount of cross-linkers (MBAAm) is only 6% of the total monomers to avoid the effect on the templates access ability and MIPs binding ability.
11. Remove unreacted monomers and oligomers.
12. To keep the wax molten.
13. To solidify the wax.
14. To remove NaOH.
15. The experiment is carried out in a dusk room.
16. Adding the $NaBH_4$ solution too fast may result in Ag precipitation.

17. The thermal decomposition of the particles is conducted in atmospheres to 600 °C.
18. The actual amount used in the assays should be very small. You should practice as much safety precaution as possible. Work in a fumehood and wear double gloves, eye protection, and a special lab coat.

Acknowledgment

This work was supported by the Fundamental Research Funds for the Central Universities in China (2015YGYL024).

References

1. Jiang S, Chen Q, Tripathy M, Luijten E, Schweizer KS, Granick S (2010) Janus particle synthesis and assembly. Adv Mater 22:1060–1071
2. de Gennes PG (1992) Soft matter. Rev Mod Phys 64:645–648
3. Walther A, Müller AHE (2013) Janus particles: synthesis, self-assembly, physical properties, and applications. Chem Rev 113:5194–5261
4. Vestal CR, Zhang ZJ (2002) Atom transfer radical polymerization synthesis and magnetic characterization of MnFe2O4/polystyrene core/shell nanoparticles. J Am Chem Soc 124:14312–14313
5. Caruso F (2001) Nanoengineering of particle surfaces. Adv Mater 13:11–22
6. Xia Y, Tang Z (2012) Chem Commun 48:6320–6336
7. Jiang S, Granick S (2008) Controlling the geometry (janus balance) of amphiphilic colloidal particles. Langmuir 24:2438–2445
8. Zimmerman SC, Wendland MS, Rakow NA, Zharov I, Suslick KS (2002) Synthetic hosts by monomolecular imprinting inside dendrimers. Nature 418:399–403
9. Yu Y, Ye L, Haupt K, Mosbach K (2002) Formation of a class of enzyme inhibitors (drugs), including a chiral compound, by using imprinted polymers or biomolecules as molecular-scale reaction. Angew Chem Int Ed 41:4459–4463
10. Hoshino Y, Koide H, Urakami T, Kanazawa H, Kodama T, Oku N, Shea KJ (2010) Recognition, neutralization and clearance of target peptides in the blood stream of living mice by molecular imprinted polymer nanoparticles: a plastic antibody. J Am Chem Soc 132:6644–6645
11. Shen X, Zhu L, Liu G, Yu H, Tang H (2008) Enhanced photocatalytic degradation and selective removal of nitrophenols by using surface molecular imprinted titania. Environ Sci Technol 42:1687–1692
12. Urraca JL, Aureliano CSA, Schillinger E, Esselmann H, Wiltfang J, Sellergren B (2011) Polymeric complements to the Alzheimer's disease biomarker β-amyloid isoforms Aβ1-40 and Aβ1-42 for blood serum analysis under denaturing conditions. J Am Chem Soc 133:9220–9223
13. Liu C, Song Z, Pan J, Wei X, Gao L, Yan Y, Li L, Wang J, Chen R, Dai J, Yu P (2013) Molecular imprinting in fluorescent particle stabilized Pickering emulsion for selective and sensitive optosensing of λ-cyhalothrin. J Phys Chem C 117:10445–10453
14. Shen X, Ye L (2011) Interfacial molecular imprinting in nanoparticle-stabilized emulsions. Macromolecules 44:5631–5637
15. Shen X, Ye L (2011) Molecular imprinting in Pickering emulsions: a new insight into molecular recognition in water. Chem Commun 47:10359–10361
16. Huang C, Shen X (2014) Janus molecularly imprinted polymer particles. Chem Commun 50:2646–2649
17. Yoshimatsu K, Reimhult K, Krozer A, Mosbach K, Sode K, Ye L (2007) Uniform molecularly imprinted microspheres and nanoparticles prepared by precipitation polymerization: The control of particle size suitable for different analytical applications. Anal Chim Acta 584:112–121
18. Ramajo L, Parra R, Reboredo M, Castro M (2009) Preparation of amine coated silver nanoparticles using Triethylenetetramine. J Chem Sci 121:83–87

Chapter 23

Surface Engineering of Nanoparticles to Create Synthetic Antibodies

Linda Chio, Darwin Yang, and Markita Landry

Abstract

Surface engineering of nanoparticles has recently emerged as a promising technique for synthetic molecular recognition of biological analytes. In particular, the use of synthetic heteropolymers adsorbed onto the surface of a nanoparticle can yield selective detection of a molecular target. Synthetic molecular recognition has unique advantages in leveraging the photostability, versatility, and exceptional chemical stability of nanomaterials. In particular, single-walled carbon nanotubes (SWNT) exhibit a large Stokes shift and near infrared emission for maximum biological sample transparency. Optical biosensors with high signal transduction and molecular specificity can be synthesized with amphiphilic heteropolymers grafted to SWNT, and discovered by high-throughput screening. Herein, we describe the development and the characterization of surface-engineered nanoparticles, or "synthetic antibodies," for protein detection.

Key words Protein detection, Sensors, Carbon nanotubes, DNA aptamers, Infrared fluorescence microscopy, Nanomaterials, Synthetic antibodies

1 Introduction

Sensors detect and relay information through a molecular recognition element and a signal transduction element. Together, molecular recognition and signal transduction constitute the governing principles and limitations by which a molecular sensor operates. For the detection of proteins, antibodies specific to the target analyte are traditionally used as the molecular recognition element, in conjunction with a signal transduction element that is either electrochemical, colorimetric, or optical. While most protein sensors utilize an electrochemical readout, these methods are often unable to detect an analyte with spatial and temporal resolutions which are necessary for biomolecular imaging [1–6].

Single-walled carbon nanotubes (SWNT) offer unique advantages in signal transduction. SWNT have a large Stokes shift exhibiting several hundreds of nanometers between excitation and

Linda Chio and Darwin Yang contributed equally to this work.

Thomas Tiller (ed.), *Synthetic Antibodies: Methods and Protocols*, Methods in Molecular Biology, vol. 1575, DOI 10.1007/978-1-4939-6857-2_23, © Springer Science+Business Media LLC 2017

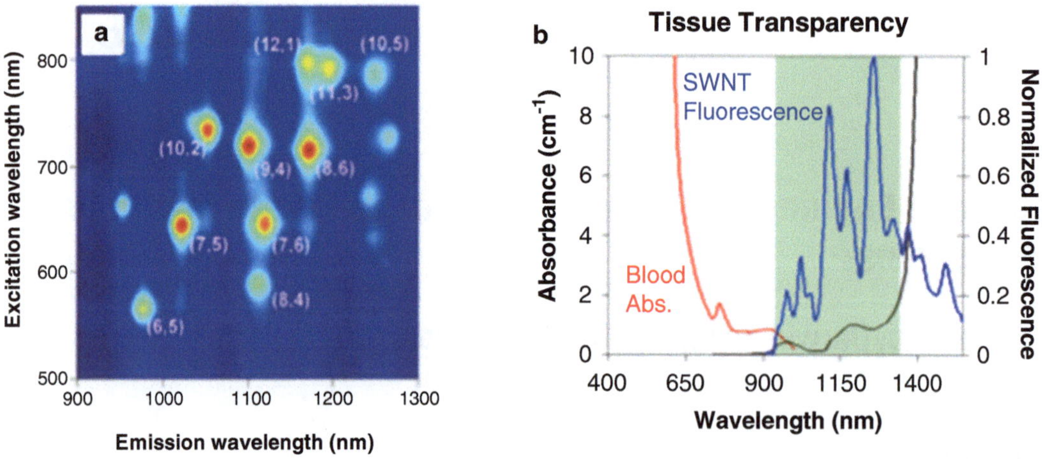

Fig. 1 Fluorescent properties of SWNT. (**a**) Excitation and emission wavelengths of various SWNT chiralities (figure reproduced with permission from [7]). (**b**) Fluorescence of SWNT occurs in a range of wavelengths in which there is an optical window into biological tissues (figure reproduced with permission from [8])

emission wavelengths, an order of magnitude greater than that of a standard fluorophore (Fig. 1a). SWNTs also have high biological sample signal transduction lengths due to their emission in the near-infrared: an optical window beyond the photon scattering range of biological materials and before the optical absorption of photons by water (Fig. 1b).

This chapter focuses on the development of protein biosensors that serve as synthetic, non-biological antibody analogues. Heteropolymers can be engineered to adsorb onto the surface of SWNT, providing the resulting polymer-SWNT hybrid with selective molecular recognition. This approach has successfully produced synthetic sensors for analytes such as riboflavin [9], nitric oxide and hydrogen peroxide, which have gone on to be used in vivo [10, 11], and dopamine [12]. Recently, sensors using this technology have been successfully implemented to produce synthetic antibodies, such as a SWNT-based sensor for the protein fibrinogen [13].

Herein, we outline the materials and methods needed to construct and analyze surface engineered SWNT biosensors for protein detection. We first outline a three-pronged approach for the discovery of synthetic antibodies through screening-based, design-based, and ratiometric nanosensor development. Next we describe the protocols used to synthesize candidate nanosensors. Depending on the stability of the heteropolymer, adsorption to SWNT nanoparticles can be achieved using probe tip sonication, bath sonication, or dialysis. We subsequently discuss the building of essential microscopy equipment (a near-infrared spectrometer, and a near-infrared epifluorescence microscope) to identify, visualize, and utilize synthetic antibodies. Finally, we discuss the screening protocols used to validate our synthetic antibodies.

2 Materials

2.1 Candidate Synthetic Antibodies: Encapsulating Single Wall Carbon Nanotubes with a Library of Different Coronas

1. Raw HiPco SWNT (Unidym).
2. NanoPure Water (Deionized water purified to a sensitivity of 18 MΩ at 25 °C).
3. Bath sonicator.
4. Biocompatible polymers, such as DNA/RNA polymers (Integrated DNA Technologies) suspended in 100 mM NaCl or phospholipid-PEG (Avanti Polar Lipids) (*see* **Note 1**).
5. 0.1 M sodium chloride (NaCl) solution: sodium chloride, NanoPure water. Weigh out 5.84 g of NaCl and dissolve in 1 L of H_2O.
6. Probe tip sonicator.
7. 3 mm probe tip (Cole-Parmer).
8. Cup-horn sonicator.
9. Surfactant, e.g., sodium dodecyl sulfate (Sigma Aldrich) (*see* **Note 2**).
10. Dialysis cartridge with proper molecular weight cutoff for retention of SWNT and polymers, and removal of monomeric surfactant.

2.2 Synthesis of a Screening Library

To ensure selectivity, a sensor must be screened against a library of analytes that is representative of the native environment of the target analyte. For example, when screening for a synthetic antibody for fibrinogen, a protein found in blood, the analyte library will consist of the most abundant proteins in human whole blood: albumin, IgG, fibrinogen, α1-antitrypsin, transferrin, haptoglobin, α2-macroglobulin, IgA, IgM, α1-acid-glycoprotein, apolipoprotein-A1, insulin, hCG, and CRP [13]. Other molecular constituents of the sensor's end-working environments should also be included in the screening analyte library.

1. Phosphate buffered saline (PBS): 137 mM NaCl, 2.7 mM KCl, 10 mM, Na_2HPO_4, 1.8 mM KH_2PO_4, pH 7.4. Weigh out 8 g NaCl, 0.2 g KCl, 1.44 g Na_2HPO_4, and 0.24 g KH_2PO_4 and dissolve in 800 mL of H_2O. Adjust the pH to 7.4 with HCl, and then add H_2O to 1 L Sterilize by autoclaving for 20 min at 15 psi on liquid cycle or by filter sterilization. Store at room temperature (*see* **Note 3**).
2. Fetal bovine serum (FBS): 10% vol PBS.
3. Protein target and other analytes.
4. 96-well plate with transparent glass bottom.

2.3 Near-Infrared Spectrometer (1D)

1. Zeiss AxioVision Inverted Microscope.
2. 785 nm photodiode laser (B&W Tek Inc.).
3. Princeton Instruments InGaAs 1D Detector.
4. PI Acton SP2500 Spectrometer.
5. Optical components: optical posts, post holders, post bases, optical table, broadband near-infrared optimized mirrors, mirror mounts.
6. Irises for beam alignment.

2.4 Near-Infrared Epifluorescence Microscope (2D)

1. Zeiss AxioVision Inverted Microscope.
2. 785 nm photodiode laser (B&W Tek Inc.).
3. Princeton Instruments InGaAs 2D Detector. (The detector to be used is dependent on the level of signal-to-noise needed for your experiments. Several InGaAs-based 2D detectors have sub-300 e/p/s dark current with or without liquid nitrogen cooling, enabling sub-10 ms single-SWNT detection, but might not be a financially feasible option. Alternative 2D detectors are given in Table 1.)
4. Optical components: post, post holder, base.
5. Irises for beam alignment.
6. Near-infrared transmissive microscope objectives.
7. Near-infrared optimized mirrors.

3 Methods

3.1 "Synthetic Antibodies" Development Outline

Surface engineered "synthetic antibodies" can be developed using a three-tiered approach: screening for a synthetic antibody, rational capture-anchor design of a synthetic antibody, or ratiometric detection of analytes (Fig. 2).

Screening Approach: Screening for a protein sensor requires the development of a candidate synthetic antibody library and screening candidate synthetic antibodies against a library of analytes. A candidate synthetic antibody library can comprise of SWNT surface-engineered with different polymers such as polynucleic acids, synthetic peptides and peptoids, amphiphilic heteropolymers, surfactants, and functionalized phospholipids. To create a selective and responsive sensor, a broad range of polymer-SWNT conjugates needs to be screened. The screening approach is the main focus of this chapter.

Design-based Approach: Design-based engineering of synthetic antibodies involves the tethering of polymers with an anchor domain and a capture element onto the surface of a SWNT [14]. Anchor domains must noncovalently and strongly adsorb to the surface of the SWNT often through hydrophobic or pi-stacking

Table 1
Comparison of 2D detectors

Manufacturer model	Photon etc. Zephir 1.7–640	Xenics cougar-640 LN	Princeton NIRvana:640	Photonic science SWIR VGA C	Photonic science SWIR VGAT
Material	InGaAs	InGaAs	InGaAs	InGaAs	InGaAs
Spectral range	0.9–1.7 μm	0.90–1.55 μm	0.9–1.7 μm	1.0–1.7 μm	1.0–1.7 μm
Array size	640 × 512	640 × 512	640 × 512	640 × 512	640 × 512
Pixel size	15 μm × 15 μm	20 μm × 20 μm	20 μm × 20 μm	25 μm × 25 μm	25 μm × 25 μm
Dynamic range	13/15 bit	24 bit	16 bit	16 bit	16 bit
FPS	120/350 HZ	1.42 HZ	110 HZ	22 HZ	22 HZ
Readout noise, gain L/H (e−)	180/45	15	120 (high gain)	380/150	380/150
Dark current	250 e/p/s	10 e/p/s	300 e/p/s	6241 e/p/s	811 e/p/s
Exposure time	1 μs to minutes	12.5 ns–53.7 s	2 μs to minutes	–	–
Full well	600 K/12 K	400 K	40 K/600 K	1.9 M/39 Ke-	1.9 M/39 Ke-
Cooling type	TEC	LN	TEC/water cooling	TEC air cooled	TEC water cooled
Operating temperature	−80 °C	−196 °C	−85 °C	−20 °C	−40 °C
Interface	Camera link	Camera link	GigE	GigE	GigE
QE	>80%	–	>85%	–	–
Pricing	60,000 USD	–	120,000 USD	32,000 USD	34,300 USD

interactions. A known strong anchor domain is a C_{30} nucleic acid sequence [15]. Capture elements have predetermined specificity to the target protein. Unwanted interactions between the capture domain and the SWNT surface could lead to unforeseeable loss of detection. This method is advantageous over a screening approach in that it leverages specific polymer sequences with predetermined affinities for protein analytes, and as such requires a smaller polymer and analyte library at the onset.

Ratiometric Approach: If both or either of the above methods provides a sensitive, but not selective, response for the protein of interest, a ratiometric detection strategy can be developed. A SWNT sensor can be created using a ratio of two distinct intensities from SWNT chiralities with different emission wavelengths. For example, the [5, 6] SWNT chirality emits at 1000 nm and the [6, 7] SWNT chirality emits at 1150 nm. A ratiometric relationship between the

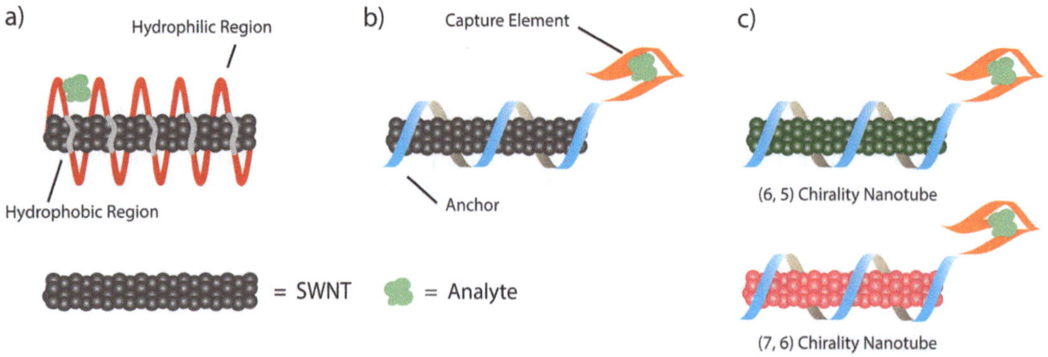

Fig. 2 Three-tiered approach to develop synthetic antibodies. (**a**) In a screening approach, a large library of amphiphilic polymers is templated by the surface of the carbon nanotube for analyte capture. (**b**) In a design-based approach, a polymer is designed with an anchor and capture element. (**c**) In a ratiometric approach, two distinct carbon nanotube chiralities provide a ratiometric response to an analyte

signal of different chiralities can provide additional information about the system by monitoring one signal as invariant to the analyte and the second signal as responsive to the analyte. Ratiometric monitoring can also eliminate baseline noise of different and often biologically complex sensing environments. While chirality-purified SWNT are available commercially, more information about isolating distinct SWNT chiralities can be found in ref. [16].

3.2 Protocol to Conjugate SWNT with Polymer Corona Phases

The method used to conjugate polymers to the surface of the carbon nanotubes depends on two parameters: the robustness of the polymer, and the affinity of the polymer for the SWNT. Probe tip sonication results in the highest SWNT-polymer conjugate yield, is completed in a few minutes, but exposes the polymer to harsh high-power conditions. This method is suitable for structurally robust polymers such as short ssDNA oligonucleotides. Bath sonication gives lower yield than probe-tip sonication, is completed in a few hours, and uses gentler sonication power conditions. Dialysis of a surfactant-suspended SWNT solution with polymer is completed in a few days' time, and involves the gentle exchange of surfactant with polymer on SWNT surface for structurally fragile polymers in which shearing is to be avoided. The three methods are compared in Table 2. Before conjugation, raw carbon nanotube samples require processing to remove impurities often found in SWNT materials (Subheading 3.2.1).

3.2.1 Processing Raw Single-Walled Carbon Nanotubes (SWNT) for Surface Engineering

Raw SWNT must be washed and processed before surface engineering. Washing the carbon nanotubes removes the organic binder, typically methanol, and residual catalyst from carbon nanotube synthesis. Initially 0.8–1.2 nm in diameter and 100 nm–1 μm in length, carbon nanotubes have a mean length of 550 nm after processing. Carbon nanotubes above ~100 nm in length retain their near-infrared fluorescence (*see* **Note 4**).

Table 2

Comparison of the three methods for synthetic antibody development

Method	Probe tip sonication	Bath sonication	Dialysis
Conditions	High-power exposure, significant heating	High-power exposure (gentler than probe tip)	Gradual exchange of surfactant for polymer, on SWNT surface
Time scale	Minutes	Hours	Days
Subheading	3.2.2	3.2.3	3.2.4
Yield	High	Low	High
Recommendations	For robust polymers (DNA and RNA)	For robust polymers prone to shearing, with high affinity for SWNT (peptide capture elements)	For fragile polymers, protein, and antibodies, with low affinity for SWNT (large dsDNA, proteins)

1. Add 200–300 mg of raw SWNT to a 50 mL Falcon tube with 45 mL of deionized water.
2. Vortex the tube at the highest setting for 2 min.
3. Place the tube in a bath sonicator for 5 min to disperse the nanoparticles.
4. Centrifuge for 20 min at 2000 × *g*. Then discard the supernatant.
5. Add 45 mL of fresh deionized water and repeat the vortexing and centrifugation steps (**steps 2** and **4**) 7 additional times.
6. Place the wet SWNT in a drying oven at 120 °C until completely dried (typically 24–48 h).
7. Grind SWNT with a mortar and pestle to homogenize SWNT length.

3.2.2 Probe Tip Sonication to Conjugate SWNT with Polymer Corona Phases

1. In a fume hood, add 1 mg of SWNT to 2 mg of polymer in 1 mL of 0.1 M sodium chloride solution in a 2 mL centrifuge tube.
2. In an ice bath, sonicate the solution with a 3 mm probe tip for 5 min at a power of 4 W (*see* **Notes 5** and **6**).
3. Benchtop-centrifuge the solution for 90 min at 16,100 × *g* (*see* **Notes 7** and **8**).
4. Pipette to collect the top 80–90% of the supernatant for further experimentation being careful not to disrupt the pellet. The pellet can be discarded (*see* **Note 9**).

3.2.3 Bath Sonication to Conjugate SWNT with Polymer Corona Phases

1. In a fume hood, add 1 mg of SWNT to 2 mg of polymer in 1 mL of 0.1 M sodium chloride solution in a 2 mL tube.
2. Bath-sonicate the solution for 1 h (*see* **Note 5**).

3. Benchtop-centrifuge the solution for 90 min at 16,100 × *g* (*see* **Notes 7 and 8**).
4. Pipette to collect and keep the top 80–90% of the supernatant, being careful not to disrupt the pellet. The pellet can be discarded (*see* **Note 9**).

3.2.4 Dialysis to Conjugate SWNT with Polymer Corona Phases

For this protocol, the carbon nanotube is first suspended with sodium dodecyl sulfate. Another surfactant may be chosen as long as it is compatible with the polymer being adsorbed, and smaller in molecular mass than the dialysis membrane chosen.

1. Add 4 g of sodium dodecyl sulfate (SDS) to 150 mL of water (*see* **Note 10**).
2. Add 60 mg of single-walled nanotubes to the aqueous solution (*see* **Note 4**).
3. Add 50 mL of water to the solution.
4. Homogenize the sample for 1 h using a homogenizer on the lowest setting. This will disperse the sample without breaking the SWNT.
5. Sonicate the sample for 10 min using a cup-horn sonicator with an amplitude of 90%.
6. Properly weigh out the sample in separate centrifuge tubes to ensure the centrifuge is balanced.
7. Ultracentrifuge the sample for 4 h at 47,000 × *g*.
8. Pipette to collect the top 80–90% of the supernatant for further experimentation being careful not to disrupt the pellet. The pellet can be discarded (*see* **Note 9**).
9. Add the polymer to the SDS-SWNT suspension to a final polymer concentration of 2 mg/mL.
10. Dialyze the mixture using an appropriate molecular weight cutoff dialysis cartridge against 2 L of water for 4–5 days. Run a concurrent control of a SDS-SWNT only suspension.
11. Change the dialysis buffer every 12 h to completely remove SDS from the SWNT surface. Dialysis is complete when the SWNT from the control precipitates out of solution (*see* **Note 11**).

3.3 Building Near-Infrared Screening and Imaging Microscopy

3.3.1 1D InGaAs Infrared Screening Spectrograph

1. Fix the Zeiss AxioVision Inverted Microscope and 785 nm laser to the optical table.
2. To align the beam, first draw the desired beam path on the optical table. Screw in post holders for the laser and shutter using Fig. 3 as a guide. Fix the laser on the corresponding post loosely so that its position may be adjusted.
3. Align the laser beam using two irises, a set distance apart. Adjust the irises to be at the same height as the microscope

port through which the beam will enter. First, adjust the tilt of the laser so that it passes through the center of the first iris. Next, adjust its height on the stability post so that the beam passes through both irises.

4. Attach post holders to the table at the corners of the beam path for placement of the mirrors. Place the mirrors on the post so that the laser hits the center of the mirror (*see* **Notes 12** and **13**).
5. Align the mirrors two at a time using the irises. Place the irises along the desired beam path ahead of two mirrors. The irises should be at the same height as before and a set distance apart from one another. Adjust the angle of the first mirror until the laser beam passes through the center of the first iris. Next, turn the second mirror until the beam is aligned with the center of the second iris. Repeat these two steps iteratively until the beam is perfectly aligned through the center of both irises.
6. Align the laser with the front and back focal plane of the objective of the inverted microscope. Similar to the previous step. Iteratively align the second mirror from the objective with the crosshair on the front focal plane, and the first mirror with the center of the eyepiece.
7. Fix the appropriate filter cube to the filter wheel of the microscope. For our setup, we use a Chroma RT—Raman 785 nm longpass set (*see* **Note 14**).

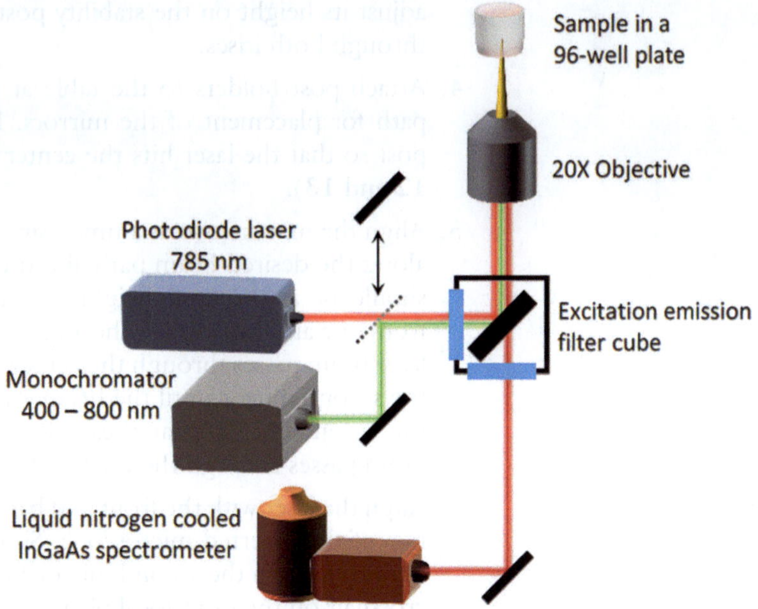

Fig. 3 Schematic of the 1D InGaAs Infrared Spectrometer for screening of synthetic antibody fluorescence response to analyte libraries

8. Ensure the laser line is collimated by reflecting the laser beam from the filter set and allowing the beam to reflect upwards toward the ceiling unperturbed. Laser beam should be collimated, or slightly divergent within a several meter-long path. Laser beam should also be approximately 1 cm in diameter as it enters the microscope port, to overfill the back of the objective (*see* **Notes 15** and **16**).
9. Place a 20× objective with near-infrared transmission capabilities in the microscope nosepiece.
10. Place the spectrometer (IsoPlane, Princeton Instruments) in line with the microscope side port and place a lens in the beam path such that the emitted light from the sample plane is directed into the entrance slit of the spectrometer. Attach a 1D InGaAs array (PyLoN-IR, Princeton Instruments) to the camera port of the IsoPlane spectrometer.

3.3.2 2D InGaAs Infrared Imaging Microscope

1. Fix the Zeiss AxioVision Inverted Microscope and 785 nm laser to the optical table.
2. To align the beam, first draw the desired beam path on the optical table. Screw in post holders for the laser and shutter using Fig. 4 as a guide. Fix the laser on the laser mounting post loosely so that its position may be adjusted.
3. Align the laser beam using two irises, a set distance apart. First, adjust the irises to be at the same height as the microscope port through which the beam will enter. Adjust the tilt of the laser so that it passes through the center of the first iris. Next, adjust its height on the stability post so that the beam passes through both irises.
4. Attach post holders to the table at the corners of the beam path for placement of the mirrors. Place the mirrors on the post so that the laser hits the center of the mirror (*see* **Notes 12** and **13**).
5. Align the mirrors two at a time using the irises. Place the irises along the desired beam path ahead of two mirrors. The irises should be at the same height as before and a distance apart from one another. Adjust the angle of the first mirror until the laser beam passes through the center of the first iris. Next, turn the second mirror until the beam is aligned with the center of the second iris. Repeat these two steps iteratively until the beam passes through the center of both irises.
6. Align the laser with the front and back focal plane of the objective of the inverted microscope. Similar to the previous step, iteratively align the second mirror from the objective with the crosshair on the front focal plane, and the first mirror with the center of the eyepiece.

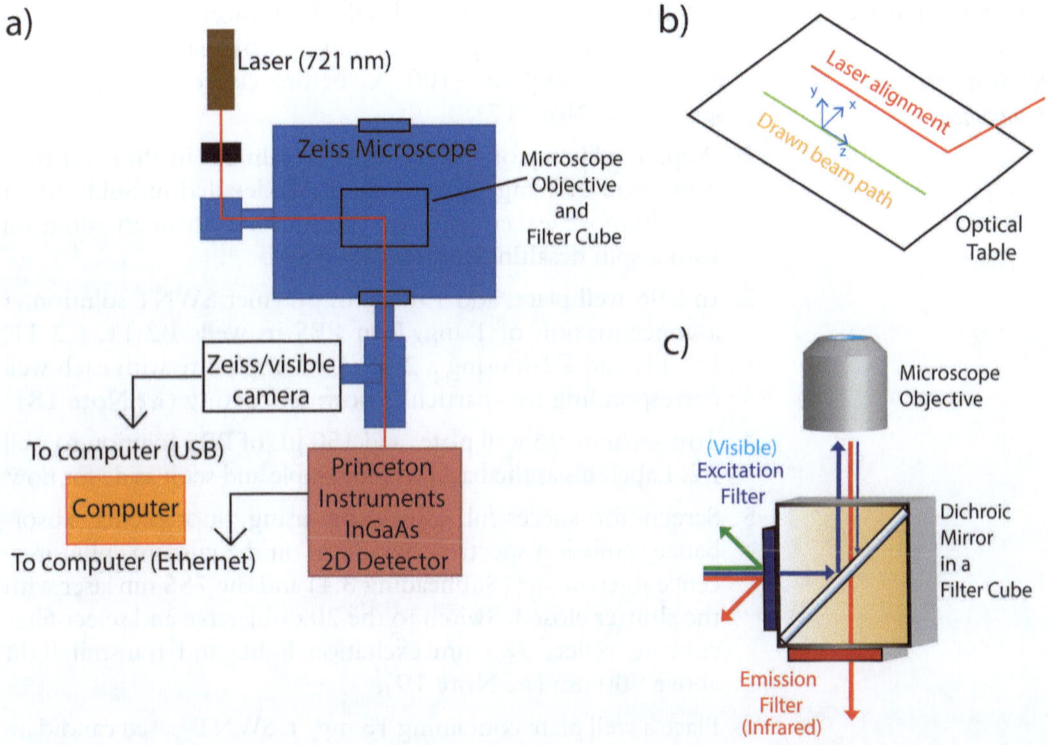

Fig. 4 (**a**) Schematic of the set up for a 2D InGaAs Infrared Screening Spectrograph to screen for synthetic antibodies. (**b**) The beam path is drawn before laser alignment. (**c**) An appropriate filter cube enables excitation with visible wavelengths and enables emission of near infrared wavelengths should be chosen for imaging of SWNT-based synthetic antibodies

7. Fix the appropriate filter cube to the filter wheel of the microscope. For our setup, we use a Chroma RT—Raman 785 nm longpass set (*see* **Note 14**).
8. Ensure the laser line is collimated by reflecting the laser beam from the filter set and allowing the beam to reflect upwards unperturbed. Laser beam should be collimated, or slightly divergent within a several meter-long path. Laser beam should also be approximately 1 cm in diameter to overfill the back of the objective (*see* **Notes 15** and **16**).
9. Place a 100× objective in the microscope nosepiece, ensuring the objective is near-infrared transmissive.
10. Put a 2-camera adapter at the exit port of the microscope Place the Princeton Instruments InGaAs 2D Detector (NIRvana 640ST) on the optical table. Mount the camera using a C-mount camera adapter to fit the exit port of the microscope and the 1-in. threaded mount of the 2D detector.
11. If a brightfield image is desired, mount a brightfield camera such as a Zeiss axiocam on the top line of the 2-camera adapter.

3.4 Spectroscopic Screening for Synthetic Antibodies

1. Fill the nIR detector (PyLoN-IR, Princeton Instruments) coolant tank with liquid N_2. Wait approximately 3 h for the detector to cool to −100 °C before collecting absorbance spectra (*see* **Note 17**).
2. Prepare a library of polymer-SWNT samples in PBS at a concentration of 1 mg/L using methods detailed in Subheading 3.1. Remove excess polymer in solution through filtration with a spin desalting column, or dialysis.
3. In a 96 well plate, add 198 μL of polymer-SWNT solution at a concentration of 1 mg/L in PBS to wells B2-11, C2-11, D2-11, and E2-8 using a 200 μL micropipette with each well corresponding to a particular polymer coating (*see* **Note 18**).
4. To a separate 96 well plate, add 150 μL of PBS solution to well B2. Label this as the background sample and set it aside for now.
5. Screen for successful suspension using fluorescence absorbance/emission spectroscopy. Turn on the custom epifluorescence microscope (Subheading 3.4) and the 785 nm laser with the shutter closed. Switch to the 20× objective and select filter cube to reflect 785 nm excitation light, and transmit light above 900 nm (*see* **Note 19**).
6. Place a well plate containing 15 mg/L SWNT-based candidate sensor on the motorized translation stage of the microscope with the first well, B2, positioned over the 20× objective. The objective should be positioned far from the bottom of the well plate.
7. Collect several sequential test spectra using an exposure time of 1000 ms. Move the objective upwards toward the bottom of the well plate slowly until maximum emission is observed. The software used for data acquisition is camera specific; however, for Princeton Instruments detectors, LightField software interfaces well with the detector. Once you achieve maximum fluorescence, keep the objective at that working distance (*see* **Note 20**).
8. Move to an empty well within your well plate. Collect a background spectrum using the exposure time from the previous step, with the objective positioned at its optimal distance from the objective.
9. Place the original sample well back onto the microscope objective. Run the automation script to collect fluorescence spectra for all wells using the same optimal exposure time. Verify that the background has been subtracted from all sample spectra. This is the initial intensity spectrum for each of your candidate sensors.
10. Remove the plate from the microscope. Using a 10 μL micropipette, add 2 μL of each analyte from the 10 mM stock solu-

tion in PBS to the corresponding polymer conjugated SWNT in the well plate. Gently stir the well plate in a circular motion while pushing the plunger in and out to thoroughly mix the solution. The final concentration of analyte in each well is 100 μM (*see* **Note 21**).

11. Place the well plate back onto the microscope. Run the automated script to scan and collect fluorescence spectra for all wells using the same exposure time as in **step 8**. These fluorescence spectra represent the fluorescence modulation incurred by each analyte.
12. Cover the well plate using a 5 cm × 5 cm sheet of Parafilm and incubate on a tabletop orbital shaker at room temperature for 1 h.
13. Remove the Parafilm and place the plate on the optical table and repeat the screening protocol to collect fluorescence spectra for each sample. These fluorescence spectra represent any time-dependent fluorescence modulation incurred on the SWNT sensor by each analyte.
14. For each well, calculate the change in fluorescence intensity at a particular wavelength upon addition of analyte, and 1-h post analyte addition (Fig. 5a). Plot these results in a heat map to determine which of your sensor-analyte pairs exhibit a significant and selective response (Fig. 5b). Candidate sensors responding selectively and strongly to the addition of a protein analyte can be used as synthetic antibodies.

3.5 Fluorescence Imaging of "Synthetic Antibodies" Using Near-Infrared Microscopy

1. Dilute polymer-SWNT conjugate solutions to a final SWNT concentration of 2 mg/L (*see* **Note 22**).
2. Clean and oxidize a glass #1.5 coverslip using a series of rinses: methanol, ethanol, acetone, and water.
3. Fix the coverslip onto the adhesive surface of a microfluidics channel slide (Ibidi, Martinsried, Germany). Only the nanotubes immobilized on the surface of the coverslip will be imaged so cleaning of this slide is not necessary.
4. Wash the channel by flowing through 50 μL of PBS solution using a micropipette.
5. Coat the surface of the coverslip with 50 μL of biotin labeled BSA (1 mg/mL) taking care not to introduce any air bubbles into the channel. Incubate for 5 min at room temperature.
6. Rinse with 100 μL PBS solution to remove any unbound BSA-Biotin.
7. Add 50 μL of 0.2 mg/mL neutravidin in PBS and let incubate for 5 min. This step sparsely populates the coverslip with SWNT binding sites.
8. Rinse again with 100 μL PBS.

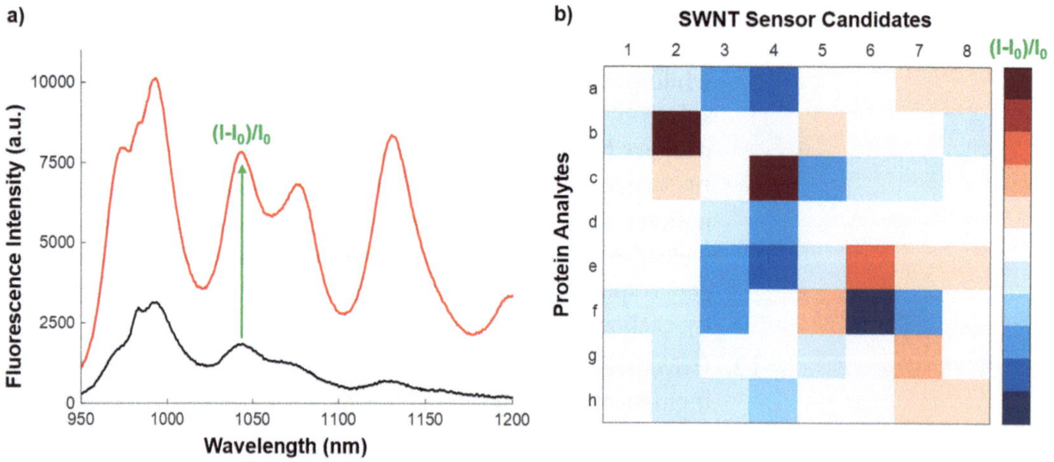

Fig. 5 Result of spectroscopic screening for a sample analyte-SWNT candidate nanosensor. (**a**) Fluorescence spectrum for a representative SWNT-polymer conjugate before (*black*) and after (*red*) addition of an analyte shown to induce fluorescence modulation. The peak at 1044 nm, for instance, can be used to calculate the sensor signal, $(I-I_0)/I_0$. (**b**) Simulated heat map of sensor response to the library of protein analytes. Candidate sensors with a significant turn-off or turn-on response to a single analyte are deemed selective synthetic antibodies, such as Protein B + candidate SWNT sensor 2, or Protein C + candidate SWNT sensor 4

9. Add 50 μL of the synthetic antibody solution to the coverslip and incubate for 5 min at room temperature. Rinse with only 50 μL PBS to prevent washing away any immobilized nanotubes (*see* **Note 23**).
10. Rinse a final time with 50 μL PBS to remove any remaining unbound synthetic antibodies. The slide is now ready to be imaged which should be done as soon as possible after completion of this step. If this is not possible, the sample can be kept refrigerated for 1–2 days.
11. Place the microscope slide on a 100× near-infrared transmissive objective with oil immersion, onto the 2D InGaAs imaging microscope with 500 ms exposure time. Slowly move the objective toward the microscope slide until surface-immobilized synthetic antibodies come into the field of view in focus (*see* Fig. 6b) (*see* **Note 24**).
12. Begin acquiring data by acquiring a sequence of frames. Make note of the frame at which you add 50 μL of protein analyte at a concentration of 10 nM. Your synthetic antibodies should respond similarly to the response recorded in the screening heat map (Fig. 5b).
13. Calculate the relative change in synthetic antibody intensity upon protein analyte addition, $((I-I_0)/I_0)$. This value will likely be higher than the value determined in Fig. 5b.

This surface can also be utilized to detect protein analytes from biological samples. For instance, one might introduce biological cells into the microscope slide and monitor the infrared response of surface-immobilized sensors as a measure of protein secretion from the cell [14]. Analyte binding can be verified by incorporating a visible single-molecule fluorescence line into the near-infrared fluorescence microscope setup as described in [17].

4 Notes

1. Polymers used generally will not exhibit affinity for a target protein prior to conjugation to SWNT.
2. Dry surfactants should be handled in a ventilated hood in case of aerosolization. Solid sodium dodecyl sulfate is flammable and should be handled in a fume hood away from sparks and flames.
3. Concentrated HCl (12 N) can be used at first to narrow the pH gap from the starting pH to the required pH. When pH is within a pH unit of the required final pH, use a series of HCl stock solutions (e.g., 6 N and 1 N) with lower ionic strengths, added dropwise, to avoid a sudden drop in pH below the required pH. Observe caution when handling caustic materials such as highly concentrated HCl.
4. Always work in a properly ventilated space, such as a fume hood or a glove box, when handling nanoparticles to avoid inhalation.
5. The SWNT initially forms small clumps in solution. If SWNT is successfully suspended by the polymer, the solution should have higher optical density and greater homogeneity.
6. Be sure that the sonicator tip is not touching the bottom or sides of the tube. Exercise caution after sonication; sonicator tip will be hot to the touch.

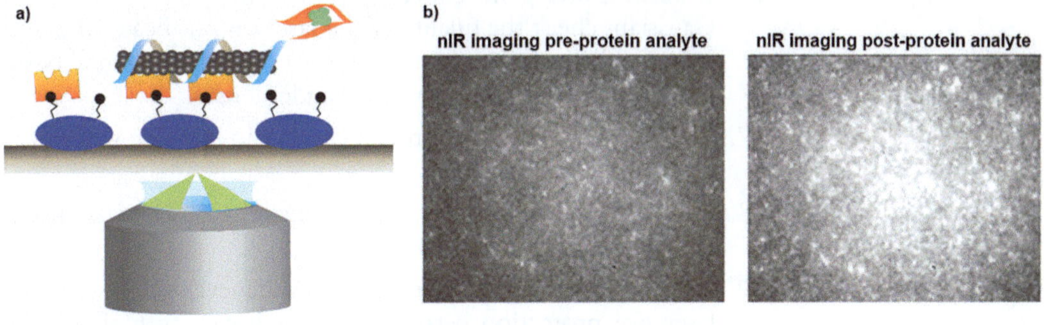

Fig. 6 Imaging of immobilized synthetic antibodies. (**a**) Schematic of excitation and emission of polymer-SWNT fixed to a coverslip. (**b**) The addition of protein analyte causes a detectable increase in the fluorescence intensity of the synthetic sensors. This can be used for the localization of analytes on a single molecule level on hour-long timescales, enabled by the theoretically infinite fluorescence photobleaching lifetime of SWNT

7. Always balance the centrifuge before operation so that the tube masses are within 0.05 g of each other.
8. If the supernatant appears clear with all SWNT in the pellet after a full 90 min of sonication, repeat the experiment and centrifuge in 5 min intervals to check stability of the polymer-SWNT suspension.
9. As soon as centrifugation ends, the pellet will begin to dissolve into the supernatant. Therefore, quick decanting is required to obtain solution of only single nanotubes, rather than tube bundles.
10. For best purity, use NanoPure or Milli-Q filtered water.
11. The complete removal of SDS from the SWNT surface allows for the adsorption of the polymer. The resulting polymer suspensions should be clear to the eye and free of SWNT aggregates.
12. When aligning the mirrors, keep in mind that the actual position of the mirror will be offset by the post.
13. Use mirrors with extended wavelength ranges that allow the passage of high wavelength beams (for the near-infrared you will need reflection of wavelengths of 700 nm and above). Some companies' standard mirrors are only optimized to manipulate light in the visible range, up to 700 nm.
14. Many filter sets have not been optimized for infrared emission. Make sure to check that the emission filter is capable of transmitting light beyond 1000 nm, specifically for the 950–1300 nm emission range of most SWNT chiralities. Check the product page of the filter cube for %Transmission using a wavelength range between 750 and 1200 nm.
15. If laser beam is smaller than 1 cm, place a beam expander in the laser alignment path along the optical table and repeat the process of iterative alignment.
16. If beam is divergent, convergent, or if the beam profile is not Gaussian, check the quality of the laser output beam, the mirrors and dichroics used for beam steering, any accidental beam path occlusions, or the beam expander lenses.
17. Thermal gloves and eye protection should be worn while handling liquid nitrogen.
18. For experiments in serum, dilute polymer-SWNT to a concentration of 5 mg/L in FBS.
19. 785 nm light is damaging to eyes. Whenever the laser is on, laser eye protection certified for this wavelength should be worn.
20. The lights in the room should be turned off while the microscope is running.

21. Protein analytes should be frozen and thawed before use. Solutions should be prepared fresh immediately before experimentation.
22. To determine polymer-SWNT concentration, take absorption spectrum of supernatant with a UV-Vis spectrometer. Concentration is roughly equal to (absorbance at 632 nm)/(0.036 L mg^{-1} cm^{-1}).
23. Longer incubation times can be used to increase SWNT adsorption. If this is required, the channel openings should be capped to prevent evaporation of the solution and formation of air bubbles.
24. If there is insufficient SWNT binding to the surface, increase the concentration of neutravidin in **step 7**.

Acknowledgments

This work was supported by Burroughs Wellcome Fund Career Award at the Scientific Interface (CASI), a Simons Foundation grant, and a Brain and Behavior Research foundation young investigator grant. D.Y. acknowledges an NSF Graduate Research Fellowship and L.C. acknowledges a LAM research fellowship. L.C. wrote the introduction, Subheadings 3.1 and 3.2. D.Y. wrote Subheadings 3.3–3.5. L.C. and D.Y. collaboratively wrote support commentary and edited the manuscript with guidance from M.L. We thank Roger Chang for insightful feedback on this work.

References

1. Cao Y, Zhu S, Yu J, Zhu X, Yin Y, Li G (2012) Protein detection based on small molecule-linked DNA. Anal Chem 84:4314–4320
2. Luo X, Freeman C, James T, Davis JJ (2013) Ultrasensitive label free electrical detection of insulin in neat blood serum. Anal Chem 85:4129–4134
3. Zhang B, Liu B, Tang D, Niessner R, Chen G, Knopp D (2012) DNA-based hybridization chain reaction for amplified bioelectronics signal and ultrasensitive detection of proteins. Anal Chem 84:5392–5399
4. Taleat Z, Cristea C, Marrazza G, Mazloum-Ardakani M, Sandulescu R (2014) Electrochemical immunoassay based on aptamer-protein interaction and functionalized polymer for cancer biomarker detection. J Electrochem Soc 717(718):119–124
5. Shi H, He X, Wang K, Wu X, Ye X, Guo Q, Tan W, Qing Z, Yang X, Zhou B (2011) Activatable aptamer probe for contrast-enhanced in vivo cancer imaging based on cell membrane protein-triggered conformation alteration. Proc Natl Acad Sci 108(10):3900–3905
6. Sriwichai S, Baba A, Phanichphant S, Shinbo K, Kato K, Kaneko F (2010) Electrochemically controlled surface plasmon resonance immunosensor for the detection of human immunoglobulin G on poly(3-aminobenzoic acid) ultrathin films. Sens Actuators B 147:322–329
7. Miyauchi Y, Chiashi S, Murakami Y, Hayashida Y, Maruyama S (2004) Fluorescence spectroscopy of single-walled carbon nanotubes synthesized from alcohol. Chem Phys Lett 387(1–3):198–203
8. Boghossian AA, Zhang J et al (2011) Near-infrared fluorescent sensors based on single-walled carbon nanotubes for life sciences application. Chem Sus Chem 4:848–863
9. Zhang J, Landry MP, Barone PW, Kim JH et al (2013) Molecular recognition using corona phase complexes made of synthetic polymers

adsorbed on carbon nanotubes. Nat Nanotechnol 8:959–968
10. Giraldo JP, Landry MP et al (2015) A ratiometric sensor using single chirality near-infrared fluorescent carbon nanotubes: application to in vivo monitoring. Small 11(32):3973–3984
11. Giraldo JP et al (2014) Plant nanobionics approach to augment photosynthesis and biochemical sensing. Nat Mater 13:400–408
12. Kruss S, Landry MP et al (2014) Neurotransmitter detection using corona phase molecular recognition on fluorescent single-walled carbon nanotube sensors. J Am Chem Soc 136:713–724
13. Bisker G et al (2016) Protein targeted corona phase molecular recognition. Nat Commn 7:10241. doi:10.1038/ncomms10241
14. Landry MP et al (2017) Single-molecule detection of protein efflux from microorganisms using fluorescent single walled carbon nanotube sensor arrays. Nat Nanotechnol. doi:10.1038/nnano.2016284
15. Landry MP et al (2015) Comparative dynamics and sequence dependence of DNA and RNA binding to single walled carbon nanotubes. J Phys Chem C Nanomater Interfaces 119(18):10048–10058
16. Liu H et al (2015) Large-scale single-chirality separation of single-wall carbon nanotubes by simple gel chromatography. Nat Commn 2:309. doi:10.1038/ncomms1313
17. Beyene AG et al (2016) Nanoparticle-templated molecular recognition platforms for detection of biological analytes. Curr Protoc Chem Biol 8:197–223. doi:10.1002/cpch.10

Chapter 24

H5N1 Virus Plastic Antibody Based on Molecularly Imprinted Polymers

Chak Sangma, Peter A. Lieberzeit, and Wannisa Sukjee

Abstract

Normally, antibodies against influenza A have been prepared from viable virus or an engineered strain in certain hosts or cultured media. Two factors concerning antibody production are obvious. The obtaining antibody that is a kind of biomolecule has to be handled carefully, e.g., to be kept in a refrigerator. Furthermore, when the virus strain is highly pathogenic, such as H5N1, antibody production has to be done carefully in a high-level biosafety lab. Here, we show how to produce an antibody against H5N1 from a polymeric material using inactivated virus which can be conducted in a low-level biosafety lab. The process is based on imprinting the whole virus on a polymer surface to form molecularly imprinted polymers (MIPs). The MIPs show some properties of H5N1 antibody as they recognize H5N1 and have some important antibody activity. The H5N1 MIPs are not to be considered biomaterial, so they can be stored at room temperature and thus do not need any special care.

Key words Plastic antibody, H5N1 influenza virus, Molecularly imprinted polymers, Surface imprinting, H5N1 MIPs, Synthetic H5N1 antibody

1 Introduction

Influenza A virus infects people worldwide every year as a seasoning flu. Occasionally, under some conditions it could become a pandemic outbreak [1, 2]. When an outbreak strain is highly pathogenic the disease can kill millions of people as it happened in 1918 from influenza A H1N1 [3–5]. Few years ago, influenza A H5N1 virus has spread mainly among poultry farms and wild birds in Asia [6–8]. It was found that H5N1 is a very highly pathogenic strain and can infect humans, although human-to-human transmission is limited [9]. Nonetheless, people fear H5N1 could evolve to become a human virus [10]. To prevent or minimize a casualty on such a global scale researchers have been working on this strain [11, 12]. Designing and making effective antibodies against H5N1 ("H5N1 antibody") is among important topics in this research area. Better understanding in the production of H5N1 antibody

Thomas Tiller (ed.), *Synthetic Antibodies: Methods and Protocols*, Methods in Molecular Biology, vol. 1575,
DOI 10.1007/978-1-4939-6857-2_24, © Springer Science+Business Media LLC 2017

and the related knowledge would benefit other disciplines such as virus diagnostics or vaccine production.

Owing to the high pathogenicity of the strain, conventional procedures from direct cell culture to reverse engineering to make H5N1 antibody from viable virus have to be carried out in a high biosafety level laboratory [13]. Methods that can produce H5N1 antibody under less restricted laboratory conditions using inactivated virus are always attractive. Based on the functionality, we can consider an antibody against a corresponding pathogen as a material that can recognize the pathogen and inhibits pathogen activity. According to this, molecularly imprinted polymers (MIPs), which are produced by imprinting an antigen on a polymer system, can be regarded as the corresponding antibody to that antigen provided the MIPs have the required properties. To get such MIPs, imprinted cavities left on a polymer surface are expected to form by self-organization of monomers induced by the antigen of interest during pre-polymerization. Here, we demonstrate how to produce MIPs as plastic antibody against H5N1. The copolymer system from four types of monomers is optimized for inactivated virus imprinting as previously reported [14]. Preparation methods are bulk-imprinting suspension copolymerization to form polymer-beads and surface imprinting by virus stamping to form thin-film plastic antibody. These materials can be applied as a platform for virus classification and virus-binding agent screening.

2 Materials

1. Phosphate buffer saline (PBS; 1×), pH 7.4: Dissolve 8.00 g of sodium chloride, 0.20 g of potassium chloride, 1.14 g of disodium hydrogen orthophosphate, and 0.20 g of potassium dihydrogen orthophosphate in 900 mL deionized water. Adjust pH to 7.4 with hydrochloric acid before making the volume up to 1 L with deionized water. Store at 4 °C.
2. Inactivated H5N1 (A/open-billed stork/Nakhonsawan/BBD0104F/04) (*see* **Note 1**).
3. Monomers: Acrylamide (AAM), methyl methacrylate (MMA), methacrylic acid (MAA), and n-vinylpyrrolidone (Merck, Kenilworth, NJ, USA) (*see* **Note 2**).
4. Initiator: Azobisisobutyronitrile (AIBN) (Sigma-Aldrich Chemical Company, St. Louis, MO, USA).
5. Cross-linking agent: Ethylene glycol dimethacrylate (EGDMA) (Merck) for plastic antibody beads. N,N′-(1,2-dihydroxyethylene)bisacrylamide (DHEBA) (Tokyo Chemical Industry; TCI, Chuo-ku, Tokyo, Japan) for surface plastic antibody.

6. Washing solution: 10% acetic acid in 10% sodium dodecyl sulfate (SDS) (*see* **Note 3**).
7. Piranha solution: 3:1 mixture of conc. sulfuric acid and 30% hydrogen peroxide.

3 Methods

3.1 Plastic Antibody Bead Synthesis

1. Mix 13 mg of AAM, 6 μL of n-vinylpyrrolidone, 6.5 μL of MMA, 10.6 μL of MAA, and 500 μL of acetone in 1.5 mL microcentrifuge tube. Add 10 μL of inactivated H5N1 virus to the mixture, follow by 1 mg of AIBN and 100 μL of EGDMA.
2. Stir with magnetic stirrer for 10 min and incubate for 20 h at 4 °C (*see* **Note 4**).
3. Polymerization of the mixture under 254 nm UV light forl h. The plastic antibody will be precipitated as white beads.
4. Grind the beads in an agate mortar to obtain powder plastic antibody. Transfer the powder to a 1.5 mL microcentrifuge tube (*see* **Note 5**).
5. Wash the plastic antibody beads with 1 mL of washing solution, 1× PBS, and deionized water by using magnetic stirrer for 10 min each time (*see* **Note 6**).
6. Dry for 15–20 h under room temperature. Grind the plastic antibody beads again with agate mortar. Keep it in a dry place until use (*see* **Note 7**).

3.2 Surface Plastic Antibody Synthesis

1. Prepare pre-polymer gel by mixing 13 mg of AAM, 6 μL of n-vinylpyrrolidone, 6.5 μL of MMA, 10.6 μL of MAA, 0.5 mg of AIBN, 48 mg of DHEBA, and 300 μL of dimethyl sulfoxide (DMSO) in a 1.5 mL microcentrifuge tube.
2. Stir the mixture with magnetic stirrer at room temperature until all components are dissolved (*see* **Note 8**).
3. Stir the mixture in a water bath at 70 °C to have it polymerized. Wait until the appearance of the pre-polymer gel (*see* **Note 9**).
4. Prepare H5N1 virus template by spreading 5 μL of virus template from stock solution over a clean 6 × 6 mm^2 glass slide (*see* **Note 10**). Spin at 1000 × *g* for 10 s with mechanical spinning. Leave it in the refrigerator for 1 h before use (*see* **Note 11**).
5. Drop 3 μL of pre-polymer gel on a desired surface (e.g., a gold electrode), spin at 1000 × *g* for 10 s with a mechanical spinning (*see* **Note 12**).
6. Gently press the glass slide of H5N1 virus template (from **step 4**) on pre-polymer gel coated on a desired surface (from **step 5**) (*see* Fig. 1), avoid the bubbles between glass slide and the imprinted surface (*see* **Note 13**).

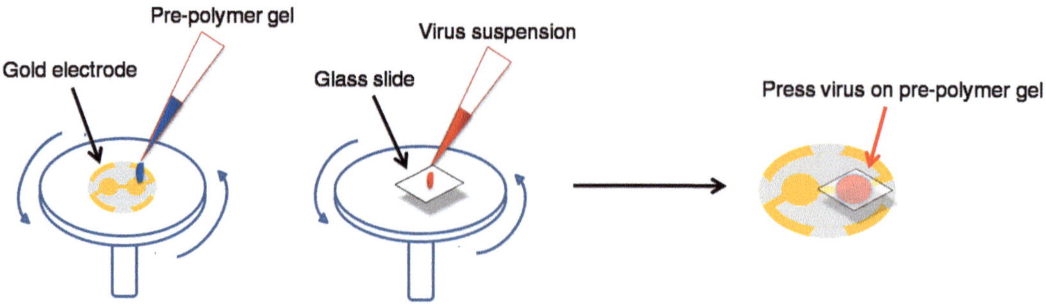

Fig. 1 A scheme showing how to make H5N1 MIPs by surface imprinting on a desired surface (in this case is a QCM electrode). First, drop and spin at 1000 × *g* the pre-polymer gel on a surface we want to have the MIPs. Then, drop and spin the H5N1 virus template on a clean glass slide. At the end, stamping the H5N1 virus template on the pre-polymer gel followed by UV-induced polymerization process

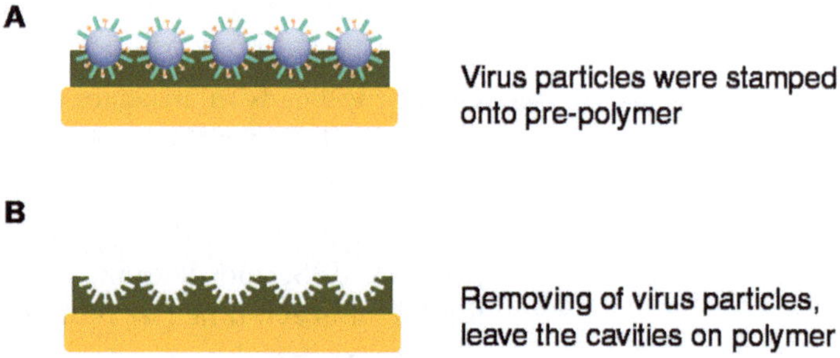

Fig. 2 H5N1 virus particles appear on thin film polymer matrix after UV light. Washing with warm deionized water can remove virus particles to obtain cavities on MIPs

7. Place under 254 nm UV light for 12–15 h to allow complete polymerization process.
8. Remove carefully the glass slide to yield a rigid copolymer thin film. This is the H5N1 MIP with template left of the surface. If the glass slide cannot be removed, soak the whole item in 55–60 °C water bath and continuously stir until the glass slide drops off (*see* **Note 14**).
9. Remove the H5N1 virus template from the MIPs surface by washing in water bath at 55–60 °C with continuous stirring approximately 1 h. Dry at room temperature before use.
10. Finally, the MIPs thin film will contain recognition sites of H5N1 virus particles (*see* **Note 15** and Fig. 2).

4 Notes

1. The use of inactivated virus sample makes it possible to do all experiments in a low-level biosafety laboratory. At present, we have no MIPs produced from viable virus. Inactivated virus

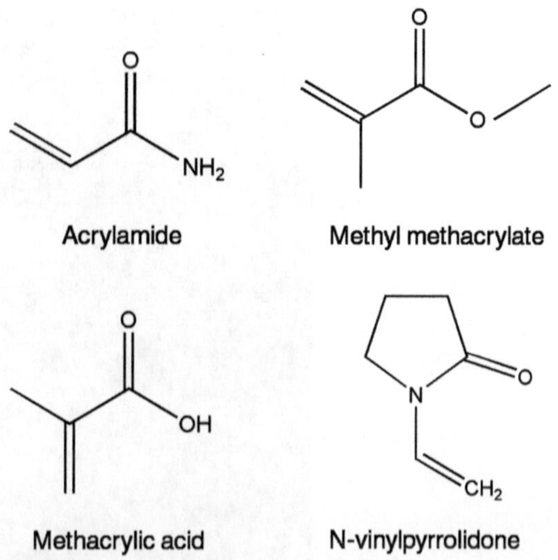

Fig. 3 Structures of functional monomer used to synthesize H5N1 plastic antibody

sample used in our work was provided by Asst. Prof. Arunee Thitithanyanont, Department of Microbiology, Mahidol University, Thailand. H5N1 virus was propagated in embryonated eggs and inactivated with β-propiolactone, followed by application of a slightly modified Centers for Disease Control and Prevention protocol [14, 15]. If other inactivating agents will be used, the resulting MIPs antibody activity could be affected. Because different inactivation could have different effects on the virus surface protein or glycosylation sugars.

2. We use monomers that can interact amino acids presented on the virus surface. Each monomer is likely to be compatible with a set of amino acid sidechains depending on polarity and shape of the monomer and particular amino acids. Monomers having polar sidechains are AAM, MMA, and MAA with amide, ester, and carboxylic sidechains, respectively, whereas n-vinylpyrrolidone is considered a nonpolar monomer (*see* Fig. 3). By mixing the monomers with virus prior to polymerization, it is assumed that a monomer will have a chance to move along the virus surface until it is hold by a suitable amino sidechain. After polymerization and cross-linking this will create an MIPs network partially optimized for the virus surface.
3. First, prepare 10% SDS solution using 10 g of SDS flakes to make 100 mL solution in deionized water. Then, use the SDS solution to dilute 10 mL of glacial acetic acid to 100 mL.
4. If it is possible, the mixture should be continuously stirred throughout an incubation time. This will allow monomers to be better organized around the virus particles.
5. Avoid the ceramic mortar because the plastic antibody will stick to its surface and thus will be hard to remove.

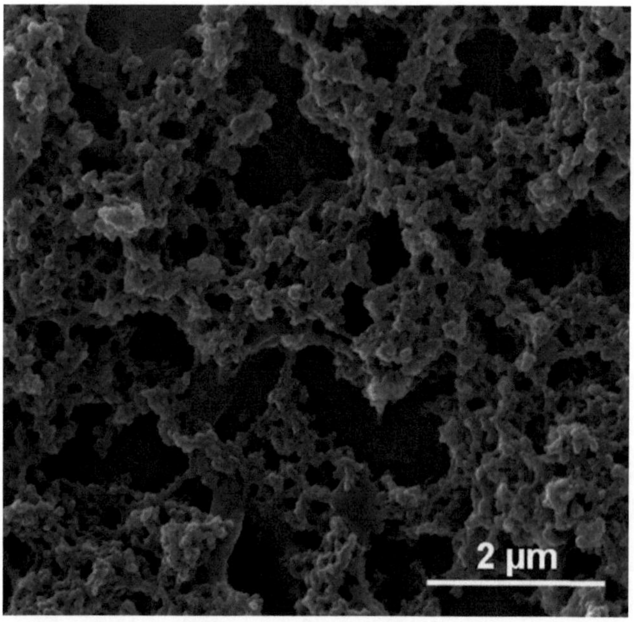

Fig. 4 Example of H5N1 bulk MIPs. Note that plastic antibody illustrates clumping morphology

6. After 10 min washing of plastic antibody beads, centrifuge at 2790 × *g* for 3 min and discard the liquid above plastic antibody beads.
7. Examples of obtained beads are shown in Fig. 4 with particle sizes between 500 and 700 nm. Further characterizations to confirm the success of H5N1 imprinting have to be done. We use virus particle counting before and after virus binding to plastic antibody beads from scanning electron microscopy (SEM) pictures to estimate our antibody affinity to H5N1. Another experiment to confirm the success imprinting is a modified agglutination assay [16]. We measure hemagglutinin (HA) titers of H5N1 left after mixing H5N1 virus sample with the plastic antibody. The decrease in virus concentration in hemagglutinating units (HAU) after binding to plastic antibody can also prove the successful H5N1 imprinting.
8. Approximately 15–20 min to obtain clear solution.
9. The mixture should to be taken outside the water bath every 20 min at the beginning and approximately 3–5 min in the final stage to obtain a gel point of the mixture. With the temperature above 65 °C, the polymerization rate becomes faster and exceeds the gel point easily. Taking out the reaction mixture regularly from a water bath during polymerization is necessary. By doing this, the reaction will slow down making it easy to observe the gel point. The gel point of pre-polymer can be checked by looking at a pre-polymer drop (5 μL sampling

from reaction mixture) on a glass slide surface. The gel point mixture will exhibit high viscosity and should not spread out.

10. Glass slides should be soaked in piranha solution for 10 min before washing two three times with deionized water and acetone. Dry at room temperature before use.
11. The template slide must be used within an hour. Avoid using the virus template where flakes of media components can be seen on a glass slide.
12. Excess pre-polymer gel outside the desired area can be removed with acetone.
13. Do not remove the glass slide at this step.
14. The stamping sample should be place in a water bath in a vertical position. This position will make the template slide slip out from the MIPs surface easier.
15. The H5N1 MIPs binding sites can be seen under atomic force microscopy (AFM). Quartz crystal microbalance (QCM) binding measurement is used to prove the virus recognition and applied as a virus-binding compound screening following the protocol in the reference [17].

Acknowledgments

This work has been supported by Royal Golden Jubilee Ph.D. Program (RGJ), Thailand Research Fund (TRF), Kasetsart University Research and Development Institute (KURDI), the National Research University Project of Thailand (NRU), and ASEAN-European Academic University Network (ASEA-UNINET).

References

1. Edwin DK (2006) Influenza pandemics of the 20th century. Emerg Infect Dis 12(1):9. doi:10.3201/eid1201.051254
2. Beveridge WI (1991) The chronicle of influenza epidemics. Hist Philos Life Sci 13(2):223–234
3. Jeffery T, David MM (2006) 1918 Influenza: the mother of all pandemics. Emerg Infect Dis 12(1):15. doi:10.3201/eid1201.050979
4. Taubenberger JK (2006) The origin and virulence of the 1918 "Spanish" influenza virus. Proc Am Philos Soc 150(1):86–112
5. Watanabe T, Kawaoka Y (2011) Pathogenesis of the 1918 pandemic influenza virus. PLoS Pathog 7(1):e1001218. doi:10.1371/journal.ppat.1001218
6. Kim H-R, Lee Y-J, Park C-K, Oem J-K, Lee OS, Kang H-M, Choi J-G, Bae Y-C (2012) Highly pathogenic avian influenza (H5N1) outbreaks in wild birds and poultry, South Korea. Emerg Infect Dis 18(3):480–483. doi:10.3201/eid1803.111490
7. Eagles D, Siregar ES, Dung DH, Weaver J, Wong F, Daniels P (2009) H5N1 highly pathogenic avian influenza in Southeast Asia. Rev Sci Tech 28(1):341–348
8. Gutiérrez RA, Naughtin MJ, Horm SV, San S, Buchy P (2009) A(H5N1) virus evolution in South East Asia. Viruses 1(3):335–361. doi:10.3390/v1030335
9. Wang H, Feng Z, Shu Y, Yu H, Zhou L, Zu R, Huai Y, Dong J, Bao C, Wen L, Wang H, Yang P, Zhao W, Dong L, Zhou M, Liao Q, Yang H, Wang M, Lu X, Shi Z, Wang W, Gu L, Zhu F, Li Q, Yin W, Yang W, Li D, Uyeki TM, Wang Y (2008) Probable limited person-to-

person transmission of highly pathogenic avian influenza A (H5N1) virus in China. Lancet 371(9622):1427–1434. doi:10.1016/S0140-6736(08)60493-6

10. Ungchusak K, Auewarakul P, Dowell SF, Kitphati R, Auwanit W, Puthavathana P, Uiprasertkul M, Boonnak K, Pittayawonganon C, Cox NJ, Zaki SR, Thawatsupha P, Chittaganpitch M, Khontong R, Simmerman JM, Chunsutthiwat S (2005) Probable person-to-person transmission of avian influenza A (H5N1). N Engl J Med 352(4):333–340. doi:10.1056/NEJMoa044021

11. Gambotto A, Barratt-Boyes SM, de Jong MD, Neumann G, Kawaoka Y (2008) Human infection with highly pathogenic H5N1 influenza virus. Lancet 371(9622):1464–1475. doi:10.1016/S0140-6736(08)60627-3

12. Baz M, Luke CJ, Cheng X, Jin H, Subbarao K (2013) H5N1 vaccines in humans. Virus Res 178(1):78–98. doi:10.1016/j.virusres.2013.05.006

13. Kang S-M, Yoo D-G, Lipatov AS, Song J-M, Davis CT, Quan F-S, Chen L-M, Donis RO, Compans RW (2009) Induction of long-term protective immune responses by influenza H5N1 virus-like particles. PLoS One 4(3):e4667. doi:10.1371/journal.pone.0004667

14. Wangchareansak T, Thitithanyanont A, Chuakheaw D, Gleeson MP, Lieberzeit PA, Sangma C (2013) Influenza A virus molecularly imprinted polymers and their application in virus sub-type classification. J Mater Chem B 1(16):2190–2197. doi:10.1039/C3TB00027C

15. Thitithanyanont A, Engering A, Ekchariyawat P, Wiboon-ut S, Limsalakpetch A, Yongvanitchit K, Kum-Arb U, Kanchongkittiphon W, Utaisincharoen P, Sirisinha S, Puthavathana P, Fukuda MM, Pichyangkul S (2007) High susceptibility of human dendritic cells to avian influenza H5N1 virus infection and protection by IFN-α and TLR ligands. J Immunol 179(8):5220–5227. doi:10.4049/jimmunol.179.8.5220

16. Eisfeld AJ, Neumann G, Kawaoka Y (2014) Influenza A virus isolation, culture and identification. Nat Protoc 9(11):2663–2681. doi:10.1038/nprot.2014.180

17. Wangchareansak T, Thitithanyanont A, Chuakheaw D, Gleeson MP, Lieberzeit PA, Sangma C (2014) A novel approach to identify molecular binding to the influenza virus H5N1: screening using molecularly imprinted polymers (MIPs). Med Chem Commun 5(5):617–621. doi:10.1039/C3MD00272A

Chapter 25

Replacement of Antibodies in Pseudo-ELISAs: Molecularly Imprinted Nanoparticles for Vancomycin Detection

Francesco Canfarotta, Katarzyna Smolinska-Kempisty, and Sergey Piletsky

Abstract

The enzyme-linked immunosorbent assay (ELISA) is a widely employed analytical test used to quantify a given molecule. It relies on the use of specific antibodies, linked to an enzyme, to target the desired molecule. The reaction between the enzyme and its substrate gives rise to the analytical signal that can be quantified. Thanks to their robustness and low cost, molecularly imprinted polymer nanoparticles (nanoMIPs) are a viable alternative to antibodies. Herein, we describe the synthesis of nanoMIPs imprinted for vancomycin and their subsequent application in an ELISA-like format for direct replacement of antibodies.

Key words Molecularly imprinted nanoparticles, Diagnostics, Assay, Analyte quantification

1 Introduction

The enzyme-linked immunosorbent assay (ELISA) is one of the most widely employed tests in diagnostics. It relies on the use of antibodies to detect and quantify the target molecule. ELISA tests can be performed in different formats: direct, indirect, competitive, and sandwich. Despite some advantages (high sensitivity and selectivity, easy operation), ELISAs exhibit several drawbacks mainly related to their high costs (due to the price of the antibodies) and poor stability of the reagents involved. In this regard, nanoMIPs thanks to their stability, cost efficiency, and easy production can be a promising alternative to antibodies in pseudo-ELISAs. Molecular imprinting describes the creation of artificial molecular recognition sites in a functional synthetic polymer by the process of forming said polymer in the presence of the target molecule (referred to as "the template") [1]. After the removal of the template, the polymer matrix will retain recognition cavities that are complementary to the template in terms of size, shape, and functionality (Fig. 1). More specifically, the three-dimensional

Thomas Tiller (ed.), *Synthetic Antibodies: Methods and Protocols*, Methods in Molecular Biology, vol. 1575,
DOI 10.1007/978-1-4939-6857-2_25, © Springer Science+Business Media LLC 2017

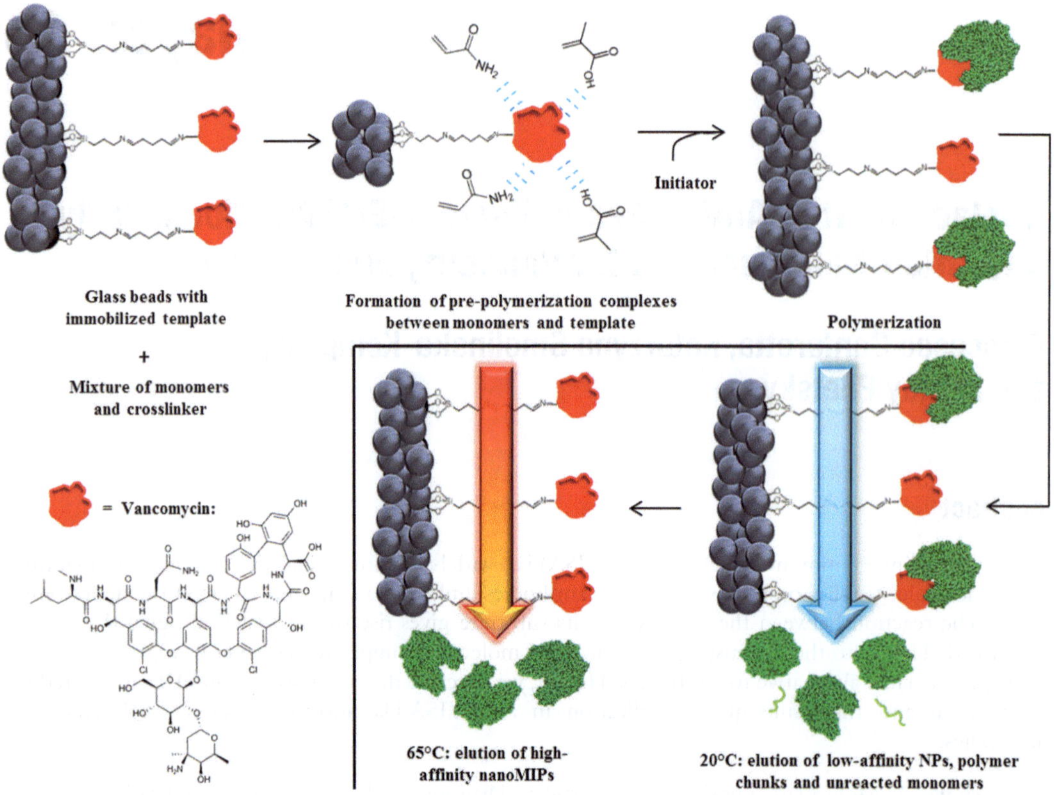

Fig. 1 Schemeofthe solid-phase synthesis of nanoMIPs. In this work, vancomycin was used as the template

arrangement of binding sites is driven by the structure of the template molecule. Therefore, the synthesized polymer is capable of rebinding the template. Compared with antibodies, the synthesis of MIPs is simpler and cheaper, and does not require involvement of animals [2]. In addition, MIPs show high stability and excellent mechanical properties and, combined with molecular modelling, they can be prepared for virtually any target [3].

However, although several papers demonstrated the applicability of MIPs in ELISAs [47], their use as Ab replacement in routine tests has not occurred yet. To date, a simple and reproducible method for coating the microplate well with MIPs is still missing. Some authors employed a MIP film to bind the target analyte, achieving a sensitivity in the micromolar range [7, 8]. Nevertheless, we do not find this approach to be particularly reproducible because preparation of high-quality MIP films depends on too many physical and chemical factors [9]. Herein, we use nanoMIPs, sometimes called "plastic antibodies," prepared by means of an innovative solid-phase approach for the detection of vancomycin in a pseudo-ELISA format, as previously demonstrated by Chianella and colleagues [10]. NanoMIPs are nanostructured polymer particles containing typically one binding site for their target molecule and capable of selectively recognizing the said target. This innovative solid-phase approach relies on the covalent immobilization of the template on a solid support (e.g., glass beads, average diameter 75–90 μm). This support bearing the template molecule is then placed in contact with the monomer mixture. After polymerization, nanoMIPs are produced around the template (Fig. 1). Then, the solid support acts as an affinity medium to isolate the high-affinity nanoMIPs from the remaining monomers, oligomers, and low-affinity polymers, which are removed by washing the beads at room temperature (20 °C). High affinity nanoMIPs are subsequently collected by increasing the temperature of the solvent (65 °C). Thanks to this affinity purification step, the produced nanoMIPs possess high affinity and specificity toward their targets and exhibit a homogeneous distribution of binding site affinities, much like monoclonal antibodies. The synthesized nanoMIPs are immobilized on the surface of microplate wells by physical adsorption. The conditions for polymer synthesis and application in pseudo-ELISA described in the present paper are generic and can be used with different types of nanoMIPs and relevant target templates, as recently demonstrated by Càceres and colleagues [11].

2 Materials

Prepare all solutions using MilliQ water (subsequently referred to as "water") and analytical grade reagents. Prepare and store all reagents at room temperature (unless indicated otherwise).

2.1 Preparation of Vancomycin Modified Solid Phase

1. Vancomycin hydrochloride.
2. Glass beads (average diameter 75–90 μm).
3. 3-aminopropyltrimethyloxysilane (APTMS).
4. Sodium hydroxide (NaOH).
5. Anhydrous toluene.
6. 250-mL sealable bottle.
7. Sintered disc filter funnel.
8. Vacuum pump.
9. 500-mL flask.

2.2 Synthesis of NanoMIPs`

1. Acrylic acid (AA).
2. *N*-isopropylacrylamide (NIPAM).
3. *N*,*N*′-methylene-bis-acrylamide (BIS).
4. *N*-tert-butylacrylamide (TBAm).
5. Ammonium persulfate (APS).

6. Tetramethylethylene-diamine (TEMED).
7. Phosphate buffered saline (PBS): 0.01 M phosphate buffer, 0.003 M potassium chloride, and 0.140 M sodium chloride, pH 7.4.
8. SPE cartridge fitted with a 20-μm porosity frit.
9. 250-mL flask.

2.3 Pseudo-ELISA with NanoMIPs Imprinted with Vancomycin

1. Bovine serum albumin (BSA), lyophilized powder, essentially globulin free, low endotoxin.
2. Horseradish peroxidase (HRP), Type II, essentially salt-free, lyophilized powder, 150–250 units/mg solid.
3. 3,3,5,5-tetramethylbenzidine, used in the form of TMB liquid substrate system for ELISA.
4. Tween 20.
5. 2-[morpholino]ethanesulfonic acid (MES).
6. Sulfuric acid min 95% $d = 1.83$, SLR, extra pure.
7. *N*-(3-Dimethylaminopropyl)-*N'*-ethylcarbodiimide hydrochloride (EDC) ≥98.0%.
8. *N*-Hydroxysuccinimide (NHS), 98%.
9. Millipore Amicon Ultra centrifuge filter unit (30 kDa MWCO).
10. Nunc™ 96-Well Microplates. Alternatively, polystyrene or polypropylene 96-Well Microplates.

2.4 Reagents

1. Washing buffer: 0.01 M PBS, (0.00268 M potassium chloride, 0.140 M sodium chloride), pH 7.4.
2. Blocking buffer: 0.1% BSA, 1% Tween 20 in washing buffer.
3. Stop solution: 0.5 M sulfuric acid in water.

3 Methods

3.1 Preparation of Vancomycin Modified Solid Phase

1. Boil the glass beads in 1 M NaOH (0.8 mL of solution per gram of glass beads) for 15 min to activate their surface, and then rinse them thoroughly with deionized water (eight times with 200 mL, for 60 g of beads), until the pH of the water/beads solution is around 6–7 (make sure that the base has been completely removed). Rinse with acetone (twice with 200 mL) and dry it at 80 °C for 3 h.
2. Prepare a solution of APTMS 2% (vol/vol) in anhydrous toluene. Incubate the dry beads in this APTMS solution for 24 h at RT in a closed container of suitable volume, using 0.4 mL of solution per gram of beads (*see* **Note 1**).

3. After incubation, decant the glass beads onto a sintered disc filter funnel and wash thoroughly with eight volumes of acetone and one volume of methanol. Dry the glass beads under vacuum (*see* **Note 2**). Alternatively, keep them on the sintered funnel connected to a vacuum source. This can be applied to all subsequent steps where vacuum is used.
4. Incubate the silanized beads in a 7% (v/v) GA solution in 0.01 M PBS (pH 7.2) for 2 h. Place them in a sintered disc filter funnel and wash them with eight volumes of deionized water under vacuum (*see* **Note 3**).
5. Incubate the glass beads in a solution of vancomycin (0.5 mg/mL) in 10 mM PBS (pH 7.4) overnight.
6. Filter the glass beads and rinse them with deionized water (8–10 volumes) in a sintered funnel under vacuum.
7. Dry the beads under vacuum and store them at 4 °C (*see* **Note 4**).

3.2 Synthesis of NanoMIPs

1. Weigh out 39 mg of NIPAm and 2 mg of BIS. Add 98 mL of water and dissolve the solids.
2. Weigh out 33 mg of TBAm in a separate vial and add 1 mL of ethanol. Once the monomer is dissolved, add this solution to the one prepared in the previous step and homogenize.
3. In a separate vial, dispense 22 μL of AAc into 1 mL of water. Then take 100 μL of this solution and add it to the solution prepared in Subheading 3.2, **step 1**. Gently swirl to homogenize.
4. Adjust the total volume of the solution to 100 mL by adding water.
5. Seal the flask, connect it to a vacuum source, and sonicate under vacuum using an ultrasonicator for 10 min. Then further bubble the solution with a gentle stream of N_2 or Ar for 20 min via a Pasteur pipette located in the bulk of the solution (Fig. 2a; *see* **Note 5**).
6. While the solution is degassing, weigh 60 g of vancomycin-derivatized glass beads into a 250-mL sealable bottle (Fig. 2b). Purge N_2 or Ar into this bottle for 5 min to ensure oxygen removal.
7. Weigh 30 mg of APS and dissolve it in 500 μL of water, then add 30 μL of TEMED.
8. Pour the degassed solution of monomers onto the glass beads, and quickly add the solution prepared in the previous step to the glass beads dispersed in the monomer mixture.
9. Purge the headspace with N_2 or Ar, then seal the vessel and incubate for 1 h at RT.

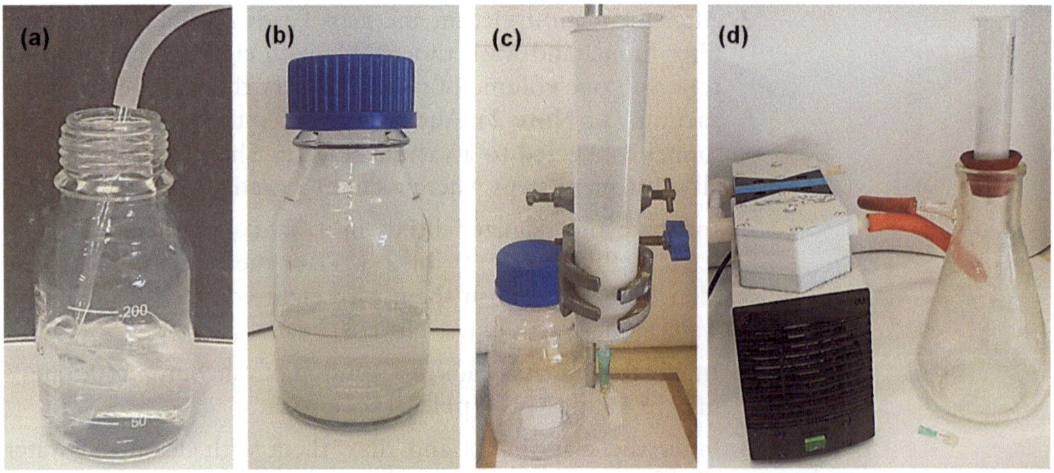

Fig. 2 Overview of the process for the production of nanoMIPs imprinted for vancomycin by solid-phase imprinting. (**a**) The polymerization mixture is purged with N_2 to remove oxygen. (**b**) Vancomycin-functionalized beads are added and the polymerization is initiated under inert atmosphere. (**c**) After polymerization, the whole content of the bottle is poured into a SPE cartridge for the subsequent (**d**) selection and elution of the nanoMIPs by means of a plunger or vacuum

10. Selection of high-affinity nanoMIPs (Fig. 1): after incubation, transfer the whole content of the vessel (both beads and solution) into a SPE cartridge fitted with a 20-μm porosity frit (Fig. 2c).
11. By means of a plunger or vacuum (Fig. 2d), remove the solution and replace it with fresh water (30 mL) at RT (*see* **Note 6**).
12. Remove low-affinity nanoMIPs by eluting the aforementioned solution at RT, replacing it with 30-mL aliquots of fresh water each time ("washing steps"). Repeat this step eight times. Stir the beads very gently every other three washing steps.
13. Close the outlet of the SPE cartridge containing the glass beads, e.g., using a female Luer cap.
14. Add 30 mL of water pre-warmed at 60 °C to the SPE cartridge containing the glass beads, stir the beads, and then place the cartridge into a water bath at 60 °C.
15. After 20 min, collect the nanoMIPs solution by means of a plunger or vacuum.
16. Add 20 mL of water pre-warmed at 60 °C and place the SPE cartridge at 60 °C for 2 min.
17. Collect the high-affinity nanoMIP solution. Repeat the previous step until ~150 mL of solution has been collected (*see* **Note 7**).

3.3 NanoMIPs Immobilization onto the Surface of Microplate Wells

1. Check the concentration of the nanoMIPs by evaporating an aliquot of the stock solution after synthesis. The recommended concentration range for nanoMIP to be used in a pseudo-ELISA for vancomycin is 0.05–0.06 mg/mL.
2. Dispense 40 μL of nanoMIP solution into each well of a 96-well Nunclon microplate, and let the solvent evaporate overnight at room temperature (*see* **Notes 8** and **9**).

3.4 Preparation of the HRP-Vancomycin Conjugate

1. Dissolve 10 mg of HRP in 1 mL 0.1 M MES buffer, pH 6. Add 0.4 mg EDC and 1.1 mg NHS, and incubate at room temperature for 15 min.
2. Dissolve 10 mg of vancomycin in 10 mL PBS at pH 7.4.
3. Immediately add the vancomycin solution to the activated HRP. Check and adjust the pH to 7.4–8.0.
4. Incubate for 2 h at room temperature.
5. To remove unreacted vancomycin and HRP, wash the conjugate by means of an Amicon Ultracentrifugal filter unit (30 kDa MWCO) seven times using 10 mL of PBS solution each time.
6. After washing, dissolve the conjugate in 2 mL of water and aliquot it into four eppendorf tubes (0.5 mL each).
7. Store the conjugate at −20 °C until further use (*see* **Note 10**).

3.5 Pseudo-ELISA with NanoMIPs Imprinted with Vancomycin

1. Dispense the nanoMIPs as in Subheading 3.3, **step 2**.
2. Wash the microplate two times with 250 μL of washing buffer per well.
3. Add 300 μL of blocking buffer per well, cover the microplate, and incubate for 1 h at room temperature.
4. Prepare various concentrations of free vancomycin solutions in water (ideal concentration range: 0.001–100 nM). Then add the same amount of conjugate (dilution 1:800), prepared in Subheading 3.4, to each of the vancomycin solutions (e.g., add 6.3 μL conjugate to 5 mL of template solution).
5. Remove the blocking buffer and wash the microplate three times with 300 μL per well of washing buffer.
6. Add 100 μL of vancomycin solutions (containing the HRP-vancomycin conjugate) to each well. Cover the microplate and incubate in the dark for 1 h at room temperature.
7. Remove the vancomycin solutions and wash the microplate three times with 300 μL of blocking solution per well.
8. Add 100 μL of TMB solution per well and incubate the microplate for 10 min at room temperature (Fig. 3).

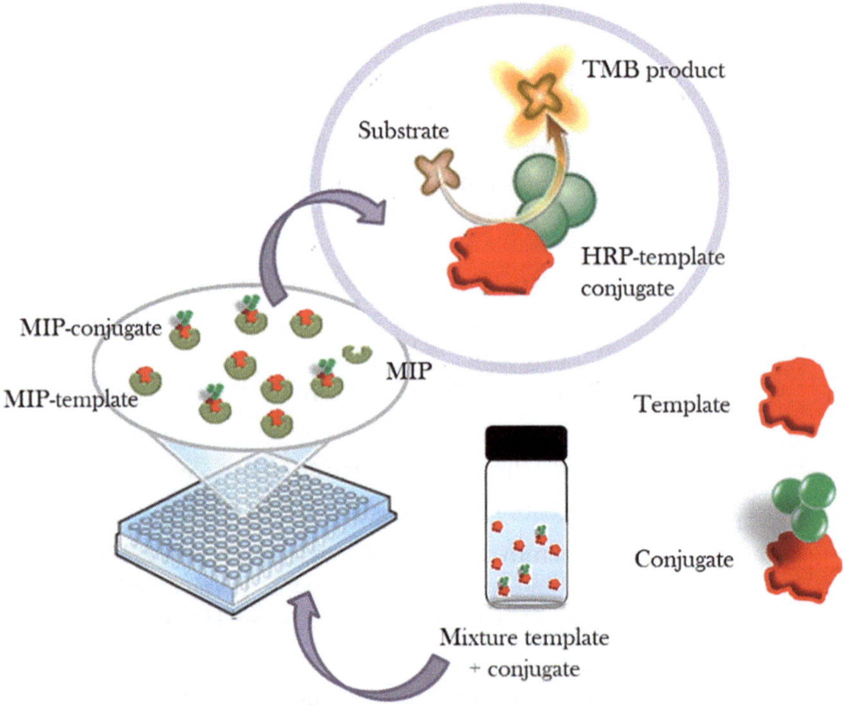

Fig. 3 Scheme of the ELISA-like assay based on nanoMIPs imprinted for vancomycin

9. Quench the reaction by adding 100 μL of stop solution per well.
10. Measure the absorbance of each well at 450 nm (*see* **Notes 11–13**).

4 Notes

1. Ensure that the glass beads are completely dry and the toluene used is anhydrous. Do not stir the beads on an orbital shaker or using a stirrer. Instead, just swirl them gently by hand from time to time, as collisions between beads may abrade their surface. This applies to all steps involving incubation of glass beads, unless otherwise specified.
2. If the glass beads are stored dry, they are stable for up to 6 months at room temperature.
3. Check the pH of the beads solution and adjust it to 7.4 if necessary prior to the addition of GA. Please note that the formation of the Schiff base is reversible, so excessive washing is not recommended. Do not store the beads at this stage, directly move to the template conjugation step.
4. The beads with immobilized vancomycin can be stored dry for 1 month. However, it is advisable to use the beads as soon as possible after preparation.

5. Make sure not to use an excessive stream of N_2 as it may cause evaporation of water.
6. It is important not to let the glass beads dry, thus leaving a water interface above the glass beads.
7. Stir the beads (by means of a spatula) every time fresh prewarmed water is added. Leave the solution at RT for 4–6 h for it to cool down, and then place it at 4 °C for storage. This stock solution containing nanoMIPs can be stored at 4 °C for a few months, depending on potential bacterial/fungal contamination issues that may arise during storage due to handling under nonsterile conditions.
8. Calculate the concentration of nanoMIPs by evaporation. If the concentration is higher than expected, dilute the solution by adding water. If the concentration is lower, increase the samples volume to be put into each well or evaporate the solvent to obtain a more concentrated solution. In general, make sure than the volume of MIP solution (40 μL) is enough to cover completely the well surface. If not, increase the volume to about 60 μL.
9. Before using the nanoMIP-coated microplate, make sure that solvent has evaporated completely. Otherwise, nanoMIPs may get wash away during the pseudo-ELISA.
10. Always use fresh conjugate. It is recommended for the conjugate not to undergo several freezing/thawing cycles.
11. If you cannot notice color development after 10 min of incubation with TMB, incubate the microplate at room temperature until the desired color intensity is reached.
12. If, despite the increase in the TMB incubation time, no color will develop in solution, it is likely that the conjugate is not working. Prepare a fresh conjugate, making sure that the HRP activity is preserved after conjugation. This could be assessed by comparison of the conjugate activity with a known concentration of free HRP.
13. If the background signal is too high, it is possible that an error in the washing steps occurred or the blocking solution is not fresh. It is recommended to store the blocking solution for no longer than 2–3 days.

References

1. Poma A, Turner AP, Piletsky SA (2010) Advances in the manufacture of MIP nanoparticles. Trends Biotechnol 28(12):629–637 Epub 2010/10/01
2. Canfarotta F, Poma A, Guerreiro A, Piletsky S (2016) Solid-phase synthesis of molecularly imprinted nanoparticles. Nat Protoc 11(3):443–455
3. Breton F, Rouillon R, Piletska EV, Karim K, Guerreiro A, Chianella I et al (2007) Virtual imprinting as a tool to design efficient MIPs for photosynthesis-inhibiting herbicides. Biosens Bioelectron 22(9–10):1948–1954

4. Surugiu I, Danielsson B, Ye L, Mosbach K, Haupt K (2001) Chemiluminescence imaging ELISA using an imprinted polymer as the recognition element instead of an antibody. Anal Chem 73(3):487–491
5. Surugiu I, Ye L, Yilmaz E, Dzgoev A, Danielsson B, Mosbach K et al (2000) An enzyme-linked molecularly imprinted sorbent assay. Analyst 125(1):13–16
6. Yonamine Y, Hoshino Y, Shea KJ (2012) ELISA-mimic screen for synthetic polymer nanoparticles with high affinity to target proteins. Biomacromolecules 13(9):2952–2957
7. Piletsky SA, Piletska EV, Bossi A, Karim K, Lowe P, Turner APF (2001) Substitution of antibodies and receptors with molecularly imprinted polymers in enzyme-linked and fluorescent assays. Biosens Bioelectron 16(9–12):701–707
8. Zhao D, Qiao X, Xu Z, Xu R, Yan Z (2013) Development of a biomimetic enzyme-linked immunosorbent assay method based on a hydrophilic molecularly imprinted polymer film for determination of olaquindox in chick feed samples. J Immunoassay Immunochem 34(1):16–29
9. Piletska EV, Guerreiro AR, Whitcombe MJ, Piletsky SA (2009) Influence of the polymerization conditions on the performance of molecularly imprinted polymers. Macromolecules 42(14):4921–4928
10. Chianella I, Guerreiro A, Moczko E, Caygill JS, Piletska EV, De Vargas Sansalvador IMP et al (2013) Direct replacement of antibodies with molecularly imprinted polymer nanoparticles in ELISA—Development of a novel assay for vancomycin. Anal Chem 85(17):8462–8468
11. Cáceres C, Canfarotta F, Chianella I, Pereira E, Moczko E, Esen C et al (2016) Does size matter? Study of performance of pseudo-ELISAs based on molecularly imprinted polymer nanoparticles prepared for analytes of different sizes. Analyst 141(4):1405–1412

Chapter 26

Cell and Tissue Imaging with Molecularly Imprinted Polymers

Maria Panagiotopoulou, Stephanie Kunath, Karsten Haupt, and Bernadette Tse Sum Bui

Abstract

Advanced tools for cell imaging are of particular interest as they can detect, localize and quantify molecular targets like abnormal glycosylation sites that are biomarkers of cancer and infection. Targeting these biomarkers is often challenging due to a lack of receptor materials. Molecularly imprinted polymers (MIPs) are promising artificial receptors; they can be tailored to bind targets specifically, be labeled easily, and are physically and chemically stable. Herein, we demonstrate the application of MIPs as artificial antibodies for selective labeling and imaging of cellular targets, on the example of hyaluronan and sialylation moieties on fixated human skin cells and tissues. Thus, fluorescently labeled MIP nanoparticles templated with glucuronic acid (MIPGlcA) and N-acetylneuraminic acid (MIPNANA) are respectively applied. Two different fluorescent probes are used: (1) MIPGlcA particles, ~400 nm in size are labeled with the dye rhodamine that target the extracellular hyaluronan on cells and tissue specimens and (2) MIP-coated InP/ZnS quantum dots (QDs) of two different colors, ~125 nm in size that target the extracellular and intracellular hyaluronan and sialylation sites. Green and red emitting QDs are functionalized with MIPGlcA and MIPNANA respectively, enabling multiplexed cell imaging. This is a general approach that can also be adapted to other target molecules on and in cells.

Key words Molecularly imprinted polymers, MIPs, Artificial antibodies, Glucuronic acid, Hyaluronic acid, Sialylation, Cell imaging, Tissue imaging, Quantum dots, Multiplexed imaging

1 Introduction

Glycosylation in biology refers mainly to the enzymatic process that attaches glycans to proteins and is a form of co-translational and posttranslational modification that the majority of proteins synthesized in the endoplasmic reticulum, undergo [1–3]. Glycosylation is critical for a wide range of biological processes in health and disease [4–6] and therefore, glycosylation sites on the cell surface are of great interest, since altered glycosylation levels or distribution are indicators of pathological conditions like infection or malignancy. Recent advances in glycobiology and cancer research have defined the key processes underlying aberrant glycosylations

Thomas Tiller (ed.), *Synthetic Antibodies: Methods and Protocols*, Methods in Molecular Biology, vol. 1575,
DOI 10.1007/978-1-4939-6857-2_26, © Springer Science+Business Media LLC 2017

with sialic acids or hyaluronan in cancer and its consequences [7–10]. Among others, overexpressed glycosylations have been associated with tumor growth, escape from apoptosis, metastasis formation and resistance to therapy. Since the polysaccharides involved in the glycosylation procedure have a highly conserved simple composition and are ubiquitously expressed in all animals that have a developed immune system, they are the so-called weak antigens. Therefore, production of antibodies that specifically recognize them is naturally difficult [11–13], and traditional immunohistochemical methods for detecting glycosylations on cells are rare.

In this context, tailor-made molecularly imprinted polymers (MIPs) are promising synthetic receptor materials. Molecular imprinting is based on a templating process at the molecular level. Monomers carrying functional groups self-assemble around a template molecule (the target or a derivative), followed by copolymerization with an excess of cross-linking monomers, which results in the formation of a polymeric mold around the template. Subsequent removal of the template reveals three-dimensional binding sites in the polymer that are complementary to the template in size, shape, and position of the functional groups [14–17]. Molecularly imprinted polymers exhibit binding properties with affinities and selectivities often comparable to those of biological antibodies but in contrast to the latter, their production is reproducible, relatively fast and economic and no animals are necessary. MIPs are potentially superior receptor materials for cell imaging applications for a number of reasons; they are physically and chemically very stable and are not degraded by proteases or denatured by solvents, and they can be engineered and tailored for a given application. In principle MIPs can be obtained for any target molecule, even when no natural receptor or antibody exists. They can be easily functionalized with fluorescent dyes [18–22], quantum dots [23–26] or other fluorescent materials like upconverting nanoparticles [27, 28] or carbon dots [29], and their size can be tuned according to the localization of the target.

In the present protocol, we apply molecularly imprinted polymers for molecular imaging of hyaluronan and sialylation sites on cells and tissue. Since molecular imprinting of entire biomacromolecules like proteins or oligosaccharides is challenging, we opted for what is commonly called the "epitope approach," which was

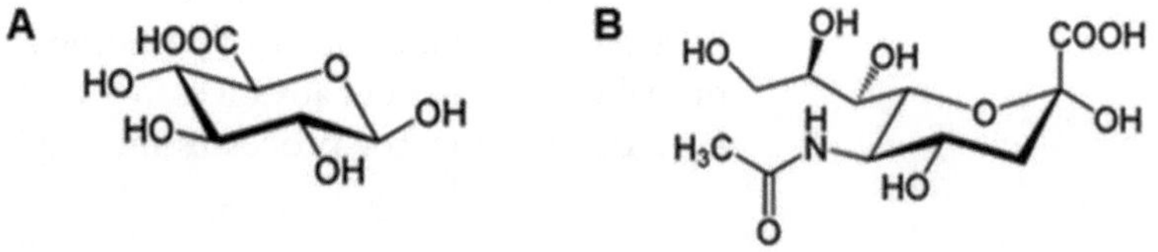

Fig. 1 Chemical structures of (**a**) glucuronic acid and (**b**) N-acetylneuraminic acid

inspired by nature [30, 31]. The monosaccharides, glucuronic acid (GlcA) and N-acetylneuraminic acid (NANA) (Fig. 1) are imprinted, and the resulting MIPs are able to bind these molecules if present and accessible, as the terminal unit on larger oligosaccharides or proteoglycans. GlcA is majorly present as a terminal moiety of hyaluronan, found extracellularly, pericellularly, and intracellularly, even within the nucleus in some cases, since nuclear staining due to the distribution of hyaluronan in nuclear clefts has previously been reported [32]. On the other hand, NANA as a terminal moiety is mostly reported to be located extracellularly as part of sialylated proteins and sphingolipids on the glycocalyx [20, 21]. MIPs in this study are synthesized in the form of dye-labeled nanoparticles using the dye rhodamine, and as MIP-coated InP/ZnS core–shell quantum dot particles. These selective fluorescent probes are then used to image human keratinocytes and skin tissues, by epifluorescence and confocal fluorescence microscopy [18, 19, 26]. Multiplexed cell imaging is also performed, demonstrating the potential of molecularly imprinted polymers, when conjugated to quantum dots of different emission colors, as a versatile multi-targeted imaging tool.

2 Materials

Small laboratory consumables, chemicals and solvents are from VWR International or Sigma-Aldrich, unless otherwise stated. All chemicals and solvents are of analytical grade. All solutions and buffers are prepared with Milli-Q water, purified using a Millipore Milli-Q system.

2.1 Cell Culture

1. Human adult low Calcium high Temperature (HaCaT) cells (Cell Lines Service, Eppelheim, Germany) (*see* **Note 1**).
2. Dulbecco's Modified Eagle Medium (DMEM) supplemented with 10% fetal bovine serum (FBS) and 1% penicillin/streptomycin.
3. 0.25% trypsin–ethylenediaminetetraacetic acid (EDTA) with phenol red.
4. Phosphate buffered saline (PBS) with 0.05% EDTA (*see* **Note 2**).
5. Cell culture consumables: 75 cm^2 sterile cell culture flasks, 12-well plates, 15 mL conical sterile polypropylene centrifuge tubes. Serological pipettes of 5, 10, and 25 mL.

2.2 Cell Sample Preparation for Imaging

1. Round glass coverslips (12 mm diameter) (Electron Microscopy Sciences, Hatfield, USA).
2. Microscope slides 76 × 26 mm.
3. PBS buffer.

4. Paraformaldehyde 3% (w/v) in PBS.
5. Glycine, 20 mM in PBS.
6. Methanol–water (1:30, v/v). Mix 180 mL of water with 6 mL of methanol to constitute a stock solution.
7. MIP suspensions:
   (a) 27 μg/mL rhodamine-labeled MIPGlcA (ultrasonicate 2.7 mg of polymer in 100 mL methanol–water (1:30, v/v)).
   (b) 60 μg/mL MIPGlcA-green QDs (ultrasonicate 3 mg of polymer in 50 mL methanol–water (1:30, v/v)).
   (c) 60 μg/mL MIPNANA-red QDs (ultrasonicate 3 mg of polymer in 50 mL methanol–water (1:30, v/v)).
8. 100 μM solution of 3,3′-dioctadecyloxacarbocyanine perchlorate (DiO) in dimethyl sulfoxide (DMSO). Dilute 10-fold a stock solution of 1 mM DiO in DMSO.
9. 1 mg/mL solution of 4′,6-diamidin-2-phenylindol (DAPI) in water.
10. Mounting medium: 0.5 mL water, 0.5 mL 1 M Tris–HCl pH 8.0, and 9 mL glycerol.
11. Transparent nail polish.

2.3 Tissue Sample Preparation for Imaging

1. Adult skin specimens (Institute of Pathology Bad Berka, Germany).
2. Cryostat microtome (Leica Biosystems, Germany).
3. Adhesive microscope slides, oven at 70 °C, cold acetone (−10 °C).
4. PBS buffer.
5. Methanol–water (1:30, v/v).
6. Rhodamine-labeled MIPGlcA (27 μg/mL in methanol–water (1:30, v/v)).
7. Rectangular glass coverslips.
8. DAPI (100 μg/mL in water). The solution is prepared as a dilution from a 1 mg/mL stock solution of DAPI in water.
9. Transparent nail polish.

2.4 Synthesis of Rhodamine-Labeled MIPGlcA by Precipitation Polymerization

1. Glass vials of volume 4 mL, equipped with screw caps and silicon septa.
2. Thermostated water bath.
3. Parafilm M.
4. Template molecule: D-glucuronic acid.
5. Functional and cross-linking monomers: 4-acrylamidophenyl (amino)methaniminium acetate (AB), methacrylamide (MAM), and ethyleneglycol dimethacrylate (EGDMA).

6. Methacryloxyethyl thiocarbonyl rhodamine B (PolyFluor 570) (Polysciences, Warrington, USA).
7. Initiator: 2,2′-azobis(2,4-dimethylvaleronitrile) (ABDV) (DuPont Chemicals, Wilmington, USA).
8. Porogen: anhydrous DMSO.
9. For template extraction, use the following solvent combinations: 1 M HCl in H_2O, methanol–acetic acid (1:9, v/v), and methanol.

2.5 Synthesis of MIP-Functionalized QDs for Multiplexed Cell Imaging

All polymerizations are carried out with a 365 nm UV-lamp (VILBER LOURMAT, 6 W). Photobleaching of the initiator dyes trapped inside the polymer particles after polymerization is done by irradiating with a fluorescent tube 18 W (MAZDAFLUOR, UK).

2.5.1 Synthesis of a Hydrophilic Cross-Linked First Shell

1. Green and red emitting InP/ZnS QDs (λ_{em} = 550 nm and λ_{em} = 650 nm, respectively). The ~20 nm green emitting QDs (green QDs) are used for the preparation of MIPGlcA-functionalized QDs, while the ~4 nm red emitting QDs (red QDs) (Sigma-Aldrich, France) are used for the preparation of MIPNANA-functionalized QDs. Prepare 1 mg/mL QD (green or red) solution in toluene, diluted from a stock solution of 5 mg/mL in toluene. Store in amber glass vials wrapped with aluminum foil at 4 °C.
2. Initiator systems: eosin Y (10 mM in DMSO–toluene (1:1, v/v)) and triethylamine (TEA) (72 mM in DMSO–toluene (1:1, v/v)) when green QDs are used. Methylene blue (10 mM in DMSO–toluene (1:1, v/v)) and TEA (72 mM in DMSO–toluene (1:1, v/v)) when red QDs are used. For this, prepare the following stock solutions:
   (a) 10 mM eosin Y solution: Dissolve 6.5 mg of eosin Y in 1 mL DMSO–toluene (1:1, v/v). Store in amber glass vials wrapped with aluminum foil at 4 °C.
   (b) 10 mM methylene blue solution: Dissolve 3.2 mg of methylene blue in 1 mL DMSO–toluene (1:1, v/v). Store in amber glass vials wrapped with aluminum foil at 4 °C.
   (c) 72 mM TEA solution: Add 10 μL TEA >99.5% as supplied to 1 mL DMSO–toluene (1:1, v/v).
3. Monomers: 2-hydroxylethyl methacrylate (HEMA) and *N,N′*-ethylenebis(acrylamide) (EbAM).
4. Porogen: DMSO–toluene (1:1, v/v).

2.5.2 Synthesis of a Second Shell Based on MIP

1. HEMA-coated green and red QDs.
2. *See* Subheadings 2.4, **item 5** and 2.4, **item 8** for functional and cross-linking monomers and porogens respectively. *See*

Subheading 2.5.1 **item 2** for the initiator systems. In the case of MIPNANA-QDs (red QDs), N-acetylneuraminic acid (NANA) is used as template.

3. For template extraction, the following solvent combinations are used: methanol–acetic acid (9:1, v/v), 100 mM NH_3 in water–methanol (7:3, v/v), water and methanol.

2.6 Imaging Acquisition and Analysis

1. All epifluorescence images are captured with a Leica DMI 6000B microscope, filter set A4, L5, TX2, N PLAN L 20.0 × 0.40 DRY, HCX PL FLUOTAR 40.0 × 0.60 DRY, HCX FLUOTAR 63.0 × 0.70 DRY and HCX FLUOTAR 100.0 × 1.30 OIL objectives with 20×, 40×, 63× and 100× magnification respectively.
2. All confocal images are captured with a Zeiss LSM 710, AxioObserver. A Plan-Apochromat 63×/1.40 OIL DIC M27 objective and 405, 488, and 543 nm lasers are used for excitation of DAPI, DiO, and rhodamine, respectively, for all images.
3. Imaging analysis software: ImageJ (National Institute of Health, USA, version 1.45 s).

3 Methods

3.1 Synthesis of Rhodamine-Labeled MIPGlcA by Precipitation Polymerization

3.1.1 Synthesis of the Functional Monomer 4-Acrylamidophenyl (Amino)Methaniminium Acetate (AB)

First, 4-acrylamidophenyl(amino)methaniminium chloride [27, 33] is synthesized.

1. In a 600 mL glass beaker, weigh 34 g (0.25 mol) of sodium acetate trihydrate and 2 g (9.6 mmol) of 4-aminobenzamidine dihydrochloride and dissolve them in 200 mL of water.
2. Cool the solution to <5 °C in an ice bath and add 4 mL (49 mmol) of acryloyl chloride dropwise, with stirring. Leave the reaction to proceed for 1 h under magnetic stirring.
3. Adjust the pH to 4.0 with 37% (w/w) hydrochloric acid; precipitation is observed.
4. Filtrate the precipitate on a Büchner funnel and redissolve the precipitate in 100 mL of water at 40 °C. Add hydrochloric acid to pH 1.0 and leave the product to crystallize overnight at 4 °C.
5. Collect the crystals by filtration and dry them in an oven at 50 °C.

   In a second step, 4-acrylamidophenyl(amino)methaniminium chloride is then converted to 4-acrylamidophenyl(amino) methaniminium acetate.

6. Suspend 1 g of 4-acrylamidophenyl(amino)methaniminium chloride in 100 mL of saturated sodium acetate solution and stir overnight.
7. Collect the product by filtration, wash with water to eliminate residual sodium acetate and dry at 50 °C in an oven (*see* **Note 3**).

3.1.2 Synthesis of MIPGlcA

1. Weigh 4.27 mg (0.022 mmol) of the template GlcA and 5.46 mg (0.022 mmol) of AB in a 4 mL glass vial. Add 1 mL of anhydrous DMSO and incubate the mixture for 1 h on a tube-rotator to form the complex GlcA-AB (*see* **Note 4**).
2. Mix together 5.6 mg (0.066 mmol) of the co-monomer MAM, 83.1 μL (0.44 mmol) of the cross-linker EGDMA, 0.0055 mmol ABDV (pipet 0.524 mL from a stock solution of 3.4 mg ABDV in 1.3 mL DMSO), 1.3 μmol of methacryloxyethyl thiocarbonyl rhodamine B (pipet 0.27 mL from a stock solution of 3 mg in 1 mL DMSO) (*see* **Note 5**) and add to the AB-GlcA complex. Synthesize a control non-imprinted polymer (NIP) in the same way but without the addition of the template (*see* **Notes 6** and **7**).
3. Sonicate the prepolymerization mixture in an ultrasonic bath for 15 min and purge for 2 min with nitrogen.
4. Seal the vials with a septum and Parafilm, place them in a water bath set at 48 °C and leave the polymerization to proceed for 18 h.
5. In order to remove the template from the polymeric matrix, wash the MIP three times with 1 M HCl, three times with methanol–acetic acid (1:9, v/v) and three times with methanol.
6. Finally, dry the particles overnight under vacuum (*see* **Note 8**).

3.2 Synthesis of MIP-Functionalized QDs for Multiplexed Cell Imaging

3.2.1 Synthesis of a Hydrophilic Cross-Linked First Shell

1. Synthesize green InP/ZnS QDs as described [34]: Add indium chloride (0.1 mmol), stearic acid (0.1 mmol), hexadecylamine (0.2 mmol) and zinc undecylenate (0.1 mmol) to 1-octadecene (2 mL). Deoxygenate the mixture repeatedly and refill with nitrogen gas to provide a water-free and oxygen-free reaction atmosphere, then heat to 270 °C with stirring. On reaching 270 °C, inject rapidly 1 mL of 0.1 M tris(trimethylsilyl) phosphine in 1-octadecene. Hold the mixture at 240 °C for 20 min, then cool to room temperature. Open the flask and add zinc diethyldithiocarbamate (0.2 mmol) and zinc undecylenate (0.2 mmol). Re-deoxygenate the mixture and place it under nitrogen gas, then heat to 180 °C for 10 min and to 240 °C for 20 min. Cool the reaction to room temperature, then add 4 mL toluene and centrifuge the solution at 2200 × *g* for 5 min. Pour off the clear QD solution and add ethanol until the QDs precipitate. Centrifuge the mixture at 2200 × *g* for 15 min. Remove the supernatant and resuspend the QDs in toluene. Repeat the precipitation with ethanol for three times to ensure the removal of synthetic residues. Finally, resuspend the QDs in toluene and store them at 4 °C in the dark until use.
2. In a 4 mL glass vial containing 16.4 mg (0.097 mmol) of EbAM, 26.5 μL (0.22 mmol) of HEMA and 100 μg green-QDs

(100 μL from the 1 mg/mL QD solution) (*see* **Note 9**) add 300 μL DMSO–toluene (1:1, v/v), 20 μL of eosin Y (10 mM in DMSO–toluene (1:1, v/v)), and 10 μL of TEA (72 mM in DMSO–toluene (1:1, v/v)). For the red QDs, follow the same procedure, but use 20 μL methylene blue (10 mM DMSO–toluene (1:1, v/v)) as the initiator dye.

3. Seal the vial with an airtight septum and sonicate in an ultrasonic bath for 15 min. Purge the mixture with nitrogen for 2 min and use a UV lamp placed at ~2 cm from the vial in order to initiate the polymerization by irradiation at 365 nm. Leave the samples to polymerize for 2 h.
4. After polymerization, transfer the content of the vial to 2 mL polypropylene microcentrifuge tubes. Subsequently, add 500 μL of DMSO–toluene (1:1, v/v) and ultrasonicate the HEMA-QDs particles, then sediment them by centrifugation for 15 min at 17,500 × *g*. Wash the particles four times with 800 μL DMSO–toluene (1:1, v/v) and twice with water.
5. Photobleach the eosin Y (or methylene blue) trapped inside the particles overnight with a fluorescent tube and, finally, dry the nanoparticles under vacuum.

3.2.2 Synthesis of a Second Shell (MIPGlcA-QDs or MIPNANA-QDs)

1. Incubate 5.46 mg AB (0.022 mmol) and 4.27 mg (0.022 mmol) GlcA (or 6.8 mg (0.022 mmol) NANA) for 1 h in 1 mL DMSO on a tube-rotator (*see* **Notes 4** and **10**).
2. Following the pre-incubation step, transfer the contents of the vial to the 4 mL glass vial containing the HEMA-QDs. Subsequently, add 80 μL (0.423 mmol) of EGDMA, 5.62 mg (0.066 mmol) MAM, 20 μL of eosin Y (10 mM) and 10 μL of TEA (72 mM). In the case of the HEMA-red QDs follow the same procedure, but use 20 μL methylene blue (10 mM in DMSO–toluene (1:1, v/v)) as the initiator dye. Synthesize a control non-imprinted polymer in the same way but without the addition of the template.
3. Seal the vial with an airtight septum and sonicate in an ultrasonic bath for 15 min. Purge the mixture with nitrogen for 2 min and use a UV lamp placed at ~2 cm from the vial in order to initiate the polymerization by irradiation at 365 nm. Leave the samples to polymerize for 2 h.
4. After polymerization, transfer the content to 2 mL polypropylene microcentrifuge tubes and wash the particles three times with methanol–acetic acid (9:1, v/v) followed by three times with 100 mM NH_3 in water–methanol (7:3, v/v), twice with water and three times with methanol.
5. Photobleach the eosin Y (or methylene blue) trapped inside the particles overnight with a fluorescent tube and, finally, dry the nanoparticles overnight under vacuum (*see* **Notes 11** and **12**).

3.3 Cell Imaging with MIPs

Prepare cell samples for imaging:

1. Human adult low Calcium high Temperature (HaCaT) cells are cultured in Dulbecco's Modified Eagle Medium high glucose with 10% FBS and 1% penicillin/streptomycin at 37 °C, 5% CO_2, and 100% humidity in 75 cm^2 sterile cell culture flasks.
2. Passage the cells when confluent. For that, add 16 mL of a solution of PBS containing 0.05% EDTA to the flask with the cells and incubate them together for 10 min at 37 °C. Aspirate the PBS solution containing EDTA. Add 2 mL of 0.25% trypsin–EDTA and incubate for 4 min at 37 °C (*see* **Note 13**). In order to stop the trypsinization, add 4 mL of culture medium and transfer the obtained cell suspension to a 15 mL conical sterile polypropylene centrifuge tube. Centrifuge for 5 min at 400 × *g*. Resuspend the cell pellet in 5 mL of culture medium and transfer 2.5 mL of the cell suspension to a new 75 cm^2 sterile cell culture flask. Add 17.5 mL medium to the new cell culture flask to get a final volume of 20 mL.
3. For the microscopic studies, the cells are cultured in 12-well plates (well diameter 22.1 mm) equipped with round glass coverslips (diameter 12 mm) (*see* **Note 14**). Pipette 100 μL of 1×10^5 suspended HaCaT cells (*see* **Note 15**) onto each coverslip. After 3 h of incubation, add 2 mL of culture medium to the cells. Afterwards, let them grow to confluency for 48–60 h.
4. Wash each coverslip with confluent HaCaT cells in the 12-well plates three times with 2 mL PBS and fix the cells at room temperature for 10 min with 600 μL paraformaldehyde 3% (w/v) in PBS (*see* **Note 16**). To stop fixation and reduce nonspecific interactions, incubate each cell sample three times with 1 mL 20 mM glycine in PBS for 20 min at room temperature (*see* **Note 17**) and finally wash them three times with 2 mL PBS.
5. Wash the cells three times with 1 mL methanol–water (1:30, v/v) and then incubate them with 1 mL of tip-sonicated polymer suspensions: 27 μg/mL rhodamine-labeled MIPGlcA (Fig. 2a) or 60 μg/mL MIPGlcA-QDs (green QDs) (Fig. 2b) or 60 μg/mL MIPNANA-QDs (red QDs) (Fig. 2c) at 37 °C for 90 min. For multiplexed cell imaging, add simultaneously 1 mL of 60 μg/mL MIPGlcA-QDs and 1 mL of 60 μg/mL MIPNANA-QDs to each well and then let them incubate at 37 °C for 90 min (*see* **Note 18**) (Fig. 2d). Afterwards, wash each fixed cell layer three times with 1 mL methanol–water (1:30, v/v) (*see* **Note 19**).
6. When quantitative analysis is not required, perform the additional staining of the cell membrane with the DiO dye, after prior MIP incubation and washing of the cells. For that, mix 200 μL FBS with 100 μL DiO (100 μM stock solution in

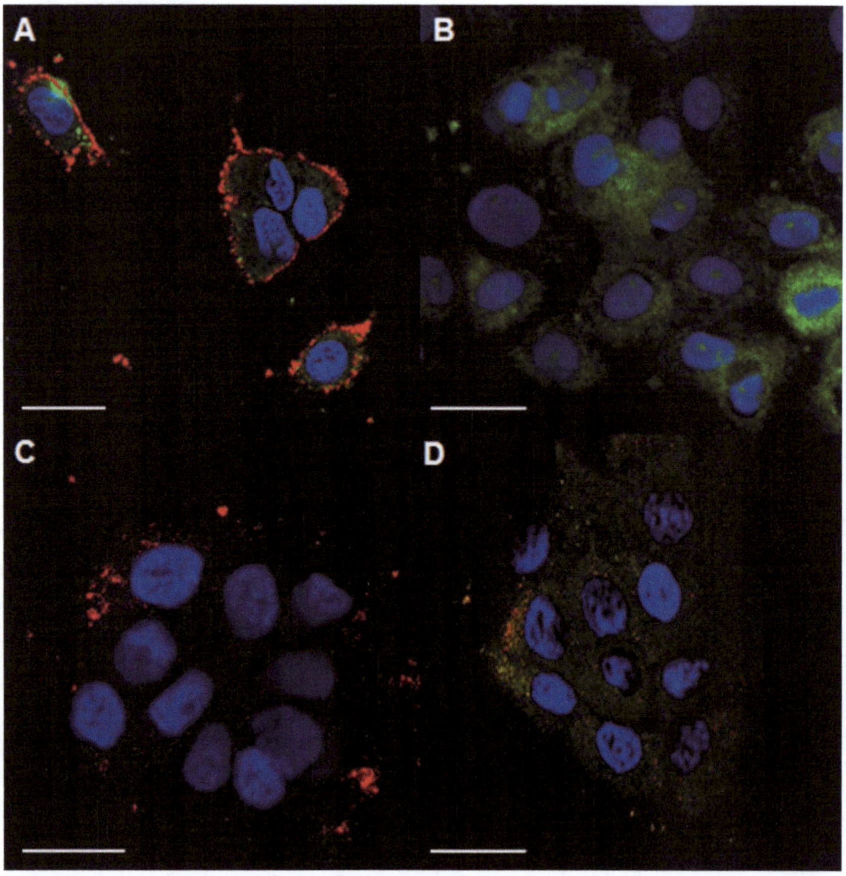

Fig. 2 Confocal images on fixated keratinocytes for the localization of (**a**) rhodamine-labeled MIPGlcA (*red*); (**b**) MIPGlcA-QDs (*green*, in some cells MIPGlcA-QDs were observed in nuclear clefts; (**c**) MIPNANA-QDs (*red*), the particles are mainly localized extracellularly and pericellularly. (**d**) Simultaneous multiplexed labeling of both GlcA and NANA by MIPGlcA-QDs (*green*) and MIPNANA-QDs (*red*). Nuclear staining is performed with DAPI (*blue*) and membrane staining in the case of (**a**) is performed with Dio (*green*). Scale bars: 25 μm. Images are reproduced with permissions from refs. 18, 19, 26

DMSO). Afterwards, add 800 μL water and use 300 μL of this solution to stain the fixed, washed cells by incubating for 30 min at room temperature (*see* **Note 20**). After staining, wash the samples three times with 1 mL water.

7. When quantitative analysis is required, do not perform cell membrane or nuclear staining. As a final step, mount the samples for fluorescence microscopy imaging on a microscope slide with 5 μL mounting medium (*see* **Note 21**).
8. For the staining of the cell nucleus, mix in a ratio 1:10 a stock solution of 1 mg/mL DAPI in water with mounting medium. Place 5 μL from this solution on a microscope slide to mount the cells on the coverslips (*see* **Note 22**).
9. After 3 min, image capturing takes place.

3.4 Tissue imaging with MIPs

1. Adult skin specimens are collected by autopsy.
2. Slice the samples in 8 μm thick sections in a cryostat microtome, transfer them to adhesive microscope slides and dry them for 30 s in an oven maintained at 70 °C to ensure good sample attachment to the glass surface. Subsequently, fix the tissue sections for 10 min with cold acetone (−10 °C) and wash them three times with PBS.
3. Prior to MIPGlcA incubation, wash the samples three times with 1 mL methanol–water (1:30, v/v) and then incubate them with 1 mL of a tip-sonicated polymer suspension of 27 μg/mL polymer in methanol–water (1:30, v/v) at 37 °C for 90 min. Afterwards, wash each tissue sample three times with 1 mL methanol–water (1:30, v/v) and mount for fluorescence microscopy in 20 μL mounting medium (Fig. 3).
4. In the cases where quantitative analysis is not required, perform additional staining of the cell nucleus with 20 μL mounting medium containing 100 μg/mL DAPI (*see* **Note 23**).

3.5 Acquisition and Analysis of Images

Quantitative cell imaging using the rhodamine-labeled MIPGlcA, MIPGlcA-QDs, or MIPNANA-QDs:

3.5.1 Epifluorescence Microscopy Imaging

1. Epifluorescence images are captured with a Leica DMI 6000B microscope, filter set A4, L5, TX2, N PLAN L 20.0 × 0.40 DRY, HCX PL FLUOTAR 40.0 × 0.60 DRY, HCX FLUOTAR 63.0 × 0.70 DRY, and HCX FLUOTAR 100.0 × 1.30 OIL objectives with 20×, 40×, 63×, and 100× magnification.

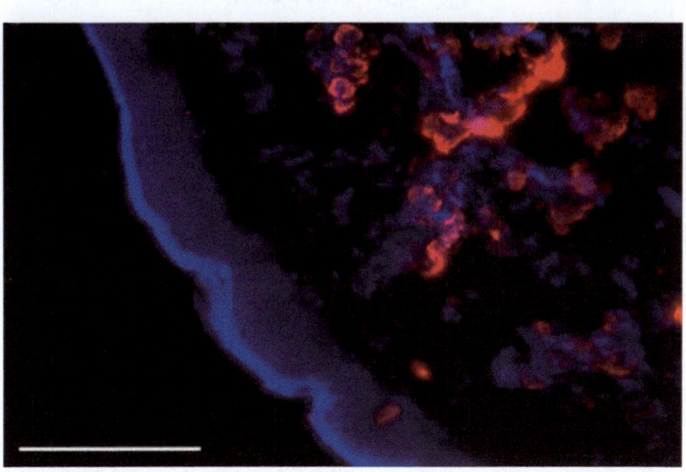

Fig. 3 Confocal image for the localization of rhodamine-labeled MIPGlcA (*red*) in human skin specimens. Nuclear staining is performed with DAPI (*blue*). The MIPGlcA particles bound to the skin tissue are mainly localized in the basal layer of the epidermis and the papillary dermis. Scale bar: 100 μm

2. Images are captured using exactly the same settings concerning light intensity and exposure time in 16-Bit Tiff format.
3. Only confluent cell layers are examined for quantification studies and when quantitative analysis is carried out, nuclear staining is not performed. From each sample, at least four images are captured with the Leica Application Suite (LAS) software and each cell sample is at least prepared in quadruplicate.
4. All fluorescence intensities are determined with ImageJ. For image analysis, if needed perform prior background subtraction in order to determine the fluorescence signal coming only from the particles. As a control for the background signal, samples with fixed cells are used and the average background signal at four different areas of each gray value image is subtracted prior to quantification. Furthermore, a slight difference in the fluorescence intensity from MIP and NIP particles is corrected during the quantification using ImageJ.

3.5.2 Confocal Microscopy Imaging

1. Confocal images are captured with a Zeiss LSM 710, AxioObserver. A Plan-Apochromat 63×/1.40 OIL DIC M27 objective and 405, 488, and 543 nm lasers are used for excitation of DAPI, DiO, and rhodamine, respectively, for all images.
2. The samples used for confocal microscopy are stained for hyaluronan using the rhodamine-labeled MIPGlcA (red), for the cell membrane using Dio (green) and for the cell nucleus using DAPI (blue). Alternatively, when staining for hyaluronan using the MIPGlcA-QDs (green) or for sialylated sites using the MIPNANA-QDs (red), do not perform cell membrane staining. Stain the cell nucleus in the latter case with DAPI (blue) (*see* **Note 24**).
3. Capture all the confocal images from the middle Z of the samples.

4 Notes

1. Successful cell culture depends on keeping the cells free from contamination by microorganisms such as bacteria, fungi, and viruses. During all cell manipulations, follow the aseptic technique in order to reduce the probability of contamination. The elements of aseptic technique are a sterile work area, good personal hygiene, sterile reagents (autoclave, spraying with 70% ethanol etc.) and media, and sterile handling.
2. Sterilize the solution of PBS containing 0.05% EDTA, in an autoclave before use.
3. The yield of the synthesis of 4-acrylamidophenyl(amino) methaniminium acetate (AB) is ~80%.

4. The stoichiometry and the association constant of the AB-GlcA complex are deduced from ^1H NMR titration studies in DMSO-d_6, yielding a 1:1 ratio and a high association constant K_a of 7.1×10^3 M^{-1}.
5. The molar ratio of methacryloxyethyl thiocarbonyl rhodamine B with respect to the other monomers was optimized to maximize the fluorescence intensity of the particles (optimal ratio: 1:0.05 AB:rhodamine). Higher dye content resulted in lower brightness due to reabsorption or energy transfer.
6. The degree of cross-linking (mol of EGDMA upon total number of mol of EGDMA and functional monomers) in the described protocol is 83%, with a molar ratio of GlcA:AB:MAM:EGDMA of 1:1:3:20 and a total concentration of monomers of 5% (mass of monomers relative to total mass of solvent and monomers). The initiator, ABDV, is used in a concentration of 0.56 mol% with respect to the total number of mol of double bonds in the polymerization mixture.
7. MIPGlcA and NIPGlcA particles with diameters of approximately 400 nm (less than 10% deviation between MIP and NIP) with good monodispersity are obtained using this precipitation polymerization protocol. The particle size is chosen to avoid possible internalization of the particles, so that only the extracellular hyaluronan is targeted.
8. The recognition properties of rhodamine-labeled MIPGlcA are evaluated by equilibrium radioligand binding assays using [^{14}C]glucuronic acid (Biotrend Chemikalien GmbH, Köln, Germany) in methanol–water (9:1, v/v) and in pure water, where maximum imprinting factors (IF_{max}) of 2.2 and 3.2 are respectively observed (IF = binding to the MIP/binding to the NIP).
9. Sonicate the 1 mg/mL QD stock solution (green or red QDs) in toluene for 15 min with a microtip (Branson Sonifier 250) prior to addition in the prepolymerization mixture.
10. The stoichiometry and the association constant of the AB-NANA complex are deduced from ^1H NMR titration studies in DMSO-d_6, yielding a 1:1 ratio and a high K_a of 41×10^3 M^{-1}.
11. The described protocol results in an average particle size of 125 ± 17 nm for both MIPGlcA-QDs and MIPNANA-QDs, enabling the staining of the intracellular glycosylation sites.
12. The recognition properties of MIP/NIPGlcA-QDs and MIP/NIPNANA-QDs are evaluated by equilibrium radioligand binding assays using [^{14}C]glucuronic acid and [6-^3H]sialic acid respectively (Biotrend Chemikalien GmbH, Köln, Germany) in water, where high specificity is observed.

13. In case there is any problem with the cell trypsinization, incubate the cells with the 2 mL of 0.25% trypsin–EDTA at 37 °C for up to maximum 10 min.
14. Prior to use, sterilize the coverslips by dipping in ethanol 70% and leave them to dry for 30 min under the hood.
15. Cell counting is performed using a disposable hemocytometer. After cell growth and trypsinization (*see* Subheading 3.3, **steps 1–2**), centrifuge the cells for 5 min at 400 × *g*. Then resuspend the cell pellet in 5 mL of culture medium and transfer the cell suspension to a 50 mL conical polypropylene centrifuge tube. Add 15 mL culture medium in order to get a final volume of 20 mL. Before the cells have a chance to settle, withdraw 500 μL of the cell suspension into an Eppendorf tube. In another Eppendorf tube, pipet 400 μL of 0.4% Trypan Blue. Subsequently, transfer 100 μL of cells into the latter tube. Mix gently by pipetting. Afterwards, take 100 μL of the Trypan Blue-treated cells and pipet them into the well of the counting chamber of the hemocytometer, allowing the capillary forces to draw it inside. Use a microscope with an objective with a 10× magnification and focus on the grid lines of the hemocytometer. Count the live, unstained cells using a hand tally counter (living cells are impermeable to Trypan Blue). In order to obtain a correct count, adapt a counting system whereby cells are only counted when they are set within a square or on the right-hand or bottom boundary line. Carry on counting until all 4 sets of 16 corners are counted. To obtain the number of viable cells/mL in the original cell suspension, take the average cell count from each of the sets of 16 corner squares. Multiply by 10^4. Multiply again by 5 to correct for the 1:5 dilution from the Trypan Blue addition.
16. The fixation of cells is based on paraformaldehyde that has low background fluorescence. Aldehyde fixatives react with amines and proteins to generate fluorescent products. Glutaraldehyde is less suitable than formaldehyde due to generation of stronger fluorescence. The simplest way to stop aldehyde-induced fluorescence is to use a fixative that does not contain an aldehyde. Carnoy, Clarke, and methacarn solutions are examples, but are used only for subsequent paraffin sectioning. The remedy to the aldehyde problem is aldehyde blocking.
17. When fixatives react and cross-link with protein molecules, lots of free aldehyde groups remain. These cell/tissue-bound free aldehyde groups will combine covalently with any amino group offered to them, including terminal and side-chain (lysine) amino groups of proteins being used as histochemical reagents, which means all antibodies, all lectins and all enzymes. This way, even a highly specific monoclonal primary antibody may bind at sites that contain basic proteins but not

the antigen of interest. This is why, aldehyde blocking is necessary. This is done by reducing the −CHO groups to $-CH_2OH$ with sodium borohydride or by feeding them small-molecule amines, e.g., glycine.

18. Incubation of the rhodamine-labeled MIPGlcA with the cells was also carried out at room temperature with and without agitation. In both cases, particle aggregation was observed rendering these samples unsuitable for cell imaging.
19. The selectivity of MIPGlcA is confirmed by competitive binding assays comparing the binding of GlcA to that of other monosaccharides such as glucose, galactose, N-acetylglucosamine, N-acetylgalactosamine and NANA, present at the terminal parts of glycoproteins or glycolipids that could potentially interfere during cell imaging; less than 1% cross-reactivity is observed. In the case of the MIPNANA, selectivity is also confirmed by competitive equilibrium binding assays using [6-^{3}H]sialic acid (Biotrend Chemikalien GmbH, Köln, Germany) which show less than 10% cross-reactivity with GlcA and negligible cross-reactivity with other terminal sugars, namely N-acetylglucosamine, glucose, galactose, and N-acetylgalactosamine. These results indicate that multiplexed cell imaging using the MIPs described in this protocol is possible.
20. Cell membrane staining with DiO was also carried out for 15 min, but better staining results are achieved with 30 min incubation.
21. In all cases, an $IF_{max} \sim 2$ is observed on the cells, verifying the imprinting effect. To further prove the specificity of the MIPs, the GlcA and NANA terminal moieties are cleaved off enzymatically with the use of hyaluronidase and neuraminidase, respectively. After the enzymatic treatment, there is no significant difference anymore between MIPs and NIPs labeling.
22. Apply transparent nail polish on the edges of the coverslips in order to ensure sample attachment to the microscope slides.
23. The MIPGlcA particles bound to the skin tissue are mainly localized in the basal layer of the epidermis and the papillary dermis. Lower amounts of particles can be found in the cornified and granular cell layer and even lower amounts in the spinous cell layer. This is in agreement with and the control results obtained with FITC-labeled hyaluronic acid binding protein applied to tissue samples from the same batch and prepared in the same way as well as to the descriptions of hyaluronan localization in the literature [35–37].
24. Confocal images for the rhodamine-labeled MIPGlcA are corrected by subtracting the rhodamine signal from the green channel as there is spectral overlap with $\lambda_{ex(rhodamine)} = 488$ nm. The subtraction was performed with ImageJ.

Acknowledgment

The authors thank the European Regional Development Fund and the Regional Council of Picardie (co-funding of equipment under CPER 2007–2013), the European Union (FP7 Marie Curie Actions, ITN SAMOSS, PITN-2013–607590), and the french embassy in Germany (postdoctoral scholarship of S.K.), for financial support. The authors thank Jörg Sänger and the Institute of Pathology Bad Berka (Germany) for providing tissue samples and for tissue imaging.

References

1. Moremen KW, Tiemeyer M, Nairn AV (2012) Vertebrate protein glycosylation: diversity, synthesis and function. Nat Rev Mol Cell Biol 13(7):448–462
2. Spiro RG (2002) Protein glycosylation: nature, distribution, enzymatic formation, and disease implications of glycopeptide bonds. Glycobiology 12(4):43R–56R
3. Bard F, Chia J (2016) Cracking the glycome encoder: signaling, trafficking, and glycosylation. Trends Cell Biol 26(5):379–388
4. Ohtsubo K, Marth JD (2006) Glycosylation in cellular mechanisms of health and disease. Cell 126:855–867
5. Rudd PM, Elliott T, Cresswell P, Wilson IA, Dwek RA (2001) Glycosylation and the immune system. Science 291:2370–2376
6. Gopaul KP, Crook MA (2006) Sialic acid: a novel marker of cardiovascular disease? Clin Biochem 39(7):667–681
7. Hascall VC, Majors AK, De la Motte CA, Evanko SP, Wang A, Drazba JA et al (2004) Intracellular hyaluronan: a new frontier for inflammation? Biochim Biophys Acta 1673: 3–12
8. Varki NM, Varki A (2007) Diversity in cell surface sialic acid presentations: implications for biology and disease. Lab Invest 87:851–857
9. Seton-Rogers S (2012) Metastasis multitasking hyaluronic acid. Nat Rev Cancer 12:228–228
10. Büll C, Stoel MA, Den Brok MH, Adema GJ (2014) Sialic acids sweeten a tumor's life. Cancer Res 74:3199–3204
11. Kawamura A, Kijima-Suda I, Sugimoto M, Itoh M, Takada K, Tomita K et al (1990) A monoclonal antibody to free N-acetylneuraminic acid. Biochim Biophys Acta 1033:201–206
12. De la Motte CA, Drazba JA (2011) Viewing hyaluronan: Imaging contributes to imagining new roles for this amazing matrix polymer. J Histochem Cytochem 59:252–257
13. Sterner E, Flanagan N, Gildersleeve JC (2016) Perspectives on anti-glycan antibodies gleaned from development of a community resource database. ACS Chem Biol 11(7):1773–1783
14. Bowen JL, Manesiotis P, Allender CJ (2013) Twenty years since *'antibody mimics'* by moelcular imprinting were first proposed: A critical perspective. Mol Imprinting 1:35–40
15. Haupt K, Linares AV, Bompart M, Tse Sum Bui B (2012) Moelcularly imprinted polymers. Top Curr Chem 325:1–28
16. Alexander C, Andersson HS, Andersson LI, Ansell RJ, Kirsch N, Nicholls IA et al (2006) Molecular imprinting science and technology: a survey of the literature for the years up to and including 2003. J Mol Recognit 19:106–180
17. Takeuchi T, Sunayama H (2015) Molecularly imprinted polymers. In: Kobayashi S, Müllen K (eds) Encyclopedia of polymeric nanomaterials. Springer, Berlin Heidelberg, p 1291
18. Kunath S, Panagiotopoulou M, Maximilien J, Marchyk N, Sänger J, Haupt K (2015) Cell and tissue imaging with molecularly imprinted polymers as plastic antibody mimics. Adv Healthc Mater 4:1322–1326
19. Panagiotopoulou M, Kunath S, Medina-Rangel PX, Haupt K, Tse Sum Bui B (2016) Fluorescent molecularly imprinted polymers as plastic antibodies for selective labeling and imaging of hyaluronan and sialic acid on fixed and living cells. Biosens Bioelectron. doi:10.1016/j.bios.2016.07.080
20. Shinde S, El-Schich Z, Malakpour A, Wan W, Dizeyi N, Mohammadi R et al (2015) Sialic acid-imprinted fluorescent core–shell particles for selective labeling of cell surface glycans. J Am Chem Soc 137:13908–13912

21. Wang S, Yin D, Wang W, Shen X, Zhu JJ, Chen HY et al (2016) Targeting and imaging of cancer cells via monosaccharide-imprinted fluorescent nanoparticles. Sci Rep 6:22757–22767
22. Ton XA, Tse Sum Bui B, Resmini M, Bonomi P, Dika I, Soppera O et al (2013) A versatile fiber-optic fluorescence sensor based on molecularly imprinted microstructures polymerized in situ. Angew Chem Int Ed 52(32):8317–8321
23. Liu RY, Guan GJ, Wang SH, Zhang ZP (2011) Core-shell nanostructured molecular imprinting fluorescent chemosensor for selective detection of atrazine herbicide. Analyst 136:184–190
24. Yang YQ, Niu H, Zhang H (2016) Direct and highly selective drug optosensing in real, undiluted biological samples with quantum-dot-labeled hydrophilic molecularly imprinted polymer microparticles. ACS Appl Mater Interfaces 8(24):15741–15749
25. Wei F, Xu G, Wu Y, Wang X, Yang J, Liu L et al (2016) Molecularly imprinted polymers on dual-color quantum dots for simultaneous detection of norepinephrine and epinephrine. Sens Actuators B Chem 229:38–46
26. Panagiotopoulou M, Salinas Y, Beyazit S, Kunath S, Duma L, Prost E et al (2016) Molecularly imprinted polymer-coated quantum dots for multiplexed cell targeting and imaging. Angew Chem Int Ed 55:8244–8248
27. Beyazit S, Ambrosini S, Marchyk N, Palo E, Kale V, Soukka T et al (2014) Versatile synthetic strategy for coating upconverting nanoparticles with polymer shells through localized photopolymerization by using the particles as internal light sources. Angew Chem Int Ed 53(34):8919–8923
28. Tang Y, Gao Z, Wang S, Gao X, Gao J, Ma Y et al (2015) Upconversion particles coated with molecularly imprinted polymers as fluorescence probe for detection of clenbuterol. Biosens Bioelectron 71:44–50
29. Xu S, Lu X (2016) Mesoporous structured MIPs@CDs fluorescence sensor for highly sensitive detection of TNT. Biosens Bioelectron 85:950–956
30. Bossi AM, Sharma PS, Montana L, Zoccatelli G, Laub O, Levi R (2012) Fingerprint-imprinted polymer: rational selection of peptide epitope templates for the determination of proteins by molecularly imprinted polymers. Anal Chem 84(9):4036–4041
31. Rachkov A, Minoura N (2001) Towards molecularly imprinted polymers selective to peptides and proteins. The epitope approach. Biochim Biophys Acta 1544(1–2):255–266
32. Tammi R, Rilla K, Pienimäki J-P, MacCallum DK, Hogg M, Luukkonen M et al (2001) Hyaluronan enters keratinocytes by a novel endocytic route for catabolism. J Biol Chem 276:35111–35122
33. Nestora S, Merlier F, Beyazit S, Prost E, Duma L, Baril B et al (2016) Plastic antibodies for cosmetics: molecularly imprinted polymers scavenge precursors of malodors. Angew Chem Int Ed 55(21):6252–6256
34. Xu S, Ziegler J, Nann T (2008) Rapid synthesis of highly luminescent InP and InP/ZnS nanocrystals. J Mater Chem 18:2653–2656
35. Tammi R, Ripellino JA, Margolis RU, Tammi M (1988) Localization of epidermal hyaluronic acid using the hyaluronate binding region of cartilage proteoglycan as a specific probe. J Investig Dermatol 90:412–414
36. Wang C, Tammi M, Tammi R (1992) Distribution of hyaluronan and its CD44 receptor in the epithelia of human skin appendages. Histochemistry 98:105–112
37. Papakonstantinou E, Roth M, Karakiulakis G (2012) Hyaluronic acid: a key molecule in skin aging. Dermatoendocrinol 4(3):253–258

ERRATUM

Synthetic Antibodies

Thomas Tiller

Thomas Tiller (ed.), *Synthetic Antibodies: Methods and Protocols*, Methods in Molecular Biology, vol. 1575, DOI 10.1007/978-1-4939-6857-2, © Springer Science+Business Media LLC 2017

DOI 10.1007/978-1-4939-6857-2_27

Contents:

On page ix, the co-author of Chapter 9, Emilie Renaud's name has been corrected. Originally the book published with the incorrect Emilie Rebaud.

Contributors:

1) On page xi, we have updated the affiliation for the author of Chapter 17, Grigory S. Filonov, to add the Cornell University information, as that is where he was working when he wrote the chapter. The contributors list has been updated as mentioned below to include both his current affiliation (Essen) and his former (Cornell):

   Grigory S. Filonov • Essen Bioscience, Ann Arbor, MI, USA (current); Department of Pharmacology, Weill Medical College, Cornell University, New York, NY, USA.

2) In page xiii, the author's name was misspelled as "Emilie Rebaud" which has now been corrected as "Emilie Renaud".

Chapter opening page:

Chapter 9: The author's name was misspelled as "Emilie Rebaud" and has now been corrected to reflect her correct name "Emilie Renaud" throughout this chapter.

Chapter 17: A footnote has been added to the chapter opening page to show the author's current and previous affiliations.

The updated online version of the original book can be found under
http://dx.doi.org/10.1007/978-1-4939-6857-2_9
http://dx.doi.org/10.1007/978-1-4939-6857-2_17

Thomas Tiller (ed.), *Synthetic Antibodies: Methods and Protocols*, Methods in Molecular Biology, vol. 1575, DOI 10.1007/978-1-4939-6857-2_27, © Springer Science+Business Media LLC 2017

INDEX

Thomas Tiller (ed.), *Synthetic Antibodies: Methods and Protocols*, Methods in Molecular Biology, vol. 1575, DOI 10.1007/978-1-4939-6857-2, © Springer Science+Business Media LLC 2017

MIX
Papier aus verantwortungsvollen Quellen
Paper from responsible sources
FSC® C105338

If you have any concerns about our products,
you can contact us on
ProductSafety@springernature.com

In case Publisher is established outside the EU,
the EU authorized representative is:
Springer Nature Customer Service Center GmbH
Europaplatz 3, 69115 Heidelberg, Germany

Printed by Libri Plureos GmbH
in Hamburg, Germany